Lecture Notes in Computer Science 839
Edited by G. Goos and J. Hartmanis

Yvo G. Desmedt (Ed.)

Advances in Cryptology – CRYPTO '94

14th Annual International
Cryptology Conference
Santa Barbara, California, USA
August 21-25, 1994
Proceedings

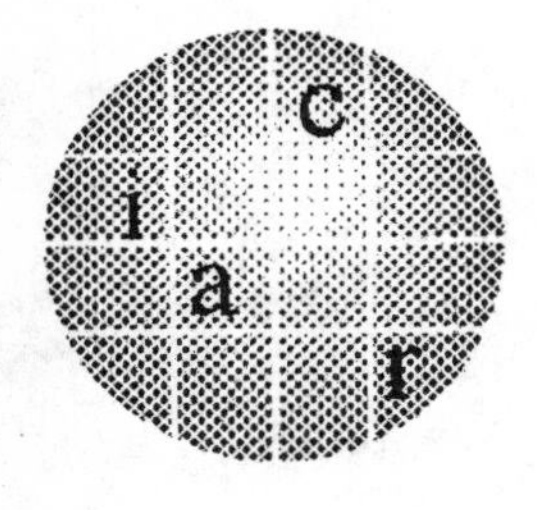

Springer-Verlag
Berlin Heidelberg New York
London Paris Tokyo
Hong Kong Barcelona
Budapest

Series Editors

Gerhard Goos
Universität Karlsruhe
Postfach 69 80
Vincenz-Priessnitz-Straße 1
D-76131 Karlsruhe, Germany

Juris Hartmanis
Cornell University
Department of Computer Science
4130 Upson Hall
Ithaca, NY 14853, USA

Volume Editor

Yvo G. Desmedt
Department of EE & CS, University of Wisconsin - Milwaukee
P. O. Box 784, Milwaukee, WI 53201, USA

CR Subject Classification (1991): E.3-4, D.4.6, G.2.1, C.2.0

ISBN 3-540-58333-5 Springer-Verlag Berlin Heidelberg New York
ISBN 0-387-58333-5 Springer-Verlag New York Berlin Heidelberg

CIP data applied for

Printed in Germany

Typesetting: Camera-ready by author
SPIN: 10475401 45/3140-543210 - Printed on acid-free paper

PREFACE

The CRYPTO '94 conference is sponsored by the International Association for Cryptologic Research (IACR), in co-operation with the IEEE Computer Society Technical Committee on Security and Privacy. It has taken place at the University of California, Santa Barbara, from August 21–25, 1994. This is the fourteenth annual CRYPTO conference, all of which have been held at UCSB. This is the first time that proceedings are available at the conference. The General Chair, Jimmy R. Upton has been responsible for local organization, registration, etc.

There were 114 submitted papers which were considered by the Program Committee. Of these, 1 was withdrawn and 38 were selected for the proceedings. There are also 3 invited talks. Two of these are on aspects of cryptography in the commercial world. The one on hardware aspects will be presented by David Maher (AT&T), the one on software aspects by Joseph Pato (Hewlett-Packard). There will also be a panel discussion on "Securing an Electronic World: Are We Ready?" The panel members will be: Ross Anderson, Bob Blakley, Matt Blaze, George Davida, Yvo Desmedt (moderator), Whitfield Diffie, Joan Feigenbaum, Blake Greenlee, Martin Hellman, David Maher, Miles Smid. The topic of the panel will be introduced by the invited talk of Whitfield Diffie on "Securing the Information Highway."

These proceedings contain revised versions of the 38 contributed talks. Each paper was sent to at least 3 members of the program committee for comments. Revisions were not checked on their scientific aspects. Some authors will write final versions of their papers for publication in refereed journals. Of course the authors bear full responsibility for the contents of their papers.

I am very grateful to the members of the Program Committee for their hard work and the difficult task of selecting roughly 1 out of 3 of the submitted papers. As has been done since 1989, submissions to CRYPTO '94 were required to be anonymous. The more recent tradition, introduced since 1991, that a Program Committee member can be the author of at most one accepted paper has been followed. Papers submitted by members of the Program Committee were sent to at least 4 referees (and, of course, no Program Committee member reviewed his or her own paper).

The following referees and external experts helped the Program Committee in reaching their decisions: Amos Beimel, Josh Benaloh, Eli Biham, Carlo Blundo, Gilles Brassard, Benny Chor, Philippe Delsarte, Yair Frankel, Atsushi Fujioka, Oded Goldreich, Dan Gordon, Thomas Hardjono, Gene Itkis, Markus Jakobsson, Burt Kaliski, Hugo Krawczyk, Kaoru Kurosawa, Eyal Kushilevitz, Susan Langford, Hendrik Lenstra, Carsten Lund, Kevin McCurley, Yi Mu, Moni Naor, Seffi Naor, Kazuo Ohta, Kevin Phelps, Jean-Jacques Quisquater, Venkatesan Ramarathnam, Jim Reeds, Ron M. Roth, Rei Safavi-Naini, Ryuichi Sakai, Doug Stinson, Jimmy Upton, Paul Van Oorschot, Scott Vanstone and Yuliang Zheng. The Program Committee appreciates their effort.

Thanks to Eli Biham for helping with postscript, Tom Cusick for being willing to provide a backup to read e-mail, Dave Rasmussen for organizing the automatic mailing facility used to distribute information, Marg Feeney and Ann Libert for secretarial work, Carlo Blundo, Giovanni Di Crescenzo, Ugo Vaccaro and William Wolfowicz for helping out at the last minute. I would also like to thank my hosts of my sabbatical year, Shimon Even, Scott Vanstone and Alfredo De Santis, where most of my work towards the conference took place. Several people have helped the General Chair with sending out the call for papers, registration, registration at the conference, etc.

Finally, I would like to thank everyone who submitted to CRYPTO '94. It goes without saying that the success of the conference depends ultimately on the quality of the submissions — CRYPTO has been and remains a leading conference in the discipline due to the high quality of the papers submitted. I am also grateful to the authors for sending me final versions of their papers for publication in these proceedings in a timely fashion.

Yvo Desmedt
Program Chair, CRYPTO '94
University of Wisconsin – Milwaukee, USA
Salerno, Italy, June, 1994.

CRYPTO '94

will take place at the University of California, Santa Barbara,
August 21–25, 1994

Sponsored by the

International Association for Cryptologic Research

in cooperation with the

IEEE Computer Society Technical Committee on Security and Privacy

General Chair

Jimmy R. Upton, Uptronics Incorporated, USA

Program Chair

Yvo Desmedt, University of Wisconsin – Milwaukee, USA

Program Committee

Tom Berson	Anagram Laboratories, USA
Don Coppersmith	IBM T. J. Watson Research Center, USA
Donald Davies	United Kingdom
Shimon Even	Technion, Israel
Amos Fiat	Tel Aviv University, Israel
Russell Impagliazzo	University of California San Diego, USA
Ingemar Ingemarsson	University of Linköping, Sweden
Mitsuru Matsui	Mitsubishi Electric Corporation, Japan
Alfred Menezes	Auburn University, USA
Andrew Odlyzko	AT&T Bell Laboratories, USA
Jennifer Seberry	University of Wollongong, Australia
Ben Smeets	Lund University, Sweden
Moti Yung	IBM T. J. Watson Research Center, USA

CONTENTS

Block Ciphers: Differential and Linear Cryptanalysis

Schemes Based on New Problems

Signatures I

Implementation and Hardware Aspects

Authentication and Secret Sharing

Zero-Knowledge

Signatures II

Combinatorics and its Applications

Number Theory

Cryptanalysis and Protocol Failures

Pseudo-Random Generation

Block Ciphers: Design and Cryptanalysis

Secure Computations and Protocols

The First Experimental Cryptanalysis of the Data Encryption Standard

Mitsuru Matsui

Computer & Information Systems Laboratory
Mitsubishi Electric Corporation
5-1-1, Ofuna, Kamakura, Kanagawa, 247, Japan
matsui@mmt.isl.melco.co.jp

Abstract. This paper describes an improved version of linear cryptanalysis and its application to the first successful computer experiment in breaking the full 16-round DES. The scenario is a known-plaintext attack based on two new linear approximate equations, each of which provides candidates for 13 secret key bits with negligible memory. Moreover, reliability of the key candidates is taken into consideration, which increases the success rate. As a result, the full 16-round DES is breakable with high success probability if 2^{43} random plaintexts and their ciphertexts are available. The author carried out the first experimental attack using twelve computers to confirm this: he finally reached all of the 56 secret key bits in fifty days, out of which forty days were spent for generating plaintexts and their ciphertexts and only ten days were spent for the actual key search.

1 Introduction

In the first paper on linear cryptanalysis [2], we introduced a new measure of linearity of S-boxes and extended it to the entire cipher structure of DES. The resultant linear approximate equations are effectively applicable to a known-plaintext attack, which proved that DES is breakable with negligible memory if 2^{47} random plaintexts and their ciphertexts are available. This is the first known-plaintext attack faster than an exhaustive key search, though the origin of linear cryptanalysis can be seen in several papers [4][5][6][7].

This paper studies an improved version of linear cryptanalysis and its application to the first successful computer experiment in breaking the full 16-round DES. We newly introduce two viewpoints; linear approximate equations based on the best (n–2)-round expression, and reliability of the key candidates derived from these equations. The former reduces the number of required plaintexts, whereas the latter increases the success rate of our attack.

In the 2^{47}-method, we established two linear approximate equations of 16-round DES using the best 15-round expression, where each equation includes one active S-box and hence recovers 7 secret key bits. This paper, however, begins with two new linear approximate equations derived from the best 14-round expression, where each equation has two active S-boxes and can recover 13 secret key bits. These equations give us, therefore, a total of 26 secret key bits, and then the remaining $56 - 26 = 30$ secret key bits are within the reach of an exhaustive search.

Moreover, we treat not only one solution of each equation but also "candidates" for the 13 secret key bits, where each candidate has its ranking of reliability such that the i-th rank represents the i-th likely solution. The aim of this approach is to give a table that relates ranking of the 26 secret key bits to that of the 13 secret key bits. This table increases the success rate of our attack at the cost of computational complexity; that is to say, if the most likely 26 key bits turn out to be wrong, we can adopt the second likely 26 key bits and search for the remaining 30 key bits again. If they are not correct either, we can try the third likely 26 key bits.

We also prove that the effectiveness of this method can be measured by DES reduced to 8 rounds. This fact enables us to experimentally determine the relationship among the number of required plaintexts, the complexity and the success rate of our attack. As a result, DES is breakable with complexity 2^{43} and success rate 85% if 2^{43} known-plaintexts are available. For another example, success rate is 10% with complexity 2^{50} if 2^{38} known-plaintexts are available.

We carried out the first experimental attack of the full 16-round DES using twelve computers (HP9735/PA-RISC 99MHz) to confirm this scenario. The program, described in C and assembly languages consisting of a total of 1000 lines, was designed to solve two equations while generating 2^{43} random plaintexts and enciphering them. We finally reached all of the 56 secret key bits in fifty days, out of which forty days were spent for generating plaintexts and their ciphertexts and only ten days were spent for the actual key search.

2 Preliminaries

We follow the notations introduced in [2]. Since our scope is a known-plaintext attack using random plaintexts, we omit the initial permutation IP, the final permutation IP^{-1}, and PC-1. The right most bit of each symbol is referred as the 0-th (lowest) bit, whereas the traditional rule defines the left most bit as the first bit [1]. The following are used throughout this paper;

P	The 64-bit data after the IP; the plaintext.
C	The 64-bit data before the IP^{-1}; the ciphertext.
P_H, P_L	The upper and the lower 32-bit data of P, respectively.
C_H, C_L	The upper and the lower 32-bit data of C, respectively.
K	The 56-bit data after the PC-1; the secret key.
K_r	The r-th round 48-bit subkey.
$F_r(X_r, K_r)$	The r-th round F-function.
$A[i]$	The i-th bit of A, where A is any binary vector.
$A[i, j, .., k]$	$A[i] \oplus A[j] \oplus ... \oplus A[k]$.

3 Principle of the New Attack

The first purpose of linear cryptanalysis is to find the following linear approximate expression which holds with probability $p \neq 1/2$ for randomly given plaintext P, the corresponding ciphertext C and the fixed secret key K :

$$P[i_1, i_2, .., i_a] \oplus C[j_1, j_2, .., j_b] = K[k_1, k_2, .., k_c], \tag{1}$$

where $i_1, i_2, .., i_a, j_1, j_2, .., j_b$ and $k_1, k_2, .., k_c$ denote fixed bit locations.

Since both sides of equation (1) essentially represent one-bit information, the magnitude of $|p - 1/2|$ expresses the effectiveness. We will refer to the most effective linear approximate expression (i.e. $|p - 1/2|$ is maximal) as the best expression and its probability as the best probability, respectively. We have found the best expression and the best probability of DES, whose results are summarized in [2] for the number of rounds varying from $n = 3$ to $n = 20$. A practical algorithm for deriving these values is described in [3].

In the 2^{47}-method, we established two equations of 16-round DES using the best 15-round expression, which holds with probability $1/2 + 1.19 \times 2^{-22}$ [2]. Our new attack, however, starts with the following two best 14-round expressions, which hold with probability $1/2 - 1.19 \times 2^{-21}$:

$$\begin{aligned} &P_L[7,18,24] \oplus C_H[7,18,24,29] \oplus C_L[15] \\ &= K_2[22] \oplus K_3[44] \oplus K_4[22] \oplus K_6[22] \oplus K_7[44] \oplus K_8[22] \oplus K_{10}[22] \oplus \\ &\quad K_{11}[44] \oplus K_{12}[22] \oplus K_{14}[22], \end{aligned} \tag{2}$$

$$\begin{aligned} &C_L[7,18,24] \oplus P_H[7,18,24,29] \oplus P_L[15] \\ &= K_{13}[22] \oplus K_{12}[44] \oplus K_{11}[22] \oplus K_9[22] \oplus K_8[44] \oplus K_7[22] \oplus K_5[22] \oplus \\ &\quad K_4[44] \oplus K_3[22] \oplus K_1[22], \end{aligned} \tag{3}$$

where P, C and K denote the plaintext, the ciphertext and the secret key of DES reduced to 14 rounds, respectively.

Then applying equations (2) and (3) to fourteen consecutive F-functions from the 2nd round to the 15th round of 16-round DES, we have the following two equations that hold with probability $1/2 - 1.19 \times 2^{-21}$ for random plaintexts and their ciphertexts (figure 1 illustrates the detailed construction of equation (4)):

$$\begin{aligned} &P_H[7,18,24] \oplus F_1(P_L, K_1)[7,18,24] \oplus C_H[15] \oplus C_L[7,18,24,29] \oplus \\ &\quad F_{16}(C_L, K_{16})[15] \\ &= K_3[22] \oplus K_4[44] \oplus K_5[22] \oplus K_7[22] \oplus K_8[44] \oplus K_9[22] \oplus K_{11}[22] \oplus \\ &\quad K_{12}[44] \oplus K_{13}[22] \oplus K_{15}[22], \end{aligned} \tag{4}$$

$$\begin{aligned} &C_H[7,18,24] \oplus F_{16}(C_L, K_{16})[7,18,24] \oplus P_H[15] \oplus P_L[7,18,24,29] \oplus \\ &\quad F_1(P_L, K_1)[15] \\ &= K_{14}[22] \oplus K_{13}[44] \oplus K_{12}[22] \oplus K_{10}[22] \oplus K_9[44] \oplus K_8[22] \oplus K_6[22] \oplus \\ &\quad K_5[44] \oplus K_4[22] \oplus K_2[22]. \end{aligned} \tag{5}$$

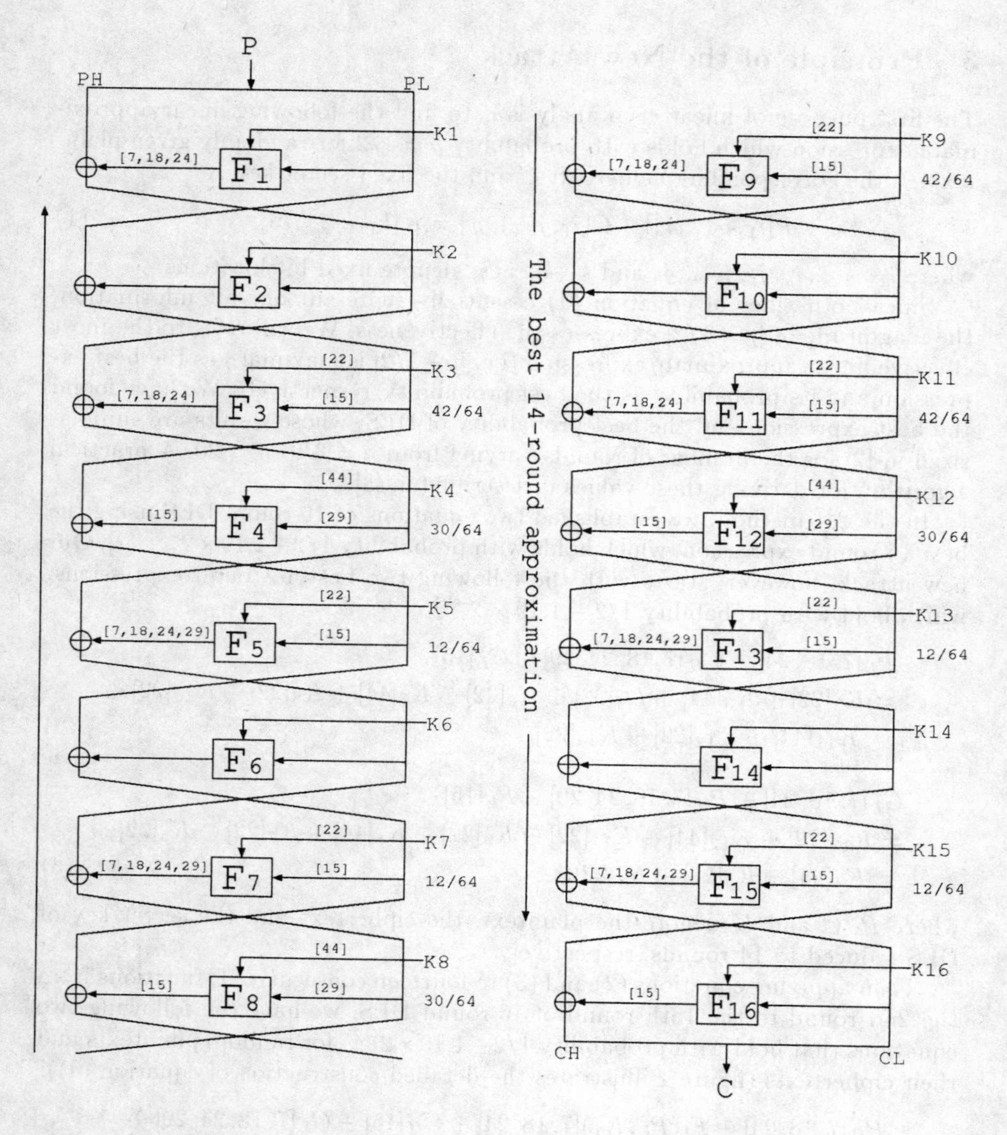

Fig. 1. New linear approximation of 16-round DES.

The first stage of our attack is to solve these equations to derive some of the 56 secret key bits. Now let us consider how much memory is required to solve them and how many secret key bits can be derived from them. For this purpose, we here define "effective text bits" and "effective key bits" of equation (4) or (5) as the text bits and the key bits which affect the left side of each equation, respectively. If an XORed value of several text/key bits affects the left side, we count as one effective text/key bit. Then the following can be easily seen:

- The effective text bits of equation (4) (13 bits):
  $P_L[11], P_L[12], P_L[13], P_L[14], P_L[15], P_L[16], C_L[0], C_L[27], C_L[28],$
  $C_L[29], C_L[30], C_L[31], P_H[7,18,24] \oplus C_H[15] \oplus C_L[7,18,24,29]$.
- The effective key bits of equation (4) (12 bits):
  $K_1[18], K_1[19], K_1[20], K_1[21], K_1[22], K_1[23],$
  $K_{16}[42], K_{16}[43], K_{16}[44], K_{16}[45], K_{16}[46], K_{16}[47]$.

- The effective text bits of equation (5) (13 bits):
  $C_L[11], C_L[12], C_L[13], C_L[14], C_L[15], C_L[16], P_L[0], P_L[27], P_L[28],$
  $P_L[29], P_L[30], P_L[31], C_H[7,18,24] \oplus P_H[15] \oplus P_L[7,18,24,29]$.
- The effective key bits of equation (5) (12 bits):
  $K_{16}[18], K_{16}[19], K_{16}[20], K_{16}[21], K_{16}[22], K_{16}[23],$
  $K_1[42], K_1[43], K_1[44], K_1[45], K_1[46], K_1[47]$.

Note that $P_H[7,18,24] \oplus C_H[15] \oplus C_L[7,18,24,29]$ and $C_H[7,18,24] \oplus P_H[15] \oplus P_L[7,18,24,29]$ represent one-bit information. This observation shows that 13 secret key bits — 12 effective key bits and one bit of the right side — can be derived from each equation using just 13 text bits. We hence obtain a total of 26 secret key bits — they are not duplicated — from equations (4) and (5) using information on 26 text bits.

Let us next consider how to solve these equations. If one substitutes an incorrect key value for K_1 or K_{16} in equation (4) or (5), the probability that the left side equals zero is expected to be closer to 1/2 (not necessarily 1/2). This leads us to maximum likelihood method in regard to key candidates; for each key candidate, we count the number of times that the left side of the equation equals zero. Then, the resultant counter value must reflect the reliability of the corresponding key candidate. We have implemented this scenario as follows:

Algorithm for breaking 16-round DES (I)

Data Counting Phase

Step 1 Prepare 2^{13} counters TA_{t_A} $(0 \leq t_A < 2^{13})$ and initialize them by zeros, where t_A corresponds to each value on 13 effective text bits of equation (4).

Step 2 For each plaintext P and the corresponding ciphertext C, compute the value 't_A' of **Step 1**, and count up the TA_{t_A} by one.

Key Counting Phase

Step 3 Prepare 2^{12} counters KA_{k_A} $(0 \leq k_A < 2^{12})$ and initialize them by zeros, where k_A corresponds to each value on 12 effective key bits of equation (4).

Step 4 For each k_A of **Step 3**, let KA_{k_A} be the sum of TA_{t_A}'s such that the left side of equation (4), whose value can be uniquely determined by t_A and k_A, is equal to zero.

Step 5 Rearrange KA_{k_A} in order of magnitude of $|KA_{k_A} - N/2|$ and rename them $\overline{KA}_{l_A}$ $(0 \leq l_A < 2^{12})$. Then, for each l_A,
- If $(\overline{KA}_{l_A} - N/2) \leq 0$, guess that the right side of equation (4) is 0.
- If $(\overline{KA}_{l_A} - N/2) > 0$, guess that the right side of equation (4) is 1.

At this stage, the key candidate corresponding to $\overline{KA}_{l_A}$ represents the l_A-th likely 13 secret key bits. The total size of required counters is $2^{13} + 2^{12}$, and the computational complexity, which depends on **Step 2** only, is $O(N)$. Note that **Step 2** is parallelizable.

Equation (5) can be also solved in the same manner, in which case we will use the notations TB_{t_B}, KB_{k_B} and $\overline{KB}_{l_B}$ instead of TA_{t_A}, KA_{k_A} and $\overline{KA}_{l_A}$. Our algorithm recovers, therefore, a total of 26 secret key bits, whose bit locations (after the PC–1) are as follows:

$$K[0], K[1], K[3], K[4], K[8], K[9], K[14], K[15], K[18], K[19], K[24], K[25], K[31],$$
$$K[32], K[38], K[39], K[41], K[42], K[44], K[45], K[50], K[51], K[54], K[55],$$
$$K[5] \oplus K[13] \oplus K[17] \oplus K[20] \oplus K[46],$$
$$K[2] \oplus K[7] \oplus K[11] \oplus K[22] \oplus K[26] \oplus K[37] \oplus K[52].$$

The next stage of our attack is to derive the remaining $56 - 26 = 30$ secret key bits. Our aim is to increase the success rate by repeating the search in order of reliability of 26 secret key bits. In other words, we want to make the following algorithm work effectively:

Algorithm for breaking 16-round DES (II)

<u>Exhaustive Search Phase</u>

Step 6 Let W_m $(m = 0, 1, 2, ...)$ be a series of candidates for the 26 secret key bits arranged in order of their reliability.

Step 7 For each W_m, search for the remaining 30 secret key bits until the correct value is found.

Now we have to describe W_m explicitly by l_A and l_B. Since the most likely candidate for the 26 key bits clearly corresponds to $\overline{KA}_0$ and $\overline{KB}_0$, we should consider this combination at first, which will be referred to as $W_0 = (\overline{KA}_0, \overline{KB}_0)$. The second likely candidates are obviously $W_1 = (\overline{KA}_0, \overline{KB}_1)$ and $W_2 = (\overline{KA}_1, \overline{KB}_0)$ with the same reliability. Then, are the next likely ones $W_3 = (\overline{KA}_0, \overline{KB}_2)$ and $W_4 = (\overline{KA}_2, \overline{KB}_0)$, or $W_3 = (\overline{KA}_1, \overline{KB}_1)$? How many candidates are needed to finish **Step 7** in reasonable time ? In the next chapter we will give a practical solution of these problems.

4 Success Rate and Complexity

We relate the problems to DES reduced to 8 rounds, which will be referred to as "8-round DES" below. Now consider the following two equations of 8-round DES derived from the best 6-round expression which holds with probability $1/2 - 1.95 \times 2^{-9}$:

$$P_H[7,18,24] \oplus F_1(P_L, K_1)[7,18,24] \oplus C_H[15] \oplus C_L[7,18,24,29] \oplus$$
$$F_8(C_L, K_8)[15] = K_3[22] \oplus K_4[44] \oplus K_5[22] \oplus K_7[22], \qquad (6)$$

$$C_H[7,18,24] \oplus F_8(C_L,K_8)[7,18,24] \oplus P_H[15] \oplus P_L[7,18,24,29] \oplus$$
$$F_1(P_L,K_1)[15] = K_6[22] \oplus K_5[44] \oplus K_4[22] \oplus K_2[22]. \quad (7)$$

Note that the left side of each equation is essentially the same as equation (4) or (5), respectively. We make use of this fact to evaluate the efficiency of our attack. The following lemma, which is an extension of lemma 4 in [2], relates the full 16-round DES to 8-round DES:

Lemma 1. *Let N be the number of given random plaintexts and p be the probability that the following equation holds:*

$$P[i_1,i_2,...,i_a] \oplus C[j_1,j_2,...,j_b] \oplus F_1(P,K_1)[u_1,u_2,...,u_d] \oplus$$
$$F_n(C,K_n)[v_1,v_2,...,v_e] = K[k_1,k_2,...,k_c]. \quad (8)$$

Assuming $|p-1/2|$ is sufficiently small, the probability that the l-th likely solution of equation (8) agrees with the real key depends on l, $u_1,u_2,...,u_d$, $v_1,v_2,...,v_e$, and $\sqrt{N}|p-1/2|$ only.

This lemma tells us that the success rate of our attack on 8-round DES with N_8 plaintexts is the same as that on 16-round DES with N_{16} plaintexts as long as the following relation holds:

$$\sqrt{N_8}|1.95 \times 2^{-9}| = \sqrt{N_{16}}|1.19 \times 2^{-21}|. \quad (9)$$

This is equivalent to

$$1.49 \times 2^{-26} \times N_{16} = N_8, \quad (10)$$

and hence 2^{43} plaintexts on 16-round DES, for instance, correspond to 1.49×2^{17} plaintexts on 8-round DES.

Note: According to the common definition of 8-round DES, which adopts eight F-functions from the first to the eighth round of 16-round DES, equations (6) and (7) yield only 23 secret key bits because three of 26 bits are duplicated. To avoid this difference from the case of 16-round DES, this paper treats the 8-round DES whose key schedule part is modified so that no secret key bit is duplicated. Our computer experiments on 8-round DES below were carried out under this condition.

We made computer experiments in solving equation (6) 100,000 times to estimate the behavior of solutions of equation (4). Figure 2 illustrates the results interpreted as the case of 16-round DES, where the ordinate (y axis) shows the probability that the ranking of a solution of equation (4) is not greater than the value of the abscissa (x axis); for example, when we solve equation (4) with 2^{43} known plaintexts, the probability that the secret key agrees with one of $\overline{KA_{l_A}}$ ($0 \le l_A < 100$) is expected to be 86%. The lowest curve represents the case where we select a key candidate randomly: namely, $y = x/2^{13}$.
Figure 3 summarizes our attack on the full 16-round DES, where the reliability of $W_m = (\overline{KA_{l_A}}, \overline{KB_{l_B}})$ has been determined in order of the magnitude of

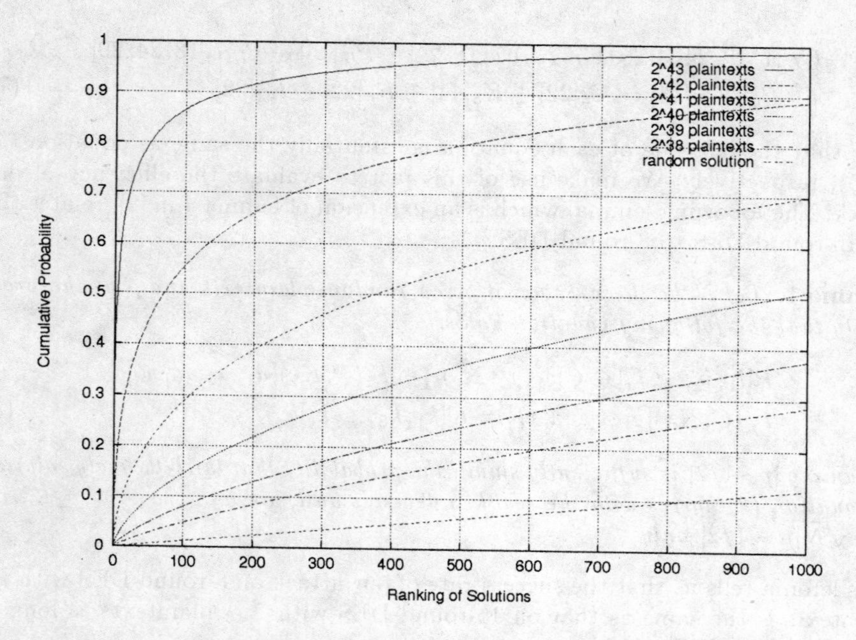

Fig. 2. Expected ranking of solutions of equation (4).

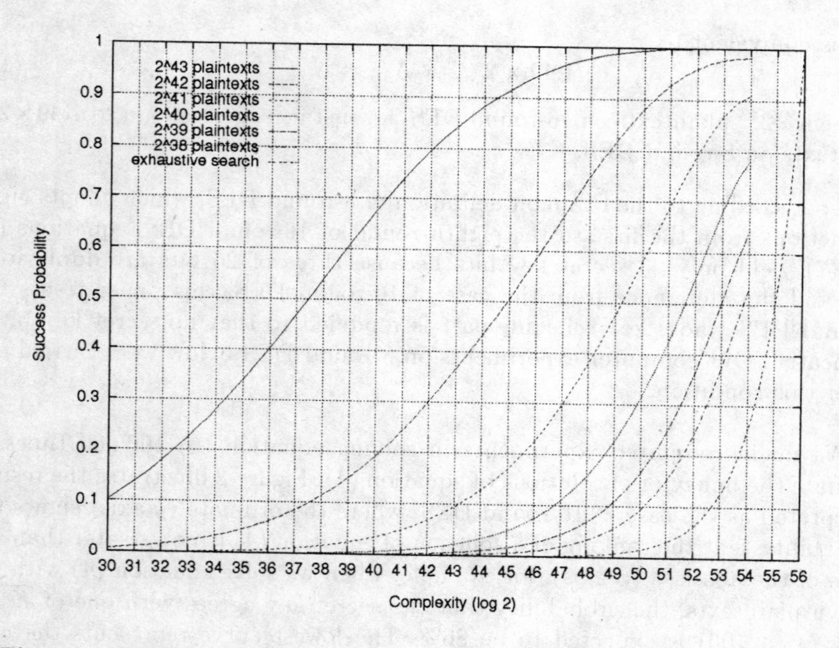

Fig. 3. Expected success rate and complexity of our attack on 16-round DES.

$(l_A+1)\times(l_B+1)$, which is the formula derived experimentally from the case of the 8-round DES. The abscissa and ordinate denote the computational complexity and the success probability, respectively. This figure tells us that when we attack the full 16-round DES with 2^{43} plaintexts, the probability that the secret key can be found within $m = 2^{13}$ (i.e. complexity $2^{30} \times 2^{13} = 2^{43}$), is expected to be 85%. For another example, the success probability is expected to be 10% with complexity 2^{50} if 2^{38} known plaintexts are available. The lowest curve represents the trivial case where we search for the 56 secret key bits exhaustively: $y = 2^{x-56}$.

5 The Computer Experiment

We made the first computer experiment in breaking the full 16-round DES on the basis of the above scenario. The program, implemented by software only, was described in C and assembly languages consisting of a total of 1000 lines. It occupies 1Mbyte in running. The main flow of the program is as follows (we use C-like notations):

```
for( i=0; i<2^43; i++ ){ /* parallelizable */

      P = Generate_Random_Plaintext();
      C = Encipher_Plaintext( P ); /* using the secret key K */

      TA[ 13bit_address_pointed_by_P_and_C ]++; /* Step 2 */
      TB[ 13bit_address_pointed_by_P_and_C ]++; /* Step 2 */
}

for( k=0; k<2^12; k++ ){ /* each value on effective key bits */
for( t=0; t<2^13; t++ ){ /* each value on effective text bits */

      if( Left_Side_of_Equation_4( t, k ) == 0 )
            KA[ k ] += TA[ t ]; /* Step 4 */
      if( Left_Side_of_Equation_5( t, k ) == 0 )
            KB[ k ] += TB[ t ]; /* Step 4 */
}
}

Rearrange_Counters( KA, KA¯ ); /* Step 5 */
Rearrange_Counters( KB, KB¯ ); /* Step 5 */

for( m=0; m<2^24; m++ ){ /* parallelizable */

      K26 = Derive_m_th_Likely_26bits( m, KA¯, KB¯ ); /* Step 6 */
      Return_Value = Search_Remaining_30bits( K26 ); /* Step 7 */
      if( Return_Value == FOUND ) exit( SUCCESS );
}

exit( FAILURE ); /* theoretically possible
                    but practically unreachable */
```

We used a sequence $\{g^0, g^1, g^2, g^3, ...\}$ for `Generate_Random_Plaintext()` routine, where g is a generator of cyclic group $GF(2^{64})^{\times}$, which is convenient for our purpose, parallel computing. `Encipher_Plaintext()` is a routine for enciphering plaintexts under a fixed key, which runs at the rate of 19Mbit/sec. On the other hand, `Search_Remaining_30bits()` is also an enciphering routine but encodes a fixed plaintext under given keys, which runs at the rate of 9Mbit/sec. `Rearrange_Counters()` is a sorting routine for a 2^{12}-dimensional array. `Derive_m_th_Likely_26bits()` can be also easily implemented using the $(l_A + 1) \times (l_B + 1)$ rule.

Calculations of both the first and last loops were carried out in parallel by 12 computers. It took 40 days to finish the first loop, where almost all time was spent for `Encipher_Plaintext()` routine. The middle loop and the sorting routine were easily executed. The last loop took 10 days and finally resulted in all of the 56 secret key bits.

6 Concluding Remarks

We have described an improvement of linear cryptanalysis and presented the first successful experiment in breaking the full 16-round DES. The topics below are remarks and possible further improvements.

- The author does not know whether **Step 1 ~ Step 5** give the best way for solving equations (4) and (5). It should be noted that we have not made use of all information available from these equations: to be concrete, when we substitute $K'_1(\neq K_1)$ and $K'_{16}(\neq K_{16})$ for K_1 and K_{16} in the left side of equation (4), the probability that the equation holds depends on only $K_1 \oplus K'_1$ and $K_{16} \oplus K'_{16}$. This fact obviously indicates more than what we have realized in this paper. Therefore if this property could be used effectively, the reliability of the solution might be improved.
- In this paper, we have solved two equations to obtain 26 key bits and then searched for the remaining 30 key bits. However, it is also possible to solve more equations to have more key bits before the search procedure (**Steps 6** and **7**). For example, there are two second best expressions that hold with probability $1/2 - 1.49 \times 2^{-21}$. Although the reliability of these solutions is lower, this loss might be recoverable by repeating the search procedure, because the number of the remaining key bits is then smaller.
- The results on figure 2 and figure 3 have been derived experimentally. If we succeed in tracing curves in figure 2 with simple functions, figure 3 can be also formalized and then a new combination rule will give more effective results instead of the $(l_A + 1) \times (l_B + 1)$ rule.

More detailed discussion including experimental data, which we have omitted due to lack of space, will appear in the full paper.

References

1. National Bureau of Standards: Data Encryption Standard. U.S. Department of Commerce, Federal Information Processing Standards **46** (1977)
2. Matsui, M.: Linear Cryptanalysis Method for DES cipher. Advances in Cryptology – Eurocrypt'93, Lecture Notes in Computer Science, Springer-Verlag **765** (1993) 386–397
3. Matsui, M.: On correlation between the order of S-boxes and the strength of DES. Pre-proceedings of Eurocrypt'94 (1994) 375–387
4. Hellman, M., Merkle, R., Schroeppel, R., Washington, L., Diffie, W., Pohlig, S., Schweitzer, P.: Results of an initial attempt to cryptanalyze the NBS Data Encryption Standard. Information Systems Laboratory, Stanford University **76-042** (1976)
5. Shamir, A.: On the security of DES. Advances in Cryptology – Crypto'85, Lecture Notes in Computer Science, Springer-Verlag **218** (1985) 280–281
6. Davies, D., Murphy, S.: Pairs and triplets of DES s-boxes. (preprint)
7. Rueppel, R.A.,: Analysis and design of stream ciphers. Springer Verlag (1986)

Linear Cryptanalysis of the Fast Data Encipherment Algorithm

Kazuo Ohta[1] and Kazumaro Aoki[2]*

[1] NTT Network Information Systems Laboratories
1-2356 Take, Yokosuka-shi, Kanagawa-ken, 238-03 Japan
[2] Department of Mathematics
School of Science and Engineering, Waseda University
3-4-1 Ookubo, Shinjuku-ku, Tokyo-to, 169 Japan

Abstract. This paper discusses the security of the Fast Data Encipherment Algorithm (FEAL) against Linear Cryptanalysis. It has been confirmed that the entire subkeys used in FEAL–8 can be derived with 2^{25} pairs of known plaintexts and ciphertexts with a success rate approximately 70% spending about 1 hour using a WS (SPARCstation 10 Model 30). This paper also evaluates the security of FEAL–N in comparison with that of the Data Encryption Standard (DES).

1 Introduction

This paper analyzes the applicability of Linear Cryptanalysis to the Fast Data Encipherment Algorithm (FEAL) [MSS88]. The structure of FEAL is similar to DES, except, for example, the permutation and the S-Boxes in F-function of DES are replaced by byte rotation and addition operation, and these differences are interesting from the viewpoint of cryptanalysis. In the Linear Cryptanalysis of FEAL, our main concerns in evaluating the security of FEAL considering the replacement of F-function and S-Boxes are: 1) how to find effective linear expressions, 2) an estimation of the success rate against the number of pairs of plaintexts and corresponding ciphertexts and the approximate probability, and 3) an estimation of the memory size and the processing amount of the attack.

2 Linear Cryptanalysis

2.1 Notations and Preliminaries

The modified FEAL and its modified F-function [MY92] are analyzed here. We use the similar notations and define the right most bit of each symbol as the 0-th bit, which is the lowest bit, as well as in the reference [M93].

2.2 Principle

Linear Cryptanalysis analyzes the probability that the following equation holds.

$$P[i_1, i_2, \ldots, i_a] \oplus C[j_1, j_2, \ldots, j_b] = S[k_1, k_2 \ldots, k_c], \tag{1}$$

where $i_1, i_2, \ldots, i_a, j_1, j_2, \ldots, j_b, k_1, k_2 \ldots, k_c$ are fixed bit locations defined by the linear expression, $(\Gamma P, \Gamma C, \Gamma K)$. The value of the right side of this equation depends only on the key values. We denote $S[k_1, k_2 \ldots, k_c]$ by S simply.

---

* A part of this research was conducted while the second author stayed at NTT Network Information Systems Laboratories as a spring intern in March of 1994.

Two kinds of probability are defined in Linear Cryptanalysis: one is $p = \mathrm{Prob}_{P,K}\{E(P,K)(\Gamma C) \oplus P(\Gamma P) = K(\Gamma K)\}$, and the other is the absolute value of probability different from a half, $p' = |p - 1/2|$. Hereafter, p' will be used as the ***probability*** of the linear expression $(\Gamma P, \Gamma C, \Gamma K)$.

2.3 Implementation Techniques

Matsui [M93] proposed the following practical implementation of an attack against DES, and captured the *effective text bits* among text information, P and C, which are essential to calculate Equation (2), and the *effective key bits* among key information, K_1 and K_n, which are essential to calculate Equation (2). Hereafter, t and k denote the number of effective text bits and the number of effective key bits, respectively.

$$P[i_1, i_2, \ldots, i_a] \oplus C[j_1, j_2, \ldots, j_b] \oplus F_1(P_L, K_1)[u_1, u_2, \ldots, u_d] \oplus F_n(C_L, K_n)[v_1, v_2, \ldots, v_e] = S. \quad (2)$$

Algorithm 1 (Counter Technique)

Step 1: Prepare 2^t counters $U_i (0 \le i < 2^t)$, where i corresponds to each value on the t effective text bits of Equation (2).

Step 2: For each plaintext and the corresponding ciphertext, compute the value 'i' of **Step 1** and count up the counter U_i by one.

Step 3: Prepare 2^k counters $T_j (0 \le j < 2^k)$, where j corresponds to each value on the k effective text bits of Equation (2).

Step 4: For each 'j' of **Step 3**, let T_j be the sum of U_i's such that the left side of Equation (2), whose value can be uniquely determined by i and j, is equal to 0.

Step 5: Let $T_{max} (T_{min})$ be the maximal(minimal) value of all $T_{i,j}$'s.
If $|T_{max} - N/2| > |T_{min} - N/2|$, then adopt the key candidate corresponding to T_{max} and guess $S = 0$ when $p > 1/2$ or 1 when $p < 1/2$.
If $|T_{max} - N/2| < |T_{min} - N/2|$, then adopt the key candidate corresponding to T_{min} and guess $S = 1$ when $p > 1/2$ or 0 when $p < 1/2$.

The computational complexity of this procedure is $O(N) + O(2^{t+k})$. The number of counters, U_i and T_j, required by this procedure is $2^t + 2^k$. If we approximate $(n-2)$-round F-functions from the second round to the $(n-1)$-th round based on an $(n-2)$-round linear expression, we call this strategy **2-Round Elimination.** **1-Round Elimination** is also defined using an $(n-1)$-round linear expression.

3 Linear Approximation of FEAL

3.1 What are the Problems

The essential differences between DES and FEAL are the structure of S-Boxes and that of F-function. More exactly, S-Boxes of DES are defined in a non-mathematical way using tables. S-Boxes of FEAL are defined mathematically using modular addition calculation with two bits left rotation. So it seems easier to find some property of S-Boxes of FEAL than that of DES. On the other hand, the eight S-Boxes in F-function of DES act in parallel more independently than four S-Boxes in F-function of FEAL which act sequentially, where the byte rotation is built in instead of the permutation of DES. Thus, it seems easier to find some semi-global property of F-function of DES than that of FEAL.

3.2 Linear Expressions of F-function

We get various linear expressions of S-Boxes approximating the addition operation with the bitwise consideration of carry propagation as was done in [CG91].

When $a + b = x$, for example, $a[i] \oplus b[i] = x[i]$ holds with probability of $2^{-(i+1)} (i \geq 0)$, $a[i, i-j] \oplus b[i] = x[i]$, $a[i] \oplus b[i, i-j] = x[i]$ and $a[i] \oplus b[i] = x[i, i-j]$ hold with probability of $2^{-(j+1)} (1 \leq j \leq i)$. Note that $a[0] \oplus b[0] = x[0]$ always holds, since there is no carry at the least significant bit in the addition operation, and this gives 15 non-trivial linear expressions of F-function with probability of 1/2, which can be always extended to 3-round linear expressions. If $j = 1$, we can make many examples with probability of 1/4 ignoring the bit position of i. This gives many local linear expressions with probability of 1/4.

Here the concatenation rule of operations inside the F-function [B94, M94] is also applicable in the same way as that between F-functions.

3.3 Linear Expressions of Reduced Round FEAL

We developed the following search algorithm to find effective 7-round linear expressions, where $(\Gamma Y_{4-r}, \Gamma X_{4-r}) = (\Gamma Y_{4+r}, \Gamma X_{4+r})$ holds for $r = 1, 2, 3$.

Algorithm 2 (Search Algorithm of 7-Round Linear Expression)

Step 1: Set $(\Gamma Y_4, \Gamma X_4) = (0, 0)$.
Step 2: Select $(\Gamma Y_2, \Gamma X_2)$ of F-function whose probability is 1/2.
Step 3: Search ΓY_3 where $(\Gamma Y_3, \Gamma X_3)$ has the probability of 2^{-2}, given $\Gamma X_3 = \Gamma Y_2$.
Step 4: Search $\Gamma X'_2$ where $(\Gamma X_2, \Gamma X'_2)$ has the probability of greater than or equal to 2^{-3}.
Step 5: Put $(\Gamma Y_1, \Gamma X_1) = (\Gamma Y_3 \oplus \Gamma X_2, \Gamma X_3 \oplus \Gamma X'_2)$. Check whether its probability is greater than or equal to 2^{-4} exhaustively, if $(\Gamma Y_3, \Gamma X_3)$ activates the same S-Boxes of F-function as $(\Gamma X_2, \Gamma X'_2)$.

We have found the following eight pairs $(\Gamma X_2, \Gamma X'_2)$ using the above algorithm and sixteen 7-round linear expressions with probability of greater than 2^{-9}. This is one of examples with probability of 1.764×2^{-9}, which is effective in our implementation described in Section 5.

Note that the middle 5-round part of the expression also has the probability of 1/8, while Biham described a 5-round linear expression with probability of 1/32 in [B94].

ΓX_2	$\Gamma X'_2$
00000100	10105050
00000100	18185858
00000100	10107878
00000100	18187070
01000000	50101010
01000000	58181818
01000000	70101818
01000000	78181010

r	ΓY_r	ΓX_r	p'_r	
P	1D000400	50101010		
1	1D000400	54111010	85×2^{-10}	1.764×2^{-9} (rounds 1–7)
2	04010000	01000000	2^{-1}	
3	1C000400	04010000	2^{-2}	
4	00000000	00000000	2^{-1}	
5	1C000400	04010000	2^{-2}	
6	04010000	01000000	2^{-1}	
7	1D000400	54111010	85×2^{-10}	
C	1D000400	50101010		

4 Discussion

4.1 Attack Strategy

Since the approximate probability of a linear expression for 6-round is larger than that for 7-round and $N = c \times p'^{-2}$ holds, the 2-Round Elimination strategy is better than 1-Round Elimination from the standpoint of the required number of pairs for attack. However, 2-Round Elimination is infeasible, since the number of effective text bits, t, and the number of effective key bits, k, satisfy $t, k = 24 \sim 30$ roughly, and the processing amount is $O(2^{42 \sim 48})$ in 2-Round Elimination where we assume $N = 2^{18}$, since Biham's linear expression for 6-round satisfies $p' = 2^{-9}$, where the Biham's iterative 4-round expression [B94] is applied to 6-round.

Let us estimate t and k for Biham's 7-round linear expression, and those for our expression for 7-round, assuming the 1-Round Elimination Technique. The processing amount of an attack using our linear expression is $O(2^{36 \sim 40})$, since $t, k = 20 \sim 24$ holds roughly and $p' = 1.149 \times 2^{-8}$. The processing amount of an attack of 1-Round Elimination using Biham's 7-round expression is $O(2^{24 \sim 30})$ using **Algorithm 1**, since $t, k = 12 \sim 15$ and $p' = 2^{-11}$. The number of counters, U_i and T_j, required by **Algorithm 1** is $2^{12 \sim 15}$, which is acceptable.

We decided to adopt the **1-Round Elimination Technique** that requires us to analyze the following equation:

$$P_L[16, 23, 25, 26, 31] \oplus C_H[31] \oplus C_L[16, 23, 25, 26] \oplus F_8(C_H \oplus C_L, K_8)[23, 25, 31] = S. \quad (3)$$

Our linear expression is effective for the later phases of an attack to derive subkeys other than those derived from the above equation.

4.2 Comparison with DES

The best expression can be obtained by an exhaustive search algorithm against DES [M93, M94]. Since the number of active S-Boxes, which are approximated with a certain masking value, at the first and last rounds of F-function is 2 and the bit length of data input to each S-Box is 6, the number of effective key bits is 12 in Equation (2).

On the other hand, since the byte rotation is built in implicitly between S-Boxes, the number of effective key bits and effective text bits seems to be $24 \sim 30$, which is larger than is true with DES. Thus 2-Round Elimination is infeasible in FEAL, which it is efficient in DES. Unfortunately, since we don't have any practical search algorithm to obtain the best expression of FEAL, there might be a better linear expression than Biham's.

How about the effective key and text bits? The closer from the right side an input bit is to a bit position related to a reference point output by the eighth F-function, the more strongly the value of the input bit determines the value of the XOR operation performed on the reference points, $F_8(C_H \oplus C_L, K_8)[23, 25, 31]$ in Equation (3). Therefore, the *effective key bits* should be subdivided into *explored key bits* and *detected key bits* for an attack against FEAL. Note that since each key bit input to an S-Box of DES influences all output bits more equally than that of FEAL, detected key bits are identical to explored key bits in DES. As a result, the treatment of effective key bits is simpler in an attack against DES than against FEAL. The similar discussion is valid in the treatment of effective text bits, which provides the number of counters, U_i. Thus there are various strategies to reduce the number of effective (explored/detected) key/text bits in an attack against FEAL.

Concerning the parameter, N, of FEAL–N, where N means the iteration number of F-function, it seems that while FEAL–32 is as secure as DES against Differential Cryptanalysis, FEAL–16 is as secure as DES from the standpoint of Linear Cryptanalysis, since the number of key bits which are explored by the attack is 12 with the 14-round linear expression with probability of 1.192×2^{-21} in DES, while it is $12 \sim 15$ with the 15-round linear expression with probability of 2^{-23} in FEAL assuming the Biham's iterative 4-round expression is applied to the 15-round case.

5 Experimentation Results

The following information was described in [OA94]:

(1) A table relating the success rate, the number of pairs of plaintexts and ciphertexts, and the effective key bits needed to solve Equation (3),
(2) How to derive the remaining values of all subkeys, and
(3) How to improve the success rate.

6 Concluding Remarks

It has been confirmed that the entire subkeys used in FEAL–8 can be derived from 2^{25} known plaintexts with a success rate approximately 70% spending about 1 hour, from 2^{26} known plaintexts with a success rate about 100% spending a little over 1 hour using a WS (SPARCstation 10 Model 30).

It seems that while FEAL–32 is as secure as the 16-round DES against Differential Cryptanalysis, FEAL–16 is as secure as it from the standpoint of Linear Cryptanalysis if we restrict ourselves to Matsui's implementation technique using Biham's linear expression.

There are several open problems:

(1) Search algorithm to obtain the best expression of FEAL,
(2) More efficient technique than **Algorithm 1**, and
(3) More efficient strategy for reducing the numbers of effective text bits and effective key bits in an attack against FEAL–8.

References

[B94] E. Biham, "On Matsui's Linear Cryptanalysis," EUROCRYPT'94

[CG91] A. Tardy-Corfdir and H. Gilbert, "A known plaintext attack of FEAL–4 and FEAL–6," CRYPTO'91

[M93] M. Matsui, "Linear Cryptanalysis Method for DES Cipher," EUROCRYPT'93

[M94] M. Matsui, "On Correlation between the Order of S-Boxes and the strength of DES," EUROCRYPT'94

[MY92] M. Matsui and A. Yamagishi, "A New Method for Known Plaintext Attack of FEAL Cipher," EUROCRYPT'92

[MSS88] S. Miyaguchi, A. Shiraishi and A. Shimizu, "Fast Data Encipherment algorithm FEAL–8," Review of Electrical Communication Laboratories, Vol. 36, No. 4, 1988

[OA94] K.Ohta and K.Aoki, "Linear Cryptanalysis of the Fast Data Encipherment Algorithm," Technical Report of IEICE Japan, ISEC94-5, May, 1994 (which contains more information than this paper.)

Differential-Linear Cryptanalysis

Susan K. Langford[1] and Martin E. Hellman

Department of Electrical Engineering
Stanford University
Stanford, CA 94035-4055

Abstract. This paper introduces a new chosen text attack on iterated cryptosystems, such as the Data Encryption Standard (DES). The attack is very efficient for 8-round DES,[2] recovering 10 bits of key with 80% probability of success using only 512 chosen plaintexts. The probability of success increases to 95% using 768 chosen plaintexts. More key can be recovered with reduced probability of success. The attack takes less than 10 seconds on a SUN-4 workstation. While comparable in speed to existing attacks, this 8-round attack represents an order of magnitude improvement in the amount of required text.

1 Summary

Iterated cryptosystems are encryption algorithms created by repeating a simple encryption function n times. Each iteration, or round, is a function of the previous round's output and the key. Probably the best known algorithm of this type is the Data Encryption Standard (DES) [6]. Because DES is widely used, it has been the focus of much of the research on the strength of iterated cryptosystems and is the system used as the sole example in this paper.

Three major attacks on DES are exhaustive search [2, 7], Biham-Shamir's differential cryptanalysis [1], and Matsui's linear cryptanalysis [3, 4, 5]. While exhaustive search is still the most practical attack for full 16 round DES, research interest is focused on the latter analytic attacks, in the hope or fear that improvements will render them practical as well. For example, linear cryptanalysis is much faster than exhaustive search, but requires an impractical 2^{43} known plaintexts. In contrast, exhaustive search requires only one known plaintext block or about 1000 bits in a ciphertext only attack. The goal of our work is therefore to reduce the amount of text required in the analytic attacks.

This paper builds on techniques from differential and linear cryptanalysis, creating an eight round attack which recovers 10 bits of key with only 512 chosen plaintexts. While the computation time is comparable to pre-existing attacks, the amount of required text is reduced by an order of magnitude. The best current

1 This author was supported by NSF grant NCR-9205663

2 Because FIPS PUB 46 specifies 16 rounds as part of the standard, strictly speaking, we should use the more cumbersome term "DES reduced to 8 rounds." While for ease of exposition we use the simpler "8-round DES" the reader should remember what is intended.

Biham-Shamir 8-round attack requires over 5,000 chosen plaintexts and Matsui's 8-round attack requires approximately 500,000 known plaintexts. In comparing our attack with these others, it should be remembered that they recover more bits of key and that Matsui's is a more desirable known plaintext attack. They also extend more efficiently to 16 rounds than ours. Our attack should therefore be viewed as providing an interesting new possibility that supplements earlier attacks when the amount of required text is at a premium. Of course, it is our hope that the attack can be extended.

2 Notation

We use FIPSPUB-46's DES numbering so that plaintext, ciphertext, and the bits of the intermediate results (L_n, R_n) are numbered from 1 to 64 reading from left to right. This numbering differs from Matsui's paper, which numbers bits from 0 to 63, reading from right to left. Similarly, in this paper the input to an S-box is taken as $(x_1, x_2, x_3, x_4, x_5, x_6)$ while Matsui uses $(x_5, x_4, x_3, x_2, x_1, x_0)$. We will use Matsui's notation in which $A[i]$ represents the ith bit of A and $A[i, j, ..., k] = A[i] \oplus A[j] \oplus ... \oplus A[k]$.

We will ignore the initial and final permutations, IP and IP^{-1}, since they have no cryptographic significance in a chosen or known text attack. Thus, we refer to (L_0, R_0), the 64 bits after IP as the plaintext and (L_n, R_n), the 64 bits before IP^{-1}, as the ciphertext. This notation differs from both Biham-Shamir and Matsui in that they take (R_n, L_n) as the ciphertext. Our notation simplifies concatenation of k-round and l-round attacks into $(k + l)$ round attacks.

3 Review of Differential and Linear Cryptanalysis

This section is included for completeness. The reader familiar with differential and/or linear cryptanalysis can omit the corresponding subsections.

3.1 Differential Cryptanalysis

The basic idea of differential cryptanalysis is that, while any single plaintext produces a ciphertext that appears random, the same is not true on a differential basis. Two chosen plaintexts, P and P^*, which XOR to a carefully chosen differential plaintext $P' = P \oplus P^*$ can encipher to two ciphertexts C and C^* such that $C' = C \oplus C^*$ takes on a specific value with non-negligible probability. As a trivial example, $P' = 0$ causes $C' = 0$ with probability 1 since $P = P^*$ implies $C = C^*$. More interestingly, Biham and Shamir found that, for 5-round DES, $P' = 405C000004000000_x$ causes $C' = 04000000405C0000_x$ with probability $\frac{1}{10485.76}$. They use this 5-round "characteristic" in an attack on 8-round DES by deciphering portions of the ciphertext (L_8, R_8) to determine when their characteristic has occurred, in which case they are able to derive a number of bits of key. Their attack is efficient because the partial deciphering of (L_8, R_8)

to tell when the characteristic has occurred depends on portions of the key small enough to allow a search. Making use of symmetries, they are able to break 8-round DES with 5,000 chosen plaintexts and 16-round DES with 2^{47} chosen plaintexts.

For the purposes of this paper, it is not necessary for the reader to understand further details of Biham and Shamir's attack. It is sufficient to be familiar with the concept of working differentially. The interested reader is referred to [1] for a complete description of Biham and Shamir's breakthrough in cryptanalysis.

3.2 Linear Cryptanalysis

A second breakthrough, linear cryptanalysis, was recently introduced by Matsui [1, 3, 4, 5]. This approach works with a known plaintext attack, as opposed to a chosen text attack. Linear cryptanalysis finds probabilistic parity relations between selected bits of the plaintext, the ciphertext, and the key. These parity relations derive from parity relations within the S-boxes that differ from the uniform 50-50 distribution and which can then be connected through multiple rounds.

Matsui was able to find useful parity relations for an arbitrary number of rounds of DES. For example, he found that for three round DES,

$$(L_0[3,8,14,25] \oplus R_0[17]) \oplus (R_3[3,8,14,25] \oplus L_3[17]) \oplus (K_1[26] \oplus K_3[26]) = 0, \quad (1)$$

with probability $p = 0.695$. In general, he uses either an $n-1$ round or $n-2$ round parity relation to attack n-round DES. He can use the 3-round relation (1) in a 4-round attack, by noting that although $K_1[26] \oplus K_3[26]$ is not known, its effect is to cause the reduced equation

$$(L_0[3,8,14,25] \oplus R_0[17]) \oplus (R_3[3,8,14,25] \oplus L_3[17]) = 0. \quad (2)$$

to be satisfied eigher with probability 0.695 or 0.305, both of which are different from 0.5, the value expected with random data. Matsui can decipher backwards through round 4, as shown in figure 1, to calculate the value of the necessary bits. This decipherment depends on six bits of key, requiring a search over only 64 values.

While we only describe Matsui's 4-round attack, the general idea follows in a straightforward manner from this example. Using the best 6-round parity relation, which holds with probability $0.5 - 1.95 * 2^{-9}$, Matsui was able to break 8-round DES with approximately 500,000 known plaintexts in less than a minute on a workstation. Similarly, using the best 14-round parity relation, which holds with probability $0.5 - 1.19 * 2^{-21}$, he was able to break 16-round DES with 2^{43} known plaintexts in 50 days using 12 HP9735 workstations. The interested reader is referred to [3, 4, 5] for more information.

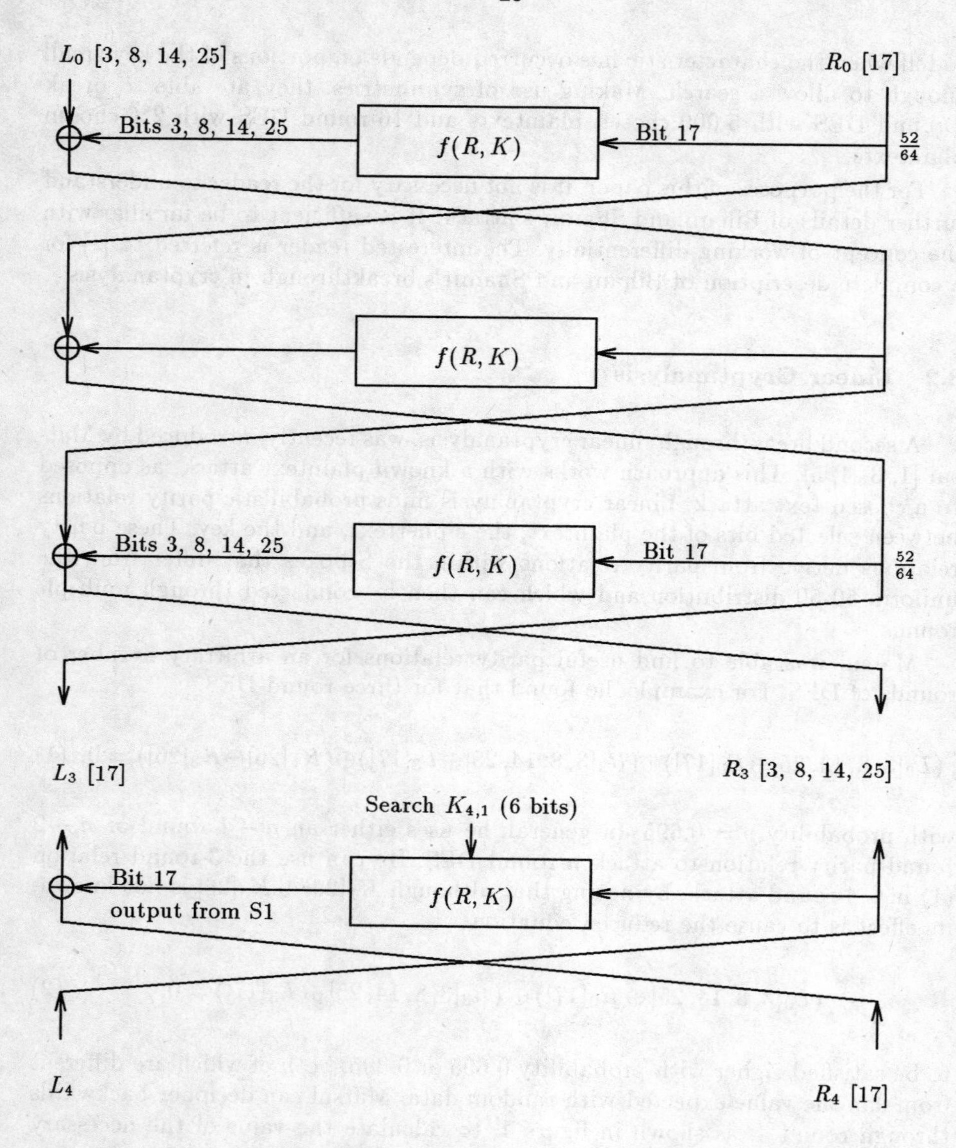

Fig. 1. Matsui's 4-round attack on DES

4 Differential-Linear Cryptanalysis

By the technique to be described in section 5, we can complement bits 2 and/or 3 of L_1 and keep the other 62 bits of (L_1, R_1) unchanged. The key observation in our attack is that this behavior in (L_1, R_1) leaves many bits of (L_4, R_4) unchanged. In particular, the input bits to Matsui's best 3-round

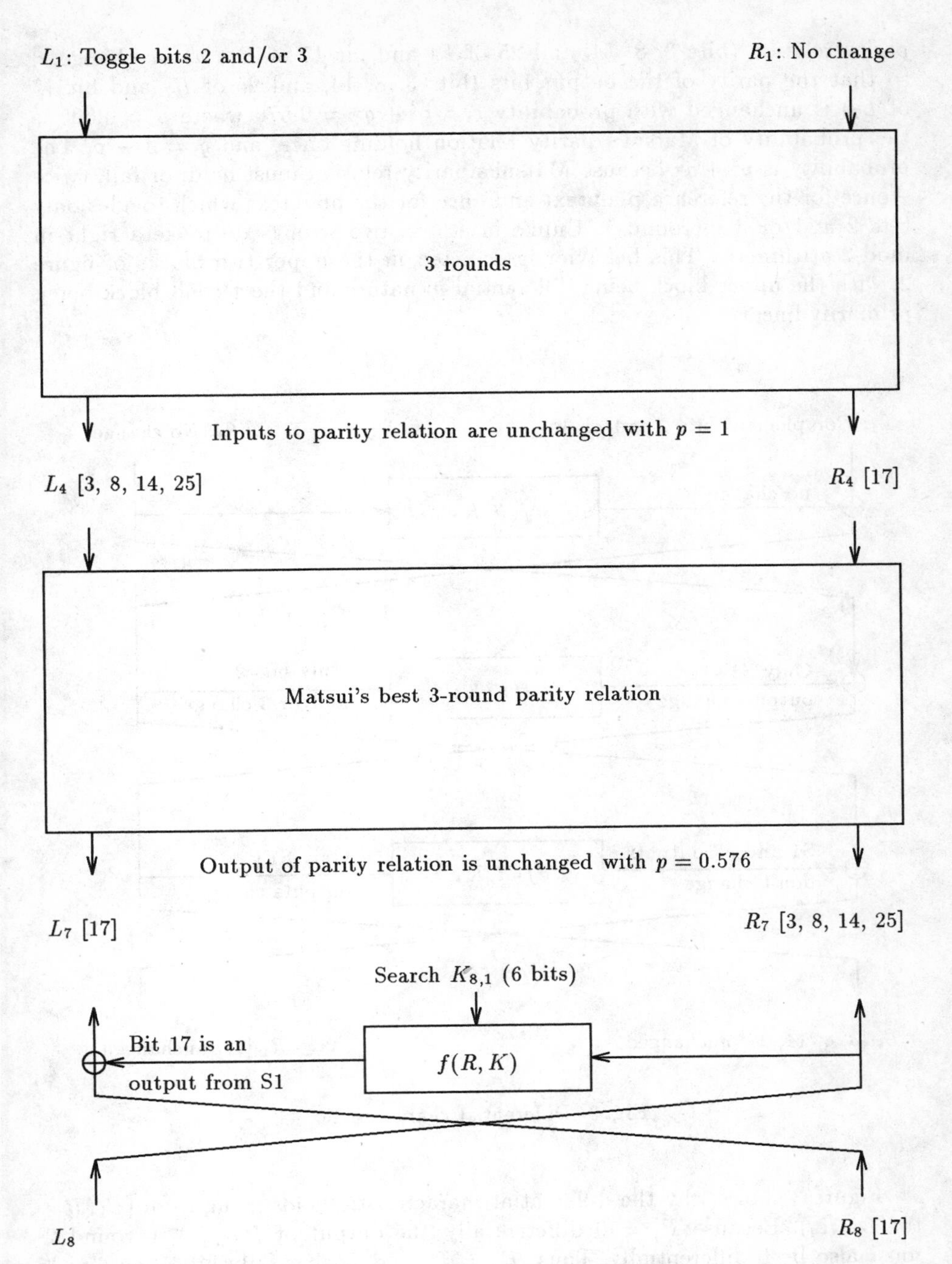

Fig. 2. Differential-Linear attack on 8-round DES

parity relation (bits 3, 8, 14, and 25 of L_4 and bit 17 of R_4) *never* change,[3] so that the parity of the output bits (bits 3, 8, 14, and 25 of R_7 and bit 17 of L_7) is unchanged with probability $r = p^2 + q^2 = 0.576$ where $p = 0.695$ is the probability of Matsui's parity relation holding once, and $q = 1 - p$. The probability is $p^2 + q^2$ because Matsui's parity relation must hold, or fail, twice – once for the reference plaintext and once for the plaintext which toggles only bits 2 and/or 3 in round 1. Unlike in ethics, two wrongs do make a right in mod-2 arithmetic. This behavior is depicted in the upper two blocks of figure 2, with the upper block being differential in nature and the second block being primarily linear.

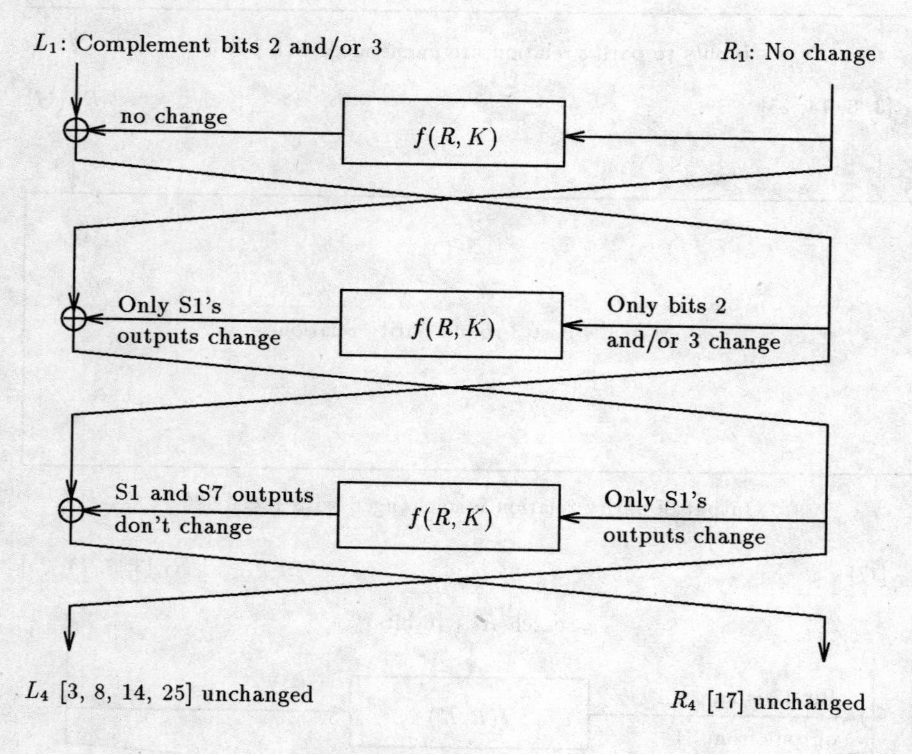

Fig. 3. Differential characteristic

Figure 3 shows why the differential characteristic holds going from (L_1, R_1) to (L_4, R_4). Because $R'_1 = 0$ differentially, the output of $f(R, K)$ in round 2 must also be 0 differentially. Thus, $R'_2 = L'_1$ and R'_2 has only bits 2 and/or 3 toggled. These two bits only affect the input of S1, so only the outputs of S1

[3] The same applies to toggling bits 10 and/or 11, but for simplicity we will not deal with that here.

can change in round 3. Since $R'_3 = L'_4$ and bits 3, 8, 14, and 25 are not outputs of S1, these bits will be unchanged in L'_4. Further, because of the E-expansion, the 4 outputs of S1 affect the inputs of 6 S-boxes in round 4. Two S-boxes will therefore be unchanged, namely S1 and S7. Bit 17 of R'_4 is the output of S1, so it will remain unchanged.

As figure 2 shows, the parity invariance to be observed occurs in round 7 with probability 0.576. Following Matsui, we decipher the two ciphertexts (L_8, R_8) and (L_8^*, R_8^*) backward through one round to get the output bits of the parity relation: bits 3, 8, 14, and 25 of R_7, and bit 17 of L_7. Bits 3, 8, 14, and 25 of R_7 are known because $R_7 = L_8$, the left half of the ciphertext. Only bit 17 of L_7 must be computed. This computation involves only S1 in round 8, so we can test the 6-bit subkey $K_{8,1}$. When the correct value of $K_{8,1}$ is used, we expect to observe parity invariance 57.6% of the time; when an incorrect value is used, the produced data is more random and we expect to observe parity invariance closer to 50% of the time.

Based on Matsui's rule of thumb that approximately $8/(r-0.5)^2$ observations are needed when r is the probability of observing a parity relation, one would expect our attack to require about 1,400 pairs of chosen plaintexts. While one would expect this number must be increased to create the desired toggling in round 1 as opposed to round 0 (the plaintext), the next section develops an approach that obtains the desired behavior while reducing the required amount of text.

5 Structures

Our attack requires plaintext pairs which toggle bits 2 and/or 3 in round 1. We produce this behavior with structures similar to those used by Biham and Shamir. Choose any reference plaintext and let $P(0)$ through $P(64)$ be the 64 plaintexts obtained by varying bits 9, 17, 23, and 31 of L_0 and bits 2 and 3 of R_0. These bits of L_0 correspond to the four outputs of S1, and bits 2 and 3 of R_0 become bits 2 and 3 of L_1, the bits to be toggled.

Bits 2 and 3 are the middle input bits to S1. Since these bits are the only bits that can change in R_0, only the outputs of S1 can change in round 1, as shown in figure 4. Because we included all 16 possibilities for these bits in the structure, if we knew the 6-bit subkey $K_{1,1}$, for each of the 64 $P(i)$'s we could choose three other $P(j)$'s which had the desired toggling in round 1. One $P(j)$ would toggle bit 2, one would toggle bit 3, and one would toggle bits 2 and 3. The 64 chosen plaintexts might therefore seem to produce $64 * 3 = 192$ differential pairs (observations), but only half of these, or 96 pairs, are distinct since (i, j) and (j, i) are the same pair.

All 96 pairs in a structure could be used to help determine $K_{8,1}$ if $K_{1,1}$ were known. Since we do not know $K_{1,1}$, we search over all values of $(K_{1,1}, K_{8,1})$. Two of the bits of $K_{1,1}$ are also part of $K_{8,1}$, so there are only 10 bits and 1024 possible subkeys to search over. Since each structure of 64 plaintexts produces 96 differential pairs and 1400 pairs are required by Matsui's rule of thumb, our

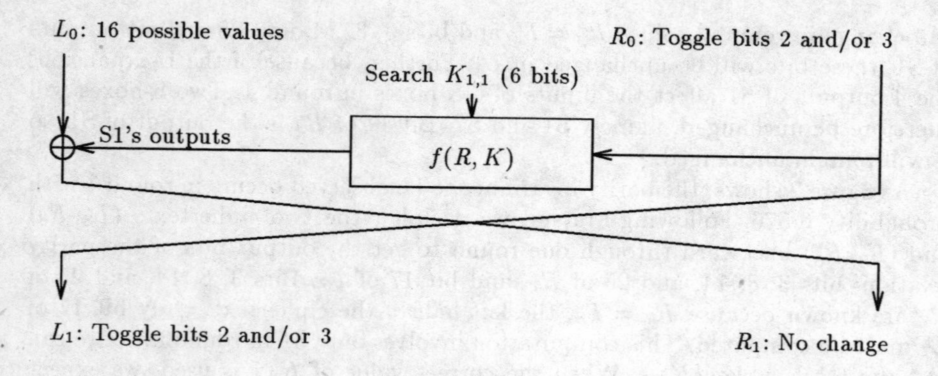

Fig. 4. First round of 8-round differential-linear attack

attack should work with approximately $64/96 * 1400 = 900$ chosen plaintexts. Our experiments find good agreement: 512 chosen plaintexts produce an 80% success rate and 768 chosen plaintexts have a 95% success rate in determining the 10 bits of key in $(K_{1,1}, K_{8,1})$. These two attacks use 8 and 12 structures of 64 chosen plaintexts respectively.

An even higher success rate can be obtained, with no increase in the number of required plaintexts, if we use ideas related to "list decoding" of error correcting codes. The most likely $(K_{1,1}, K_{8,1})$ is tried first in the semi-exhaustive search over the 46 remaining key bits and, if it does not work (which happens one time in five for the 512 chosen plaintexts attack), the next most likely value is tried, etc. With a list of two, this method increases the average computation by only 20%, while increasing the probability of success to 87%. A list of size 8 increases the probability of success to 95%.

6 Additional Bits of Key

While a conservative definition of security would regard a cryptosystem as broken when even one bit of key can be recovered, the 10 bits recovered in our attack leaves a large semi-exhaustive search over 2^{46} keys. However, we can recover additional bits of key using other, lower probability 3-round parity relations in rounds 5 to 7, in place of Matsui's optimal 3-round parity relation. For example, bits 5, 15, 21, 27, and 63 of round 4 also remain unchanged when bits 2 and/or 3 of round 1 are toggled. Therefore, we can use the relation

$$(L_4[5, 15, 21, 27] \oplus R_4[31]) \oplus (R_7[5, 15, 21, 27] \oplus L_7[29]) = 0. \quad (3)$$

Differentially, equation (3) holds with probability 0.527, instead of 0.576 for Matsui's optimal relation. Use of this parity relation requires searching over $(K_{1,1}, K_{8,6})$ instead of $(K_{1,1}, K_{8,1})$, thereby recovering the six additional bits of key $K_{8,6}$.

The lower probability of the second relation increases the probability of error. The 768 chosen plaintexts which had a 95% success rate on the first ten bits of key, have an 85% success rate for all sixteen bits of key. The idea of list decoding, mentioned above, can be applied here with even greater success.

7 Work in Progress

The above analysis treats each differential pair as if it were independent of all other pairs. Since each structure of 64 chosen plaintexts yields 96 differential pairs, this assumption is clearly not true. However, the close agreement between the predicted and experimental results shows that the effect of dependence is small for the eight round attack.

We have begun a more precise analysis based on the fact that in each structure of 64 chosen plaintexts, there are 16 sets of four plaintexts. These four plaintexts differ only in bits 2 and/or 3 of (L_1, R_1), and therefore have the same input to Matsui's parity relation. Starting from the probability of the parity relation, we can calculate the probability that each set of four texts will have a particular set of four output parity bits. When we ran our eight round attack using the results from this more complex analysis, we obtained only a one percent improvement in the probability of success. Although the more complex analysis makes only a small difference in the eight round attack, it might be more useful in attacks on a larger number of rounds. Such work is in progress.

References

1. E. Biham and A. Shamir, *Differential Cryptanalysis of the Data Encryption Standard*, Berlin: Springer-Verlag, 1993.
2. W. Diffie and M. Hellman, "Exhaustive Cryptanalysis of the NBS Data Encryption Standard," *Computer*, vol.10, no. 6, June 1977, pp. 74–84.
3. M. Matsui, "Linear Cryptanalysis Method for DES Cipher," *Advances in Cryptology–EUROCRYPT '93 Proceedings*, Berlin: Springer-Verlag, 1994, to appear.
4. M. Matsui, "Linear Cryptanalysis of DES Cipher (I)," *Journal of Cryptology*, to appear.
5. M. Matsui, "The first experimental cryptanalysis of the Data Encryption Standard," *Advances in Cryptology–Crypto '94 Proceedings*, Springer-Verlag, to appear.
6. National Bureau of Standards, *Data Encryption Standard*, U.S. Department of Commerce, FIPS pub. 46, January 1977.
7. M. Wiener, "Efficient DES Key Search," *Advances in Cryptology– Crypto '93 Proceedings*, Springer-Verlag, to appear.

Linear Cryptanalysis Using Multiple Approximations

Burton S. Kaliski Jr. and M.J.B. Robshaw

RSA Laboratories
100 Marine Parkway
Redwood City, CA 94065, USA

Abstract. We present a technique which aids in the linear cryptanalysis of a block cipher and allows for a reduction in the amount of data required for a successful attack. We note the limits of this extension when applied to DES, but illustrate that it is generally applicable and might be exceptionally successful when applied to other block ciphers. This forces us to reconsider some of the initial attempts to quantify the resistance of block ciphers to linear cryptanalysis, and by taking account of this new technique we cover several issues which have not yet been considered.

1 Introduction

Matsui and Yamagishi [6] introduced the idea of *linear cryptanalysis* in 1992 in an attack on FEAL [10]. The techniques used in this attack were refined by Matsui and used with dramatic effect on DES [7] in a theoretical attack on the full 16-round DES requiring 2^{47} known plaintext/ciphertext pairs [4]. After further work an experiment was performed during which the key used in the full 16-round version of DES was recovered using 2^{43} known plaintext/ciphertext pairs [9].

The most notable feature about linear cryptanalysis is that it is a known plaintext attack rather than a chosen plaintext attack (differential cryptanalysis [1] is a chosen plaintext attack) and as such poses more of a practical threat to a block cipher. At present, however, a successful linear cryptanalytic attack on DES still requires a large quantity of known plaintext.

In this paper we consider an extension to the linear cryptanalytic attack [4, 5] using multiple linear approximations. This offers a slight improvement in the efficiency of an attack on DES but more importantly, it is generally applicable and in certain circumstances it might well be extremely effective in reducing the amount of data required by a cryptanalyst for a successful attack on a block cipher using linear cryptanalysis.

Our paper is organized as follows. We briefly describe the technique of linear cryptanalysis, and then we present an adaptation of these methods which allows us to use multiple linear approximations. After providing theoretical estimates for the performance of our techniques we present experimental evidence that supports our claims. We then consider some of the implications of our work and draw our conclusions.

2 Linear Cryptanalysis

We shall assume familiarity with the original paper by Matsui [4] which introduced the linear cryptanalytic attack on DES; we shall also refer to the more recent paper due to Matsui [5].

The basic idea behind linear cryptanalysis is to find some linear approximation to the action of the iterated block cipher which connects together, in one expression, some bits of the plaintext $P_{i_1} \ldots P_{i_a}$, ciphertext $C_{j_1} \ldots C_{j_b}$ and key $K_{k_1} \ldots K_{k_u}$. We shall write $P_{i_1} \oplus \ldots \oplus P_{i_a}$ as $P[\chi_P]$ and we can write a single linear approximation as

$$P[\chi_P] \oplus C[\chi_C] = K[\chi_K]. \tag{1}$$

If equation 1 is correct with probability $p = \frac{1}{2} + \epsilon$ for randomly chosen plaintext and a fixed key, then we say that it has *bias* ϵ. By collecting *known* plaintext/ciphertext pairs the cryptanalyst can make a guess for the value of $K[\chi_K]$; provided $\epsilon \neq 0$ the guess becomes more reliable as the cryptanalyst collects more plaintext/ciphertext pairs.

We note that the expected value of the left side will be $\frac{1}{2} + \epsilon$ if $K[\chi_K] = 1$ and $\frac{1}{2} - \epsilon$ if $K[\chi_K] = 0$. Under certain heuristic assumptions the cryptanalyst is attempting to distinguish a distribution with mean $\frac{1}{2} + \epsilon$ and variance $\frac{1}{4} - \epsilon^2$ from one with mean $\frac{1}{2} - \epsilon$ and variance $\frac{1}{4} - \epsilon^2$ [2]. The cryptanalyst has to take sufficiently many plaintext/ciphertext pairs to be confident that this distinction is being correctly made and so the smaller the value of ϵ, the more plaintext/ciphertext pairs are required to give the same level of confidence that the identification is correct.

The basic algorithm that allows the cryptanalyst to deduce one bit of key information from a single linear approximation is *Algorithm 1* in [4, 5].

Algorithm 1

Suppose that equation 1, $P[\chi_P] \oplus C[\chi_C] = K[\chi_K]$, is correct with probability $p = \frac{1}{2} + \epsilon$.

Step 1 Let T be the number of plaintext/ciphertext pairs such that the left side of equation 1 is equal to 0 and let N be the total number of pairs.

Step 2 If $T > N/2$
- then guess $K[\chi_K] = 0$ (when $\epsilon > 0$) or 1 (when $\epsilon < 0$),
- else guess $K[\chi_K] = 1$ (when $\epsilon > 0$) or 0 (when $\epsilon < 0$).

Of more practical importance is *Algorithm 2* [5] which allows a cryptanalyst attacking DES to recover up to 13 bits of key information in the full 16-round DES. More generally, for an r-round Feistel cipher we approximate $(r-2)$ iterations of the F-function from the second to the $(r-1)^{\text{th}}$ round using some linear approximation while guessing the subkey bits that are relevant to the first and final rounds. To keep the number of candidates small, it is advisable to consider linear approximations that involve guessing few subkey bits in the first and r^{th}

round. In the case of DES, this is conveniently achieved by guessing the subkeys relevant to a single S-box in the outer rounds.

Following Matsui we write the parity of the relevant output bits from the F-function in round one, which is dependent on subkey K_1 and plaintext block P_L, as $F_1(P_L, K_1)[\chi_{F_1}]$; we use similar notation for the output of the F-function in round r. The relevant linear approximation can be written as follows:

$$P[\chi_P] \oplus C[\chi_C] \oplus F_1(P_L, K_1)[\chi_{F_1}] \oplus F_r(C_L, K_r)[\chi_{F_r}] = K[\chi_K]. \quad (2)$$

Suppose that equation 2 is correct with probability $p = \frac{1}{2} + \epsilon$.

Algorithm 2

Step 1 Let $K_1^{(g)}$ $(g = 1, 2, \ldots)$ and $K_r^{(h)}$ $(h = 1, 2, \ldots)$ be possible candidates for K_1 and K_r respectively. Then for each pair $(K_1^{(g)}, K_r^{(h)})$, let $T_{g,h}$ be the number of plaintexts such that the left side of equation 2 is equal to 0 when K_1 is replaced by $K_1^{(g)}$ and K_r by $K_r^{(h)}$. Let N be the total number of plaintexts.

Step 2 Let T_{max} be the maximum value and T_{min} be the minimum value of all $T_{g,h}$'s.

- If $|T_{max} - N/2| > |T_{min} - N/2|$, adopt the key candidate corresponding to T_{max} and guess $K[\chi_K] = 0$ (when $\epsilon > 0$) or 1 (when $\epsilon < 0$).
- If $|T_{max} - N/2| < |T_{min} - N/2|$, adopt the key candidate corresponding to T_{min} and guess $K[\chi_K] = 1$ (when $\epsilon > 0$) or 0 (when $\epsilon < 0$).

Algorithm 2 provides a guess for the relevant subkey bits of K_1 and K_r together with the additional bit of subkey information obtained from the linear approximation. Matsui [5] gives both experimental and theoretical justification for the belief that by taking $8\epsilon^{-2}$ known plaintext/ciphertext pairs *Algorithm 2* is correct with high probability.

To obtain more subkey bits, another linear approximation can be used or, for a Feistel cipher, the roles of the ciphertext and plaintext can be reversed and *Algorithm 2* applied again.

In the particular case of DES, Matsui [5] uses the second approach to obtain 26 bits of subkey information in total. Fortuitously, the key schedule in DES ensures that these 26 bits of subkey information correspond to 26 distinct bits of key information. The remaining unknown 30 bits of key can be derived using an exhaustive search.

3 Multiple Linear Approximations

There are many different linear approximations to a block cipher over a given number of rounds. Suppose we have n linear approximations which involve the same key bits but differ in the plaintext and ciphertext bits that they use.

Matsui's *Algorithm 1* can be used on any of the n individual linear approximations to define a statistic T_i, $1 \leq i \leq n$, with a certain bias and variance;

the success of *Algorithm 1* depends on both. However, we now show that it is possible to use more than one linear approximation at the same time. In short, we work harder on the information we already have rather than calling for more plaintext/ciphertext pairs.

Suppose the ith linear approximation for $1 \leq i \leq n$ has the following form:

$$P[\chi_P^i] \oplus C[\chi_C^i] = K[\chi_K]. \tag{3}$$

For the sake of analysis, we will suppose that each bias ϵ_i is positive; changes can easily be made to the data collecting phase to ensure that this condition holds.

Algorithm 1M

Step 1 For $1 \leq i \leq n$ let T_i be the number of plaintext/ciphertext pairs such that the left side of equation 3 is equal to 0. Let N denote the total number of plaintexts.

Step 2 For some set of weights $a_1, \ldots a_n$ where $\sum_{i=1}^{n} a_i = 1$ calculate

$$U = \sum_{i=1}^{n} a_i T_i.$$

Step 3 If $U > \frac{N}{2}$ then guess $K[\chi_K] = 0$, else guess $K[\chi_K] = 1$.

The analysis below will show that we have introduced a new statistic U (defined in terms of the T_i, $1 \leq i \leq n$) which has a bias comparable to the T_i's but a reduced variance. Because of this reduced variance the attack requires fewer known plaintexts. We will also see in Lemma 2 that the success rate of *Algorithm 1M* is optimized if the weights are defined as $a_i = \epsilon_i / \sum \epsilon_i$.

3.1 Analysis of *Algorithm 1M*

For each linear approximation define x_i to be a random variable whose value is 0 when the left side of the ith linear approximation is equal to 1 and 1 when the left side is equal to 0. By making this definition we see that sum of x_i for the ith linear approximation over all plaintext/ciphertext pairs in the experiment is equal to T_i.

We observe that for a good block cipher the value of the left sides of two different linear approximations will be essentially independent since the two left sides differ by a combination of plaintext and ciphertext bits. This observation is supported by the experimental work presented in Section 4, and leads to the following assumption:

Assumption 1 *For all i and j with $i \neq j$, $x_i = x_j$ with probability $\frac{1}{2}$, where the probability is taken over randomly chosen plaintexts.*

We now establish the values for the mean and variance of the statistic U.

Lemma 1. *Let* $y = \sum_{i=1}^{n} a_i x_i$ *with* $\sum_{i=1}^{n} a_i = 1$. *Then under Assumption 1 we have* $E[y] = \frac{1}{2} + \sum_{i=1}^{n} a_i \epsilon_i$ *if* $K[\chi_K] = 0$ *and* $E[y] = \frac{1}{2} - \sum_{i=1}^{n} a_i \epsilon_i$ *if* $K[\chi_K] = 1$. *The variance satisfies*

$$\sigma_y^2 = \sum_{i=1}^{n} \frac{a_i^2}{4} - \left(\sum_{i=1}^{n} a_i \epsilon_i \right)^2 .$$

Proof. See Appendix A. □

Following Matsui and noting that each value of the statistic $U = \sum a_i T_i$ can also be expressed as a sum of independent values $y = \sum a_i x_i$, we will make the following assumption which is consistent with the experimental results presented in Section 4.

Assumption 2 *The distribution of the statistic* $U = \sum_{i=1}^{n} a_i T_i$ *can be accurately modeled using a normal distribution.*

Our goal is to maximize the distance between $N/2$ and $E[U] = NE[y]$ in terms of the standard deviation $\sigma_U = \sqrt{N}\sigma_y$. This increases our success rate for a fixed N, or alternatively, allows us to use a smaller N while maintaining the same level of success.

Lemma 2. *Under Assumptions 1 and 2, with the statistic* U *defined as in Algorithm 1M, the distance* $|N/2 - E[U]|/\sigma_U$ *is maximized for a given* N *when the weights* a_i *are proportional to the biases* ϵ_i.

Proof. See Appendix A. □

The maximum distance is easily calculated as $2\sqrt{N}\sqrt{(\sum \epsilon_i^2)/(1 - 4\sum \epsilon_i^2)}$. Adopting the conventional notation of $\Phi(\cdot)$ for the normal cumulative distribution function [2], this leads to the following theorem:

Theorem 3. *Under Assumptions 1 and 2, the success rate of Algorithm 1M, with optimal weights* a_i, *is*

$$\Phi\left(2\sqrt{N} \sqrt{\frac{\sum_{i=1}^{n} \epsilon_i^2}{1 - 4\sum_{i=1}^{n} \epsilon_i^2}} \right) .$$

Proof. Follows directly. □

When $\sum \epsilon_i^2$ is small, we approximate the success rate as $\Phi(2\sqrt{N}\sqrt{\sum \epsilon_i^2})$, a generalization of Matsui's single-approximation success rate of $\Phi(2\sqrt{N}\epsilon)$.

As an illustration of the improvement possible, suppose that we have n linear approximations all involving the same key bits and all having the same bias ϵ. Then using *Algorithm 1M* and N' known plaintext/ciphertext pairs the success rate is given by $\Phi\left(2\sqrt{N'}\sqrt{\sum \epsilon_i^2}\right) = \Phi\left(2\sqrt{N'}\sqrt{n}\epsilon\right)$. Using *Algorithm 1* with a single linear approximation, the success rate is $\Phi\left(2\sqrt{N}\epsilon\right)$. We see that if $N' = N/n$, then there is an n-fold reduction in the amount of data and yet the two algorithms have the same success rate.

It is somewhat artificial to assume that there are n linear approximations involving the same key bits, all correct with the same probability. Later we will outline some extensions which allow us to relax these conditions on the use of multiple linear approximations. However, we can summarize our results so far by saying that the use of several carefully chosen linear approximations concurrently will lead to a reduction in the amount of data required for a linear cryptanalytic attack.

3.2 *Algorithm 2M*

We now introduce an extension to *Algorithm 2* [4, 5] which uses multiple linear approximations. For an r-round Feistel cipher we approximate $(r-2)$ iterations of the F-function from the second to the $(r-1)^{\text{th}}$ round using n linear approximations while we still make guesses for the subkey bits needed to extend through the first and final rounds. Note that to keep the number of candidates small, the approximations should all involve the same guessed subkey bits in round one, as well as in round r. Following the notation established for *Algorithm 2* we can write the ith linear approximation as follows:

$$P[\chi_P^i] \oplus C[\chi_C^i] \oplus F_1(P_L, K_1)[\chi_{F_1}^i] \oplus F_r(C_L, K_r)[\chi_{F_r}^i] = K[\chi_K]. \qquad (4)$$

We will again suppose, without loss of generality, that each bias ϵ_i is positive.

Step 1 Let $K_1^{(g)}$ $(g = 1, 2, \ldots)$ and $K_r^{(h)}$ $(h = 1, 2, \ldots)$ be possible candidates for K_1 and K_r respectively. Then for each pair $(K_1^{(g)}, K_r^{(h)})$ and each linear approximation i let $T_{g,h}^i$ be the number of plaintexts such that the left side of equation 4 is equal to 0 when K_1 is replaced by $K_1^{(g)}$ and K_r by $K_r^{(h)}$. Let N be the total number of plaintexts.

Step 2 Let $a_i = \epsilon_i / \sum_{i=1}^{n} \epsilon_i$. Calculate, for each g, h,

$$U_{g,h} = \sum_{i=1}^{n} a_i T_{g,h}^i$$

Step 3 Let U_{max} be the maximum value and U_{min} be the minimum value of all $U_{g,h}$'s.

- If $|U_{max} - N/2| > |U_{min} - N/2|$, adopt the key candidate corresponding to U_{max} and guess $K[\chi_K] = 0$.
- If $|U_{max} - N/2| < |U_{min} - N/2|$, adopt the key candidate corresponding to U_{min} and guess $K[\chi_K] = 1$.

Algorithm 2M is a generalization of *Algorithm 1M* and we expect it to be successful for essentially the same reasons. The main issue is that we are replacing the statistic defined using *Algorithm 2* with one that has a smaller variance. We know that the variance is reduced for the correct key guess; the complicating issue is what happens to the statistics $U_{g,h}$ for incorrect key guesses. By Lemma 1 the use of multiple linear approximations ensures that the variance of any statistic, even those for incorrect key guesses, will be reduced. This gives us one reason to expect a higher success rate with *Algorithm 2M* for a given number of plaintext/ciphertext pairs.

Another reason is due to a cancellation property. Let $P^i_{g,h}$ be the probability for a random plaintext/ciphertext pair that an incorrect key guess gives the same answer as a correct key guess; that is, the probability that

$$F_1(P_L, K_1^{(g)})[\chi^i_{F_1}] \oplus F_r(P_L, K_r^{(h)})[\chi^i_{F_r}] = F_1(P_L, K_1)[\chi^i_{F_1}] \oplus F_r(P_L, K_r)[\chi^i_{F_r}].$$

Then it is easy to show that

$$E[U_{g,h}] = \frac{N}{2} + N \sum_{i=1}^{n} a_i \epsilon_i \delta^i_{g,h}$$

where $\delta^i_{g,h}$ is a "correlation coefficient" defined as $\delta^i_{g,h} = 2P^i_{g,h} - 1$. This coefficient takes values between -1 and $+1$.

The greater the magnitude of the difference between $E[U_{g,h}]$ and $N/2$, the greater the likelihood that the key guess $(K_1^{(g)}, K_r^{(h)})$ is adopted. However, since a correlation coefficient $\delta^i_{g,h}$ can take both positive and negative values for an incorrect key guess, but is always 1 for the correct key guess, the correct guess is at a significant advantage. We will explore this phenomenon more closely in subsequent work.

Part of Section 4 contains results on the performance of *Algorithm 2M*.

4 Experimental Results

In this section we report the findings from several experiments performed on small-round versions of DES. We chose small-round versions because we wish to present confirmation of the performance of our algorithms and we feel that these examples are particularly illustrative.

4.1 Confirming the assumptions and *Algorithm 1M*

The aim of the first set of experiments was to substantiate the assumptions made during Section 3.1 and to test the theoretical predictions made about the use of multiple linear approximations. Although the experiments were conducted on a small scale, they provide good corroborating evidence for the validity of both the assumptions and the theoretical results. It is reasonable to assume that these results would extend to a larger scale.

We used a five-round version of DES with two linear approximations identified using the notation in [5] as $-$`ACD`$-$ and $-$`DCA`$-$. Each linear approximation holds with bias $\epsilon = 25 \times 2^{-12} \approx 6.104 \times 10^{-3}$. Let K_i denote the 48-bit subkey used at round i, P_L the low 32 bits of plaintext and C_L the low 32 bits of ciphertext. Then we have

$$P_1 : P_L[7,18,24,29] \oplus C_L[7,18,24] = K_2[22] \oplus K_3[44] \oplus K_4[22]$$
$$P_2 : P_L[7,18,24] \oplus C_L[7,18,24,29] = K_2[22] \oplus K_3[44] \oplus K_4[22]$$

First we consider Assumption 1. Using 2,000,000 plaintext/ciphertext pairs and linear approximations P_1 and P_2 we found that the outcome of the two relations agreed 999,351 times and disagreed the remaining 1,000,649 times; this gives us some confidence that Assumption 1 is reasonable.

We next completed 100 attempts at the linear cryptanalysis of five-round DES using *Algorithm 1M* on an increasing number of plaintext/ciphertext pairs; these numbers were chosen so that the number of plaintext/ciphertext pairs was roughly $\frac{1}{8}\epsilon^{-2}, \frac{1}{4}\epsilon^{-2}, \frac{1}{2}\epsilon^{-2}$ and ϵ^{-2}. The success rates achieved in practice using P_1 and P_2 individually and then jointly are presented below, together with the results predicted by the theory for *Algorithm 1* using individual linear approximations and for *Algorithm 1M* using multiple linear approximations.

	experimental results				*theoretical predictions*			
number of pairs	3,356	6,711	13,422	26,844	3,356	6,711	13,422	26,844
P_1	81%	86%	94%	99%	76%	84%	92%	98%
P_2	75%	88%	92%	99%	76%	84%	92%	98%
using P_1 *and* P_2	92%	95%	98%	100%	84%	92%	98%	100%

These results confirm Theorem 3: it is clear that the use of two linear approximations gives success rates comparable to individual linear approximations with half as many plaintext/ciphertext pairs.

4.2 Confirming *Algorithm 2M*

The aim of the second set of experiments was to confirm the expected behavior of *Algorithm 2M*. The experiments consisted of attacking a seven-round version of DES by using the five-round linear approximations P_1 and P_2 from the previous experiment for rounds two through six, and guessing the 12 subkey bits used in the first and seventh rounds. The results are presented below. We note that the same success rate achieved using either of the single linear approximations on their own could be achieved by using both linear approximations together with half as many plaintext/ciphertext pairs.

	experimental results			
number of pairs	13,422	26,844	53,688	107,376
P_1	3%	2%	17%	51%
P_2	1%	1%	15%	45%
using P_1 *and* P_2	4%	13%	46%	94%

5 Effectiveness and Extensions

In an attack on DES it is perhaps surprising how little advantage is gained at present using multiple linear approximations. An exhaustive search reveals that there are 10,006 14-round linear approximations involving a single S-box at each round with bias $|\epsilon_i| > 10^{-8}$. We found that $\sum_{i=1}^{10,006} \epsilon_i^2 \approx 1.23 \times 10^{-11}$. With our present techniques we can use only a small fraction of these approximations, giving only a minor improvement in the number of plaintext/ciphertext pairs required.

Even if we could use all these linear approximations at the same time, it would only result in a reduction by a factor of about 38 in the number of plaintext/ciphertext pairs, since for the "best" linear approximation, $\epsilon^2 \approx 3.22 \times 10^{-13}$. Future research will undoubtedly reveal whether techniques exist using multiple linear approximations which can begin to deliver such improvements.

Note that there may well be different linear approximations involving different subkey bits which, because of the key schedule used in the block cipher, correspond to the same key-bits in the user-provided key. In such a case these linear approximations would each provide a guess for the same bit of key information, and could be used together with *Algorithm 1M* or *2M*. Unfortunately the existence of such linear approximations depends closely on the key schedule in the cipher.

Our aim is to remove the assumption that all the linear approximations use the same subkey bits. Our approach is to modify the techniques we have, without regard for the key schedule, to use good linear approximations which potentially involve different subkey bits. At this point we present only an extension to *Algorithm 1M*.

Suppose we have n linear approximations and the ith linear approximation has the form

$$P[\chi_P^i] \oplus C[\chi_C^i] = K[\chi_K^i]. \tag{5}$$

To consider the approximations together, we must first guess, for each j, $2 \le j \le n$, whether $K[\chi_K^j]$ and $K[\chi_K^i]$ are equal or not. There are at most 2^{n-1} guesses and for each guess we obtain n linear approximations of the form

$$P[\chi_P^i] \oplus C[\chi_C^i] \oplus \Delta^i = K[\chi_K^1] \tag{6}$$

where $\Delta^1 = 0$ and Δ^i for $i > 1$ depends on the guess.

At this point, for each guess, we can determine by *Algorithm 1M* whether $K[\chi_K^1] = 0$ or 1, where the statistic T_i in *Algorithm 1M* is taken from the left side of equation 6. Note that in practice one would not repeat *Algorithm 1M* for each guess; instead one would consider up to 2^{n-1} ways of combining the statistics T_i for $1 \le i \le n$.

This approach does not determine which guess is correct, but for each guess it does give an estimate for the value of $K[\chi_K^1]$. In effect this halves the exhaustive search space. We need only have a high success rate for the value of $K[\chi_K^1]$ for the correct guess. Analysis similar to that for *Algorithm 1M* suggests that fewer plaintext/ciphertext pairs are required for a given success rate than if we only used a single approximation.

As an example of this approach, consider the following linear approximations to six-round DES. Following Matsui's notation, let P_1 represent the linear approximation `A-ACD-` while P_2 represents `E-DCA-`. These hold with bias 3.81×10^{-3} and 3.05×10^{-3} respectively.

$$P_1 : P_H[7,18,24,29] \oplus P_L[15] \oplus C_L[7,18,24] = K_1[22] \oplus K_3[22] \oplus K_4[44] \oplus K_5[22] \quad (7)$$

$$P_2 : P_H[7,18,24] \oplus P_L[12,16] \oplus C_L[7,18,24,29] = K_1[19,23] \oplus K_3[22] \oplus K_4[44] \oplus K_5[22] \quad (8)$$

We can use both linear approximations in *Algorithm 1M* while guessing whether the right sides are equal or not. The results of this experiment are provided below. The amount of data in these experiments was chosen so that the number of plaintext/ciphertext pairs was roughly $\frac{1}{8}\epsilon^{-2}$, $\frac{1}{4}\epsilon^{-2}$, $\frac{1}{2}\epsilon^{-2}$ and ϵ^{-2} where $\epsilon = 3.81 \times 10^{-3}$. The success rates provided in the table below are for the correct guess, which in our experiment was that the right sides are equal.

More analysis of this technique will be presented in future work.

	experimental results				*theoretical predictions*			
number of pairs	8,590	17,180	34,360	68,720	8,590	17,180	34,360	68,720
P_1	81%	82%	92%	100%	76%	84%	92%	98%
P_2	73%	84%	86%	96%	71%	79%	87%	95%
using P_1 *and* P_2	89%	93%	99%	100%	82%	90%	96%	99%

6 Implications

If we consider DES-like ciphers generally, we can imagine situations where the use of multiple linear approximations might well be very significant. If an S-box has b output bits, then there are potentially $2^b - 1$ useful linear approximations using the same input mask, and hence the same subkey bits. We note that there are no useful variants of the trivial one-round linear approximation (denoted as – by Matsui).

The Feistel structure of a DES-like cipher means that we can consider $2^b - 1$ variations on the left half of the plaintext and $2^b - 1$ variations on the right half which would provide alternative linear approximations on exactly the same subkey bits. Since the trivial one-round linear approximation has no useful variations, the best linear approximations (which we would expect to use the trivial one-round linear approximation) might not allow consideration of all these variations.

One of the most interesting cases is a DES-like block cipher with larger S-boxes. Our results show that the extra linear approximations admitted by the use of larger S-boxes might increase the vulnerability of the block cipher to some form of linear cryptanalysis using multiple linear approximations. More research is required to assess the significance of this development but we note that it should be an important consideration; after the work of O'Conner [8] larger S-boxes have been proposed as a way of increasing the resistance of a block cipher to differential cryptanalysis [3].

Another consequence of our work is that in assessing the resistance of a block cipher to linear cryptanalysis it is not sufficient to consider the best linear approximation in isolation; one must also take account of other linear approximations which might be used at the same time. We suggest that some of the preliminary work in assessing the practical security of a cipher against linear cryptanalytic attack [3] should be broadened since these estimates of strength are made under the assumption that the best attack is completed using a single linear approximation. Under this assumption a lower bound on the complexity of a linear cryptanalytic attack on a Feistel cipher is obtained by considering (i) the bias of the best non-trivial one-round linear approximation and (ii) the number of rounds in the cipher.

By considering a theoretical best linear approximation one overlooks the possibility that several good linear approximations could be used concurrently to obtain a more efficient attack. We suggest that a more useful measure of practical security against linear cryptanalytic attack should also consider the use of multiple linear approximations.

7 Conclusions

We have presented an extension to the basic linear cryptanalytic attack which offers an improvement in the number of known plaintext/ciphertext pairs required for the linear cryptanalysis of a block cipher. While its effectiveness in an attack on DES is at present somewhat limited, it is a general technique which might have excellent results in the cryptanalysis of other less well designed block ciphers.

Importantly, we note that the use of larger S-boxes, which is sometimes recommended as a way of increasing the security of DES-like block ciphers, might in certain circumstances facilitate the use of linear cryptanalysis with multiple linear approximations. More research is needed to ascertain quite how significant a threat this might eventually be.

We believe that the use of multiple linear approximations is an important cryptanalytic tool and one which should be considered both in the design of block ciphers and in any attempt to provide a theoretical bound on the resistance of block ciphers to linear cryptanalysis.

References

1. E. Biham and A. Shamir. *Differential Cryptanalysis of the Data Encryption Standard.* Springer-Verlag, New York, 1993.
2. A.W. Drake. *Fundamentals of Applied Probability Theory.* McGraw-Hill Book Company, New York, 1967.

3. L.R. Knudsen. Practically secure Feistel ciphers. In *Proceedings of Cambridge Security Workshop, December 1993*, Springer-Verlag, Berlin, To appear.
4. M. Matsui. Linear cryptanalysis method for DES cipher. In T. Helleseth, editor, *Advances in Cryptology — Eurocrypt '93*, pages 386–397, Springer-Verlag, Berlin, 1994.
5. M. Matsui. Linear cryptanalysis of DES cipher (I). January 1994. Preprint.
6. M. Matsui and A. Yamagishi. A new method for known plaintext attack of FEAL cipher. In R.A. Rueppel, editor, *Advances in Cryptology — Eurocrypt '92*, pages 81–91, Springer-Verlag, Berlin, 1992.
7. National Institute of Standards and Technology (NIST). *FIPS Publication 46-2: Data Encryption Standard.* December 30, 1993. Originally issued by National Bureau of Standards.
8. L. O'Conner. On the distribution of characteristics in bijective mappings. In T. Helleseth, editor, *Advances in Cryptology — Eurocrypt '93*, pages 360–370, Springer-Verlag, Berlin, 1994.
9. E. Okamoto. Personal communication. March 1994.
10. A. Shimizu and S. Miyaguchi. Fast data encipherment algorithm FEAL. In D. Chaum and W.L. Price, editors, *Advances in Cryptology — Eurocrypt '87*, pages 267–280, Springer-Verlag, Berlin, 1988.

A Proof of Lemmas 1 and 2

We need the following additional lemma for our proof of Lemma 1.

Lemma 4. *Under Assumption 1, $E[x_i x_j]$ is given by*

$$E[x_i x_j] = E[x_i]E[x_j] - \epsilon_i \epsilon_j .$$

Proof. Clearly $x_i x_j = 1$ if, and only if, $x_i = x_j = 1$. Thus $E[x_i x_j] = \frac{1}{2} E[x_i | x_i = x_j]$, since $x_i = x_j$ with probability $\frac{1}{2}$. If $K[\chi_K] = 0$, we know that

$$\frac{1}{2} E[x_i | x_i = x_j] + \frac{1}{2} E[x_i | x_i \neq x_j] = \frac{1}{2} + \epsilon_i$$

and

$$\frac{1}{2} E[x_j | x_i = x_j] + \frac{1}{2} E[x_j | x_i \neq x_j] = \frac{1}{2} + \epsilon_j .$$

It is not hard to solve for the individual expectations

$$E[x_i | x_i = x_j] = E[x_j | x_i = x_j] = \frac{1}{2} + \epsilon_i + \epsilon_j ,$$

so that

$$E[x_i x_j] = \frac{1}{4} + \frac{\epsilon_i + \epsilon_j}{2} = \left(\frac{1}{2} + \epsilon_i\right)\left(\frac{1}{2} + \epsilon_j\right) - \epsilon_i \epsilon_j$$

and the result follows. Analysis is similar if $K[\chi_K] = 1$. □

Proof of Lemma 1

Proof. It is straightforward to verify that $E[y] = \frac{1}{2} \pm \sum a_i \epsilon_i$ depending on $K[\chi_K]$. The calculation of the variance σ_y^2 is more involved.

$$
\begin{aligned}
\sigma_y^2 &= E[y^2] - E[y]^2 \\
&= E\left[\left(\sum_{i=1}^{n} a_i x_i\right)^2\right] - (E[y])^2 \\
&= E\left[\sum_{i=1}^{n} a_i^2 x_i^2 + 2\sum_{i=1}^{n}\sum_{j=i+1}^{n} a_i a_j x_i x_j\right] - \left(\sum_{i=1}^{n} a_i E[x_i]\right)^2 \\
&= \sum_{i=1}^{n} a_i^2 E\left[x_i^2\right] + 2\sum_{i=1}^{n}\sum_{j=i+1}^{n} a_i a_j E[x_i x_j] \\
&\quad - \sum_{i=1}^{n} a_i^2 \left(E[x_i]\right)^2 - 2\sum_{i=1}^{n}\sum_{j=i+1}^{n} a_i a_j E[x_i]E[x_j] \\
&= \sum_{i=1}^{n} a_i^2 \left(E[x_i^2] - (E[x_i])^2\right) + 2\sum_{i=1}^{n}\sum_{j=i+1}^{n} a_i a_j \left(E[x_i x_j] - E[x_i]E[x_j]\right) \\
&= \sum_{i=1}^{n} \frac{a_i^2}{4} - \sum_{i=1}^{n} a_i^2 \epsilon_i^2 + 2\sum_{i=1}^{n}\sum_{j=i+1}^{n} a_i a_j \left(E[x_i x_j] - E[x_i]E[x_j]\right).
\end{aligned}
$$

Under Assumption 1 we established in Lemma 4 that

$$E[x_i x_j] = E[x_i]E[x_j] - \epsilon_i \epsilon_j .$$

This then gives us

$$
\begin{aligned}
\sigma_y^2 &= \sum_{i=1}^{n} \frac{a_i^2}{4} - \sum_{i=1}^{n} a_i^2 \epsilon_i^2 - 2\sum_{i=1}^{n}\sum_{j=i+1}^{n} a_i a_j \epsilon_i \epsilon_j \\
&= \sum_{i=1}^{n} \frac{a_i^2}{4} - \left(\sum_{i=1}^{n} a_i \epsilon_i\right)^2 .
\end{aligned}
$$

□

Proof of Lemma 2

Proof. The distance $|N/2 - E[U]|/\sigma_U$ is maximized when $(N/2 - E[U])^2/\sigma_U^2$ is maximized. Expanding $E[U]$ and σ_U^2, we wish to maximize

$$
\begin{aligned}
\frac{(N/2 - NE[y])^2}{N\sigma_y^2} &= N\frac{\left(\sum_{i=1}^{n} a_i \epsilon_i\right)^2}{(1/4)\sum_{i=1}^{n} a_i^2 - \left(\sum_{i=1}^{n} a_i \epsilon_i\right)^2} \\
&= N\frac{4}{\left(\sum_{i=1}^{n} a_i^2\right)/\left(\sum_{i=1}^{n} a_i \epsilon_i\right)^2 - 4}.
\end{aligned}
$$

It is easy to show that this is maximized when $(\sum a_i\epsilon_i)^2/(\sum a_i^2)$ is maximized; note that the denominator must be positive since σ_y^2 must be. We now give an upper bound of $\sum \epsilon_i^2$ on this fraction, by contradiction.

Suppose that the fraction exceeds the upper bound. Then we must have

$$(\sum_{i=1}^{n} a_i\epsilon_i)^2 > (\sum_{i=1}^{n} a_i^2)(\sum_{i=1}^{n} \epsilon_i^2).$$

Expanding the terms, we have

$$\sum_{i=1}^{n}\sum_{j=1}^{n} a_i\epsilon_i a_j\epsilon_j > \sum_{i=1}^{n}\sum_{j=1}^{n} a_i^2\epsilon_j^2.$$

Thus we must have, for some i and j,

$$2a_i\epsilon_i a_j\epsilon_j > a_i^2\epsilon_j^2 + a_j^2\epsilon_i^2,$$

or $(a_i\epsilon_j - a_j\epsilon_i)^2 < 0$, which is a contradiction.

The upper bound is achieved when $a_i\epsilon_j - a_j\epsilon_i = 0$ for all i and j, i.e., when a_i/ϵ_i is constant. □

Hashing with SL_2

Jean-Pierre Tillich and Gilles Zémor

Ecole Nationale Supérieure des Télécommunications
Network Department
46 rue Barrault, 75634 Paris Cedex 13, France
tillich@inf.enst.fr, zemor@res.enst.fr

Abstract. We propose a new family of hash functions based on computations over a finite field of characteristic 2. These functions can be computed quickly, detect small modifications of the input text, and their security is equivalent to a precise mathematical problem. They rely on the arithmetic of the group of matrices SL_2, and improve upon previous functions based on the same strategy.

1 Introduction

We focus on the problem of designing easily computable cryptographic hash functions. Such a function H should map the set of variable length texts over an alphabet $\mathcal{A}$, to a set of (short) fixed length texts that are named *hashcodes.*

$$H : \mathcal{A}^* \longrightarrow \mathcal{A}^n$$

A hash function should have the following properties :

- It should be easily (i.e. quickly) computable.
- It should be computationally difficult to find "collisions", i.e. two texts having the same hashcode. (This is sometimes known as the strong collision criterion).

Hash functions are widely used in numerous cryptographic protocols, and a lot of work has already been put into devising adequate hashing schemes. Despite that, numerous propositions have been shown to be insecure, and the security of those which remain unbroken remains formally unproved.

Following ideas introduced in [12] (see also [13]), we elaborate here on a design principle which enables one to obtain the following unconditional security property. *Small modifications of the input text are always detected.* More precisely we will present here a new hash algorithm which meets this property, and which displays several other attractive features. In particular, it improves upon previous proposals of this kind [12] [11].

- The algorithm can be easily implemented in software by using only basic operations, namely addition in a finite field of characteristic 2, with 2^n elements $\mathbf{F}_{2^n}$ (with say n in the range 130-170), which allows fast computations.
- It is particularly well suited to parallelisation, and to precomputations.
- The security of the scheme we propose is equivalent to a precise mathematical problem, for which there exist several results in favor of its difficulty.

The hash algorithm we propose can be described as follows.

Defining Parameter. An irreducible polynomial $P_n(X)$ of degree n in the aforementioned range.
Algorithm. Let A and B be the following matrices.

$$A = \begin{pmatrix} X & 1 \\ 1 & 0 \end{pmatrix} \qquad B = \begin{pmatrix} X & X+1 \\ 1 & 1 \end{pmatrix}$$

Define the mapping

$$\begin{aligned} \pi : \{0,1\} &\to \{A, B\} \\ 0 &\mapsto A \\ 1 &\mapsto B \end{aligned}$$

The hashcode of binary message $x_1 x_2 \ldots x_k$ is just the matrix product

$$\pi(x_1)\pi(x_2)\ldots\pi(x_k)$$

where computations are made in the quotient field $\mathbf{F}_{2^n} = \mathbf{F}_2[X]/P_n(X)$ of 2^n elements. The hashcode is thus some element of the group $SL_2(\mathbf{F}_{2^n})$ of 2×2 matrices with determinant 1 over $\mathbf{F}_{2^n}$. We need $3n+1$ bits to encode the hashed value (that is 390-510 bits).

A design strategy lies behind the construction of this function. It is based on associating to such a function a Cayley graph, and by exploiting the fact that security is related to graphical parameters. We will elaborate on this in the next section. Provable properties of the hash algorithm using this strategy will be given in section 3.

2 A Design Strategy. Graph-theoretic Issues

2.1 The General Construction

We have based the design of our hash functions on the following general scheme.

Defining Parameter. A finite group G, and a set of generators $\mathcal{S}$ of the same size as the text alphabet $\mathcal{A}$. Choose a function $\pi : \mathcal{A} \to \mathcal{S}$ which defines a one-to-one correspondence between $\mathcal{A}$ and $\mathcal{S}$.
Algorithm.

The hashcode of the text $x_1x_2 \dots x_k$ is just the group element

$$\pi(x_1)\pi(x_2)\dots\pi(x_k)$$

Motivation for this construction is twofold : the hash functions display a concatenation property, and one can associate to such a scheme a Cayley Graph, several parameters of which are relevant to security (namely its girth and expanding properties).

Concatenation property. If x and y are two texts, then their concatenation xy has hashed value $H(xy) = H(x)H(y)$. This clearly allows an easy parallelisation of the scheme, and precomputations when parts of the message are known in advance.

Parameters of the associated Cayley graph. We can associate to this scheme the Cayley graph $\mathcal{C}(G, \mathcal{S})$: its vertex set is G and there is a directed edge from g_1 to g_2 if and only if $g_1^{-1}g_2 \in \mathcal{S}$. The following parameters are of fundamental importance when studying the security of the hash function.

The girth of the graph. We shall use the following definition of the *directed girth.*

Definition 2.1 — *Call the* directed girth *of a graph $\mathcal{G}$, the largest integer ∂ such that given any two vertices v and w, any pair of distinct directed paths joining v to w will be such that one of those paths has length (i.e. number of edges) ∂ or more.*

It is readily seen that for the associated Cayley graph $\mathcal{C}(G, \mathcal{S})$, this notion is translated into the following property of the hash function.

Proposition 2.2 — *If we replace k consecutive symbols of a text*

$$x = x_1x_2\dots x_i \boxed{x_{i+1}\dots x_{i+k}} x_{i+k+1}\dots x_t$$

by a string of h consecutive symbols so that the resulting text

$$x' = x_1x_2\dots x_i \boxed{y_{i+1}\dots y_{i+h}} x_{i+k+1}\dots x_t$$

has the same hashed value, then $\sup(k, h) \geq \partial$.

In other words, if we can obtain $\mathcal{C}(G, \mathcal{S})$ with a large ∂, we protect against local modifications of the text.

Expanding properties. A desirable feature of any hash function is the equidistribution of the hashed values. This property can be guaranteed if the associated Cayley graph $\mathcal{C}(G, \mathcal{S})$ satisfies

Proposition 2.3 — *If $\mathcal{C}(G, \mathcal{S})$ is a Cayley graph such that the gcd of its cycle lengths equals 1, then for the corresponding hash function, the distribution of hashed values of texts of length n tends to equidistribution when n tends to infinity.*

This is proved by classical graph-theoretic (or Markov Chain) methods by studying the successive powers A^n of the adjacency matrix of the graph. To prove that this property occurs in practice, we need to evaluate the speed with which equidistribution is achieved. The best results will be achieved if the Cayley graph $\mathcal{C}(G, \mathcal{S})$ sufficiently resembles a random graph. This can be obtained for graphs with a high *magnifying* or *expansion* coefficient (for more details see [12]).

2.2 The Choice of SL_2.

The groups $SL_2(\mathbf{F}_q)$ of 2×2 matrices of determinant one over a finite field $\mathbf{F}_q$ seem to us to be a promising choice for devising quality hash functions. There are several reasons for this. By choosing simple matrices for generators, one obtains fast hash functions: this is because multiplication by such a matrix, i.e. processing one bit of text, amounts to a few additions in $\mathbf{F}_q$: in this paper, we focus on the case when $q = 2^n$ which provides the fastest computations of all. It is comparatively easy to obtain Cayley graphs over those groups that have large girths, [7]: [8] is a record-breaking construction. Cayley graphs over $SL_2(\mathbf{F}_q)$ tend to display good expanding properties: this is justified both theoretically [9] and experimentally [6].

2.3 On the Difficulty of Finding Collisions.

Another attractive feature of the general hashing scheme is that we can express clearly in mathematicals terms the problem of finding collisions. It may readily be checked that the problem reduces to finding two strings of generators (elements of $\mathcal{S}$) such that the corresponding products coincide in G ; i.e. find $s_1, s_2, \ldots s_n$, $\sigma_1, \sigma_2, \ldots \sigma_m \in \mathcal{S}$ such that

$$s_1 s_2 \ldots s_n = \sigma_1 \sigma_2 \ldots \sigma_m$$

equivalently

$$s_1 s_2 \ldots s_n \sigma_m^{-1} \sigma_{m-1}^{-1} \ldots \sigma_1^{-1} = 1. \tag{1}$$

So we see that finding a collision is equivalent to finding factorisations of the form (1). Now it can be argued that there are always trivial factorisations of the form (1) in any finite group (e.g. $s^{|G|} = 1$, for any $s \in \mathcal{S}$). But we must note that only $n \approx \log |G|$ bits are needed to express hashed values, so that we can choose groups of large cardinality (e.g. $|G| = 2^{500}$) for which trivial factorisations involving $N \sim |G|$ elements are useless as actual forgeries (because no text has 2^{500} bits !). This means, broadly speaking, that the strong collision criterion is satisfied whenever it is computationally difficult to find short factorisations of the form (1). This problem is known to be potentially difficult. For instance Jerrum [5] considers the problem of finding the shortest factorisation of an arbitrary element of an arbitrary permutation group over some set of generators, and proves that it is Pspace-complete. This provides additional motivation for pursuing this group-theoretic strategy.

Moreover when we choose $G = SL_2(\mathbf{F}_q)$ as explained above, it seems that the search for short factorisations in G yields challenging and apparently difficult problems, see e.g. [1]. Some work has already been devoted to solve this problem, and the main results are the following probabilistic algorithms.

1. *Subgroup attacks.* A hashing scheme based on matrix computations has been devised in the past [2] with a group of too small a size for which a probabilistic attack was devised in [3]. It is based on the search for texts whose hashcode falls into a subgroup. For our choice of q this algorithm turns out to be inefficient, because all subgroups of $SL_2(\mathbf{F}_q)$ have index larger than $q + 1$ (theorem 3.2).
2. *Density attacks.* A preliminary version of a group theoretic hash function was shown to be insecure in [11]. It was based on the group $SL_2(\mathbf{Z}_p)$ and the two generators $C = \left(\begin{smallmatrix} 1 & 1 \\ 0 & 1 \end{smallmatrix}\right)$ and $D = \left(\begin{smallmatrix} 1 & 0 \\ 1 & 1 \end{smallmatrix}\right)$. The key to the forgery was the following: it is readily seen that a short factorisation over $SL_2(\mathbf{Z}_p)$ of the identity produces collisions (by insertion into any message M). To find such a factorisation, the strategy is to reduce the problem to factorising in an infinite group, in this case $SL_2(\mathbf{Z})$. Look for a matrix U of $SL_2(\mathbf{Z})$ which reduces modulo p to the identity, and can be expressed as a product of C's and D's. In this case this simply means that U should have non-negative coefficients. There is an effective algorithm to obtain the factorisation of U. To have an effective forgery, one must have a way of finding such a matrix U whose factorisation into C's and D's is short. A probabilistic algorithm that does this is given in [11]: it is based on the fact that the set of matrices of $SL_2(\mathbf{Z})$ with non-negative coefficients is a "dense" subset of $SL_2(\mathbf{Z})$. To protect against such attacks, one should choose sets of generators $\mathcal{S}$ that generate sufficiently sparse submonoids of the infinite groups associated to $SL_2(\mathbf{F}_q)$.

3 A Specific Hash Function

In this section we turn to the specific hash function associated to $G = SL_2(\mathbf{F}_q)$ with $q = 2^n$ and with the generators A and B mentioned in the introduction. We address the following topics related to its security, i.e.

- What is the set of hashcodes?
- Why does this hash function detect local modifications of the input.
- Why does this hash function protect against density attacks.

3.1 The Set of Hashcodes

In order to show that the set of all hashcodes is not too small we need to clarify what is the subset (actually the subgroup) of $SL_2(\mathbf{F}_{2^n})$ which is generated by A and B. More precisely we show here that this subset is the whole group: this is the following theorem

Theorem 3.1 —

$$< A, B >= SL_2(\mathbf{F}_{2^n})$$

(where $< A, B >$ denotes the group generated by A and B)

From that it follows that the set of hashcodes has $2^n(2^{2n} - 1)$ elements (since $|SL_2(\mathbf{F}_{2^n})| = 2^n(2^{2n} - 1)$), which is sufficiently large to avoid direct probabilistic attacks, which use the birthday paradox for instance. Moreover, it is straightforward to check that we can code efficiently these hashcodes by using just $3n + 1$ bits.

Theorem 3.1 is proven by using some known results about the subgroups of $SL_2(\mathbf{F}_{2^n})$. For our purposes, we use the classification obtained by Dickson (see [4]), and which can also be found in [10] (see theorem 6.25 p.412-413)

Theorem 3.2 — *All possible proper subgroups of $SL_2(\mathbf{F}_{2^n})$ are the following:*

(a) Abelian subgroups.
(b) Dihedral subgroups of order $2d$, where d divides $2^n + 1$ or $2^n - 1$.
(c) The alternating groups A_4, or A_5, or the symmetric group S_4.
(d) The upper triangular subgroup, its subgroups, and their conjugates.
(e) $SL_2(\mathbf{F}_{2^m})$, where m is a divisor of n, (and its conjugates).

It is straightforward to check that none of these cases can occur.
Case (a) is impossible since $A.B \neq B.A$.
Case (b) cannot occur, because neither A, nor B is of order 2.
Case (c) is checked with a little computation: for $n > 2$, A and B generate a subgroup whose order is larger than those of the groups of (c).

Case (d) uses a different approach. Let us note that all the matrices of the upper triangular group (or a conjugate of this group) have a common eigenvector. Since A and $S = A^{-1}B = \begin{pmatrix} 1 & 1 \\ 0 & 1 \end{pmatrix}$ have no common eigenvector, A and B can not generate only the upper triangular group (or its conjugates) or its subgroups.
Case (e) is settled by noting that X cannot belong to a subfield $\mathbf{F}_{2^m}$ of $\mathbf{F}_{2^n}$, hence A does not belong to $SL_2(\mathbf{F}_{2^m})$. A and B can not generate a conjugate $U^{-1}SL_2(\mathbf{F}_{2^m})U$ of this group ($U = \begin{pmatrix} a & b \\ c & d \end{pmatrix} \in SL_2(\mathbf{F}_{2^n})$) since this would imply

$$U.A^{-1}.B.U^{-1} \in SL_2(\mathbf{F}_{2^m}) \tag{2}$$

$$U.A^{-2}.B.A.U^{-1} \in SL_2(\mathbf{F}_{2^m}) \tag{3}$$

(2) implies that a^2 and d^2 belong to $\mathbf{F}_{2^m}$, hence a and d belong to $\mathbf{F}_{2^m}$. (3) implies that c^2 and b^2 belong to $\mathbf{F}_{2^m}$, hence c and b belong to $\mathbf{F}_{2^m}$. This in turn implies that U belongs to $SL_2(\mathbf{F}_{2^m})$, which would imply that A belongs to this same subgroup, which is impossible.

Moreover, by using the above classification of subgroups and similar arguments, we can show (details will be archived in a technical report) that the Cayley graph corresponding to our hash function satisfies the condition of proposition 2.3, so we obtain an even stronger property on the hashcodes, that is :

Proposition 3.3 — *The distribution of hashcodes of length l tends to equidistribution when l tends to infinity.*

3.2 Protection Against Local Modifications

We have seen in section 2 that if the Cayley graph associated to our scheme has a girth larger than ∂ then any modification of a text obtained by replacing k consecutive bits by h other bits will necessarily change the hashcode if k and h are less than ∂.

The relevance of our proposition lies in the fact that the girth for the associated graph of the scheme we propose is large :

Theorem 3.4 — *The girth of the Cayley graph $\mathcal{G}(SL_2(\mathbf{F}_{2^n}), A, B)$ is larger than n.*

In other words, for n chosen as explained in the introduction we are *sure* to detect any modification of up to $n = 130 - 170$ consecutive bits of the input text.
Proof.
In the Cayley graph setting, the girth is just the minimum value l for which there exist two different strings of A's and B's $s_1, s_2, \ldots s_l$, and $\sigma_1, \sigma_2, \ldots \sigma_m$ (with $m \leq l$) such that

$$s_1 s_2 \ldots s_l = \sigma_1 \sigma_2 \ldots \sigma_m$$

If we consider now A and B as elements of $SL_2(\mathbf{F}_2[X])$, then the two products $s_1 s_2 \ldots s_l$ and $\sigma_1\sigma_2\ldots\sigma_m$ differ in this infinite group (see lemma 3.5). Since $\mathbf{F}_{2^n}$ is defined as the quotient $\mathbf{F}_2[X]/P_n(X)$, where $P_n(X)$ is some irreducible polynomial over $\mathbf{F}_2$ of degree n, it follows that the only way these products will coincide over $SL_2(\mathbf{F}_{2^n})$, is that one of the matrices $s_1 s_2 \ldots s_l$ or $\sigma_1\sigma_2\ldots\sigma_m$ (computed over $SL_2(\mathbf{F}_2[X])$) has one of its entries equal to a polynomial of degree $\geq n$. It is easy to check that this can happen only if $l \geq n$ or $m \geq n$. □

Lemma 3.5 — *For two different strings of A's and B's , $s_1, s_2, \ldots s_l$, and $\sigma_1, \sigma_2, \ldots \sigma_m$ the products (computed over $SL_2(\mathbf{F}_2[X])$) $s_1 s_2 \ldots s_l$ and $\sigma_1\sigma_2\ldots\sigma_m$ are different.*

Proof.
It is straightforward to show by induction that a product $\sigma_1\sigma_2\ldots\sigma_m$ takes the form $\begin{pmatrix} P_m(X) & Q_{m-1}(X) \\ P_{m-1}(X) & Q_{m-2}(X) \end{pmatrix}$ if $\sigma_m = A$ (The P's and Q's denote polynomials whose degree is indicated by the subscript), or $\begin{pmatrix} P_m(X) & Q_m(X) \\ P_{m-1}(X) & Q_{m-1}(X) \end{pmatrix}$ if $\sigma_m = B$. Hence two products $s_1 s_2 \ldots s_l$ and $\sigma_1\sigma_2\ldots\sigma_m$ can be equal over $SL_2(\mathbf{F}_2[X])$ only if $m = l$, and $\sigma_m = s_l$. By simplifying on the right by $\sigma_m = s_l$, and by iterating this argument we obtain the lemma. □

3.3 Protection Against Density Attacks

A density attack in the case of this particular hash function takes the following form.

1. find a matrix U of $SL_2(\mathbf{F}_2[X])$ which is equal to the identity modulo $P_n(X)$ (where $P_n(X)$ is the irreducible polynomial of degree n which defines the field $\mathbf{F}_{2^n}$ as $\mathbf{F}_2[X]/P_n(X)$).
2. express this matrix U as a product of A and B in $SL_2(\mathbf{F}_2[X])$ (if possible).
3. this factorisation becomes a factorisation of the identity when computed over $SL_2(\mathbf{F}_{2^n})$.
4. one can deduce from it a message whose hashcode is the identity and which may be inserted in every message M without changing the hashed value of M.

Point 1 is rather easy to solve (in the same way as over $SL_2(\mathbf{Z})$), nevertheless the key of our proposition is to make point 2 infeasible. What makes this last problem so difficult is that there are very few matrices of $SL_2(\mathbf{F}_2[X])$, which can be expressed as a product of A and B. More precisely

Theorem 3.6 — *Define the set*

$$E_m = \{U = \begin{pmatrix} P(X) & Q(X) \\ R(X) & S(X) \end{pmatrix} \in SL_2(\mathbf{F}_2[X]) / degP, Q, R, S \leq m\}.$$

We have

$$\frac{|\{U \in E_m\ ,\ U \textit{ is a product of } A \textit{ and } B\}|}{|E_m|} = O(\frac{1}{2^m})$$

Sketch of the proof:
This follows from

$$|\{U \in E_m\ ,\ U \text{ is a product of } A \text{ and } B\}| = 2^{m+1} - 2$$

(because of the form of products of A and B, see the proof of lemma 3.5) and

$$|E_m| = O(2^{2m})$$

(this comes from a counting argument giving the number of pairs of coprime polynomials of bounded degree). □

The above theorem shows that even if we can find a matrix U satisfying point 1 by random search methods, then since it will have an entry which is a polynomial of degree at least n (see proof of theorem 3.4), U will have a very small probability of being a product of A and B (that is $O(\frac{1}{2^n})$).

4 Concluding Remarks

We have defined a new family of hash functions based on computations in $SL_2(\mathbf{F}_{2^n})$. These functions improve upon previous schemes [12], which were defined over the group $SL_2(\mathbf{Z}_p)$. First, computing time is substantially speeded up. It requires at most a few shifts and XOR's of 150-bit quantities per message bit. Because of the concatenation property, precomputations and fast multiplications in fields of characteristic 2 can also be used. Second, we can prove an explicit security property which speaks against the density attacks that have shown the first attempt of [13] to be insecure.

We have chosen the size parameter q so that $\sqrt{q}$ stays too large with respect to computing power. This is to protect against the subgroup approach used in conjunction with birthday attacks.

References

1. Babai, L., Kantor, W. M., Lubotsky, A: Small-diameter Cayley graphs for finite simple groups. Europ. J. of Combinatorics **10** (1989) 507–522
2. Bosset, J: Contre les risques d'altération, un système de certifications des informations. 01 Informatique (1977)
3. Camion, P: Can a fast signature scheme without secret key be secure ? In proc. AAECC (1987) Springer-Verlag Lec. N. Comp. Sci. 228 pp. 187–196
4. Dickson, L: Linear groups with an exposition of the Galois field theory. Dover New York 1958
5. Jerrum, M. R: The complexity of finding minimum length generator sequences. Theoretical Computer Science **36** (1985) 265–289

6. J.Lafferty, Rockmore, D: Numerical investigation of the spectrum for certain families of cayley graphs. In 1992 DIMACS Workshop on expanders graphs (1993) D. S. in Disc. Math., T. C. S. in Discrete Mathematics, and T. C. Science, Eds. pp. 63–74
7. Margulis, G. A: Explicit constructions of graphs without short cycles and low density codes. COMBINATORICA **2** (1982) 71–78
8. Margulis, G. A: Explicit group-theoretical constructions of combinatorial schemes and their application to the design of expanders and concentrators. Problemy Peredachi Informatsii **24** (1988) 51–60
9. Sarnack, P: Some applications of modular forms. Cambridge University Press 1990
10. Suzuki, M: Group theory, volume I. Springer-Verlag 1982
11. Tillich, J.-P., Zémor, G: Group-theoretic hash functions. In First French-Israeli workshop on algebraic coding (1994) Springer-Verlag Lec. N. Comp. Sci. 781 pp. 90–110
12. Zémor, G: Hash functions and Cayley graphs. To appear in Designs, Codes and Cryptography
13. Zémor, G: Hash functions and graphs with large girths. In EUROCRYPT 91 (1991) LNCS 547 Springer-Verlag pp. 508–511

Design of Elliptic Curves with Controllable Lower Boundary of Extension Degree for Reduction Attacks

Jinhui Chao[1], Kazuo Tanada[2] and Shigeo Tsujii[1]

[1] Chuo University, 1-13-27, Kasuga, Bunkyo-ku, Tokyo, 112, Japan
[2] Tokyo Institute of Technology, 2-12-1, Ookayama, Meguro-ku, Tokyo, 152, Japan

Abstract. In this paper, we present a design strategy of elliptic curves whose extension degrees needed for reduction attacks have a controllable lower boundary, based on the complex multiplication fields method of Atkin and Morain over prime fields.

1 Introduction

In recent years, elliptic curves have been used to define a new category of discrete logarithm problems, in hope to build new one-way functions instead of the existing cryptographic functions [1][2] [3].

An elliptic curve over a field K, E/K is defined by the Weierstrass canonical form

$$E/K \ : y^2 + a_1xy + a_3y = x^3 + a_2x^2 + a_4x + a_6 \quad (a_1, a_3, a_2, a_4, a_6 \in K). \tag{1}$$

When $\mathrm{char}(K) \neq 2, 3$, E/K can be transformed by an isomorphism to a form of

$$E/K \ : y^2 = x^3 + ax + b \quad (a, b \in K). \tag{2}$$

The discrete logarithm problem over an elliptic curve E/K is to find $x \in \boldsymbol{Z}$ such that for P, $Q \in E/K, Q = xP$. Hereafter we will assume that $\mathrm{char}(K) = p$.

The above problems are expected to provide a new cryptographic function with stronger integrity and have been applied to build cryptosystems. Until now, two algorithms are known as attacks on the problems: the Baby-step-Giant-step algorithm[4] and the MOV reduction [5].

The first method by Shanks costs $O(\sqrt{\#E(K)} \log \#E(K))$ of fully exponential time. Its fast versions, e.g. Pohligh-Hellman's algorithm[6] reduced the computation to order of the root of the maximum prime factor of $\#E(K)$. If the maximum prime factor is smaller than $\log \#E(K)$, it costs $O((\log \#E(K))^2)$ and becomes a very powerful method.

The second algorithm by Menezes, Okamoto and Vanstone uses the Weil pairing to embed the discrete logarithm problems over $E(\boldsymbol{F}_q)$ into the classic discrete logarithm problems over certain extension of the ground field $\boldsymbol{F}_{q^k}$, which then can be solved by efficient algorithms such as Adleman's index algorithm[7] of subexponential time. This approach works when the extension degree of the ground field required for a well-defined embedding is very low.

To defend the elliptic-curves-based cryptosystems against the first attack, the order of elliptic curves $\#E(K)$ has to contain a large prime factor. As to the second attack, it is known that for a class of elliptic curves called super-singular curves, the reduction can be fulfilled with extension of 6 degree of the ground field[5]. The super-singular curves however are in fact very few. For the ordinary or non-super-singular curves, it is shown in [8] that with high probability, the extension degrees of the ground field needed for reduction attack on random curves are an exponential function of $char(\boldsymbol{F}) = p$. However, this is an asymptotic conclusion and not directly applicable to a fixed prime p [8]. In cryptosystem design practice, it would be desirable to find some strategy to control the lower bound of the extension degree for particular curves defined over fixed fields.

In this paper, we consider about the extension degree of ground fields which is needed to reduce discrete logarithm problems over elliptic curves to discrete logarithm problems over finite fields. Then we show an algorithm to design elliptic curves over primary fields which can control the extension degree for the reduction attack. This algorithm is based on the complex multiplication field method by Atkin and Morain[9][10]. (Both the algorithms can also be generalized over extension fields[11].)

2 Current Design Methods of Curves

Below we review popular design methods of elliptic curves.

(1) Method using Schoof's algorithm[12] : This method selects randomly curves until a desired group structure is found. The order of the curve is calculated by Schoof's algorithm. Koblitz used the following algorithm to design curves of prime orders and the extension degree for the reduction attack to be larger than $(\log p)^2$[8].

[Koblitz's algorithm]

1. Choose a curve $E/\boldsymbol{F}_p$ randomly;
2. Calculate $N = \#E(\boldsymbol{F}_p)$ by Schoof's algorithm;
3. Check that if N is a prime, if not, go back to 1;
4. Check that if $p^j \not\equiv 1 \bmod N (1 \leq j \leq (\log p)^2)$ if not, go back to 1.

In this algorithm, the order calculation part is of most costly, which requires $O((\log p)^8)$ computations by Schoof's algorithm. This becomes awkward when p is large. Recently, progresses have appeared in development of fast order calculation algorithms. However, the order-counting problem seems to be difficult for curves with arbitrary orders. One way to avoid this difficulty is to choose an order first which is "easy" in certain sense and with desired cryptographic property, then build a curve with such order. This is the method by Atkin and Morain.

(2) Method using complex multiplication field : It is implemented by Atkin and Morain to build curves of assigned order with complex multiplication[9] [10].

[Atkin, Morain's algorithm]

1. Determine an order $\#E(\boldsymbol{F}_p) = m$ such that $t = p+1-m$ satisfies $|t| < 2\sqrt{p}$;
2. Calculate d such that $t^2 - 4p = c^2 d$;
3. Calculate the class equation $H_d(x)$;
4. Solve $H_d(x) \equiv 0 \bmod p$ to find a root j_0;
5. Build a curve with j_0 as its j-invariant.

Theoretically, one can produce curves over the prime field with arbitrary order with this algorithm. The most demanding part of the algorithm is the step 3, 4. Since the degree of $H_d(x)$ equals the class number of d, $h(d)$ which is of $O(\sqrt[4]{p})$, the calculation of $H_d(x)$ and the solution of it over $\boldsymbol{F}_p$ by Berlekamp's[13] or Rabin's[14] algorithms require computations of exponential time.

Thus, this algorithm can only be used for small class number cases. Under this condition, curves with the order equals to the characteristic or contains a large prime factor are built in [15][16]. The curve used in [15] is interesting because it can resist any reduction attacks. However, there is only one isogeny class of such curves over a prime field. (Although much richer isogeny classes of p-divisible curves exist over extension fields [17]). Besides, in order to make the class number small, one has to restrict the prime p to meet certain conditions. For the same reason, isogeny classes of curves are also restricted. On the other hand, it seems that to build curves without using Schoof's algorithm could be computationally attractive.

3 Design of Curves with Controlled Lower Boundary of Extension Degree against Reduction Attack

To control the extension degree for the reduction attack, we choose the following strategy, i.e., to specify a lower boundary B of the extension degree for the reduction attack, then design a curve with the order satisfies this lower boundary.

First we consider the extension degree for the reduction attack on non-p-divisible curves.

The condition for any well-defined reduction to $\boldsymbol{F}_{q^k}$ with $m = \#E(\boldsymbol{F}_q)$ is that

$$m \mid q^k - 1 \tag{3}$$

or

$$q^k \equiv 1 \bmod m \tag{4}$$

By Euler's theorem, the minimum k satisfies (4) must a factor of $\varphi(m)$. Thus, take the primary factorization of $\varphi(m)$, one can find the minimum factor satisfy (4), then find the minimum k. However, the primary factorization of $\varphi(m)$ is then necessary, which could become a new computational burden.

Now we give a condition of order m for the extension degree to be larger than B.

Theorem 1. *Assume an elliptic curve over $\boldsymbol{F}_q$ has order m. Denote the extension degree for arbitrary reduction as k. If $\varphi(m)/2$ is B-nonsmooth and $q^2 \not\equiv 1 \bmod m$ then*

$$k \geq B.$$

Here n is B-nonsmooth means that n has not primary factors less than B.

Proof : As before, by Euler's theorem, k divides $\varphi(m)$. If $\varphi(m)/2$ is B-nonsmooth, then $k = 2$ or $k \geq B$. Thus, if $q^2 \not\equiv 1 \bmod m$, then $k \geq B$. QED

Corollary 2. *Let $\#E(\boldsymbol{F}_q) = m$, $l \mid m$ (l : a prime), and the extension degree for any reduction as k. If $(l-1)/2$ is B-nonsmooth and $q^2 \not\equiv 1 \bmod l$ then*

$$k \geq B$$

QED

Therefore, curves are to be designed to satisfy the following condition.
zh
Condition : $l \mid \#E(\boldsymbol{F}_p)$ and $(l-1)/2$ is B-nonsmooth.
zh
We need then a method to assign order of a curve. There is currently only one method for this purpose, the one with complex multiplication fields, or Atkin and Morain's algorithm.

Below, we show an algorithm over primary fields to build the curves satisfy the above condition based on the Atkin and Morain's algorithm.

[Algorithm]

1. Choose a large prime l such that $(l-1)/2$ to be B-nonsmooth and $(\frac{d}{l}) = 1$, choose also $d < 0$ with small $h(d)$;
2. Choose t, c, s such that $(t-2)^2 = 4sl + c^2 d$, $(t \not\equiv 0, 2 \bmod l)$;
3. Check if $p = sl + t - 1$ is a prime , if not, go back to 2. ;
4. Calculate the class equation $H_d(x)$ and solve $H_d(x) \equiv 0 \bmod p$ to find a root j_0 ;
5. Define a curve with j-invariant as j_0.

In this way, curves of order sl are derived.

In Step1, to search for B-nonsmooth l need about $(\log l)(\log B)$ primality tests. $(\frac{d}{l}) = 1$ holds in probability of 1/2. Once $(\frac{d}{l}) = 1$ is true, there are plenty of solutions for Step2. Assuming p is random, Step 3 will repeat about $\log p$ times to pass the check. (In simulation it seems quite easy.)

[Example]

1. Choose prime $l = 21838143759917965991093122527538322503$ and $d = -43$; where $l - 1 = 2 * 10\,91907\,18799\,58982\,99554\,65612\,63769\,16251$
2. Choose $t = 5472\,72782\,79345\,38832$, $s = 4$;
3. Obtain $p = 87\,35257\,50396\,71864\,01909\,97683\,89498\,68843$;
4. From $H_d(x) = x + 960^3$, we have $j_0 = -960^3$, which defines a curve as

zh

$$\begin{aligned} y^2 &= x^3 + ax + b \bmod p \qquad (5) \\ a &= 29\,71431\,93700\,48984\,66387\,07954\,89768\,65095 \\ b &= 9\,30797\,87665\,24631\,56378\,60591\,36653\,79551 \\ p &= 87\,35257\,50396\,71864\,01909\,97683\,89498\,68843 \end{aligned}$$

This curve has its order and the extension degree for any reduction as

$$\begin{aligned} \#E(\boldsymbol{F}_p) &= 4 * 21\,83814\,37599\,17965\,99109\,31225\,27538\,32503 \\ B &= 10\,91907\,18799\,58982\,99554\,65612\,63769\,16251 \end{aligned}$$

References

1. Miller, V. S.: Use of elliptic curves in cryptography. Advances in Cryptology-CRYPTO'85, Lecture Notes in Computer Science, **218** (1986) 417-426
2. Koblitz, N. : Elliptic Curve Cryptosystems. Math. Comp. **48** (1987) 203-209
3. Menezes, A. J.: *Elliptic Curve Public Key Cryptosystems* Kluwer Academic Publishers (1993)
4. Knuth, D. E.: *The art of computer programming*, Sorting and searching. vol. 3, Addison Wesley (1973)
5. Menezes, A. J., Vanstone, S., Okamoto T.: Reducing elliptic curve logarithms to logarithms in a finite field. Proc. of STOC'91 (1991) 80-89
6. Pohlig, S. C., Hellman, M. E.: An improved algorithm for computing logarithm over $GF(p)$ and its cryptographic significance. IEEE Trans. Information Theory, **IT-24**, 1 (1978) 106-110,
7. Adleman, L. M.: A subexponential algorithm for the discrete logarithm problem with applications to cryptography. Proc. of IEEE 20th Symp. on Foundations of Comp. Sci. (1979) 55-60
8. Koblitz, N.: Elliptic curve implementation of zero-knowledge blobs. Journal of Cryptology, vol. 4, No. 3 (1991) 207-213,
9. Atkin, A. O. L., Morain F.: Elliptic curves and primality proving. Research Report 1256, INRIA, June (1990)
10. Morain, F., Building cyclic elliptic curves modulo large primes. Advances in Cryptology -EUROCRYPT'91, Lecture Notes in Computer Science, **547** (1991) 328-336
11. Chao, J., Tanada, K., Tsujii S.: On secure elliptic curves against the "reduction attack" and their design strategy. Proc. of SCIS'94 (1994) 10A, IEICE Tech. Report, ISEC93-100, 29-37

12. Schoof, R.: Elliptic curves over finite fields and the computation of square roots mod p. Math. Comp. **44** (1985) 483-494
13. Berlekamp, E. R.: *Algebraic coding theory.* MacGraw–Hill (1968)
14. Rabin, M. O.: Probabilistic algorithm in finite fields. SIAM J. on Comput., Vol.9, No.2 (1980) 273-280
15. Miyaji, A.: On ordinary elliptic curve cryptosystems. Advances in Cryptology-ASIACRYPT'91, Lecture Notes on Computer Science **739** (1991) 460-469
16. Miyaji, A.: Fast elliptic curve cryptosystems, Technical Report of IEICE (1993) COMP93-25
17. Chao, J., Ikemoto, H., Tanada, K., Tsujii, S.: On Discrete Logarithm Problems over elliptic curves with p-divisible groups. Proc. of Joint Workshop on Information Security and Cryptography, JW-ISC93 (1993) 99-104
18. Koblitz, N.: Constructing elliptic curve cryptosystems in characteristic 2. Advances in Cryptology - CRYPTO'90, Lecture Notes in Computer Science, **537** (1990) 156-167

Cryptographic Protocols Based on Discrete Logarithms in Real-quadratic Orders

Ingrid Biehl Johannes Buchmann Christoph Thiel

Fachbereich Informatik
Universität des Saarlandes
Postfach 151150, 66041 Saarbrücken, Germany

Abstract. We generalize and improve the schemes of [4]. We introduce analogues of exponentiation and discrete logarithms in the principle cycle of real quadratic orders. This enables us to implement many cryptographic protocols based on discrete logarithms, e.g. a variant of the signature scheme of ElGamal [8].

1 Introduction

1.1 Motivation

The security of many cryptographic protocols (see for example [7], [8], [12]) is based on the difficulty of solving the *discrete logarithm problem* (DL-problem) in the multiplicative group $GF(p)^*$ of prime fields $GF(p)$ of characteristic $p > 0$.

Recently, Gordon [9] has shown that under reasonable assumptions the discrete DL-problem in GF$(p)^*$ can be solved in expected time

$$L_p[1/3, c] = exp((c + o(1)) \cdot (\log p)^{1/3} \cdot (\log\log p)^{2/3})$$

by means of the number field sieve (NFS), thereby lowering the best known asymptotically upper bound considerably. Experience with similar integer factoring algorithms shows that the NFS can be expected to be practical (see [5], [1]). It is therefore by no means clear that the discrete logarithm problem remains difficult in the future and one must search for other problems that can serve as basis for one-way and trapdoor one-way functions (see for example [4], [10], [11]). It would be useful to have some sort of hierarchy of difficult problems. If one of the problems turns out to be easy one can use the next difficult one that remains intractable. A first step in this direction is to employ algebraic number fields (see [3]) as a source for computationally hard problems. In [4], [13] it is shown how to use the infrastructure of the cycles of reduced ideals in real quadratic orders to implement the Diffie-Hellman key exchange protocol. Breaking that scheme was shown to be at least as hard as factoring. Also, it was the first case of a Diffie-Hellman-implementation which is not based on the arithmetic of a finite abelian group. That application, however, looks rather restricted since it only solves the problem of key management. In this paper we generalize and improve the scheme of [4]. We introduce analogoues of exponentiation and discrete logarithms in the principle cycle of real quadratic orders. This enables us to implement many cryptographic protocols which are based on discrete logarithms, e.g. the ElGamal scheme [8], and we argue that computing generalized discrete logarithms is at least as hard as factoring.

1.2 Discrete Logarithms in Real-quadratic Orders

Let $\mathcal{D}$ be the set of all $\Delta \in \mathbb{Z}_{>0}$ which are no squares in $\mathbb{Z}$ and satisfy $\Delta \equiv 0, 1 \bmod 4$. Let $\Delta \in \mathcal{D}$. Then $\mathcal{O}_\Delta = \mathbb{Z} + \frac{\Delta+\sqrt{\Delta}}{2}\mathbb{Z}$ is the *real quadratic order of discriminant* Δ. The *field of fractions* of $\mathcal{O}_\Delta$ is $\mathrm{K}_\Delta = \mathbb{Q}(\sqrt{\Delta})$, its multiplicative group is denoted by K_Δ^*. For $\alpha \in \mathrm{K}_\Delta$, $\alpha = (x + y\sqrt{\Delta})/(2z)$ with $x, y, z \in \mathbb{Z}$, $z > 0$ and $\gcd(x, y, z) = 1$, we set $\overline{\alpha} = (x - y\sqrt{\Delta})/(2z)$.

A finitely generated $\mathcal{O}_\Delta$-module $A \subseteq \mathrm{K}_\Delta$, $A \neq \{0\}$ is called *fractional ideal* of $\mathcal{O}_\Delta$. It can be written as $A = q\left(\mathbb{Z} + \frac{b+\sqrt{\Delta}}{2a}\mathbb{Z}\right)$, where $a, b \in \mathbb{Z}, q \in \mathbb{Q}$, $a, q > 0$, and $c = (b^2 - \Delta)/(4a) \in \mathbb{Z}$. The numbers a and q are uniquely determined whereas b is unique modulo $2a$. To make the representation unique we choose $-2a + \sqrt{\Delta} < b < \sqrt{\Delta}$ if $a < \sqrt{\Delta}$, and $-a < b \leq a$ if $a > \sqrt{\Delta}$. If $q = 1$ then A is called *normal* . We will only use normal ideals and write $a = a(A)$, $b = b(A)$, $c = c(A)$. We also define $\gamma(A) = \frac{b+\sqrt{\Delta}}{2a}$, so $A = \mathbb{Z} + \gamma(A)\mathbb{Z}$. Fractional ideals A, B of $\mathcal{O}_\Delta$ are called *equivalent* if there is $\alpha \in \mathrm{K}_\Delta^*$ such that $A = \alpha B$. The number α is called *generator of A relative to B*. If $B = \mathcal{O}_\Delta$ then A is called *principle* and α is called *generator* of A. The equivalence classes of the above equivalence relation are called *ideal classes.*

In cryptosystems based on real quadratic orders, the analogue of the finite field GF(p) is the set of reduced principal ideals. An ideal A of $\mathcal{O}_\Delta$ is called *reduced* if $\gamma(A) > 1$ and $-1 < \overline{\gamma(A)} < 0$. If A is reduced, then $a(A), |b(A)| < \sqrt{\Delta}$. Thus, the set $\mathcal{R}_\Delta$ of reduced principle ideals of $\mathcal{O}_\Delta$ is finite.

Shanks [14] discovered that $\mathcal{R}_\Delta$ resembles a cyclic group. This enables us to formulate a discrete logarithm problem in $\mathcal{R}_\Delta$. If A, B are ideals of $\mathcal{O}_\Delta$, α a generator of A relative to B then the set of all generators of A relative to B is $\{\eta\alpha : \eta \text{ is a unit of } \mathcal{O}_\Delta\}$. There is a unique unit ϵ in $\mathcal{O}_\Delta$ with $\epsilon > 1$ such that any other unit η in $\mathcal{O}_\Delta$ can be written as $\eta = \pm\epsilon^k$ with $k \in \mathbb{Z}$. Hence the generators of A relative to B are of the form $\alpha' = \pm\epsilon^k\alpha$, $k \in \mathbb{Z}$, and we have $\log|\alpha'| = \log|\alpha| + k\mathrm{R}_\Delta$ where $\mathrm{R}_\Delta = \log\epsilon$ is the *regulator* of $\mathcal{O}_\Delta$. The residue class $\log|\alpha| + \mathrm{R}_\Delta\mathbb{Z}$ is called *distance* from A to B. That distance is denoted by $\delta(A, B)$. If $A = \mathcal{O}_\Delta$ then we briefly write $\delta(B)$. On a circle of circumpherence R_Δ fix a point for $\mathcal{O}_\Delta$. Then $B \in \mathcal{R}_\Delta$ corresponds to a point on that circle whose distance from $\mathcal{O}_\Delta$ is $\delta(B)$. The set $\mathcal{R}_\Delta$ forms a cyclic graph. In that graph, each reduced principle ideal has a *left neighbor* and a *right neighbor*.

We explain exponentiation in $\mathcal{R}_\Delta$. We associate with $k \in \mathbb{R}$ a pair (A, c) consisting of the nearest reduced principal ideal A to k and the distance from k to A. That nearest ideal is defined by the property that there is $\alpha \in \mathcal{O}_\Delta$, with

$(1/\alpha)\mathcal{O}_\Delta = A$ and such that $|\log|\alpha| - k| < |\log|\alpha'| - k|$ for all $\alpha' \in \mathcal{O}_\Delta$. The distance from k to A is $c = \log|\alpha| - k$. In case that A is not uniquely determined by those conditions we make A unique by requiring c to be positive. Let m be a positive integer. Let $A \in \mathcal{R}_\Delta$ and c a real number. Let $\delta \in \delta(A)$. The mth power of (A, c) is a pair (B, d) where B is the reduced principal ideal which is nearest to $x = m(\delta + c)$ and d is the distance from x to B. This power is independent of the choice of δ and is denoted by $\exp((A, c), m)$. It is easy to see that if $A \in \mathcal{R}_\Delta$, $c \in \mathbb{R}$, $k, m \in \mathbb{Z}_{>0}$ then we have $\exp(\exp((A, c), k), m) = \exp(\exp((A, c), m), k)$. This statement enables us to implement many cryptographic protocols. The difficulty of inverting exp follows from the following theorem which can be proved using ideas similar to those explained in [4].

Theorem 1. *There is a probabilistic polynomial time reduction of factoring integers to inverting* exp.

2 Computing Powers

Let c be a complex number, and let $q \in \mathbb{Z}_{>0}$. An *approximation of precision* q to c is a number $\hat{c} \in 2^{-(q+1)}\mathbb{Z}[i]$ such that $|c - \hat{c}| < 2^{-q}$.

We now present the algorithm EXP. Let $A \in \mathcal{R}_\Delta$, $c \in \mathbb{Q}$, $x \in \mathbb{Z}_{>0}$, $\delta \in \delta(A)$, and $\exp((A, c), x) = (B, d)$. Given A, c, x, and a sufficiently large precision $p \in \mathbb{Z}_{>0}$ EXP determines B or its left neighbor and an approximation d of precision p to the distance between B and $x(d + c)$. Since there might be two ideals in $\mathcal{R}_\Delta$ of almost equal distance from $x(\delta + c)$, it is, in general, not clear, whether B itself can be computed in polynomial time.

Algorithm 2 (EXP).

Input: $\Delta \in \mathcal{D}$, $p \in \mathbb{Z}$, $p \geq \lceil 3 + \log \Delta \rceil$, $A \in \mathcal{R}_\Delta$, $c \in \mathbb{Q}$, $x \in [0, \ldots, \lceil\sqrt{\Delta}\rceil]$.
Output: *As specified above*

$C := \mathcal{O}_\Delta;\ p' := p + 3 \cdot \lceil \log \Delta \rceil + 3$	(1)
IF $x = 1$	(2)
THEN $\xi := 1;\ C := A;\ m := 0;\ \ell := c$	(3)
ELSE *Compute the binary decomposition* $x = \sum_{i=0}^{m} x_i 2^i$ *with* $m = \lceil \log x \rceil$, $x_i \in \{0, 1\}$ *for* $0 \leq i \leq m$; $d = 0;\ \ell := 0$	(4)
FOR $i = 0, 1, \ldots, m$	(5)
$\gamma_i := 1$	(6)
IF $x_i = 1$ *THEN* $C := CA;\ \ell := \ell + d$	(7)
$(A, \gamma_i) := REDUCE(\Delta, (A)^2);\ d := 2d + APPROX(\Delta, \gamma_i, p')$	(8)
$(C, \gamma) := REDUCE(\Delta, C);\ \ell := -\ell + cx - APPROX(\Delta, \gamma, p')$	(9)
$\alpha := TARGET(\Delta, C, \ell)$	(10)
$B := (1/\alpha)C;\ d := \ell - APPROX(\Delta, \alpha, p')$	(11)

EXP determines its result by means of binary exponentiation. The basic operation in the set of reduced principal ideals is ideal multiplication which can be performed in quadratic time followed by ideal reduction which also requires quadratic time. Reduction of ideals is performed using the procedure REDUCE which on input of the discriminant Δ and an ideal B of $\mathcal{O}_\Delta$ returns a reduced ideal in the ideal class of B and a generator γ of B relative to C. If we compute a reduced principal ideal C in the class of the product of two reduced principal ideals B and B' using REDUCE then $\delta(C) = \delta(B) + \delta(B') + \log|\gamma|$. While carrying out binary exponentiation in EXP we therefore accumulate the error term $\log|\gamma|$ in the variable ℓ. This is done using APPROX which given Δ, a number γ in K_Δ and a precison constant q finds in linear time an approximation of precison q to $\log|\gamma|$. After executing the FOR loop we have found an ideal C which is close to $x\delta + \ell$. Procedure TARGET then determines a generator of C relative to the real output ideal B. TARGET uses the procedures which were introduced in [6]. Its correctness follows from the arguments presented there. We will show in the full paper that the running time of EXP is cubic just as exponentiation in GF(p).

Algorithm 3 (TARGET).

Input: $\Delta \in \mathcal{D}$, *a reduced principal ideal* A *of* $\mathcal{O}_\Delta$, $s \in \mathbb{Q}_{>0}$.
Output: *A minimum* α *that is closest to* s *or the left neighbor of it.*

$p := \lceil 3 + \log \Delta \rceil$	*(1)*
$\xi := CLOSE(\Delta, A, s);\ e := s - APPROX(\Delta, \xi, p)$	*(2)*
$\eta := NEAREST(\Delta, (1/\xi) \cdot B, e, p);\ \alpha := \xi \cdot \eta;\ r := APPROX(\Delta, \alpha, p) - s$	*(3)*
$IF\ r \geq 0\ THEN\ \alpha := LMIN(\Delta, B, \alpha)$	*(4)*

2.1 Cryptographic Applications of EXP

We need another simple but important procedure MULT which replaces the multiplication in GF(p). Let A, B be reduced principal ideals, $a, b \in \mathbb{Q}$, and let I be the reduced principal ideal such that $\delta(I)$ is as close as possible to $\delta(A)+\delta(B)+a+b$. Given (A, a), (B, b) and a precision $p \in \mathbb{Z}_{>0}$ MULT determines a pair (C, c) where $C = I$ or C is the left neighbor of I and c is an approximation of precision p to the absolute smallest representative of $\delta(C)-\delta(A)-\delta(B)-a-b$. MULT can be easily implemented using the techniques of [6]. Details will be presented in the full paper.

Using MULT and EXP one can implement many cryptographic protocols. Clearly, the Diffie-Hellman key exchange protocol can be implemented using those techniques. Here we present a variant of the ElGamal sigature scheme [8]. Suppose that Alice wants to be able to sign messages. She chooses a real quadratic discriminant Δ and a reduced principal ideal A of $\mathcal{O}_\Delta$ for which she knows a good approximation a of a representative of $\delta(A)$. That can be realised by randomly choosing a number $a' \in \mathbb{R}$ and by a procedure CLOSEIDEAL which given a' and a precision p computes in polynomial time A and a such that

A is an ideal closest to a' or its left neighbor and a is an approzimation of some $\delta \in \delta(A)$ of precision p. Both Δ and A are made public. Also, the precision in the following computation is $p = \lceil 3 + \log \Delta \rceil$.

To sign the message $m \in \mathbb{Z}$, $m \in \{1, 2, \ldots, \lfloor \sqrt{\Delta} \rfloor\}$ Alice randomly chooses $r \in \{1, 2, \ldots, \lfloor \sqrt{\Delta} \rfloor\}$ and determines $(X, x) = \mathrm{EXP}((\mathcal{O}_\Delta, 1), r, p)$. The distance of X is very close to r. The number x is the corresponding correction term. Next, Alice computes $b = (m - a)/r$, $b_1 = \lfloor b \rfloor$, $b_2 = b - b_1$. Also, she determines a reduced principal ideal C which is close to rb_2 and the corresponding correction term c. The signature is $((X, x), b_1, (C, c), m)$. In order for Bob to verify the signature he computes $(Y_1, y_1) = \mathrm{EXP}((X, x), b_1, p)$, $(Y_2, y_2) = \mathrm{MULT}((A, 0), (Y_1, y_1), p)$, $(Y_3, y_3) = \mathrm{MULT}((Y_2, y_2), (C, c), p)$. It is easy to see that Y_3 must be the reduced principal ideal which is closest to m or its left neighbor. If this is incorrect then the signature is incorrect. As in the original ElGamal scheme, there is currently no other way of breaking this scheme than inverting exp and finding r or a. Again, a formal version of this protocol will be presented in the full version.

References

1. Bernstein, D.J., Lenstra, A.K.: A general number field sieve implementation. In: A. K. Lenstra, H. W. Lenstra, Jr. (Eds.) The Development of the Number Field Sieve (LNM 1554) (1993), Springer-Verlag, pp. 103-126
2. Buchmann, J., Thiel, C., Williams, H.C.: Short representation of quadratic integers. To appear in Proc. of CANT 1992
3. Buchmann, J.: Number Theoretic Algorithms and Cryptology. Proc. of FCT'91 (LNCS 529) (1991), Springer-Verlag, pp.16-21
4. Buchmann, J., Williams, H.C.: A Key Exchange System Based on Real-quadratic Fields. Proc. of CRYPTO'89 (LNCS 435) (1989), Springer-Verlag, pp. 335-343
5. Buchmann, J., Loho, J., Zayer, J.: An Implementation of the General Number Field Sieve. Proc. of CRYPTO'93 (LNCS 773) (1993), Springer-Verlag, pp. 159-165
6. Biehl, I., Buchmann, J.: Algorithms for quadratic orders. To appear in Proc. of Symposia in Applied Mathematics (1993)
7. Diffie, W., Hellman, M.: New directions in Cryptography. IEEE Trans. Inform. Theory 22 (1976), pp. 472-492
8. ElGamal, T.: A public key cryptosystem and a signature scheme based on discrete logarithms. IEEE Trans. Inform. Theory 31 (1985), pp. 469-472
9. Gordon, D.: Discrete Logarithms in $GF(p)$ Using the Number Field Sieve. Siam Jour. on Discrete Math. 6 (1993), pp. 124-138
10. Koblitz, N.: Elliptic curve cryptosystems. Math. Comp. 48 (1987), pp. 203-209
11. Miller, V.: Use of Elliptic Curves in Cryptography. Proc. of CRYPTO'85 (LNCS 218) (1986), Springer-Verlag, pp. 417-426
12. National Institute of Standards and Technology. The Digital Signature Standard, proposal and discussion. Comm. of the ACM, 35 (7), Juli 1992, pp. 36-54
13. Scheidler, R., Buchmann, J., Williams, H.C.: Implementation of a Key Exchange Protocol Using Real Quadratic Fields. Proc. of EUROCRYPT'90 (LNCS 473) (1990), Springer-Verlag, pp. 98-109
14. Shanks, D.: The infrastructure of a real quadratic field and its applications. Proc. of the 1972 Number Theory Conference, Boulder (1972), pp. 217-224

Designated Confirmer Signatures and Public-Key Encryption are Equivalent

Tatsuaki Okamoto

NTT Laboratories
Nippon Telegraph and Telephone Corporation
1-2356 Take, Yokosuka-shi, Kanagawa-ken, 238-03 Japan
Email: okamoto@sucaba.ntt.jp

Abstract. The concept of designated confirmer signatures was introduced by Chaum [Cha94] to improve a shortcoming of undeniable signatures. The present paper formalizes the definition of designated confirmer signatures and proves that a designated confirmer signature scheme is equivalent to a public-key encryption scheme with respect to existence. In addition, the paper proposes practical designated confirmer signature schemes which are more efficient in signing than the previous scheme [Cha94].

1 Introduction

The concept of *undeniable signatures* was proposed by Chaum et al. [Cha90, CA89]; the recipient of a signature cannot misuse the signature and the signer cannot subsequently deny the signature. Unfortunately, for many practical applications undeniable signatures have one major shortcoming compared to normal (self-authenticating) digital signatures. Since undeniable signatures rely on the signer cooperating in subsequent confirmations of the signature, if the signer should become unavailable, such as might be expected in the case of a default on the agreement authorized by the signature, or should refuse to cooperate, then the recipient cannot make use of the signature.

The concept of *designated confirmer signatures* was introduced by Chaum [Cha94] to solve this weakness of undeniable signatures. It involves three parties: the signer, recipient, and confirmer. In designated confirmer signature schemes, if the signer is unavailable to confirm the signature, the confirmer, previously designated by the signer, can confirm the signature for the recipient.

In [Cha94], however, no formal definition (i.e., no rigorous concept) of designated confirmer signatures was given, and only an example of designated confirmer signature based on the RSA scheme was proposed. As mentioned in [Cha94], the remaining problems were as follows:

- give a formal definition (i.e., rigorous concept) of designated confirmer signatures.
- construct a designated confirmer signature based on a more general assumption. (i.e., find a sufficient assumption which is as weak as possible.)
- clarify what assumption is essential for constructing a designated confirmer signature. (i.e., find a necessary assumption which is as strong as possible.)
- propose more efficient constructions than [Cha94].

The present paper solves all these problems.

- This paper gives the first formal definition (i.e., rigorous concept) of designated confirmer signatures.
- This paper ultimately answers both the questions on necessary and sufficient assumptions. It proves that *designated confirmer signatures exist if and only if public-key encryption [GM84] exists.* That is, it shows that the existence of public-key encryption is the necessary and sufficient assumption for constructing designated confirmer signatures.
- This paper proposes practical designated confirmer signature schemes that are more efficient in signing than the previous scheme [Cha94]. The proposed schemes are based on three move identification protocols, while the previous scheme is based on the RSA scheme. Two typical constructions are shown, one of which utilizes the Schnorr scheme [Sch91] as a three move protocol, while the other uses instead the extended Fiat-Shamir scheme [GQ88, OhOk88] as a three move protocol.

The theoretical part of our results can be considered from the following viewpoint. In the theoretical research fields of cryptography, the relationships of (computational) cryptographic primitives have been investigated extensively for the latest ten years [1] and many typical cryptographic primitives have been classified into two classes: one-way function family (OWF) and encryption-decryption function family (EDF). OWF consists of the primitives that are equivalent to one-way functions with respect to existence. EDF consists of the primitives that seem to require the encryption-decryption property as well as the one-way property. Impagliazzo and Rudich's result [IR89] seems to imply that several primitives such as secret key agreement may essentially require the encryption-decryption property in addition to the one-way property. In other words, EDF seems to be exclusive to OWF. OWF includes digital signatures [NY89, Rom90], pseudo-random generators [ILL89, Has90], and bit-commitment [Nao90]. EDF includes secret key agreement, public-key encryption, and oblivious transfer. (Here note that the security of these primitives have been formally defined, and that, for example, a digital signature scheme means a digital signature scheme which is existentially secure against adaptive chosen message attacks [GMRi88].)

The following natural question is suggested: To which class, OWF or EDF, do undeniable signatures and designated confirmer signatures belong? Boyar et al [BCDP90] gave an answer to this question: undeniable signatures exist if and only if digital signatures exists (i.e., if and only if one-way functions exist [NY89, Rom90]). That is, undeniable signatures belong to OWF. Hence, the other remaining problem has been to determine to which class, OWF or EDF, designated confirmer signatures belong.

The present paper answers this problem. It shows that designated confirmer signatures belong to EDF. This result seems to be somewhat surprising since designated confirmer signatures belong to a class different from the class of undeniable signatures, although designated confirmer signatures were introduced as a variant of undeniable signatures. (On the contrary, the previous result that undeniable signatures belong to the same class as digital signatures belong to

[1] The relationships of physical (i.e., information theoretical, or unconditionally secure) cryptographic primitives have been also investigated (e.g., [Oka93]).

[BCDP90] is less surprising, since undeniable signatures were introduced as a variant of digital signatures.)

This paper is organized as follows. Section 2 formalizes designated confirmer signatures and proves the main theorem that implies the equivalence of the designated confirmer signatures and public-key encryption. Section 3 proposes practical designated confirmer signature schemes that are more efficient in signing than the previous scheme.

2 Theoretical Results

2.1 Definitions

In this section, we formally define the "designated confirmer signature", as a variant of the definition of the "undeniable signature" by [BCDP90]. In the definition of designated confirmer signatures, a "designated confirmer" is introduced in addition to the properties of the undeniable signatures.

Definition 1. [Secure designated confirmer signature scheme] A *secure* "designated confirmer signature" scheme is $(G_S, G_C, Sign, Conf_{(S,V)}, Conf_{(C,V)})$ such that the following conditions hold:

1. **Key generation (G_S, G_C):**
   Let S be a signer, and C be a designated confirmer. G_C is a probabilistic poly-time algorithm which, on input 1^n (the security parameter), outputs a pair of strings, (C's secret-key, C's public-key), which is denoted by $G_C(1^n) = (G1_C(1^n), G2_C(1^n))$.
   G_S is a probabilistic poly-time algorithm which, on input strings 1^n and C's public-key $\in G2_C(1^n)$, outputs a pair of strings, (S's secret-key, S's public-key), which is denoted by $G_S(1^n, G2_C(1^n)) = (G1_S(1^n, G2_C(1^n)), G2_S(1^n), G2_C(1^n))$. Hereafter, however, for simplicity, $G_S(1^n, G2_C(1^n)) = (G1_S(1^n, G2_C(1^n)), G2_S(1^n), G2_C(1^n))$ will be just written by $G_S(1^n) = (G1_S(1^n), G2_S(1^n))$. The probability is taken over G_C's and G_S's coin tosses. Note that, as a variant, the input of G_S can be restricted only to 1^n. (Then, a slight modification of the definitions of the signing and privacy is required in some cases, although it is fairly easy. See Note of this definition.)
2. **Signing ($Sign$):**
   $Sign$ is a probabilistic poly-time algorithm which, on input strings 1^n, m (message), C's public-key $\in G2_C(1^n)$, and S's (secret-key, public-key) $\in G_S(1^n)$, outputs a string ("designated confirmer signature"), which is denoted by $Sign(1^n, m, G_S(1^n))$ (shortly by $Sign(m)$). The probability is taken over $Sign$'s coin tosses. Let $\Sigma_S(m)$ be the set of $Sign(m)$.
3. **Confirmation and disavowal ($Conf_{(S,V)}, Conf_{(C,V)}$):**
   $Conf_{(S,V)}$ is an interactive proof [GMRa89] between S and V, which, on common input strings 1^n, m, s (the presumed signature of m), S's public-key $\in G2_S(1^n)$, and C's public-key $\in G2_C(1^n)$, outputs either 0 ("true") or 1 ("false"). Here, signer S is the prover with an auxiliary input, S's secret-key $\in G1_S(1^n)$, and V is the verifier. For all m, for any constant c, and for sufficiently large n,

$$\Pr(Conf_{(S,V)}(1^n, m, s, G2_S(1^n), G2_C(1^n)) = 0) > 1 - 1/n^c,$$

if $s = \Sigma_S(m)$, and

$$\Pr(Conf_{(S,V)}(1^n, m, s, G2_S(1^n), G2_C(1^n)) = 1) > 1 - 1/n^c,$$

otherwise. The probability is taken over the coin tosses of S and V.
$Conf_{(C,V)}$ is an interactive proof between C and V, which, on common input strings 1^n, m, s, S's public-key $\in G2_S(1^n)$, and C's public-key $\in G2_C(1^n)$, outputs either 0 ("true") or 1 ("false"). Here, designated confirmer C is the prover with an auxiliary input, C's secret-key $\in G1_C(1^n)$, and V is the verifier. For all m, for any constant c, and for sufficiently large n,

$$\Pr(Conf_{(C,V)}(1^n, m, s, G2_S(1^n), G2_C(1^n)) = 0) > 1 - 1/n^c,$$

if $s = \Sigma_S(m)$, and

$$\Pr(Conf_{(C,V)}(1^n, m, s, G2_S(1^n), G2_C(1^n)) = 1) > 1 - 1/n^c,$$

otherwise. The probability is taken over the coin tosses of C and V.

4. **Security:**
   - **[Security for signers: Unforgeability against any adaptively chosen message attacks]** Let F be a probabilistic poly-time forging algorithm which, on input strings 1^n, S's public-key $\in G2_S(1^n)$, C's public-key $\in G2_C(1^n)$, and C's secret-key $\in G1_C(1^n)$, can request and receive S's signatures of polynomially-many adaptively chosen messages $\{m_i\}$, can request the execution of $Conf_{(S,F)}$ for polynomially-many adaptively chosen strings (either true signature strings or fake signature strings), and finally outputs a pair of strings (m, s). Then, for all such F, for any constant c, for all sufficiently large n, the probability that F outputs (m, s) for which either $Conf_{(S,V)}$ or $Conf_{(C,V)}$ outputs 0 is less than $1/n^c$. The probability is taken over the coin tosses of G_S, G_C, *Sign*, S, V and F.
   - **[Security for confirmers]** Let A be a probabilistic poly-time attacking algorithm which, on input strings 1^n, S's public-key $\in G2_S(1^n)$, S's secret-key $\in G1_S(1^n)$, and C's public-key $\in G2_C(1^n)$, can request the execution of $Conf_{(C,A)}$ for polynomially-many strings (either true signature strings or fake signature strings) adaptively chosen by A, and finally execute $Conf_{(A,V)}$ which is accepted by V, where V is a honest verifier. Then, for all such A, for any constant c, for all sufficiently large n, the probability that A succeeds in the above-mentioned attack is less than $1/n^c$. The probability is taken over the coin tosses of G_S, G_C, C, V and A.
5. **Privacy (Untransferability):**
   - **[Signature simulator]** A *signature simulator*, relative to a verifier V, is a probabilistic polynomial time algorithm, which, when given a message m, outputs
     $Simulator(m) = (Fake(m), FakeT_{(S,V)}(Fake(m), m),$
     $FakeT_{(C,V)}(Fake(m), m)),$
     where $Fake(m)$ is a fake signature of m, $FakeT_{(S,V)}(Fake(m), m)$ is a simulated transcript of verifying conversation between the signer S and V proving that $Fake(m)$ is a valid signature, and $FakeT_{(C,V)}(Fake(m), m)$ is a simulated transcript of verifying conversation between the designated confirmer C and V proving that $Fake(m)$ is a valid signature.

- **[Signature oracle]** The *signature oracle* O, relative to a verifier V, receives a message m as input and outputs
$Oracle(m) = (Sign(m), ValidT_{(S,V)}(Sign(m), m),$
$ValidT_{(C,V)}(Sign(m), m)),$
where $Sign(m)$ is a string chosen randomly from valid signatures of m, $ValidT_{(S,V)}(Sign(m), m)$ is a transcript chosen randomly from true verifying conversations between the true signer S and V proving that $Sign(m)$ is a valid signature, and $ValidT_{(C,V)}(Sign(m), m)$ is a transcript chosen randomly from true verifying conversations between the true designated confirmer C and V proving that $Sign(m)$ is a valid signature.
- **[Privacy (Untransferability)]** Let D be a polynomial time distinguisher, which is allowed to choose a message m, obtain some valid signatures of messages in set M', with $m \notin M'$, and interact with the true signer in verifying (and denying) the validity of the signatures of messages in set M'. Let $D(s, T_{(S,V)}, T_{(C,V)}, m)$ denote the output of a distinguisher D when its input is the possible signature s for message m, and possible transcripts $T_{(S,V)}$ and $T_{(C,V)}$. Let n be the security parameter.
Then, for any verifier V, there exists a signature simulator relative to V such that, for any polynomial time distinguisher D, and for any constant c, the following holds for n sufficiently large:

$$|\Pr(D(Oracle(m), m) = 1) - \Pr(D(Simulator(m), m) = 1)| < 1/n^c.$$

Note: When the input of G_S is restricted only to 1^n, as a variant, a slight modification of the definitions of the signing and privacy is required in some schemes, as mentioned in the key generation. For example, $Sign(m)$ consists of two parts: one part ($Sign_1$) depends on C's public-key but not on m, and the other part ($Sign_2$) depends on m. Then, $Sign_1$ can be considered to be a part of S's public key, although it is not published before. (See the note in Step 1 of the signing protocol in the [If part] of the proof of Theorem 3.) Then, in the privacy definition, $Sign_2$ is considered to be $Sign(m)$ instead.

There are three kinds of definitions of a "secure" public-key encryption scheme [GM84], and these three definitions have been proven to be equivalent [MRS88]. Here, we adopt the definition based on the indistinguishability.

Definition 2. [Secure public-key encryption scheme] A *secure* "public-key encryption" scheme is (G, E, D) such that the following conditions hold:

1. **Key generation (G):**
G is a probabilistic poly-time algorithm which, on input 1^n (the security parameter), outputs a pair of strings, (secret-key, public-key), which is denoted by $G(1^n) = (G1(1^n), G2(1^n))$. The probability is taken over G's coin tosses.
2. **Encryption (E):**
E is a probabilistic poly-time algorithm which, on input strings 1^n, m (plaintext), public-key $\in G2(1^n)$, outputs a string ("ciphertext") c, which is denoted by $E_{G2(1^n)}(m)$. The probability is taken over E's coin tosses.
3. **Decryption (D):**

D is a probabilistic poly-time algorithm which, on input strings 1^n, ciphertext $c = E_{G2(1^n)}(m)$, (secret-key, public-key) $\in G_S(1^n)$, outputs a string, which is denoted by $D_{G(1^n)}(c)$. For any m, for any constant c, and for sufficiently large n,

$$\Pr(D_{G(1^n)}(E_{G2(1^n)}(m)) = m) > 1 - 1/n^c.$$

The probability is taken over D's coin tosses.

4. **Security:**
A public-key encryption scheme (G, E, D) is *secure* if for any polynomial sequence of random variables $X_n = (X_n^{(1)}, X_n^{(2)})$, for any polynomial time machine A, for any constant c and for any sufficiently large n

$$\Pr(X_n = (m_0, m_1)) \cdot (|\Pr(A((m_0, m_1), E_{G(1^n)}(m_0)) = 1) - \Pr(A((m_0, m_1), E_{G(1^n)}(m_1)) = 1)|) < 1/n^c.$$

The probability is taken over $X_n^{(1)}$ and $X_n^{(2)}$'s distributions and coin tosses of A, G and E.

2.2 Equivalence of Designated Confirmer Signature and Public-Key Encryption

Theorem 3. [Main Theorem]
A secure designated confirmer signature scheme exists if and only if a secure public-key encryption scheme exists.

Sketch of Proof:

[If part:]

First, we assume the existence of a secure public-key encryption scheme.

Let $E_{e_C}(m)$ be a secure public-key encryption of a plaintext m for receiver C (e_C is C's public-key), where only C can decipher m by $m = D_{d_C}(E_{e_C}(m))$ (d_C is C's secret-key).

Then, an ordinary signature scheme which is existentially secure against adaptive chosen message attacks [GMRi88] (hereafter, we will call this signature scheme simply the "secure" signature scheme) exists, since a secure signature scheme exists if and only if a one-way function exists [NY89, Rom90], and a one-way function exists if a secure public-key encryption scheme exists (use the key generation function of the public-key encryption scheme to construct a one-way function).

Let $\sigma_S(m)$ be the set of ordinary secure signatures of m generated by signer S, and $V_S(m, s)$ be a verification boolean function for S's signature, where $V_S(m, s) = 0$ iff $s \in \sigma_S(m)$ and $V_S(m, s) = 1$ iff $s \notin \sigma_S(m)$.

We will now explain a designated confirmer protocol based on a secure public-key encryption scheme and ordinary secure signature scheme. Let S be a signer, C be a confirmer, and V be a verifier. The designated confirmer protocol consists of the signing protocol by S, confirmation protocol between S and V, confirmation protocol between C and V.

First the key generation and signing protocol between S and V is as follows:

Protocol: (Key generation and signing: designated confirmer signature)

Step 1 (Key generation) C generates a pair of keys, (e_C, d_C), and publishes e_C as C's public-key. S publishes a public function V_C as S's public-key for the verification of S's signature. S also publishes $P_{S,C} = E_{e_C}(K_C)$ for each designated confirmer C.
Note: Instead $P_{S,C}$ can be transmitted along with S's signature (i.e., certificate) in $\sigma_S(P_{S,C})$.

Step 2 (Signing) S generates a designated confirmer signature $Sign(m)$ of message m such that

$$Sign(m) = BC(s, h_{K_C}(m)),$$

where $s \in \sigma_S(m)$, BC is a secure bit commitment function [Nao90], and h is a pseudo-random function [GGM84]. S sends $(m, Sign(m))$ to V.

Note: A secure bit commitment function exists, if a secure public-key encryption scheme exists, since a secure bit commitment function exists when a one-way function exists [Nao90, ILL89, Has90]. A pseudo-random function exists, if a secure public-key encryption scheme exists. (This is also from the reduction to a one-way function [GGM84].)

The confirmation and disavowal protocol between S and V is as follows:

Protocol: (Confirmation and disavowal between S and V: designated confirmer signature)

Step 1 S determines, on input (m, Z), whether $Z \in \Sigma_S(m)$ or not, by the BC opening of Z with $h_{K_C}(m)$ (S can open the BC, since S knows K_C, i.e., $h_{K_C}(m)$).

Step 2 When S proves the validity of $Z = Sign(m) \in \Sigma_S(m)$, S proves to V that there exists (s, K_C, r) satisfying $V_S(m, s) = 0$, $Sign(m) = BC(s, h_{K_C}(m))$, and $P_{S,C} = E_{e_C}(K_C)$ with a zero-knowledge interactive proof for any NP problem [BCC88, IY87, GMW86], where r is a random string which is used for generating $E_{e_C}(K_C)$ from K_C. Note that such a zero-knowledge interactive proof exists since it is a poly-time predicate that (s, K_C, r) satisfies $V_S(m, s) = 0$, $Sign(m) = BC(s, h_{K_C}(m))$, and $P_{S,C} = E_{e_C}(K_C)$.

Step 3 When S proves that $Z \notin \Sigma_S(m)$ is an invalid signature of m, S proves to V either one of the followings with a zero-knowledge interactive proof [BCC88, IY87, GMW86]:
- there exists (K_C, r) such that $P_{S,C} = E_{e_C}(K_C)$ and the BC opening of Z with $h_{K_C}(m)$ is unsuccessful (i.e., $Z \neq BC(*, h_{K_C}(m))$).
- there exists (s', K_C, r) such that $P_{S,C} = E_{e_C}(K_C)$, $Z = BC(s', h_{K_C}(m))$ and $V_S(m, s') = 1$.

The confirmation and disavowal protocol between C and V is as follows:

Protocol: (Confirmation and disavowal between C and V: designated confirmer signature)

Step 1 C calculates $K_C = D_{d_C}(P_{S,C})$. C determines, on input (m, Z), whether $Z \in \Sigma_S(m)$ or not, by the BC opening of Z with $h_{K_C}(m)$.

Step 2 When C proves the validity of $Z = Sign(m) \in \Sigma_S(m)$, C proves to V that there exists (s, K_C, t) satisfying $V_S(m, s) = 0$, $Sign(m) = BC(s, h_{K_C}(m))$, and $G_t(1^n) = (e_C, d_C)$, $K_C = D_{d_C}(P_{S,C})$, with a zero-knowledge interactive proof [BCC88, IY87, GMW86], where t is a random string for key generation algorithm G to generate keys (e_C, d_C).

Step 3 When C proves that $Z \notin \Sigma_S(m)$ is an invalid signature of m, C proves to V either one of the followings with a zero-knowledge interactive proof [BCC88, IY87, GMW86]:
- there exists (K_C, t) such that $G_t(1^n) = (e_C, d_C)$, $K_C = D_{d_C}(P_{S,C})$, and the BC opening of Z with $h_{K_C}(m)$ is unsuccessful.
- there exists (s', K_C, t) such that $G_t(1^n) = (e_C, d_C)$, $K_C = D_{d_C}(P_{S,C})$, $Z = BC(s', h_{K_C}(m))$ and $V_S(m, s') = 1$.

Now, we show that the above-mentioned scheme satisfies the conditions for a secure designated confirmer signature scheme.

1. **Confirmation and disavowal:**
   - If $Z \in \Sigma_S(m)$, there exists (s, K_C, r) such that $V_S(m, s) = 0$, $Z = BC(s, h_{K_C}(m))$, and $P_{S,C} = E_{e_C}(K_C)$. Then, signer S knows (s, K_C, r) and prove verifier V that Z is S's valid signature of message m.
   If $Z \notin \Sigma_S(m)$, either one of the two cases described in the above protocol occurs. Then, S can prove that Z is not S's valid signature of message m.
   - Similarly, C, given (m, Z), can prove verifier V $Z \in \Sigma_S(m)$ and $Z \notin \Sigma_S(m)$ correctly.
2. **Security:**
   - If we assume that the proposed designated confirmer signature scheme does not satisfy the security condition for signers, then we can easily show that the underlying ordinary signature in $\sigma_S(m)$ is not *secure*. This is contradiction. Note that the zero-knowledge property of $Conf_{(S,V)}$ is used in this part. (Here, note that forger F can distinguish a true signature and false signature since F knows C's secret key.)
   - The security condition for confirmers is satisfied, from the zero-knowledge property of $Conf_{(C,V)}$. Note that attacker A can distinguish a true signature and false signature since A knows S's secret key.
3. **Privacy (Untransferability):**
   A signature simulator, *Simulator*, relative to V can be constructed as follows:
   - *Simulator* selects a random message a and calculates $E_{e_C}(a)$ as $Fake(m)$.
   - Let M_1 and M_2 be zero-knowledge simulators of confirmation protocols between S and V, and C and V, respectively. *Simulator* runs M_1 and sets $FakeT_{(S,V)}(Fake(m), m)$ = the output of M_1. *Simulator* also runs M_2 and sets $FakeT_{(C,V)}(Fake(m), m)$ = the output of M_2.

   From the definition of secure public-key encryption, $Fake(m)$ can be easily shown to be indistinguishable from $Sign(m)$. From the definition of zero-knowledge, $FakeT_{(S,V)}(Fake(m), m)$ and $FakeT_{(C,V)}(Fake(m), m)$ are indistinguishable from $ValidT_{(S,V)}(Sign(m), m)$, and $ValidT_{(C,V)}(Sign(m),$

m)), respectively. Following the well known Hybrid argument of [GM84], we conclude that $Oracle(m)$ and $Simulator(m)$ are indistinguishable.

[Only if part:]

We assume the existence of a secure designated confirmer signature scheme.

Then, the public-key encryption scheme can be constructed using the designated confirmer signature scheme as follows:

Protocol: (Public-key encryption)

Step 1 Key generation: The public-key of the designated confirmer in the underlying designated confirmer signature scheme is used for the public-key, e, of the encryption scheme. The corresponding secret-key of the confirmer is used for the secret-key, d, of the encryption scheme.

Step 2 Encryption: Suppose a plaintext, b, is one bit (0 or 1). An arbitrary message m is selected, and the other necessary parameters (signer's secret and public keys) for a signer in the underlying designated confirmer signature scheme is generated.
When $b = 0$, $E_e(b)$ is a valid designated confirmer signature, $Sign(m)$, of m along with the generated signer's parameters. When $b = 1$, $E_e(b)$ is a fake signature $Fake(m) \notin \Sigma_S(m)$, which is generated by the signature simulator $Simulator$, along with the signer's parameters. Here, note that $Fake(m) \notin \Sigma_S(m)$ can be checked by signer's confirmation protocol. Here, the public-key of the designated confirmer is used as e.

Step 3 Decryption: Execute the confirmation and disavowal protocol, $Conf_{(C,V)}$, for $E_e(b)$ as a designated confirmer signature of m. If the output of the protocol, $Conf_{(C,V)}$, is valid, $D_d(E_e(b))$ is 0. Otherwise, $D_d(E_e(b))$ is 1. Here, the secret-key of the designated confirmer is used as d.

Note: If the length of the plaintext is k bits, repeat the above procedure of encryption and decryption k times. (Note that the parameters of the signer can be shared.)

Now, we show that the above-mentioned scheme satisfies the conditions for a secure public-key encryption scheme.

First, from the property of the confirmation protocol of the designated confirmer, $D_d(E_e(b)) = b$ with overwhelming probability.

Next, we show that the above-mentioned public-key encryption scheme is *secure*. For simplicity of description, here we assume that a ciphertext is one bit. Then, $E_e(0)$ (= $Sign(m) \in \Sigma_S(m)$, m, public parameters) and $E_e(1)$ (= $Fake(m) \notin \Sigma_S(m)$, m, public parameters) are indistinguishable, since $Oracle(m)$ including $Sign(m)$ and $Simulator(m)$ including $Fake(m)$ are indistinguishable from the privacy condition of Definition 1.

□

Boyar et al [BCDP90] introduced the concept of convertible undeniable signatures as a variant of the undeniable signatures. The present paper shows that a convertible designated confirmer signature scheme can be constructed similarly, and that it is equivalent to the public-key encryption with respect to the existence.

Corollary 4. *A secure convertible designated confirmer signature scheme exists if and only if a secure public-key encryption scheme exists.*

Sketch of Proof:

[If part:] In addition to the protocols in Theorem 3 of the designated confirmer signature scheme, the *conversion* protocol [BCDP90] between C (or S) and V is as follows:

Protocol:

Step 1 C (or S) sends K_C to V.

Step 2 V calculates s by opening the bit commitment $Sign(m) = BC(s, h_{K_C}(m))$ through $h_{K_C}(m)$, and checks whether $V_S(m, s) = 0$ holds.

[Only if part:]

Same as this part of Theorem 3.

□

3 Practical Constructions

This section introduces a new type of practical constructions. The new schemes are more efficient in signing than Chaum's scheme [Cha94]. They are based on three move identification protocols such as Feige-Fiat-Shamir [FFS88], Schnorr [Sch91], the extended Fiat-Shamir [GQ88, OhOk88], and the modified Schnorr [Oka92], while Chaum's scheme is based on the RSA scheme. That is, many constructions are possible (e.g., FFS type, Schnorr type, etc.). Among these three move protocols, [Sch91] and [Oka92] are based on the discrete logarithm problem, while [FFS88] and [GQ88, OhOk88] are based on the factoring problem.

This section first gives general description which is common among these constructions is given. Next, two examples are given: one is based on the Schnorr scheme, and the other is based on the extended Fiat-Shamir scheme.

The advantage of the proposed schemes based on the discrete logarithm type protocols [Sch91, Oka92] compared to the Chaum scheme [Cha94] is:

- If the preprocessing technique is used in the signing stage, the time taken for signing is much shorter than with the Chaum scheme. That is, in the Chaum scheme, the running time for signing is at least as same as that for the RSA scheme (i.e., very slow), while, in the proposed scheme, the running time for signing after the preprocessing is negligible.
- The security of this construction depends on only one arithmetical problem, i.e., the discrete logarithm problem, while the Chaum scheme depends on two arithmetical problems, i.e., the discrete logarithm problem and factoring problem (if either one is breakable, the scheme is breakable, although our scheme is breakable only if the discrete logarithm problem is tractable). Apart from this security advantage, our scheme has some practical merits. One is that all required arithmetical procedures can be executed using the same modulus p and q. Another is that an elliptic curve variant can be constructed, which has practical merits such as shorter data size and less computational complexity.

- The [Oka92] variant of our scheme is provably secure (unforgeable against chosen message attacks) under fairly weak assumptions. On the contrary, the Chaum scheme depends on a very strong assumption that the RSA scheme with hash functions is unforgeable against chosen message attacks.

The advantage of the proposed schemes based on the factoring type protocols [FFS88, GQ88, OhOk88] over the Chaum scheme is:

- Even if the preprocessing technique is not used in the signing stage, signing is faster than with the Chaum scheme.
- Our schemes are provably secure (unforgeable against chosen message attacks) under fairly weak assumptions [FFS88, OhOk88, Oka92].

3.1 Basic Protocol

Let (A, D) be a three move identification protocol such that first S (prover) sends a message $x = A_1(w)$ to D (verfier), D sends e to S, S sends $y = A_2(w, e, s)$ to V, and finally D checks the validity of (x, e, y) by checking $x = D(e, y, a)$. Here w is random coin flips of S, a is S's public key, and s is S's secret key. H denotes a one-way hash function. (Theoretically, H should be an ideal random function, or a correlation-free one-way hash function [Oka92].)

We assume that signer S utilizes the functions (A_1, A_2, D) of a three move protocol (A, D), and S's public and secret keys are (a, s). Let $b = g^u \bmod p$ be confirmer C's public key and u be C's secret key. Here, p is a prime, and q is also a prime which divides $p - 1$. The order of g in the multiplicative group of Z_p^* is q.

First the signing protocol between S and V is as follows:

Protocol: (Signing and confirmation between S and V)

Step 1 S generates a designated confirmer signature (d, e, y) of message m such that

$$d = g^r \bmod p,\ e = (b^r \bmod p) \oplus H(m, x),\ x = A_1(w),\ y = A_2(w, e, s).$$

Here, $r \in_R Z_q$, and w is a random number. S sends $(m, (d, e, y))$ to V.

Step 2 S and V calculate

$$z = e \oplus H(m, D(e, y, a)).$$

S proves to V that $\log_g d = \log_b z$ in a zero-knowledge manner (without revealing r). Several efficient perfect zero-knowledge protocols have been known [Cha90, BCDP90].

The confirmation protocol between C and V is as follows:

Protocol: (Confirmation between C and V)

Step 1 C receives $(m, (d, e, y))$ from S or V. C and V calculate

$$z = e \oplus H(m, D(e, y, a)).$$

Step 2 C proves to V that $\log_g b = \log_d z$ in a zero-knowledge manner (without revealing u).

The conversion protocol between C and V is as follows:

Protocol: (Conversion by C)
Step 1 C calculates

$$z = e \oplus H(m, D(e, y, a)).$$

C proves to V that $\log_g b = \log_d z$ C sends (l_1, l_2, k) to V as follows:

$$l_1 = g^t \bmod p,\ l_2 = d^t \bmod p,\ k = t + H(l_1, l_2)u \bmod q.$$

Step 2 V checks whether

$$g^n \equiv l_1 b^{H(l_1,l_2)} \pmod p,\ d^n \equiv l_2 z^{H(l_1,l_2)} \pmod p$$

holds.

3.2 Example Based on Schnorr

Here, the Schnorr identification protocol [Sch91] is used as (A, D) such that first S (prover) sends a message $x = g^w \bmod p$ $(= A_1(w))$ to D (verfier), D sends $e \in Z_L$ to S, S sends $y = w + es \bmod q$ $(= A_2(w, e, s))$ to V, and finally D checks the validity of (x, e, y) by checking $x = g^y a^e \bmod p$ $(= D(e, y, a))$. Here $w \in_R Z_q$, $a = g^{-s} \bmod p$. (a: S's public key, s: S's secret key).

Let $b = g^u \bmod p$ be confirmer C's public key and u be C's secret key.

First the signing protocol between S and V is as follows:

Protocol: (Signing and confirmation between S and V)
Step 1 S generates a designated confirmer signature (d, e, y) of a message m such that

$$d = g^r \bmod p,\quad e = (b^r \bmod p) \oplus H(m, x),$$
$$x = g^w \bmod p,\quad y = w + es \bmod q.$$

S sends $(m, (d, e, y))$ to V.
Step 2 S and V calculate

$$z = e \oplus H(m, g^y a^e \bmod p).$$

S proves to V that $\log_g d = \log_b z$ in a zero-knowledge manner (without revealing r) by using [Cha90, BCDP90].

The confirmation protocol between C and V and the conversion protocol between C and V can be shown similarly.

3.3 Example Based on the Extended Fiat-Shamir

Here, the extended Fiat-Shamir identification protocol [GQ88, OhOk88] is used as (A, D) such that first S (prover) sends message $x = w^L \bmod n$ $(= A_1(w))$ to D (verfier), D sends $e \in Z_L$ to S, S sends $y = ws^e \bmod n$ $(= A_2(w, e, s))$ to V, and finally D checks the validity of (x, e, y) by checking $x = y^L a^e \bmod n$ $(= D(e, y, a))$. Here $w \in_R Z_n$, $a = 1/s^L \bmod n$, and $n = PQ$ (P, Q: primes). (a: S's public key, s: S's secret key).

Let $b = g^u \bmod p$ be confirmer C's public key and u be C's secret key.

First the signing protocol between S and V is as follows:

Protocol: (Signing and confirmation between S and V)

Step 1 S generates a designated confirmer signature (d, e, y) of a message m such that

$$d = g^r \bmod p, \quad e = (b^r \bmod p) \oplus H(m, x),$$

$$x = w^L \bmod n, \quad y = ws^e \bmod n.$$

Here, $r \in_R Z_q, w \in_R Z_n$. S sends $(m, (d, e, y))$ to V.

Step 2 S and V calculate

$$z = e \oplus H(m, y^L a^e \bmod n).$$

S proves to V that $\log_g d = \log_b z$ in a zero-knowledge manner (without revealing r) by using [Cha90, BCDP90].

The confirmation protocol between C and V and the conversion protocol between C and V can be shown similarly.

4 Conclusion

This paper has formalized the definition of designated confirmer signatures and has proven that a designated confirmer signature scheme is equivalent to a public-key encryption scheme with respect to existence. This paper also presented practical designated confirmer signature schemes which are more efficient in signing than the previous scheme.

Acknowledgments

The author wishes to thank anonymous referees for their useful suggestions, and is grateful to Kazuo Ohta for valuable discussions.

References

[BCC88] Brassard, G., Chaum, D., and Crépeau, C.: Minimum Disclosure Proofs of Knowledge. J. Computer and System Sciences, **37** (1988) 156–189

[BCDP90] Boyar, J., Chaum, D., Damgård, I., Pedersen, T.: Convertible Undeniable Signatures. Proc. of Crypto'90, LNCS **537**, Springer–Verlag, (1991) 189–205

[Cha90] Chaum, D.: Zero-Knowledge Undeniable Signatures. Proc. of Eurocrypto'90, LNCS **473**, Springer–Verlag, (1991) 458–464

[Cha94] Chaum, D.: Designated Confirmer Signatures. Proc. of Eurocrypt '94, LNCS, Springer–Verlag (to appear)

[CA89] Chaum, D., van Antwerpen, H.: Undeniable Signatures. Proc. of Crypto'89, LNCS **435**, Springer–Verlag, (1990) 212–216

[DH76] Diffie, W., Hellman, M. E.: New Directions in Cryptography. IEEE Trans. Information Theory, **22**, 6, (1976) 644–654

[ElG85] ElGamal, T.: A Public-Key Cryptosystem and a Signature Scheme Based on Discrete Logarithms. IEEE Trans. Information Theory, **31**, 4, (1985) 460–472

[FFS88] Feige, U., Fiat, A., Shamir, A.: Zero-Knowledge Proofs of Identity. J. of Cryptology, **1**, 2 (1988) 77–94

[GGM84] Goldreich, O., Goldwasser, S., Micali, S.: How to Construct Random Functions. J. of ACM, **33**, 4 (1984) 792–807

[GL89] Goldreich, O., Levin, L.: A Hard-Core Predicate for any One-way Function. Proc. of STOC'89 (1989) 25–32

[GM84] Goldwasser, S., Micali, S.: Probabilistic Encryption. J. Computer and System Sciences, **28**, 2 (1984) 270–299

[GMRa89] Goldwasser, S., Micali, S., Rackoff, C.: The Knowledge Complexity of Interactive Proof Systems. SIAM J. Comput., **18**, 1 (1989) 186–208

[GMRi88] Goldwasser, S., Micali, S., Rivest, R.: A Digital Signature Scheme Secure Against Adaptive Chosen-Message Attacks. SIAM J. Comput., **17**, 2 (1988) 281–308

[GMW86] Goldreich, O., Micali, S., Wigderson, A.: Proofs that Yield Nothing But their Validity and a Methodology of Cryptographic Protocol Design. Proc. FOCS (1986) 174–187

[GQ88] Guillou, L. C., Quisquater, J.J.: A Practical Zero-Knowledge Protocol Fitted to Security Microprocessor Minimizing Both Transmission and Memory. Proc. of Eurocrypt'88, LNCS **330**, Springer–Verlag (1988) 123–128

[Has90] Håstad, J.: Pseudo-Random Generators under Uniform Assumptions. Proc. of STOC (1990) 395–404

[ILL89] Impagliazzo, R., Levin, L., Luby, L.: Pseudo-Random Number Generation from One-Way Functions. Proc. of STOC (1989) 12–24

[IR89] Impagliazzo, R., Rudich, S.: Limits on the Provable Consequence of One-Way Permutations. Proc. of STOC (1989) 44–61

[IY87] Impagliazzo, R., Yung, M.: Direct Minimum-Knowledge Computations. Proc. of Crypto'87, LNCS **293**, Springer–Verlag (1988) 40–51

[MRS88] Micali, S., Rackoff, C., Sloan, B.: The Notion of Security of Probabilistic Cryptosystems. SIAM J. Comput., **17**, 2 (1988) 412–426

[Nao90] Naor, M.: Bit Commitment Using Pseudo-Randomness. Proc. of Crypto'89, LNCS **435**, Springer–Verlag, (1990) 128–136

[NY89] Naor, M., Yung, M.: Universal One-Way Hash Functions and Their Cryptographic Applications. Proc. of STOC (1989) 33–43

[OhOk88] Ohta, K., Okamoto, T.: A Modification of the Fiat-Shamir Scheme. Proc. of Crypto'88, LNCS **403**, Springer-Verlag (1990) 232-243

[Oka92] Okamoto, T.: Provably Secure and Practical Identification Schemes and Corresponding Signature Schemes. Proc. of Crypto'92, LNCS **740**, Springer–Verlag, (1993) 31–53

[Oka93] Okamoto, T.: On the Relationship among Cryptographic Physical Assumptions. Proc. of ISAAC'93, LNCS **762**, Springer–Verlag, (1993) 369–378

[Rom90] Rompel, J.: One-Way Functions are Necessary and Sufficient for Secure Signature. Proc. of STOC (1990) 387–394

[Sch91] Schnorr, C. P.: Efficient Signature Generation by Smart Cards. J. of Cryptology, **4**, 3 (1991) 161–174

Directed Acyclic Graphs, One-way Functions and Digital Signatures

(Extended Abstract)

Daniel Bleichenbacher and Ueli M. Maurer

Institute for Theoretical Computer Science
ETH Zürich
CH-8092 Zürich, Switzerland
Email: {bleichen,maurer}@inf.ethz.ch

Abstract. The goals of this paper are to formalize and investigate the general concept of a digital signature scheme based on a general one-way function without trapdoor for signing a predetermined number of messages. It generalizes and unifies previous work of Lamport, Winternitz, Merkle, Even et al. and Vaudenay. The structure of the computation yielding a public key from a secret key corresponds to a directed acyclic graph $\mathcal{G}$. A signature scheme for $\mathcal{G}$ can be defined as an antichain in the poset of minimal verifyable sets of vertices of $\mathcal{G}$ with the naturally defined computability relation as the order relation and where a set is verifyable if and only if the public key can be computed from the set.

1 Introduction

Lamport [5] proposed a so-called one-time signature scheme based on a general one-way function (OWF), i.e., a function f that is easy to compute but computationally infeasible to invert, for suitable definitions of "easy" and "infeasible". Lamport's scheme for signing a single bit is set up by choosing as the secret key two strings x_0 and x_1 at random and revealing as the public key the pair $\langle f(x_0), f(x_1)\rangle$. The signature for bit b is x_b. For signing longer messages, several instances of this scheme can be used.

Motivated by Lamport's approach, many researchers have subsequently proposed more efficient one-time signature schemes. The goals of this paper are to formalize the concept of a signature scheme based on any OWF for signing a predetermined number of messages, and to present several results on the number and size of messages that can be signed with a given scheme. In contrast to Rompel's result [9] showing that a signature scheme can be obtained from any OWF, the emphasis of this paper is on efficiency and on a unified description of the general idea rather than on rigorously proving the security of schemes with respect to a certain intractability assumption.

In addition to the general interest in a class of intriguing graph-theoretic problems, our motivations for considering the design of signature schemes based on OWFs are as follows. First, there is a severe limitation on the diversity of mathematical problems (such as factoring integers or computing discrete logarithms in certain finite groups) that can at present be used as the bases for a digital signature scheme. Therefore an alternative design approach with a much larger degree of freedom in choosing the underlying cryptographic function appears to be of interest. Second, for applications where only few messages need to be signed, schemes based on an arbitrary one-way function have the potential of being computationally more efficient than presently-used number-theoretic schemes, but their disadvantage is that each public key can only be used for signing a predetermined number of messages. Moreover, even if these schemes turn out to be of limited interest as regular digital signature schemes, they do have applications in other contexts such as on-line/off-line signatures [3] and the signature schemes of [1].

The number (i.e., diversity) of messages that can be signed by the Lamport scheme with r public-key pairs is 2^r. Using the same secret key and public key, but allowing as signatures all subsets of cardinality r of the set of $2r$ public-key components, the number of messages can be improved to $\binom{2r}{r}$, which is optimal [11]. These sets are compatible because computing a signature from a different signature requires the inversion of the OWF for at least one value.

Note that the size of the secret key of such a scheme can be reduced significantly by generating all the secret-key components in a pseudo-random fashion from a single secret key S. Similarly, the public key can be reduced to a single value P by applying a one-way hash function to the list of public-key components.

A generalization of the Lamport scheme attributed by Merkle to Winternitz [6] is to apply the OWF to two secret key components iteratively a fixed number of times, resulting in a two-component public key. Meyer and Matyas [7] proposed as a further improvement to use more than two chains of function evaluations: they observed that a one-time signature scheme for a message space of size $K!$ can be obtained from a scheme with K chains of length K each, by allowing as signatures all combinations of K nodes containing one node in each chain such that at each level there is one of these nodes. This scheme was generalized further in [3] and later in [12] to a scheme with l chains of length k where the signatures consist of one node in each chain such that the total sum of the levels of these nodes (within their chains) is constant. It can be proved that this strategy yields the maximal number of signatures for such a computation structure.

The described schemes can only be used to sign a single message. Merkle [6] proposed the so-called tree-authentication scheme for signing several messages consecutively with a single public key P.

2 One-time signature schemes based on directed acyclic graphs

In this paper, vertices and sets of vertices of a graph are denoted by small and capital letters, respectively, and graphs, posets as well as sets of sets of vertices are denoted by calligraphic letters. We summarize some well-known definitions and results on partially ordered sets (poset). A poset is defined as a set with an antisymmetric, transitive and reflexive order relation, denoted $\leq$. Two elements x and y of a poset $\mathcal{Z} = (Z, \leq)$ are *comparable* if and only if $x \leq y$ or $y \leq x$ and they are *incomparable* otherwise. A subset $U \subseteq Z$ is a *chain* if every pair of elements of U is comparable, and it is called an *antichain* if every pair of elements of U is incomparable. A chain (antichain) is called *maximal* if it is not a subset of another chain (antichain).

Definition 1. The *width* of a poset $\mathcal{Z}$, denoted $w(\mathcal{Z})$, is the maximal cardinality of an antichain.

Definition 2. For a poset $\mathcal{Z} = (Z, \leq)$, a function $r : Z \longrightarrow \mathbb{N}$ is called a *representation function* of $\mathcal{Z}$ if for all distinct $x, y \in Z$, $x \leq y$ implies $r(x) < r(y)$.

Let B be a suitable large set (e.g., the set of 64, 96 or 128-bit strings) and let $f_1, f_2, \ldots$ with $f_i : B^i \longrightarrow B$ be a list of one-way functions, where f_i takes as input a list of i values in B and produces as output a single value in B. Consider a scenario in which a secret key S consisting of u values $s_1, \ldots, s_u \in B$ is chosen at random, and a sequence of values $s_{u+1}, s_{u+2}, \ldots, s_t$ is computed from $s_1, \ldots, s_u$ by applications of the one-way functions f_i. More precisely, for $u+1 \leq j \leq t$, s_j is the result of applying an appropriate OWF to a subset U_j (of appropriate size) of $\{s_1, \ldots, s_{j-1}\}$, where the order of the arguments is assumed to be fixed but is irrelevant for the further discussion. Some of these computed values will not be used as input to a OWF and are published as the public key P. Signatures consist of appropriately chosen subsets of $\{s_1, \ldots, s_t\}$.

In the following we need to distinguish between the structure of the described computation for setting up a digital signature scheme and the particular values resulting for a particular choice of the secret key. Consider a directed graph $\mathcal{G} = (V, E)$ with vertex set $V = \{v_1, \ldots, v_t\}$, where v_i corresponds to the value s_i, and with edge set E containing the edge (v_i, v_j) if and only if s_i is an input to the OWF resulting in s_j. Hence the value corresponding to v_j can be computed from the values corresponding to the predecessors of v_j, and it functionally depends on the value s_k (corresponding to v_k) if and only if there exists a directed path from v_k to v_j.

In such a graph the secret key set and the public key set correspond to the sets of vertices with in-degree 0 and out-degree 0, respectively. The graph $\mathcal{G}$ is assumed to be known publicly and can be used by all users, but the values corresponding to the vertices for a user's particular secret key are kept secret by the user, except those values corresponding to the public key. A signature scheme assigns a signature pattern, i.e. an appropriate subset of vertices, to every message in the message space. A user's signature for a given message consists of the values (for that user's secret key) corresponding to the vertices in the signature pattern for that message, when the computation according to $\mathcal{G}$ is performed for that user's secret key. The set of signature patterns must satisfy certain conditions discussed below.

Definition 3. For a given directed acyclic graph (DAG) $\mathcal{G} = (V, E)$, the *secret key pattern* $S(\mathcal{G}) \subset V$ and the *public key pattern* $P(\mathcal{G}) \subset V$ are defined as the sets of vertices with in-degree 0 and out-degree 0, respectively. Let X be a subset of V. A vertex v is defined recursively to be *computable* from X if either $v \in X$ or if v has at least one predecessor and all predecessors are computable from X. A set Y is computable from X if every element of Y is computable from X. Note that V and hence every subset of V is computable from the secret key $S(\mathcal{G})$. A set $X \in V$ is called *verifyable* (with respect to the public key) if $P(\mathcal{G})$ is computable from X. A verifyable set X is *minimal* if no subset of X is verifyable. Two minimal verifyable sets (MVS) X and Y are *compatible* if neither X is computable from Y nor Y is computable from X. A set of MVSs is compatible if they are pairwise compatible.

Remarks.
(1) Of course, the OWFs used for evaluating different vertices can be different, as long as a function together with the order of the arguments is uniquely specified for each vertex.
(2) As mentioned before, the secret key components can be generated in a pseudo-random manner from a single secret key. We can hence extend $\mathcal{G}$ by introducing an extra vertex s_0 (the real secret key) and edges form s_0 to the vertices $s_1, \ldots, s_u$ (using the convention that when two vertices in $\mathcal{G}$ have the same set of predecessors, then the two OWFs used in the corresponding computation steps are different and unrelated). Similarly, one can without much loss of generality restrict the discussion to graphs with only one public-key component because a list of public-key components could be hashed using a secure cryptographic hash function.
(3) Because messages can be hashed prior to signing, it suffices to design signature schemes for a message space corresponding to the range of a secure cryptographic hash function, for instance the set of 128-bit strings.

The computability relation on the set of MVSs of a graph is transitive, anti-symmetric and reflexive, and hence the set of MVSs of a graph $\mathcal{G}$, denoted $\mathcal{W}(\mathcal{G})$, forms a poset $(\mathcal{W}(\mathcal{G}), \leq)$ with computability as the order relation, i.e., we have $X \leq Y$ for $X, Y \in \mathcal{W}(\mathcal{G})$ if and only if X is computable from Y. Note that two MVSs of $\mathcal{G}$ are compatible if and only if they are incomparable in $(\mathcal{W}(\mathcal{G}), \leq)$.

Definition 4. The *associated poset* of DAG $\mathcal{G}$ is the poset $(\mathcal{W}(\mathcal{G}), \leq)$ of minimal verifyable subsets of $\mathcal{G}$.

In order to remove a possible source of confusion it should be pointed out that a DAG in which every edge (x, y) is the only path from x to y has itself the structure of a poset and $\mathcal{W}(\mathcal{G})$ is the poset of cutsets in this poset. However, we have avoided the term "cutset" for the signature patterns because this term has a different meaning for graphs.

Definition 5. A *one-time signature scheme* $\mathcal{A}$ for an acyclic directed graph $\mathcal{G} = (V, E)$ is an antichain of the associated poset $\mathcal{W}(\mathcal{G})$.

The important parameters of a one-time signature scheme $\mathcal{A}$ for a graph $\mathcal{G} = (V, E)$ are the number $|V|$ of vertices (which is equal to the sum of the size of the secret key and the number of function evaluations required for computing a public key from a secret key), the number $|\mathcal{A}|$ of signatures which must be at least equal to the size of the message space, and the maximal size of signatures, $\max_{X \in \mathcal{A}} |X|$.

Example: Figure 1 shows a graph for which it is especially easy to design a signature scheme. At each level of the graph, the verifier is given one of two unknown values, where a message bit determines which one is given. This scheme allows to sign 1 bit per three vertices, i.e., 1/3 bit per vertex. More efficient schemes will be discussed later.

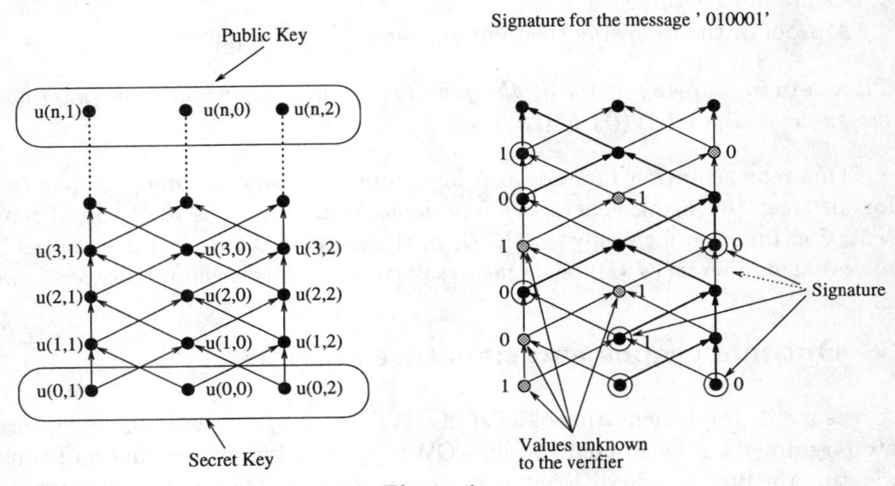

Figure 1.

The following interesting problems are now well-motivated. First, for a given graph $\mathcal{G}$ to find a large (ideally a maximal-sized) antichain in the associated poset. Note that $w(\mathcal{W}(\mathcal{G}))$ denotes the maximal size of such an antichain. Second, for a given size of the message space to find a graph with few (ideally the minimal number of) vertices allowing the construction of a one-time signature scheme. Third, both problems should be treated with a constraint on the maximal size of signatures, and also for a generalized definition of a signature scheme for signing several rather than only one message.

The maximal-sized antichain of a poset can be determined by a flow algorithm whose running time is polynomial in the number of elements of the poset. This method is only feasible for graphs of less than about 30 vertices. For larger graphs, a very useful technique for determining lower bounds on the size of the maximal antichain is based on representation functions.

It follows from the definition of a representation function r of a poset $\mathcal{Z} = (Z, \leq)$ that for any $x \neq y$, $r(x) = r(y)$ implies that x and y are incomparable. Hence for any representation function r of the associated poset $(\mathcal{W}(\mathcal{G}), \leq)$ of a given DAG $\mathcal{G}$ and for any integer k, the set $\{U \in \mathcal{W}(\mathcal{G}) : r(U) = k\}$ is a one-time signature scheme. Let

$$\beta(\mathcal{G}, r) = \max_k(|\{U \in \mathcal{W}(\mathcal{G}) : r(U) = k\}|)$$

be the maximal cardinality of these sets.

In order to find good signature schemes for a given graph, we need to find a good representation function, that is one with a large maximal coefficient. For a given DAG $\mathcal{G}$ let $c_{\mathcal{G}} : \mathcal{W}(\mathcal{G}) \rightarrow \mathbb{N}$ be the function defined by

$$c_{\mathcal{G}}(U) := |\{v : v \notin U \text{ and } v \text{ is computable from } U\}|$$

i.e. $c_{\mathcal{G}}(U)$ is the cardinality of the set of vertices of $\mathcal{G}$ that are computable from U but are not contained in U.

A proof of the following theorem appears in the full paper.

Theorem 6. *For any DAG $\mathcal{G}$, the function $c_{\mathcal{G}}$ is a representation function of the associated poset $\mathcal{W}(\mathcal{G})$ of $\mathcal{G}$.*

This representation function can be computed easily for many graphs (e.g., for all trees) by using generating functions. Moreover $c_{\mathcal{G}}$ is an optimal representation function for many graphs $\mathcal{G}$, in the sense that $\beta(\mathcal{G}, c_{\mathcal{G}})$ is equal to the maximal number $w(\mathcal{W}(\mathcal{G}))$ of signatures patterns, but this is not true in general.

3 Optimal graphs and signature schemes

A reasonable implementation of a list of OWFs $f_1, f_2, f_3 \ldots$ with one, two, three, etc. arguments is by implementing a OWF f_2 with two arguments and implementing the function f_1 with one argument as $f_1(x) = f_2(x, x)$ and the functions f_i for $i \geq 3$ as $f_i(x_1, \ldots, x_i) = f_2(f_{i-1}(x_1, \ldots, x_{i-1}), x_i)$. The function f_2 can for instance be implemented by applying DES in an appropriate mode, but much more efficient implementations of good candidate OWFs are possible.

In the described implementation based on a function f_2, the graph could be considered to consist only of vertices with fan-in 1 or 2. In the sequel we discuss the problem of maximizing the number of signature patterns for a given number n of vertices under this fan-in restriction. Let $\nu(n)$ be the maximal number of MVSs obtainable for a graph with n vertices and let $\mu(n)$ be the maximal number of compatible MVSs for a graph with n vertices, i.e., let

$$\nu(n) = \max\{|\mathcal{W}(\mathcal{G})| : \mathcal{G} = (V, E) \text{ with } |V| = n\}$$
$$\mu(n) = \max\{w(\mathcal{W}(\mathcal{G})) : \mathcal{G} = (V, E) \text{ with } |V| = n\},$$

where $\mathcal{G}$ has fan-in at most 2 and public key of size 1. We will now derive concrete and asymptotic results on $\mu(n)$.

For a DAG $\mathcal{G} = (V, E)$ we define $\mathcal{R}_{\mathcal{G},l}$ to be the graph consisting of l unconnected identical copies of $\mathcal{G}$. The poset of MVSs of $\mathcal{R}_{\mathcal{G},l}$ consists of all l-tuples $(S_1, \ldots, S_l)$ for which S_i is a MVS of the i-th copy of $\mathcal{G}$.

Let $r_{\mathcal{G}}$ be any representation function of $\mathcal{W}(\mathcal{G})$ such that there exist $S_1, S_2 \in \mathcal{W}(\mathcal{G})$ where $r_{\mathcal{G}}(S_1) - r_{\mathcal{G}}(S_2) = 1$. We define the representation function r of $\mathcal{W}(\mathcal{R}_{\mathcal{G},l})$ by $r(S) = \sum_{i=1}^{l} r_{\mathcal{G}}(S_i)$ for an MVS $S = (S_1, \ldots, S_l) \in \mathcal{W}(\mathcal{R}_{\mathcal{G},l})$.

Theorem 7. *For the representation function r defined above we have*

$$\lim_{l \to \infty} \beta(\mathcal{R}_{\mathcal{G},l}, r) \frac{\sqrt{l}}{m^l} = \frac{1}{\sigma\sqrt{2\pi}}$$

where $m = |\mathcal{W}(\mathcal{G})|$ and σ is the standard deviation of $r_{\mathcal{G}}(S)$ if S is chosen uniformly from $\mathcal{W}(\mathcal{G})$.

Proof sketch. Let Y be the random variable defined by $Y = (r_{\mathcal{G}}(S) - E[r_{\mathcal{G}}(S)])/\sigma$ where S is chosen uniformly from $\mathcal{W}(\mathcal{G})$. The distribution of Y is a lattice distribution with span $1/\sigma$, $E[Y] = 0$ and $E[Y^2] = 1$. Now we can apply Theorem 3 of [4, p.490] to complete the proof. □

It should be mentioned that $\beta(\mathcal{R}_{\mathcal{G},l}, r) = O(m^l/\sqrt{l})$ is satisfied for any choice of $r_{\mathcal{G}}$. Theorem 7 implies the following result which is proved in the full paper.

Corollary 8. $\lim_{n \to \infty} \dfrac{\log_2 \mu(n)}{n} \geq \sup_m \dfrac{\log_2 \nu(m)}{m+1}.$

We have found a DAG $\mathcal{G}$ with 26 vertices whose associated poset has 5004 vertices. Thus $\mathcal{R}_{\mathcal{G},l}$ contains $O(5004^l/\sqrt{l})$ signatures patterns. In order to combine l copies of the graph in a tree with fan-in 2 we need a tree with only $l-1$ additional vertices. Therefore, there exists a sequence of graphs with n vertices which allows to sign $\alpha n - O(\log n)$ bits, i.e., α bits per vertex, where $\alpha = \log_2(5004)/27 = 0.4551$. This result cannot be achieved by trees as the following theorem demonstrates.

Theorem 9. *For every tree $\mathcal{T}$ with n vertices,*

$$|\mathcal{W}(\mathcal{T})| \leq 2^{\gamma(n+1)} \quad \text{where } \gamma = \log_2(685/216)/4 \approx 0.4162. \tag{1}$$

In other words, no tree with n vertices allows to sign more than $\gamma(n+1)$ bits. On the other hand for $n \geq 1$ there exists a tree $\mathcal{T}_n$ with n vertices with

$$w(\mathcal{W}(\mathcal{T}_n)) \geq \frac{2^{\delta n}}{3\sqrt{n} + 3/2} \quad \text{where } \delta = \log_2(101)/16 \approx 0.4161. \tag{2}$$

Note that the upper and lower bounds of Theorem 9 are extremely close to each other. By refined arguments, γ can be reduced and δ increased slightly so that they agree in the first 8 decimal digits, but nevertheless it remains to prove that there exists a constant which is both upper and lower bound.

4 Concluding remarks

One problem with the schemes discussed in this extended abstract is that the signatures are relatively long. An interesting problem is to devise schemes with small signature patterns. One such scheme based on a forest of chains is discussed in [12]. Our definitions can be extended in this direction and also to include schemes for signing a fixed number of messages (rather than only one). Such generalizations will be discussed in a forthcoming paper.

Acknowledgements

The authors would like to thank Martin Perewusnyk, Adi Shamir, and Roger Wattenhofer for many inspiring discussions on the topic of this paper.

References

1. J.N.E. Bos and D. Chaum, Provably unforgeable signatures, *Advances in Cryptology – CRYPTO '92* (E. Brickell, ed.), Lecture Notes in Computer Science, vol. 740 , Springer Verlag, 1993, pp. 1-14.
2. N. de Brujin, C. A. van Ebbenhorst Tengebergen, and D. R. Kruyswijk, "On the set of divisors of a number," *Nieuw Arch. Wisk*, vol. 23, pp. 191–193, 1952.
3. S. Even, O. Goldreich and S. Micali, On-line/off-line digital signatures, *Advances in Cryptology – CRYPTO '89*, Lecture Notes in Computer Science, vol. 435 (G. Brassard, ed.), Springer Verlag, 1990, pp. 263-275.
4. W. Feller, *An Introduction to Probability Theory and Its Applications*, vol. II., second corrected printing, Wiley & Sons, 1966.
5. L. Lamport, Constructing digital signatures from a one-way function, Technical Report SRI Intl. CSL 98, 1979.
6. R. Merkle, A certified digital signature, *Advances in Cryptology – CRYPTO '89*, Lecture Notes in Computer Science, vol. 435 (G. Brassard, ed.), Springer Verlag, 1990, pp. 218-238.
7. C. Meyer and S. Matyas, *Cryptography – a new dimension in computer data security*, John Wiley & Sons, Inc., 1982.
8. R.L. Rivest, A. Shamir, and L. Adleman, A method for obtaining digital signatures and public-key cryptosystems, *Communications of the ACM*, Vol. 21, No. 2, pp. 120-126, 1978.
9. J. Rompel, One-way functions are necessary and sufficient for secure signatures, *Proc. 22nd ACM Symp. on Theory of Computing (STOC)*, 1990, pp. 387-394.
10. C.P. Schnorr, Efficient identification and signatures for smart cards, Advances in Cryptology – Crypto '89, Lecture Notes in Computer Science, vol. 435 (G. Brassard, ed.), Springer-Verlag 1990, pp. 239-252.
11. E. Sperner, Ein Satz über Untermengen einer endlichen Menge, *Mathematische Zeitschrift*, vol. 27, pp. 544–548, 1928.
12. S. Vaudenay, One-time identification with low memory, *Proc. of EUROCODE '92*, Lecture Notes in Computer Science, Springer Verlag. CISM Courses and Lectures, No. 339, International Centre for Mechanical Sciences, P. Camion, P. Charpin and S. Harari (eds.), Springer-Verlag, pp. 217–228.

An Identity-Based Signature Scheme with Bounded Life-Span

Olivier Delos
Dept of Computer Sc. (INGI)
Place Sainte-Barbe, 2
B-1348 Louvain-la-Neuve, Belgium
E-mail: delos@info.ucl.ac.be
Fax: + 32 10 45 03 45

Jean-Jacques Quisquater
Dept of Elec. Eng. (DICE)
Place du Levant, 3
B-1348 Louvain-la-Neuve, Belgium
E-mail: jjq@dice.ucl.ac.be
Fax: +32 10 47 86 67

Abstract

The aim of this paper is to present a signature scheme in which the ability to sign messages of a signer is limited to a fixed number k of signatures. It is an identity-based signature scheme in which each signature can be used only once. We called such schemes "bounded life-span". It is based on mental games and it uses zero-knowledge tools. A validation center is needed to initialize this identity-based scheme. A credential center is used to insure the unicity and the bounded life-span aspects. It allows delegation and numerous practical applications.

1 Introduction

There are a lot of situations where the receiver of a message needs assurances concerning its non-alteration (accidental or voluntary), *i.e.* the authenticity (integrity) of the message and of its origin. The sender must be able to "sign" a message in such a way that any alteration of the message will be immediately revealed by the "signature". Furthermore, if the signature cannot be forged then this will also authenticate the sender.

Diffie and Hellman first introduced the concept of "digital signature" [DH76]. Since their paper, this concept has been the subject of numerous researches [GMR88]. We distinguish undeniable [CvA90], convertible [BCDP91], unconditionally secure, fail stop [WP90], blind, group [vH92], and multi-signatures [DF92, DQ94].

The aim of this paper is to present a signature scheme in which the ability to sign messages of a signer is limited to a fixed number k of signatures. It is an identity-based signature scheme in which each signature can be used only once. We called such schemes "bounded life-span". It is based on mental games [SRA81] and it uses zero-knowledge tools [Sh85, GMRa89, BGKW88].

Cryptography using one-time pad or key is known to be the most secure.

There exist similar aimed authentication schemes described in [dWQ90] but their signature adaptations are too complex. There exist also schemes based on non zero-knowledge techniques as described in [Vau93].

We combine a Guillou-Quisquater signature scheme [GQ89b, DQ94] with an instance of Lamport's scheme [L81]. Our scheme limits the signing power of users of the system to a fixed number of signatures.

Our scheme may also be used as a delegation scheme as it is explained in this paper. Practical applications are easily derived from our scheme (*e.g.* to implement payment systems).

The one-time signature is an old concept attributed to Lamport-Diffie and improved in [Mer79]. Other versions which are Public Key Systems oriented are described in [GMR88, BM88, NY89]. Our contribution is to link the original idea of Lamport together with zero-knowledge schemes in an efficient way. It is a practical identity-based scheme whose goal is the limitation of the number of acknowledged signatures.

2 Bounded Life-span Signatures

We now outline our bounded life-span signature scheme, *i.e.* an identity-based signature scheme in which the right to produce acknowledged signatures is limited. Each user has a different but fixed identity I, which is validated once by the authority at the beginning of the system. The following desirable properties apply to such a scheme :

1. Signatures must be used only once, or it will be easily *detected* that a signature was reused,
2. Bounded life-span aspect should be revealed while sending out the signature.
3. No secret information should be revealed,
4. The representation of a user's identity must be fixed at all times, throughout his lifetime in the system.

2.1 Initialization of the System

2.1.1 First Step

A trusted center computes a public composite modulus $n = p \cdot q$ whose factors are strong primes. [1] These are kept secret. The center chooses also a public prime exponent v. The pair $(v\,;n)$ is made public.

Each user of the system has an identity I^*. [2] We assume that the identity string J is built from I^* with some added redundancy. This considerably enhances the intractability of an identity fraud. The center computes the secrets as follows : a redundancy function Red is applied to the original identity I^* such that the left and right parts of the result match a particular pattern. The Red identity J is then used to extract a secret number D using the signing function S of the center [GQ89b]. The center then issues the signer with this secret number $D \in Z_n$. [3] For modulus n and an identity I^* half the length of n, the center thus computes :

[1] A prime p is said to be *strong* if $p = 2\,p' + 1$ where p' is also a prime. A product of strong primes seems to be in general more difficult to factorize [McC90].

[2] To prevent misuse of authentications, we must append to the identity I a particular flag which restricts the use of I to either only authentication schemes, or only signature schemes [DQ94]. This gives I^*.

[3] This secret number D may be stored in a tamper-resistant device (Smart Card) [GUQ91].

$$I^* \longrightarrow J = Red(I^*) \quad \longrightarrow D = S(J)$$
$$\text{where } D^v \cdot J \equiv 1 \mod n$$

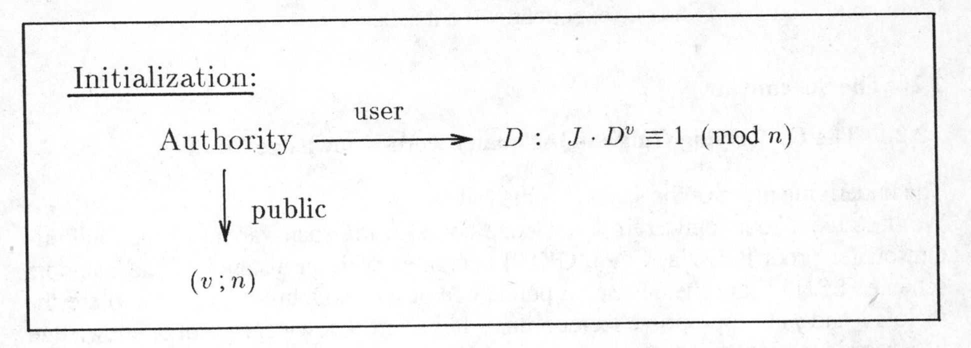

After the initialization of all users, the trusted center can be removed. It will not take part anymore in the system other than as a judge to settle eventual conflicts.

2.1.2 Second Step

Consider the following instance of the Lamport scheme [L81] :

Let $F(x) = x^v \mod n$. Here the function F is a one-way function with trapdoor the secret factors of the modulus n. From x_0 we get

$$x_i = F(x_{i-1}), \text{ for } i = 1, 2, \ldots, k$$

That is, it is hard to compute x_{k-1} from x_k under the RSA assumption, *i.e.* if n is the product of two distinct primes p and q, and if v is coprime to $\phi(n)$, it is conjectured that extracting a vth root modulo n is as difficult as factoring n. Clearly $x_i \equiv (x_0)^{v^i}$, $i = 1, 2, \ldots, k$. The number x_0 is chosen randomly, as long as the modulus n.

A "credential center" issues each signer with a different and unique (not already used) secret pair (x_k, x_0). Only the signer is able to compute the values $\{x_i\}_{1 \leq i < k}$ from the secret x_0 and nobody can compute x_{i-1} from x_i, unless one knows the secret factors of n.

The credential center registers in a public directory a pair (J, x_k) for each user having an identity I^* and wishing to take part in this system. The indexes i of the public x_i will be decreased each time a signature is registered. The x_i will be published in this directory in the decreasing order following their use by the signer.

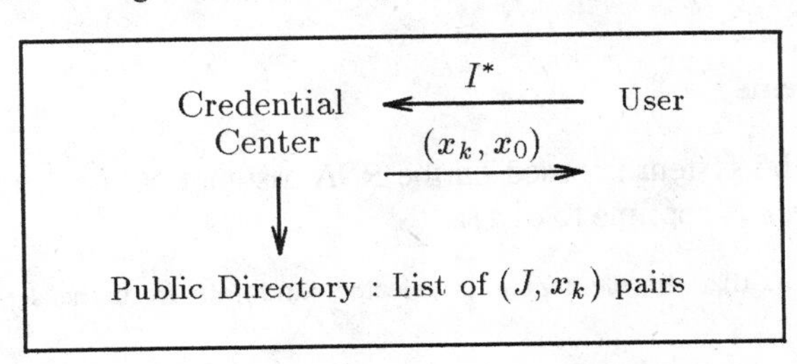

This secret distribution from the credential center to a legitimate user may be achieved by using a bi-authenticated secret message transmission as described in [DQ].

Secrets are stored in signer's smart card. No secret is needed by the verifier. There is only one modulus n and one exponent v in our scheme.

2.2 The Scheme

2.2.1 The Underlying Guillou-Quisquater Zero-Knowledge Proof

The underlying proof of the scheme is the following :
We describe a four round zero-knowledge bounded life-span variant of the Guillou-Quisquater proof [GQ89a, CP94, GK89] combined with an instance of the Lamport scheme [L81]. This scheme uses a public composite modulus $n = p \cdot q$, where the factors p and q are appropriate secret primes. It is a zero-knowledge proof of knowledge of vth residues, where v is an appropriate public exponent [FFS88, GK89]. Its security is based on the RSA assumption if the exponent v is coprime to $\phi(n)$ [Bu93]. The prover and the verifier are respectively denoted by P and V. The prover has a current public pair (J, x_m) registered in the public directory :

GQ Bounded Life-span Zero-Knowledge Proof :

Input : $(J, x_m; v, n)$.
Language : $L_{c_1c_2} = \{(J, x_m; v, n) \mid v$ is a prime, $c_1|n| < v < c_2|n|$ and J is a vth residue in $Z_n\}$ with $0 < c_1 < c_2$.
Protocol : To prove that $(J, x_m; v, n) \in L_{c_1c_2}$, we use the following four rounds :

- $V \rightarrow P : \{w_i = l_i^2 \bmod n\}_{i=1}^{c_1|n|}$, with $0 < l_i < n/2$.
- $P \rightarrow V : T_m = r_m^v \mod n$, where $r_m \in_R Z_n$. [4]
- $V \rightarrow P : \{l_i\}_{i=1}^{c_1|n|}$ and P constructs $d_m = [d_{m_1}, \ldots, d_{m_{c_1|n|}}]_2$, where $d_{m_i} = \frac{(l_i/n)+1}{2}$.
- $P \rightarrow V : t_m = r_m \cdot (D \cdot x_{m-1})^{d_m} \mod n$, where $D^v \cdot J \equiv 1 \pmod{n}$.

Verification: $T_m \stackrel{?}{\equiv} t_m^v \cdot (J \cdot x_m^{-1})^{d_m} \pmod{n}$ (and $T_m \neq 0$).

Note that we could use a three round bounded life-span variant of the Guillou-Quisquater proof but this protocol is not zero-knowledge though practically secure. Nevertheless a three round **zero-knowledge** bounded life-span variant of the Guillou-Quisquater proof can be obtained by iterating the protocol.

2.2.2 The Scheme

The security of the system is based on the RSA assumption, *i.e.* on the difficulty of factoring n (since v is coprime to $\phi(n)$).

[4] $a \in_R A$ means that the element a is selected randomly from the set A with uniform distribution.

The verifier has access to the public values n, v and to the public pairs (J, x_k) registered in the public directory.

Let us assume that $x_k, x_{k-1}, \ldots, x_{m+1}, x_m$ have already been used by the user having identity J, with $m \geq 1$. The corresponding public "key" of this user is now registered in the public directory as (J, x_{m+1}).

The protocol may be divided in the following steps. To send out a signature using x_{m-1} as a secret with bounded life-span, the signer performs the following :

Bounded Life-span Signature Protocol :

Input : $(J, x_{m+1} ; v, n)$

Signer :

1. Reveals (J, x_m).
2. Picks $r_m \in_R Z_n$, and computes $T_m = r_m^v \mod n$.
3. Computes the hashing value $d_m = h(T_m, M|I^*)$, where $M|I^*$ is the message to be signed. [5]
4. Computes the final witness $t_m = r_m \cdot (D \cdot x_{m-1})^{d_m} \mod n$.
5. Sends the verifier the signature $\mathsf{Sgn}(T, t, M|I^*)$. [6]

Verifier :

6. Knows $(J, x_{m+1} ; v, n)$ and (J, x_m) and checks that :
   1. $x_{m+1} \stackrel{?}{=} x_m^v \mod n$, [7]
   2. $d_m \stackrel{?}{=} h(t_m^v \cdot (J \cdot x_m^{-1})^{d_m} \mod n, M|I^*)$.

   It is easy to see that if the signer follows the protocol, the verification will be valid. Indeed,

$$\begin{aligned} t_m^v \cdot (J \cdot x_m^{-1})^{d_m} &= (r_m^v \cdot D^{d_m v} \cdot x_{m-1}^{d_m v}) \cdot J^{d_m} \cdot x_m^{-d_m} \mod n \\ &= r_m^v \cdot (D^v J)^{d_m} \cdot (x_{m-1}^v \cdot x_m^{-1})^{d_m} \mod n \\ &= r_m^v \mod n \end{aligned}$$

   Using this protocol the verifier is convinced with overwhelming probability that the signer knows $D = S(J)$ (and thus is an authorized entity) and that he knows the secret x_{m-1}. This x_{m-1} which is used but not revealed at the end of the protocol will be actually revealed during the next signing operation.

7. The verifier sends the revealed (J, x_m) with the corresponding signature to the credential center in order to update the public directory.

---

[5] In this paper, $a|b$ means that a is concatenated with b.

[6] Note that x_m may be revealed as a part of the signature. **Sgn** may merely denote concatenation.

[7] Actually, x_{m+1} is obtained from the public directory.

Schematically,

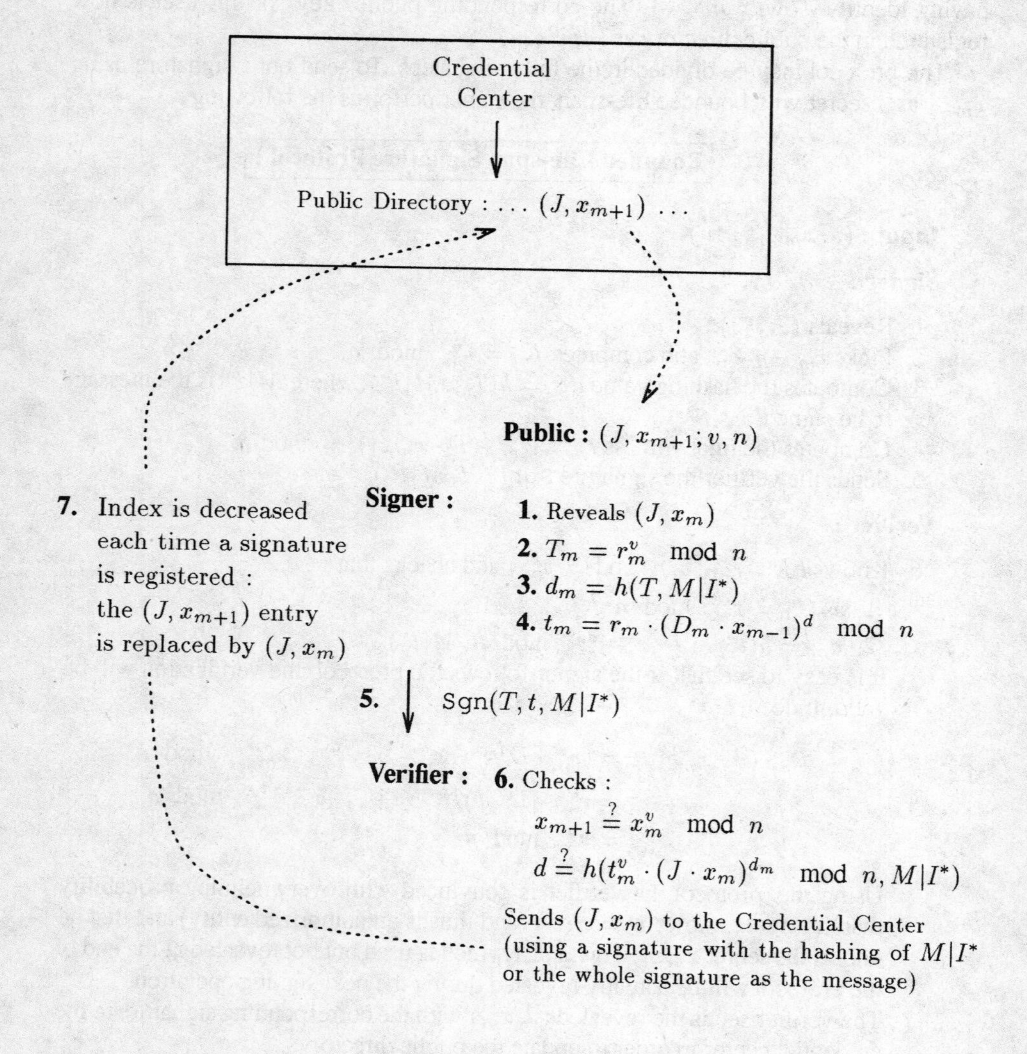

The verifier sends the new public pair (J, x_m) to the credential center as an update request of the public directory. This request is composed of the new entry and either the hashing of the signed message concatenated with the identity string I^* of the signer or the whole signature. This request is signed by the verifier using a classical Guillou-Quisquater signature scheme using D as the secret number.

The bounded life-span secrets x_i may be related to one exclusive credential center (service provider, *etc*). But this deals with management.

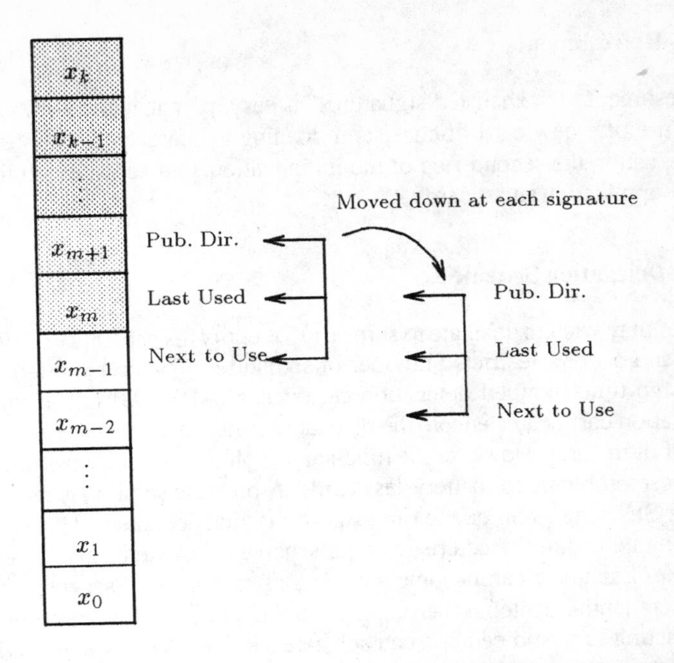

Actually the number of available signatures is $k - 1$ if x_0 is used and $k - 2$ if not.

3 Analysis and Remarks

3.1 What Happens if a Signature is Replayed ?

Using for a second time the same value x_m to produce a signature implies a second verification. If the previous verifier was honest the public directory was updated and the second signature will be rejected. At the same time of sending the signature, the signer may publish the value x_m to prevent a reusing by the verifier.

A crooked signer could sign different messages with the same Lamport secret number to different verifiers. But only one signature, the one which will be registered by the corresponding credential center, will be accepted by the verifier.

How could a judge settle an *a posteriori* conflict between two signatures using the same secret(s) but involving two different messages ? If two verifiers are involved in the conflict, the answer is simple : positive decision in favour of the (first) verifier who has signed the update request of the Public Directory. The same solution is adopted if there is only one verifier involved in the conflict since the signed update request of the Public Directory contains the (hashing of the) involved message.

The (J, x_{m+1}) entry in the Public Directory means that the secrets $x_k, x_{k-1}, \ldots, x_m$ have already been used and x_{m-1} is the next secret to be used. Excepted the credential center, nobody can successfully request the update of the Public Directory without the help of the legitimate user.

3.2 Refreshment

Refreshment of "exhausted signatures" is easy, practical and does not require the manufacture of a new card. It consists in loading the card with a new secret pair (x'_k, x'_0) by repeating the second step of the initialization. Our scheme is fully compatible with multi-application smart cards.

3.3 Delegation Scheme

A user may wish to delegate to some trusted entity his capability of signing for a limited time and/or for a restricted number of signatures. The user's smart card can construct and sign time-stamped delegation certificates [ABKL93] to the trusted entity. These delegation certificates enable the delegated entity to convince others that he acts on the behalf of the user. However the time-stamp solution requires a secure time source. This can pose problems for battery-less cards. A possible solution is nevertheless to use the Haber-Stornetta technique to time-stamp a digital document [HS91].

An interesting characteristic of our scheme is its extension to a *delegation signature scheme*. Each user can become a credential "sub-center" accepted (recognized) by the main credential center. When x_{m-1} is the next secret Lamport number to be used, a user, acting as a "sub-center", can delegate his signature power by giving to another user a secret x_i with $1 \leq i \leq m-1$ (using [DQ]). The delegation may concern either any legitimate user or a particular user having an identity string J related to the identity string of the "sub-center" (J contains a particular pattern class introduced by the redundancy function) *without endangering* neither his secret number D nor the secrets x_l, with $1 \leq l < i$.

The identity of the delegate could be related to the delegator. Either the delegator sends an update request to the credential center concerning the identity of its public entry : (J, x_{m+1}) becoming (J, J', x_{m+1}). Or at the initialization of the system, identity classes have been designed on the basis of the identity of candidate delegators. The first solution gives a better protection against stolen secrets (though this should not be a problem by using smart cards).

This delegation scheme is transitive and "suspensive" in the sense that the sub-center must wait the use of x_i before using x_{i-1} otherwise the delegate could not use his available signature. However this revocation possibility gives a full control to the delegator.

3.4 General Remarks

- Actually the credential center is **not** an authority or a trusted center in the sense that he does not (need to) know the factorization of the modulus n. The only part he plays in this scheme is to issue secret Lamport pairs (x_k, x_0) and to manage the Public Directory. Trust is then limited to his application.
- To prevent the fraudulent use of values of k greater than the maximal authorized value, a kind solution is to insert this value in one "field" of the redundant identity J which is signed by the Authority.

- The credential center (and the sub-center or delegator in a delegation scheme) is able to revoke "licenses", signature power, the ability to sign of the signer, once granted.
- How to choose the Lamport numbers :
  A new Lamport pair must never have been used before. Since there could exist many independent credential centers, such numbers should include a particular redundant pattern related to the credential center. Or the simple fact that "registration" is the key word suffices : the credential center is the only entity mastering the **public** directory; this implies a control from the users.
- Performances : From the signer point of view, the costs are roughly the same than for a classic Guillou-Quisquater signature scheme. The verification process is roughly 50% more expansive.
- The protocol is on-line. An off-line version would be possible provided we could have confidence in smart cards.
- Note that this scheme simplifies and generalizes the scheme presented in [dWQ90].
- Would not it be more simple to use a Guillou-Quisquater signature in parallel with an instance of the Lamport scheme, as $F(x) : x^v \mod n$. Doing this we loose the interconnection between the Lamport values and the user's secret $S(J)$. It prevents unauthorized, illegitimate Lamport instances; x_0 must be chosen and validated by the credential center.
  Moreover acting this way the secret numbers must be disclosed before the approval of the signature and then before using the service granted by this approval. In case of (server) (communication) failure some odd instances of secret numbers may be reused. On the other hand, in our scheme, secret numbers are revealed the next time a signature is performed. This also provide the verifier with a proof that the service was granted.
  No secret information is revealed during transactions; when x_{m-1} is revealed it may be public.

3.5 Applications

3.5.1 Pre-payment Cards

One type of application leads to the following setting : there is one credential center which is the only one verifier. That means practically that any seller can manage his own payment system allowing his clients to use prepaid cards to buy food, things or services (including computer network services).

When the credential center is also the only one verifier, it may fix an appropriate lifetime to individual signature, *i.e.* it may allow replays of a signature for a fixed period of time or a fixed number of occurrences.

A similar or derived application may allow several verifiers to be connected to the same credential center : prepaid cards for parking in private areas, or prepaid access to services.

3.5.2 Electronic Cash

Another application deals with the possibility to withdraw money thanks to banking cards. A banking card could be loaded at a bank counter with the equivalent of a fixed number of bank notes. This loading operation could occur regularly or only a single time on the same card. The physical bank notes could then be withdrawn at any ATM in connection with the bank (even abroad). A practical application could be a (nice) alternative to traveller's checks.

3.5.3 Pay-TV

Another type of implementation may be achieved by means of an hierarchization of the management performed by the credential center. The setting is as follows : One main application center, say a cable TV organization, is the only authority able to load smart cards [8] with (x_k, x_0) secret pairs related to one particular (set of) user(s) and the user's decoder. The decoder plays both the role of the verifier and of the public directory manager described in our scheme.

Only a particular set of users are allowed to use a particular decoder. This set may be distinguished from the others by using a special class of redundancy in I^* identity strings.

Or in an original setting, the decoder sends request to the main application center which manages the public directory and delegation allows any card user to use any decoder.

4 Conclusion

We outlined a signature scheme which limits the right of the signer to produce acknowledged signatures. It is based on the Guillou-Quisquater zero-knowledge proof and Lamport authentication scheme. It is practical and flexible. Our scheme allows delegation and refreshment is easy.

A trusted center initializes the system and will not take part any more other than as a judge to settle eventual conflicts.

A credential center issues each signer with a different and unique (not already used) set of k bounded life-span secrets which are to be used to produce signatures. This center manages the bounded life-span property of the signature scheme.

Actually the credential center is **not** an authority or a trusted center in the sense that he does not (need to) know the factorization of the modulus n.

The credential center (and the sub-center or delegator in a delegation scheme) is able to revoke "licenses", signature power, the ability to sign of a user, once granted.

5 Acknowledgments

We would like to thank Mike V. D. Burmester for discussions and comments about this work.

[8] Multi-applications smart cards.

References

[ABKL93] M. Abadi, M. Burrows, C. Kaufman and B. Lampson. Authentication and delegation with smart cards. *Science of Computer Programming*, N ° 21, pp. 93–113, Elsevier, 1993.

[BM88] M. Bellare and S. Micali. How to sign given any trapdoor function. *Proceedings of the 20th Symposium on Theory of Computing, STOC'90*, pp. 427–437.

[Bu93] M. V. D. Burmester. Recent developments in efficient Zero-Knowledge proofs. *Talk given at the Université Catholique de Louvain*, June 1993.

[BCDP91] J. Boyar, D. Chaum, I. Damgard and T. Pedersen. Convertible Undeniable Signatures. *Advances in cryptology, Proceedings of CRYPTO '90, Lecture Notes in Computer Science*, N° 537, pp. 189–205, Springer-Verlag, 1991.

[BD89] M. V. D. Burmester and Y. G. Desmedt. Remarks on Soundness of Proofs. *Electronic letters*, pp. 1509–1510, Vol. 25, N ° 22, 26th October 1989.

[BGKW88] M. Ben-Or, S. Goldwasser, J. Killian and A. Wigderson. Multi-prover interactive proofs : How to remove intractability assumptions. *Proceedings of the twentieth annual ACM Symp. Theory of Computing, STOC'88*, pp. 113–131, May 2–4,1988.

[CP94] G. do Crescenzo and G. Persiano. Round-optimal perfect zero-knowledge proofs. *Information Processing Letters*, pp. 93–99, Vol. 50, N ° 2, 22 April 1994.

[CvA90] D. Chaum and H. van Antwerpen. Undeniable Signatures. *Advances in cryptology, Proceedings of CRYPTO '89, Lecture Notes in Computer Science*, N ° 435, pp. 212–216, Springer-Verlag, 1990.

[DF92] Y. Desmedt and Y. Frankel. Shared Generation of Authenticators and Signatures. *Advances in cryptology, Proceedings of CRYPTO '91, Lecture Notes in Computer Science*, N ° 576, pp. 457–469, Springer-Verlag, 1992.

[DH76] W. Diffie and M. E. Hellman. New Directions in Cryptography. *IEEE Transactions on Information Theory*, Vol. IT-22, N° 6, pp. 644–654, 1976.

[DQ94] O. Delos and J.-J. Quisquater. Efficient multi-signature schemes for cooperating entities. *Proceedings of French-Israeli Workshop on Algebraic Coding, Lecture Notes in Computer Science*, N° 781, pp. 63–74 , Springer-Verlag, 1994.

[DQ] O. Delos and J.-J. Quisquater. Biauthentication and secret message transmission. *Manuscript UCL 1994.*

[dWQ90] D. de Waleffe and J.-J. Quisquater. Better login protocols for computer networks. *Proceedings of ESORICS '90*, pp. 163–172, October 1990.

[FFS88] U. Feige, A. Fiat and A. Shamir. Zero-knowledge proofs of identity. *Journal of Cryptology*, 1(2), pp. 77–94, 1988.

[GDQ89] L. C. Guillou, M. Davio and J.-J. Quisquater Public-key techniques: Randomness and Redundancy. *Cryptologia*, Vol. 13, N ° 2, pp. 167–189, April 1989.

[GK89] O. Goldreich and H. Krawczyk. On the Composition of Zero-Knowledge Proof Systems. Technical Report N° 570 of Technion, 1989.

[GMR88] S. Goldwasser, S. Micali and R. Rivest. A digital signature scheme secure against adaptative chosen-message attacks. *Siam J. Comput.*, 1988, Vol. 17, pp. 281–308.

[GMRa89] S. Goldwasser, S. Micali and C. Rackoff. The Knowledge Complexity of Interactive Proof Systems. *Siam J. Comput.*, 1989, Vol. 18, N° 1, pp. 186–208.

[GQ88a] L. C. Guillou and J.-J. Quisquater. Efficient digital public-key signatures with shadow. *Advances in cryptology, Proceedings of CRYPTO '87, Lecture Notes in Computer Science*, N° 304, p. 223, Springer-Verlag, 1988.

[GQ89a] L.C. Guillou and J.-J. Quisquater. A practical zero-knowledge protocol fitted to security microprocessor minimizing both transmission and memory. In C. G. Günther, editor, *Advances in Cryptology, Proceedings of EUROCRYPT '88, Lecture Notes in Computer Science*, N° 330, pp. 123–128, Springer-Verlag, 1988.

[GQ89b] L.C. Guillou and J.-J. Quisquater. A "paradoxical" identity-based signature scheme resulting from zero-knowledge. *Advances in Cryptology, Proceedings of CRYPTO '88, Lecture Notes in Computer Science*, N ° 403, pp. 216–231, Springer-Verlag, 1989.

[GUQ91] L. C. Guillou, M. Ugon and J.-J. Quisquater. The Smart Card: A Standardized Security Device Dedicated to Public Cryptology. *Contemporary Cryptology : The Science Information Integrity*, edited by G. J. Simmons, IEEE Press, 1991.

[HS91] S. Haber and W.S. Stornetta. How to Time-Stamp a Digital Document. *Advances in Cryptology, Proceedings of CRYPTO '90, Lecture Notes in Computer Science*, N ° 537, pp. 437–455, Springer-Verlag, 1991.

[L81] L. Lamport. Password Authentication With Insecure Communication. *Comm. of ACM*, Vol. 24, N° 11, pp. 770–772, Nov. 1981.

[McC90] K. Mc Curley. Odd and ends from cryptology and computational number theory. *Cryptology and computational number theory*, edited by C. Pomerance, AMS short course, pp. 145–166, 1990.

[Mer79] R. C. Merkle. A Certified Digital Signature. *Advances in Cryptology, Proceedings of CRYPTO '89, Lecture Notes in Computer Science*, N ° 435, pp. 218–238, Springer-Verlag, 1989.

[NY89] M. Naor and M. Yung. Universal One-way Hash Functions and their Cryptographic Applications. *Proceedings of the 21 st Symposium on Theory of Computing, STOC'89*, pp. 33–43, 1989.

[Q87] J.-J. Quisquater. Secret distribution of keys for public-key system. *Advances in cryptology, Proceedings of CRYPTO '87, Lecture Notes in Computer Science*, N ° 293, pp. 203–208, Springer-Verlag, 1987.

[R80] M. O. Rabin. Probabilistic algorithms for testing primality. *Journal on Number Theory*, Vol. 12, pp. 128–138, 1980.

[Sh85] A. Shamir. Identity-based cryptosystems and signatures schemes. *Advances in cryptology, Proceedings of CRYPTO '84, Lecture Notes in Computer Science*, N ° 196, pp. 47–53, Springer-Verlag, 1985.

[SRA81] A. Shamir, R. Rivest and L. Adleman. Mental Poker. *The Mathematical Gardner*, edited by D. A. Klarner, Wadsworth International, 1981.

[Vau93] S. Vaudenay. Mémoire de Magistère de Mathématiques Fondamentales et Appliquées et d'Informatique. GRECC, Laboratoire d'Informatique de l'Ecole Normale Supérieure, Paris, 1993.

[vH92] E. van Heijst. Special Signature Schemes. Thesis for the degree of Doctor at the Eindhoven University of Technology (The Netherlands), July 1992.

[WP90] M. Waidner and B. Pfitzmann. The Dining Cryptographers in the Disco : Unconditional Sender and Recipient Untraceability with computationally Secure Serviceability. *Advances in cryptology, Proceedings of EUROCRYPT '89, Lecture Notes in Computer Science*, N ° 434, p.690, Springer-Verlag, 1990.

More Flexible Exponentiation with Precomputation

Chae Hoon Lim and Pil Joong Lee

Department of Electrical Engineering, Pohang University of Science and Technology (POSTECH), Pohang, 790-784, KOREA

Abstract. A new precomputation method is presented for computing g^R for a fixed element g and a randomly chosen exponent R in a given group. Our method is more efficient and flexible than the previously proposed methods, especially in the case where the amount of storage available is very small or quite large. It is also very efficient in computing $g^R y^E$ for a small size E and variable number y, which occurs in the verification of Schnorr's identification scheme or its variants. Finally it is shown that our method is well-suited for parallel processing as well.

1 Introduction

The problem of exponentiating fast in a given group (usually Z_N, N a large prime or a product of two large primes) is very important for efficient implementations of most public key cryptosystems (hereafter it is assumed w.l.o.g. that the computation is performed over Z_N and thus multiplication denotes multiplication mod N). A typical method for exponentiation is to use the binary algorithm, known as the square-and-multiply method [1]. For 512 bit modulus and exponent, this method requires 766 multiplications on average and 1022 in the worst case. The signed binary algorithm [2-3] can reduce the required number of multiplications to around 682 on average and 768 in the worst case.

On the other hand, using a moderate amount of storage for intermediate values, the performance can be considerably improved again. Knuth's 5-window algorithm [1,4] can do exponentiation in about 609 multiplications on average, including the on-line precomputation of 16 multiplications. The fastest known algorithm for exponentiation is the windowing method based on addition chains, where we can use bigger windows such as 10 [4] and need more storage for intermediate values [5]. Though finding the shortest addition chain is an NP-complete problem [6], it is reported [4] that, by applying heuristics, an addition chain of length around 605 can be computed.

These general methods can be used for any cryptosystems requiring exponentiation such as RSA [7] and ElGamal [8]. However, in many cryptographic protocols based on the discrete logarithm problem, we need to compute g^R for a fixed base g but for a randomly chosen exponent R. Thanks to the fixed base element, a precomputation table can be used to reduce the number of multiplications required, of course at the expense of storage for precomputed values.

At Eurocrypt'92, Brickell et al. [9] proposed such a method for speeding up the computation of g^R (called the BGMW method, for the convenience of reference). Their basic strategy is to represent an exponent R in base b, that is, $R = d_{t-1}b^{t-1} + \cdots + d_1 b + d_0$ where $0 \leq d_i < b$ $(0 \leq i < t)$, and precompute all powers $g_i = g^{b^i}$. Then g^R can be computed by $\prod_{i=0}^{t-1} g_i^{d_i} = \prod_{d=1}^{b-1}(\prod_{d_i=d} g_i)^d$. Using a basic digit set for base b, they extended the basic scheme so that the computation time can be further decreased while the storage required is increased accordingly.

In this paper, we propose another precomputation method for fast exponentiation. Our method is a generalization of the simple observation that if an n-bit exponent R is divided into two equal blocks (i.e., $R = R_1 \times 2^{n/2} + R_0$) and $g_1 = g^{2^{n/2}}$ is precomputed, then g^R can be evaluated in a half of the time required by the binary method in the worst case ($\frac{7}{12}$ on average) by computing $g_1^{R_1} g^{R_0}$. It will be seen that the proposed method is more efficient, especially when the storage available is very small or quite large, and also more flexible, giving a wide range of time-memory tradeoffs, than the BGMW method. The case of using a small amount of storage is of great importance for an application to smart cards having limited storage and computing power, but the BGMW method is not so efficient for this case.

Another advantage of the proposed method is its efficiency in computing $g^R y^E$, where y is not fixed and the size of E is much less than that of R, which is needed for the verification of Schnorr's identification and signature scheme [10] or its variants, e.g., Brickell-McCurley's scheme [11] and Okamoto's scheme [12]. Note that, for this kind of computation, representing exponents in non-binary power base may considerably increase the on-line computational load (see section 4). Finally we show that the proposed method is also well-suited for parallel processing.

Throughout this paper, we will use g as a fixed element of Z_N and R as an n-bit random exponent over $[0, 2^n)$. We denote by $|S|$ the bit-length of S for an integer S or the cardinality of S for a set S. We also denote by $\lceil x \rceil$ the smallest integer not less than x and by $\lfloor x \rfloor$ the greatest integer not greater than x.

2 Review of Previous Work : BGMW Method

In this section, we briefly review the BGMW method, the precomputation method proposed by Brickell, Gordon, McCurley and Wilson at Eurocrypt'92. For more details, see the original paper [9].

A set of integers D is called a *basic digit set* for base b if any integer can be represented in base b using digits from the set D [13]. Suppose that we can choose a set M of multipliers and a parameter h for which

$$D(M, h) = \{mk \mid m \in M, 0 \leq k \leq h\} \tag{1}$$

is a basic digit set for base b. Then an n-bit exponent R can be represented as

$$R = \sum_{i=0}^{t-1} d_i b^i, \quad d_i = m_i k_i \in D(M, h). \tag{2}$$

With this representation of R, g^R can be computed by

$$g^R = \prod_{i=0}^{t-1} g^{m_i k_i b^i} = \prod_{k=1}^{h} (\prod_{k_i=k} g^{m_i b^i})^k = \prod_{k=1}^{h} c_k^k. \quad (3)$$

Therefore, if we precompute and store powers g^{mb^i} for all $i < t$ and $m \in M$, then g^R can be computed in at most $t + h - 2$ multiplications using about $t|M|$ precomputed values by the following algorithm.

```
u := ∏_{k_i=h} g^{m_i b^i};
v := u;
for w := h - 1 to 1 step -1
    u := u * ∏_{k_i=w} g^{m_i b^i};
    v := v * u;
return(v);
```

It is easily seen that the number of multiplications performed by the above algorithm is $t + h - 2$ in the worst case ($t - h$ multiplications for computing products of the form $\prod_{k_i=w} g^{m_i b^i}$ for $w = 1, \cdots, h$ and $2h - 2$ multiplications for completing the for-loop) and $\frac{b-1}{b}t + h - 2$ on average (For a randomly chosen exponent, $\frac{t}{b}$ digits are expected to be zero.).

The most obvious example for D is the base b number system ($M = \{1\}$, $h = b - 1$, $t = \lceil \log_b (2^n - 1) \rceil$). For a 512 bit exponent, the choice of $b = 26$ minimizes the expected number of multiplications. This basic scheme requires 127.8 multiplications on average, 132 in the worst case, and storage for 109 precomputed values. More convenient choice of base will be $b = 32$, since then the digits for the exponent R can be computed without radix conversion by extracting 5 bits at a time. With this base, the required number of multiplications is increased only by one for the average case and remains unchanged for the worst case. Though the basic scheme is the obvious choice in the case where the storage available is small, its performance is considerably degraded as the number of storage is going down below 109. This means that the BGMW method does not provide an efficient way to perform the computation when the storage available is very small.

Brickell et al. also presented several schemes using other number systems to decrease the number of multiplications required, of course using more storage for precomputed values. One of the extreme examples is to choose the set M as $M_2 = \{m | 1 \le m < b, \omega_2(m) = 0 \bmod 2\}$, where $\omega_p(m)$ is the highest power of p dividing m. Then, for $1 \le d_i < b$, we have $d_i = m$ or $2m$ for some $m \in M_2$ (i.e., $h = 2$). Thus g^R can be computed in t multiplications on average and $\frac{b-1}{b}t$ multiplications in the worst case, with the storage to $|M_2| \lceil \log_b (2^n - 1) \rceil$ values. For example, taking $b = 256$ ($t = 64$, $|M_2| = 170$), we can achieve an average of 63.75 multiplications with 10880 precomputed values. Two tables that Brickell et al. presented in [9] are given in the appendix A for the purpose of comparison with our results.

3 The Proposed Method

We now present our method for fast evaluation of g^R using a precomputation table. Let R be an n-bit exponent for which we want to compute g^R. We first divide the exponent R into h blocks R_i, for $0 \le i \le h-1$, of size $a = \lceil \frac{n}{h} \rceil$ and then subdivide each R_i into v smaller blocks $R_{i,j}$, for $0 \le j \le v-1$, of size $b = \lceil \frac{a}{v} \rceil$ as follows (see figure 1):

$$R = R_{h-1} \cdots R_1 R_0 = \sum_{i=0}^{h-1} R_i 2^{ia},$$

$$R_i = R_{i,v-1} \cdots R_{i,1} R_{i,0} = \sum_{j=0}^{v-1} R_{i,j} 2^{jb}. \tag{4}$$

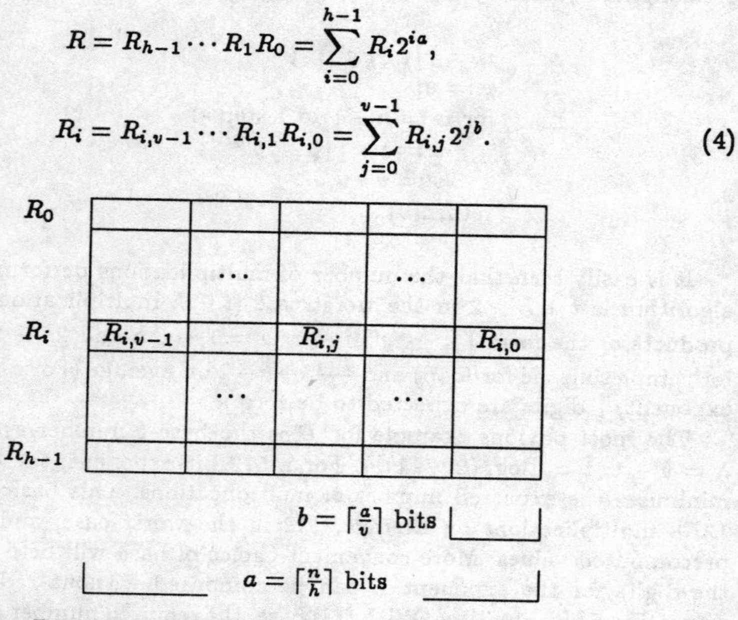

Figure 1 : Division and arrangement of an n-bit exponent R

Let $g_0 = g$ and define g_i as $g_i = g_{i-1}^{2^a} = g^{2^{ia}}$ for $0 < i < h$. Then, using the equations (4), we can express g^R as

$$g^R = \prod_{i=0}^{h-1} g_i^{R_i} = \prod_{j=0}^{v-1} \prod_{i=0}^{h-1} \left(g_i^{2^{jb}} \right)^{R_{i,j}} . \tag{5}$$

If we let $R_i = e_{i,a-1} \cdots e_{i,1} e_{i,0}$ be the binary representation of R_i $(0 \le i < h)$, then $R_{i,j}$ $(0 \le j < v)$ is represented in binary as

$$R_{i,j} = e_{i,jb+b-1} \cdots e_{i,jb+k} \cdots e_{i,jb+1} e_{i,jb}.$$

Therefore the expression (5) can be rewritten as follows :

$$g^R = \prod_{k=0}^{b-1} \left(\prod_{j=0}^{v-1} \prod_{i=0}^{h-1} g_i^{2^{jb} e_{i,jb+k}} \right)^{2^k} \tag{6}$$

Next suppose that the following values are precomputed and stored for all $1 \leq i < 2^h$ and $0 \leq j < v$.

$$G[0][i] = g_{h-1}^{e_{h-1}} g_{h-2}^{e_{h-2}} \cdots g_1^{e_1} g_0^{e_0},$$
$$G[j][i] = (G[j-1][i])^{2^b} = (G[0][i])^{2^{jb}}. \tag{7}$$

Here the index i is equal to the decimal value of $e_{h-1} \cdots e_1 e_0$. Then, using the precomputed values of (7), we can rewrite the expression (6) as

$$g^R = \prod_{k=0}^{b-1} \left(\prod_{j=0}^{v-1} G[j][I_{j,k}] \right)^{2^k}, \tag{8}$$

where $I_{j,k} = e_{h-1,bj+k} \cdots e_{1,bj+k} e_{0,bj+k}$ $(0 \leq j < b)$, which corresponds to the k-th bit column of the j-th block column in the figure 1. Now it is straightforward to compute g^R using the expression (8) by the ordinary square-and-multiply method as follows :

```
Z := 1;
for k := b − 1 to 0 step -1
    Z := Z * Z;
    for j := v − 1 to 0 step -1
        Z := Z * G[j][I_{j,k}];
return(Z);
```

We next count the number of multiplications required by the above algorithm. Here we have to note that the $(v-1)$-th blocks in the figure 1 may not be full of b bits. In fact, they are of $bv - a$ bit size. Thus the number of terms to be multiplied together in the inner for-loop is $v+1$ for the first $bv - a$ rounds and $v+2$ for the remaining $b - bv + a$ rounds. Therefore, the total number of multiplications required is at most v(bv-a)+(v+1)(b-bv+a)-2 = a+b-2 in the worst case. Since we may assume that the probability of $I_{j,k}$ being zero is $\frac{1}{2^h}$ and there are a occurrences of $I_{j,k}$ in the above algorithm, the expected number of multiplications is given by $\frac{2^h-1}{2^h}a + b - 2$. Of course, this performance is achieved with storage for $(2^h - 1)v$ precomputed values.

In the above, we assumed that the exponent R is partitioned into hv blocks of almost equal size and that these hv blocks are arranged in a $h \times v$ rectangular shape. In most cases, such partitions and arrangements yield better performance than others for a given amount of storage, but sometimes this may not be the case. For example, consider two configurations shown in the figure 2 below, where a 512-bit exponent is partitioned and arranged in two different ways. The first configuration corresponds to the case we analyzed in the above and results in the performance of 118.78 multiplications on average (122 in the worst case) with storage for 155 values. On the other hand, with the second configuration, we can do the exponentiation in 117.13 multiplications on average (119 in the worst case) using storage for 157 values. This shows that we had better choose the second configuration.

19	21	21	21	21
19				
18				
18				
18				

28	31	31

5×5 configuration $\quad$ $5 \times 1|6 \times 2$ configuration

Figure 2 : Two different configurations with almost the same storage requirement

For the configuration of type $h_1 \times v_1|h_2 \times v_2$ where $h_1 < h_2$, we can easily derive general formulae for the worst/average-case performance. Let b_1 and b_2 be the size of partitioned blocks in $h_1 \times v_1$ and $h_2 \times v_2$ respectively. For better performance, b_2 must be greater than or equal to b_1 and can be obtained in the same way as b in the $h \times v$ configuration. Thus we get $b_2 = \lceil \frac{n}{h_1v_1+h_2v_2} \rceil$ and $b_1 = \lceil \frac{n-b_2h_2v_2}{h_1v_1} \rceil$. Now the worst-case number of multiplications required for this configuration can be directly obtained from the formula for the $h \times v$ configuration. by replacing a by $h_1v_1+h_2v_2$ and b by b_2 respectively, This results in $b_1v_1+b_2(v_2+1)-2$ multiplications in the worst case. Similarly, the expected number of multiplications can be shown to be $\frac{2^{h_1}-1}{2^{h_1}}b_1v_1 + \frac{2^{h_2}-1}{2^{h_2}}b_2v_2 + b_2 - 2$. This performance can be achieved with the storage for $(2^{h_1}-1)v_1+(2^{h_2}-1)v_2$ precomputed values. We can easily see that no configurations other than the two types, $h \times v$ and $h_1 \times v_1|h_2 \times v_2(h_2 = h_1+1)$, yield better performance for a given amount of storage.

The number of multiplications and storage requirements for a 512 bit modulus are summarized in tables B.1 and B.2 in the appendix B, for a 160-bit and 512-bit exponent respectively. Note that not only is the proposed method simpler, but it also achieves better performance than the BGMW method. In particular, due to its effectiveness over a wide range of storage, our method is flexibly applicable to various computing environments according to the amount of storage available. For example, to speed up the computation by smart cards, we may choose the configuration of 4×2. Then for 512 bit modulus and exponent the computation of g^R can be done in 182 multiplications on average with 1920 bytes of storage. On the other hand, when a relatively large amount of storage is available, we can choose, for example, the configuration of 7×4, achieving 90.42 multiplications on average with about 32 Kbytes of storage.

4 Speeding up Identification and Signature Verifications

Based on the discrete logarithm problem, a lot of identification and digital signature schemes have been developed (e.g., [10-12]). In all these schemes, along with a few modular multiplications the prover (or the signer) needs to compute g^R for a random R, which can be efficiently performed by the method described in section 3. On the other hand, to validate the prover's identity or the signature, the verifier needs to perform the computation of the form $g^R y^E$ where y corresponds to the public key of the prover (or the signer) and thus varies in each run of the protocol. The size of E typically lies between 20 and 40 in identification schemes and around 80 in the corresponding signature schemes. This section investigates the performance of the proposed method for computing $g^R y^E$.

Let t be the size of E. It is clear that if $t \leq b$, then $g^R y^E$ can be computed in $a+b+t-2$ multiplications in the worst case and $\frac{2^h-1}{2^h}a+b+0.5t-2$ on average. In case of $t > b$, we can either proceed as above or do the computation after partitioning E into smaller blocks. The first case yields the performance of $a+2t-2$ multiplications in the worst case and $\frac{2^h-1}{2^h}a+1.5t-2$ on average. However, if t is much larger than b, the performance can be further improved by dividing E into smaller blocks.

Thus, for more general formulae, suppose that E is partitioned into u blocks of almost equal size (Consider the whole configuration for computing $g^R y^E$ as $u \times 1|h \times v$). Let c be the bit-length of the partitioned blocks (i.e., $c = \lceil \frac{t}{u} \rceil$). Then, we first have to compute $y^{2^{kc}}$ for $k = 1, 2, \cdots, u-1$, and each product of their possible combinations, which all together requires $(u-1)c+2^u-u-1$ multiplications. For the range of t we are interested in, i.e., up to $t = 80$, u takes on at most 3. Now, if $c \leq b$, then at most c additional multiplications are sufficient in the worst case ($\frac{(2^u-1}{2^u}c$ on average). Therefore the total number of multiplications required in this case is $a+b+uc+2^u-u-3$ in the worst case and $\frac{2^h-1}{2^h}a+b+\frac{u2^u-1}{2^u}c+2^u-u-3$ on average. Similarly, for the case of $c > b$ we can easily show that the number of multiplications is $a+(u+1)c+2^u-u-3$ in the worst case and $\frac{2^h-1}{2^h}a+\frac{(u+1)2^u-1}{2^u}c+2^u-u-3$ on average.

With the proposed method, the Schnorr-like identification and/or signature schemes can be made more practical for smart card implementations. For example, with a 512-bit modulus, 160-bit exponents and $t = 30$, the verification condition can be checked in 80.5 multiplications on average, if 1920 bytes of storage are available (4×2 configuration). Similarly, a signature with $t = 80$ can be verified in 144.13 multiplications on average using the same amount of storage. This is a considerable speedup only with a very small amount of storage, compared with the binary method requiring 246.5 multiplications for $t = 30$ and 259.0 multiplications for $t = 80$ on average. Moreover, identification or signature verifications are usually performed in much more powerful terminals capable of being equipped with a large amount of memory. In such an environment, we may adopt the 8×2 configuration and thus can perform, on average, identity verifications in 60.2 multiplications for $t = 30$ and signature verifications in 126.6 multiplications for $t = 80$, using about 32 Kbytes of storage.

Further improvement with additional communication : Small additional communication can considerably reduce the number of multiplications for computing $g^R y^E$ again. That is, the verifier can save the on-line computational load for preparing $y_k = y^{2^{kc}}$ for $k = 1, 2, \cdots, u-1$, if they are precomputed and stored by the signer (or the prover), since y is a fixed number to him, and then transmitted together with other data. For example, for the signature scheme with $t = 80$, if the signer sends 2 additional 512 bit blocks y_1, y_2 where $y_1 = y^{2^{27}}$ and $y_2 = y_1^{2^{27}}$, together with a signature for message, then the signature verification can be done in 90.13 multiplications on average with the 4×2 configuration. Therefore, 54 multiplications can be saved only with the increase of 128 bytes of communication. This corresponds to about a 3-fold speedup on average over the binary method which requires 259 multiplications on average.

For comparison, it is worth mentioning that the BGMW method is less efficient for the computation of the form $g^R y^E$ in either case we considered above. In case of no additional communication, if the exponents are represented in non-binary power base, more computations are needed in performing the on-line precomputation required for y^E. When addtional communication is allowed, more precomputed values must be transmitted due to the use of small base.

The above method of combining precomputation and additional communication can be used to speed up the verification of the digital signature standard (DSS) [14] as well. In DSS, we have to perform the computation of the type $g^R y^E$ with $|R| = |E| = 160$ and thus without additional communication we can gain no advantage with precomputation. However, if the signer sends 3 additional blocks $\{y_1, y_2, y_3\}$ where $y_i = y_{i-1}^{2^{40}}$ for $i = 1, 2, 3$ and if the verifier adopts the 4×2 configuration, then the signature can be verified in 124 multiplications on average. This is more than a 2-fold improvement over the binary method which requires 279 multiplications on average, only with 1920 bytes of storage and 192 bytes of additional communication (for a 512 bit modulus).

signature generation				
schemes	n / t	binary	4 × 2	8 × 2
Schnorr	160 / 80	318 / 238	58 / 55.5	28 / 27.9
DSS	160 / 160	318 / 238	58 / 55.5	28 / 27.9
BM	512 / 80	1022 / 766	190 / 182.0	94 / 93.8

signature verification				
schemes	n / t	binary	4 × 2	8 × 2
Schnorr	160 / 80	319 / 259	89 / 85.3	69 / 67.7
DSS	160 / 160	319 / 279	129 / 124.0	109 / 106.4
BM	512 / 80	1023 / 787	221 / 211.8	125 / 123.5

Table 1 : Worst/average performance w/ 3 block additional commun.

The table 1 shows the number of multiplications required for signature generation and verification in three signature schemes (Schnorr [10], DSS [14] and

Brickell-McCurley [11]), under the assumption that the signer sends additionally 3 precomputed values for his public key together with a signature, as mentioned above. Here we only take into account the number of multiplications for exponentiation operations, neglecting some other necessary operations such as reduction mod q and multiplicative inverse mod q where q is a prime of about 160 bit size. Two configurations of 4×2 and 8×2 are taken as examples, since the former is suitable for smart card applications and the latter for more general applications with a relatively large amount of storage available. For comparison, the performance of the binary method is also presented.

5 Parallel Processing

With multiple processors, the proposed method can be parallelized, much more efficiently than the BGMW method, by assigning the j-th processor to the j-th column of the $h \times v$ configuration (see figure 1). That is, if v processors are available, then the j-th processor can be assigned to compute

$$\prod_{k=0}^{b-1} (G[j][I_{j,k}])^{2^k} \tag{9}$$

in the expression of

$$g^R = \prod_{j=0}^{v-1} \prod_{k=0}^{b-1} (G[j][I_{j,k}])^{2^k}, \tag{10}$$

where we assume that each processor stores in its local memory $2^h - 1$ precomputed values. The computation of each processor can be completed in at most $2(b-1)$ multiplications. After then, we need $\lceil log_2 v \rceil$ multiplications in addition to produce the final result. Therefore, the total number of multiplications is $2(b-1) + \lceil log_2 v \rceil$.

Table B.3 in the appendix B shows the required number of multiplications for 160/512 bit exponents, according to the number of processors and the storage needed per processor. Note that only with a small number of processors the performance can be greatly improved. For example, for 512-bit modulus and exponent, we can compute g^R in 32 multiplications , when 4 processors are available and each processor has a local storage for 255 precomputed values (about 16 Kbytes). With more processors, say 16, the exponentiation can be done in 10 multiplications with the same storage requirement.

The described parallel computation can be more efficiently implemented by a special-purpose hardware. For example, with 8 pairs of multiply and squaring circuits together with read-only memory for 120 precomputed values (4×8 configuration), we can compute g^R with $|R| = 512$ in 18 multiplications and 15 squarings. If we use storage for 2040 values (8×8 configuration) with the same circuits, then the computation can be done in 10 multiplications and 7 squarings.

6 Conclusion

We have proposed a new method for fast exponentiation with precomputation. The proposed method is very simple but achieves better performance than the BGMW method [9]. Our method is also preferable since it is flexibly applicable to various computing environments due to its wide range of time-storage tradeoffs. In particular, using the proposed method, we can substantially speed up the computation by smart cards with only a very small amount of storage. We also showed that the proposed method can also speed up the computation of the form $g^R y^E$ with y variable. This can make much more practical the Schnorr-type identification and signature scheme, since the verifier as well as the prover (signer) can gain great computational advantage with a moderate amount of storage. Finally we presented how the proposed algorithm can be parallelized. Such parallel processing may be useful in high performance server machines with multiple processors.

References

1. D.E.Knuth, *The art of computer programming, Vol.2 : Seminumerical algorithms*, second Edition, Addison-Wesley (1981).
2. J.Jedwab and C.J.Mitchell, "Minimum weight modified signed-digit representations and fast exponentiation," *Elect. Let.* 25 (17), 1171-1172 (1989).
3. C.N.Zhang, "An improved binary algorithm for RSA," *Computers Math. Applic.* 25 (6), 15-24 (1993).
4. J.Bos and M.Coster, "Addition chain heuristics," In *Advances in Cryptoloy-Crypto'89, Lecture Notes in Computer Science 435*, (edited by G.Brassard), pp.400-407, Springer-Verlag (1990).
5. J.Sauerbrey and A.Dietel, "Resource requirements for the application of addition chains modulo exponentiation," In *Proc. Eurocrypt'92*, Balatonfured, Hungary (1992).
6. P.Downey, B.Leony and R.Sethi, "Computing sequences with addition chains," *Siam J. Comput.* 3, 638-696 (1981).
7. R.L.Rivest, A.Shamir and L.Adleman, "A method for obtaining digital signatures and public-key cryptosystems," *Commun. ACM*, 21 (2), 120-126 (1978).
8. T.ElGmal, "A public key cryptosystem and a signature scheme based on the discrete logarithm," *IEEE Trans. Inform. Theory* 31 (4), 469-472 (1985).
9. E.F.Brickell, D.M.Gordon, K.S.McCurley and D.Wilson, "Fast exponentiation with precomputation," In *Proc. Eurocrypt'92*, Balatonfured, Hungary (1992).
10. C.P.Schnorr, "Efficient signature generation by smart cards," *J. Cryptology* 4 (3), 161-174 (1991).
11. E.F.Brickell and K.S.McCurley, "An interactive identification scheme based on discrete logarithms and factoring," *J. Cryptology* 5 (1) 29-39, (1992).
12. T.Okamoto, "Provably secure and practical identification schemes and corresponding signature schemes," In *Proc. Crypto'92*, Santa Barbara, CA (1992).
13. D.W.Matula, "Basic digit sets for radix representation," *J.ACM* 29, 1131-1143 (1982).
14. A proposed Federal information processing standard for digital signature standard (DSS), *Federal Register* 56 (169), 42980-42982 (1991).

A The Performance of the BGMW Method

Table A.1 : Selected parameters for a 160-bit exponent

b	M	h	storage	worst / average
13	$\{1\}$	12	45	54 / 50.25
19	$\{\pm1\}$	9	76	45 / 43.00
29	$\{\pm1, \pm2\}$	9	134	41 / 39.83
36	$\{\pm1, 9, \pm14, \pm17\}$	7	219	37 / 36.11
36	M_3	3	620	32 /31.14
64	M_2	2	1134	27 / 26.58
128	M_3	3	1748	24 / 23.82
256	M_2	2	2751	21 / 20.92

Table A.2 : Selected parameters for a 512-bit exponent

b	M	h	storage	worst / average
26	$\{1\}$	25	109	132 / 127.81
45	$\{\pm1\}$	22	188	114 / 111.91
53	$\{\pm1, \pm2\}$	17	362	106 / 104.28
67	$\{\pm1, \pm2, \pm23\}$	16	512	100 / 98.72
64	M_3	3	3096	87 / 85.66
122	M_3	3	5402	75 / 74.39
256	M_2	2	10880	64 / 63.75

B The Performance of the Proposed Method

Table B.1 : Selected parameters for a 160 bit exponent

configuration	storage	worst / average
2×2	6	118 / 98.00
$2 \times 1 \| 3 \times 1$	10	94 / 82.00
3×2	14	79 / 72.25
$3 \times 1 \| 4 \times 1$	22	67 / 62.69
4×2	30	58 / 55.50
4×3	45	52 / 49.50
5×2	62	46 / 45.00
$4 \times 1 \| 5 \times 2$	77	44 / 42.63
5×3	93	41 / 40.00
5×4	124	38 / 37.00
$5 \times 1 \| 6 \times 2$	157	36 / 35.44
6×3	189	34 / 33.58
6×4	252	32 / 31.58
$6 \times 1 \| 7 \times 2$	317	30 / 29.75
7×3	381	29 / 28.82
7×4	508	27 / 26.82
7×6	762	25 / 24.82
8×4	1020	23 / 22.92
8×7	1785	21 / 20.92
$8 \times 1 \| 9 \times 4$	2299	20 / 19.96

Table B.3 : Performance on parallel processing ($|R| = 160/512$)

np \ sp	15	31	63	127	255	511	1023
2	39 / 127	31 /103	27 / 85	23 / 73	19 / 63	17 / 57	15 / 51
4	20 / 64	16 / 53	14 / 44	12 / 38	10 / 32	- / 30	8 / 26
6	15 / 45	13 / 37	11 / 31	9 / 27	- / 23	7 / 21	- / 19
8	11 / 33	9 / 27	- / 23	7 / 21	- / 17	- / -	5 / 15
16	8 / 18	6 / 16	- / 14	- / 12	- / 10	- / -	4 / -
32	7 / 11	5 / -	- / 9	- / -	- / 7	- / -	- / -

* np = no. of processors (v), sp = storage / processor ($2^h - 1$)

Table B.2 : Selected parameters for a 512 bit exponent

configuration	storage	worst / average
2×2	6	382 / 318.00
$2 \times 1 \| 3 \times 1$	10	306 / 267.63
3×2	14	255 / 233.63
$3 \times 1 \| 4 \times 1$	22	218 / 204.38
4×2	30	190 / 182.00
4×3	45	169 / 161.00
5×2	62	153 / 149.78
5×3	93	136 / 132.78
$5 \times 2 \| 6 \times 1$	125	126 / 123.50
$5 \times 1 \| 6 \times 2$	157	106 / 104.66
6×3	189	113 / 111.66
6×4	252	106 / 104.66
$6 \times 1 \| 7 \times 2$	317	101 / 100.20
7×3	381	97 / 96.42
7×4	508	91 / 90.42
7×5	635	87 / 86.42
$7 \times 1 \| 8 \times 3$	892	81 / 80.68
8×4	1020	78 / 77.75
8×5	1275	75 / 74.75
8×6	1530	73 / 72.75
8×8	2040	70 / 69.75
9×5	2555	67 / 66.89
9×6	3066	65 / 64.89
$9 \times 6 \| 10 \times 1$	4089	62 / 61.90
$9 \times 1 \| 10 \times 5$	5626	59 / 58.94
$9 \times 1 \| 10 \times 7$	7672	57 / 56.95
$10 \times 2 \| 11 \times 4$	10234	54 / 53.97
$10 \times 2 \| 11 \times 6$	13305	52 / 51.98

A Parallel Permutation Multiplier for a PGM Crypto-chip

Tamás Horváth[1], Spyros S. Magliveras[2] and Tran van Trung[1]

[1] Institute for Experimental Mathematics, University of Essen, Ellernstrasse 29, 45326 Essen, Germany
[2] Department of Computer Science and Engineering, University of Nebraska, Lincoln NE, 68588-0115, U.S.A.

Abstract. A symmetric key cryptosystem, called PGM, based on *logarithmic signatures* for finite permutation groups was invented by S. Magliveras in the late 1970's. PGM is intended to be used in cryptosystems with high data rates. This requires exploitation of the potential parallelism in composition of permutations. As a first step towards a full VLSI implementation, a parallel multiplier has been designed and implemented on an FPGA (Field Programmable Gate Array) chip. The chip works as a co-processor in a DSP system. This paper explains the principles of the architecture, reports about implementation details and concludes by giving an estimate of the expected performance in VLSI.

1 Introduction

A symmetric key cryptosystem PGM based on *logarithmic signatures* for finite permutation groups was invented by by S. Magliveras in the late 1970's. The system was described in [1], and its statistical and algebraic properties were studied in [2], [3], [4]. Recent significant results have been obtained on closely related material by S.A. Vanstone and by M. Qu [5]. Here we include only a short description of PGM.

Let G be a finite permutation group of degree n. A *logarithmic signature* for G is an ordered collection $\alpha = \{A_i : i = 1, \ldots, s\}$ of ordered subsets, $A_i = \{a_{i,0}, \ldots, a_{i,r_i-1}\}$ of G, such that each element $g \in G$ can be expressed uniquely as a product of the form

$$g = q_s \cdot q_{s-1} \cdots q_2 \cdot q_1 \qquad q_i \in A_i \tag{1}$$

The A_i are called the *blocks* of α and the vector of block lengths $(r_1, \ldots, r_s)$ is called the *type* of α. A logarithmic signature is called *tame* if the factorization in equation (1) can be achieved in time polynomial in the degree n of G; it is called *supertame* if the factorization can be achieved in time $O(n^2)$. A logarithmic signature is called *wild* if it is not tame. In [4] the authors describe how a logarithmic signature α induces an efficiently computable bijection $\hat{\alpha} : Z_{|G|} \rightarrow G$. The inverse of $\hat{\alpha}$ is efficiently computable only if α is tame. *Basic* system PGM is described as follows: For a given pair of tame logarithmic signatures, α, β, the *encryption* transformation $E_{\alpha,\beta}$ is the mapping $E_{\alpha,\beta} = \hat{\alpha}\hat{\beta}^{-1} : Z_{|G|} \rightarrow Z_{|G|}$.

The corresponding *decryption* transformation is obtained by reversing the order of the pair of logarithmic signatures, i.e. $D_{\alpha,\beta} = E^{-1}_{\alpha,\beta} = E_{\beta,\alpha} = \hat{\beta}\hat{\alpha}^{-1}$.

To effect the fastest possible PGM encryption and decryption operations, one must compute efficiently products of permutations as in equation (1). Unlike multiplication of integers, composition of two permutations is inherently parallelizable. Hence, we can achieve fast computation of $\hat{\alpha}$ and its inverse by designing a permutation multiplier which takes advantage of this property of permutation composition. In this paper we describe a design for such a permutation multiplier, as a first step towards a full VLSI implementation of PGM.

2 Principles of multiplication in parallel

For easy understanding, we shall explain the principles by means of a simple example. We consider permutations of degree 4, and represent them in *cartesian* form, $\pi = [\pi(0), \pi(1), \pi(2), \pi(3)]$. This form is particularly convenient for representing permutations in hardware, where a vector register of length n is used to represent a permutation of degree n. For example, $\pi = [3, 2, 1, 0]$ is our notation for the permutation $\pi = (0\ 3)(1\ 2)$ as the product of disjoint cycles. In general, this representation needs $nlog_2n$ bits to store a permutation of degree n. Throughout the example, we define five input operands to work with, namely: $\iota = [0, 1, 2, 3]$ (the identity permutation), $\alpha = [2, 3, 0, 1]$, $\beta = [1, 3, 2, 0]$, $\gamma = [3, 2, 0, 1]$ and $\delta = [1, 2, 3, 0]$.

The multiplication unit is in essence a crossbar switching network. A 4x4 switching matrix is depicted in Figure 1. The matrix has three input ports, labeled A, B and C respectively, and one output port named Q. Ports B and C are connected to the vertical lines in the matrix, whereas A and Q to the horizontal lines. At the cross-points of vertical and horizontal lines reside the *switching cells*, each consisting of a *transfer gate* and *cell-logic*. The cell-logic controls the transfer gate. If the gate is open, which is denoted by a dot ($\bullet$) in the Figures, it allows the signal to propagate from the vertical line onto the corresponding horizontal line. A closed gate does not influence the signal on the horizontal line. Multiplication takes place in two phases:

(a) In the first, so-called *setup phase* the contents of A and B appear on the horizontal and vertical lines respectively. At each cross-point, the corresponding cell-logic unit compares the horizontal input to the vertical input and saves the result of the comparison: 1 in case of a match, and 0 otherwise. (See Figure 1(a))

(b) In the second, so-called *pass-through phase*, the transfer gates at cells where a match was found, open. The C operand is placed onto the vertical lines and is transferred via the open gates onto port Q as the result. (See Figure 1(b))

It is now relatively easy to see that the result Q can be expressed in terms of the other operands as $Q = (A \circ B^{-1}) \circ C$, where $\circ$ denotes *composition* of permutations. We verify the result Q when $A = \alpha = [2, 3, 0, 1]$, $B = \beta = [1, 3, 2, 0]$ and $C = \gamma = [3, 2, 0, 1]$ as in Figure 1: $([2, 3, 0, 1] \circ [1, 3, 2, 0]^{-1}) \circ [3, 2, 0, 1] = [2, 3, 0, 1] \circ [3, 0, 2, 1] \circ [3, 2, 0, 1] = [0, 2, 1, 3]$.

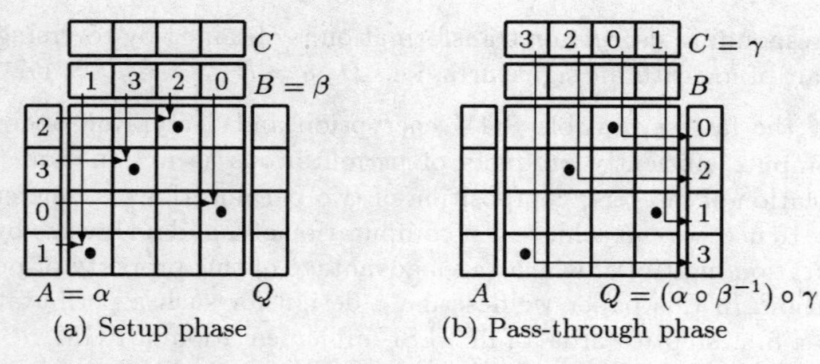

(a) Setup phase (b) Pass-through phase

Fig. 1. Principle of multiplication in the switching matrix.

We remark here that the partial product $\pi = \alpha \circ \beta^{-1}$ is implicitly stored in the state of the transfer gates, and can be retrieved by passing $C = \iota$ through the matrix. Furthermore, it is possible to compute several products with the same first operand π, without setting up the matrix again. This kind of operation we call *continuous mode*. By dedicating separate lines to A, B, C and Q respectively, it becomes possible to overlap in time the pass-through phase of a multiplication and the setup phase of the next one. This two-stage *pipelining* is shown in Figure 2. The state of the gates is always changed at the end of the phases, thus pass-through operations can take place using the previous setup.

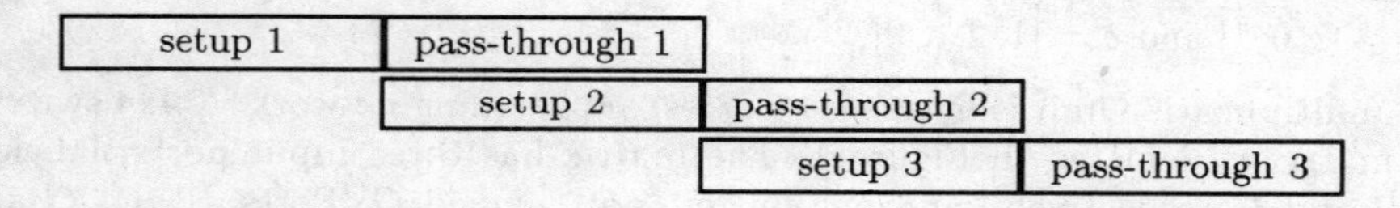

Fig. 2. Pipelining setup and pass-through stages.

A particular case of the pipelined operation is used in the implementation. The contents of port Q are fed back as input to port A, within the same phase. By letting $B = \iota$ constantly, the matrix evaluates products of the form $\pi_1 \circ \pi_2 \ldots \circ \pi_n$ in exactly n cycles, i.e. without losing cycles, by merely loading back partial products as input operands. This mode of operation is called *feedback mode*. The computation of $\alpha \circ \gamma \circ \delta = [2,3,0,1] \circ [3,2,0,1] \circ [1,2,3,0] = [1,2,0,3]$ can be followed in Figure 3.

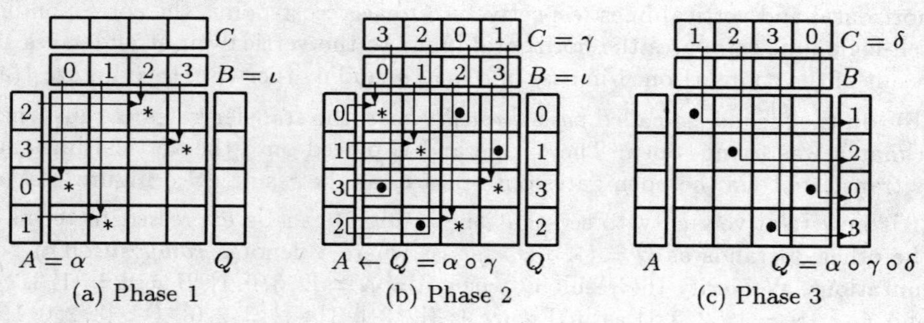

(a) Phase 1 (b) Phase 2 (c) Phase 3

Fig. 3. Feeding back Q to A to compute $\alpha \circ \gamma \circ \delta$

3 Implementation details

As a first step towards a VLSI implementation of PGM, a hybrid hardware-software prototype has been developed based on a Texas Instruments 320C30 DSP processor. Multiplication of permutations is effected in the permutation co-processor chip, which is connected to the DSP system via a 16 bit peripheral bus, called DSPLINK. The DSP accesses the co-processor through I/O instructions. The co-processor is an XC3190 FPGA (*Field Programmable Gate Array*), a product of the Xilinx Corporation. The FPGA is a perfect prototyping tool, in view of the flexibility it affords for design changes. However, the achievable complexity is rather low, only a few thousand gate equivalents. This constraint limits the degree n of permutations that are processed on the chip to $n = 16$.

In order to be able to carry out one setup or pass-through operation in each cycle, the operands have to be led through the crossbar network in parallel, i.e. needing $log_2 n$ lines per operand. For practical applications n should be at least 32, requiring thus at least $5.2^5 = 160$ lines. Although a fully parallel implementation may still be feasible on a VLSI chip, we follow a different approach. The vectors of first, second, etc. bits of the n elements in the permutation are sent through the crossbar serially, in $log_2 n$ cycles. This principle reduces dramatically the total number of lines needed, the complexity of the cells, and hence the overall chip area. Due to shorter lines, propagation delays shorten considerably, too. We estimate the performance of a serialized multiplier to be about 50% that of a fully parallel one. This seems to be a good trade-off between price and speed.

Let us now take a closer look at the FPGA multiplier. The circuitry belonging to one cell is depicted in Figure 4. As a convention, the vector of least significant bits (LSBs) is processed first, followed by the other bit layers in order of significance. The cells function as follows:

- Essentially, the XOR gate compares the corresponding bits of the operands A and B, received from the neighboring horizontal and, respective vertical lines. The result of a bit comparison is AND-ed with the accumulated result of previous comparisons, and is reflected by the state of ACCUFF. The output of the AND function becomes the new accumulated result, and is written into ACCUFF at the end of the cycle due to a low-high transition on the global ACCU clock net.

- During the first cycle of the setup phase a global signal, called INIT, is activated. This makes the cells ignore the accumulated result, and simply enter the output of the XOR gate into ACCUFF.

- At the end of the last cycle a transition occurs in FIRE, the second global clock net, which causes the final result to be entered into FIREFF. The output of this flip-flop controls GATE, a tri-state buffer, thus a new setup also comes into effect at this moment.

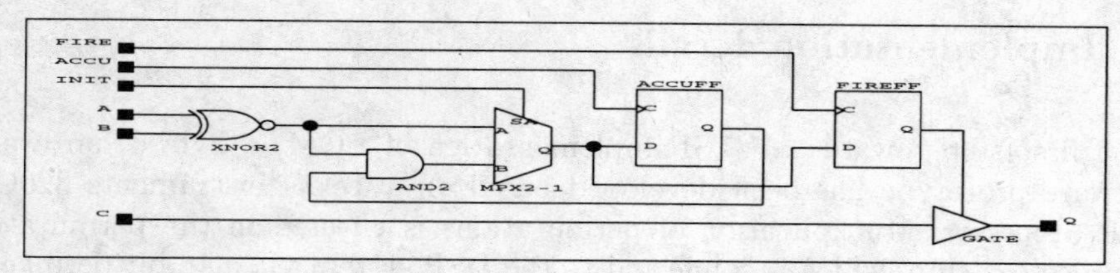

Fig. 4. The logic scheme of a cell.

The data-path of the permutation chip, reduced to 4 bit vector length, can be seen in Figure 5. Ports A and Q of the multiplier array are unified as port AQ on the left edge. Similarly, ports B and C are fused to form port BC on the top edge. The external pins of AQ and BC are also connected together on the embedding card to form one data bus for connecting to DSPLINK. The identity operand I is *hard-wired* on the chip in units ICODE$_n$ All signals controlling the ports and cells, are generated in unit CTRLOGIC.

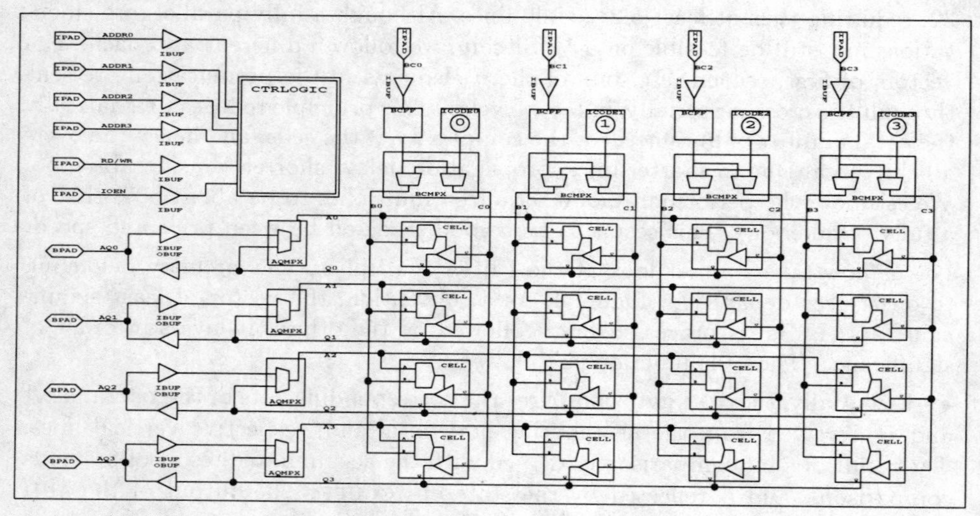

Fig. 5. The data path of the chip, reduced to vector length 4.

The 4 bit address bus of DSPLINK presents the instruction code to the chip, thus instruction and data are transferred at the same time. An instruction set of 6 elements has been defined to control the assembly. Because of space limitations we can not go into their semantics here.

4 Conclusions

A permutation multiplier chip has been developed, verified by simulation, attached to a DSP system and successfully tested by means of a simple DSP program. The processing speed is satisfactory, 100ns for one cycle.
In our implementation the degree of processed permutations was set to $n = 16$.

This is of course too small for practical applications. Nevertheless, we consider this prototyping work an important step towards a full VLSI implementation of PGM. Our multiplier architecture can be easily extended to larger n, and be quickly transferred to larger scale technology.

For future work we plan to complete the DSP implementation so as to gain more insight into the actual processing and storage requirements of the PGM algorithm. Afterwards we intend to augment the permutation matrix with other hardware units to embrace the entire algorithm with fast, special-purpose hardware.

References

1. S. S. Magliveras, A cryptosystem from logarithmic signatures of finite groups, In *Proceedings of the 29'th Midwest Symposium on Circuits and Systems*, Elsevier Publishing Company (1986), pp 972–975.
2. S. S. Magliveras and N. D. Memon, The Linear Complexity Profile of Cryptosystem PGM, *Congressus Numerantium*, Utilitas Mathematica, **72** (1989), pp 51–60.
3. S. S. Magliveras, N. D. Memon and K.C. Tam, Complexity tests for cryptosystem PGM, *Congressus Numerantium*, Utilitas Mathematica, **79** (1990), pp 61–68.
4. S. S. Magliveras and N. D. Memon, Algebraic Properties of Cryptosystem PGM, in *Journal of Cryptology*, **5** (1992), pp 167 –183.
5. M. Qu and S. A. Vanstone, Factorizations of elementary Abelian p-groups and their cryptographic significance, to appear in J. of Cryptology.

Cryptographic Randomness from Air Turbulence in Disk Drives

Don Davis,[1] Ross Ihaka,[2] and Philip Fenstermacher[3] *

[1] Openvision Technologies, 1 Main St. Cambridge, MA 02142
[2] University of Auckland, Mathematics Dept, Auckland, NZ
[3] 18A Forest St. Cambridge, MA 02138

Abstract. A computer disk drive's motor speed varies slightly but irregularly, principally because of air turbulence inside the disk's enclosure. The unpredictability of turbulence is well-understood mathematically; it reduces not to computational complexity, but to information losses. By timing disk accesses, a program can efficiently extract at least 100 independent, unbiased bits per minute, at no hardware cost. This paper has three parts: a mathematical argument tracing our RNG's randomness to a formal definition of turbulence's unpredictability, a novel use of the FFT as an unbiasing algorithm, and a "sanity check" data analysis.

1 Introduction

Secure PRNG design commonly rests on computational complexity [2, 5, 6, 13, 24], but none of the underlying problems has been proven to be hard. Specialized hardware can provide naturally random physical noise, but has disadvantages: dedicated devices tend to be expensive; natural noise tends to be biased and correlated; hardware failure can silently suppress randomness; and physical randomness is only an article of faith. Our random number generator, which is based on disk-speed variations,[4] addresses each of these problems. Timing data are very low-cost, easily whitened, reliable, and mathematically noisy.

I/O randomness is well-known in cryptography [17], and a spinning-disk RNG was used even 50 years ago [11]. Still, our approach is subtly novel, because a disk drive combines three important features most economically. First, the OS detects and reports disk faults, so that silent randomness failures are unlikely. Second, unlike most other I/O devices, the disk can be secured from outside influence and measurement. Last, nonlinear dynamics gives us an *a priori* mathematical argument for our generator's randomness. This has not been possible for other noise sources, which rely on *a posteriori* statistical measurements.

This paper has three parts. First, we trace the disk's speed-variations to air turbulence, and we show why these variations can show only short-term correlations. Second, we show that the FFT removes bias and correlations from the disk's timing-data. Third, we describe our "sanity check" analysis of some timing-data and the resulting random numbers.

---

* Affiliations during this work: MIT Project Athena, MIT Stat. Ctr., MIT LCS, resp.

[4] Disk drives use brushless DC motors [10, 18], so these speed variations are independent of the AC line frequency.

2 Turbulence in Disk Drives

In this section, we review studies demonstrating turbulent air flow in disk drives. Oversized mockups have clarified the various turbulent flow regimes inside a disk drive, and have shown how rotational speed, disk spacing, and cooling flow affect the flow [1]. The apparatus was a stack of 1-meter glass disks, spun in a water tank at low speeds (5 – 60 rpm). A close-fitting cylindrical shroud enclosed the disks, and 50 cm.-diameter hubs separated them, to closely model typical modern disk drives of various sizes. Dye and bubbles made the flows visible, and a rotating video camera recorded the results. Our disk's speed and configuration were similar to those studied.

Turbulence arose at the read/write heads and their support arms, in Coriolis circulation between the disk surfaces, and in Taylor-Couette flow at the disks' rims. Crucially, the T-C flow pumped turbulence into the Coriolis flow. Numerical simulation of disk flow showed similar turbulence patterns [21], and yielded an estimate of 3% for the consequent fluctuations in the windage torque. This is clearly large enough to influence the disk's speed.

Spectral measurements of the fluid velocity showed both sharp peaks and broadband features [1], reflecting *weak turbulence*: very noisy motion with a periodic component [3]. The spectra were taken at various rotational speeds, but the peaks always contained only a small proportion of the spectral power, rising only a factor of 2–3 above the white-noise background. This broadband spectral component was maximized at Reynolds numbers near those commonly found in disk drives.

The classic Taylor-Couette (T-C) flow experiment models a disk drive's dominant turbulence pretty well: a tall cylinder spins inside a fluid-filled glass sleeve, which displays the fluid's toroidal convection. Laser-Döppler velocimetry experiments have precisely measured the fluctuating convection in T-C flow [9]. The flow changed from periodic to quasiperiodic, and then *abruptly* to weakly turbulent, as the fluid's velocity was gradually increased. This development was consistent with a formal model of weak turbulence in simple quasiperiodic systems:

Theorem 1 (Newhouse, Ruelle, Takens, 1978 [14]). *"Let v be a constant vector field on the torus $T^n = \mathbb{R}^n/\mathbb{Z}^n$. If $n \geq 3$, every C^2 neighborhood of v contains a vector field v' with a strange Axiom A attractor. If $n \geq 4$, we may take C^∞ instead of C^2."*

Here, the torus T^n does not represent the toroidal T-C vortices directly, but is a simple dynamical system's phase space. "Axiom A"[5] refers to a formal definition of dynamical systems that show a close mixture of periodic and turbulent behavior. (For the definition of Axiom A flows and attractors, see [15].)

[5] The finite-dimensional Axiom A formalism can't apply *directly* to Navier-Stokes infinite-dimensional phase flows. The machinery of inertial manifolds, though, has shown that bounded-velocity flows have finite-dimensional attractors [20].

Even weak turbulence is sufficiently random for our purposes, because it shows sensitive dependence on initial conditions (SDIC). (Theorem 1's Axiom A result implies SDIC.) Somewhat formally, a phase-space flow $f_t : S \longrightarrow S$ has the SDIC property if $\exists$ an attractor $A \subset S$ s.t. $\forall x \in A$, $\forall$ small $U \ni x$, the diameter of $f_t(U)$ increases exponentially with time [15]. Informally, to completely forecast a system that shows SDIC, we must specify its parameters and initial conditions with infinite precision; measurement limitations limit the forecast to short-term accuracy [3]. Thus, *it is not computational complexity, but information losses in measurement, that prevent effective prediction in such physical systems.*

3 Converting Access-Times to Random Numbers

In Section 2, we showed that the disk's speed variations show sensitive dependence on initial conditions (SDIC). Even so, the disk's access-times are still strongly structured, biased, and correlated, so they clearly cannot directly simulate a tossed coin. Some solutions to this problem [4, 23] make assumptions that don't fit our noise source. Semi-randomness [16, 22, 8] successfully formalizes imperfect randomness, but is restrictively pessimistic because it requires two sources. This motivates our use of the FFT. In this part of our paper, we show that the FFT is a good unbiasing algorithm, and that SDIC justifies this use of the FFT. In presenting these results we follow Brillinger [7].

Let $(X_1, \ldots, X_k)$ be a vector of random variables. The *joint cumulant* of kth order, $cum\,(X_1, \ldots, X_k)$, is defined as the coefficient of $i^k t_1 \ldots t_k$ in the Taylor series expansion of the logarithm of the characteristic function of $(X_1, \ldots, X_k)$ about the origin. For a stationary time-series X_t with $E\,|X_t|^k < \infty$, we define the *joint cumulant function* of order k to be

$$c_{X \ldots X}(t_1, \ldots, t_{k-1}) = cum(X_{t_1}, \ldots, X_{t_{k-1}}, X_0) \tag{1}$$

The cumulant is thus the mean of a polynomial function of k staggered copies of the time-series X_t, and its parameters t_i describe the copies' offsets. The cumulants represent the dependencies present in X_t. The requirement that these dependencies fall off over time is known as a *mixing* condition. Our use of the FFT as an unbiasing algorithm rests on the following mixing assumption.

Assumption 2. *The time series X_t possesses moments of all orders and its cumulant functions satisfy*

$$\sum_{t_1} \cdots \sum_{t_{k-1}} |c_{X \ldots X}(t_1, \ldots, t_{k-1})| < \infty. \tag{2}$$

In our case, the disk-timing data X_t are bounded, so the cumulants do exist.

Define the *power spectrum* and the *finite Fourier transform*, respectively, as

$$f_{XX}(\lambda) = \sum_{t=-\infty}^{\infty} |c_{XX}(t)|\, e^{-i\lambda t} < \infty \tag{3}$$

$$d_V^T(\lambda) = \sum_{t=0}^{T-1} X_t e^{-i\lambda t} \tag{4}$$

where V denotes a T-vector $(X_0, \ldots, X_{T-1})$. Then

Theorem 3 (Brillinger, 1981 [7]). *Let X_t be a stationary time series which satisfies Assumption 2, and let $\lambda_1, \ldots, \lambda_k$ be distinct values in the interval $[0, 2\pi]$ s.t. $\lambda_j \neq 0, \pi, 2\pi$. Then as $T \to \infty$, the values $d_V^T(\lambda_j)$ converge asymptotically to independent (complex) normal random variables with mean 0 and variances equal to $2\pi T f_{XX}(\lambda_j)$.*

As an immediate consequence it is clear that the phase angles $\phi_j = arg\left(d_V^T(\lambda_j)\right)$ are asymptotically independent and uniformly distributed on the interval $[0, 2\pi]$.

Theorem 3 is a generalization of the Law of Large Numbers. Like the L.L.N., it grants perfect normality only in the limit as $T \to \infty$, so the spectral distributions are only approximately normal. Another price of the theorem's generality is that the output distribution converges only pointwise to the desired joint normal. Lacking convergence-rates, we must measure how well the spectra approach normality, so that we can choose a practical spectrum-length T.

To feed the FFT, we filtered and decimated our raw disk-periods to remove some obvious quantization structure. We discarded the FFT's predictable spectral lines at $\lambda_0, \lambda_{T/2}, \lambda_{T/2+1}, \ldots, \lambda_T$. Note that if we take from this algorithm more bits than we feed into it, we run the risk that the mapping $arg(d_{()}^T) : V \to [0, 2\pi)^{T/2}$ can be inverted, even if we only keep part of the spectrum as random numbers.

We claim that disk access-times satisfy Assumption 2. No statistical test can justify such a claim, so we will argue instead that the cumulants' decay follows from SDIC. The N.R.T. theorem ensures exponentially-damped autocorrelation [14], but says nothing about higher-order correlations.

The mixing condition is formally very different from SDIC, but their meanings are similar: loosely, both put limits on how measurements can aid prediction of the system's long-term behavior:

- SDIC means that *measured* initial conditions are insufficient;
- mixing means polynomial functions of past measurements are insufficient.

To see that SDIC implies our mixing condition, suppose that a cumulant of order k fails to decay, so that arbitrarily long-term dependencies can exist among k measurements of the time series. Then $n < k$ early measurements can suffice to predict some function of $k - n$ later measurements' values, for some arbitrarily large separations between the two groups. The n early samples, though, all have limited accuracy, and cannot specify the underlying system's state with infinite precision. Thus, we've contradicted the SDIC property: the first n samples specify a range of initial conditions, whose consequences are recurrently and significantly parallel.

4 Statistical Analysis

In this section we summarize two data analyses, as "sanity checks." We tested:

- our measured access-times, to ensure that noise was present.
- the RNG's product, for several deviations from randomness.

We call these measurements sanity checks, because our argument for the disk's value as a noise-source actually rests on the mathematical properties of the disk's air turbulence, and not on our observations. These tests' failure would have disproven our claim that the disk's motion reflects turbulent flow.

For our measurements, we used an IBM RT/PC desktop workstation and a Micropolis 1320 series 40 Mb hard disk with nonremovable 5.25 inch media. A permanent-magnet brushless DC motor turns the disk spindle at a nominal rate of 3600 r.p.m. The motor's phase-locked loop stabilizes the rate to $\pm$ 0.03%, which amounts to a positional accuracy of 5 μsec. [19].

The workstation's operating system was MIT Project Athena's port of 4.3bsd UNIX, with machine-dependent routines from the IBM Academic Information Systems release. For our tests, we debugged the UNIX kernel's `microtime()` subroutine, and we modified the disk-scheduler software to record the time at each disk-access' initiation and completion. We were particularly careful to avoid disturbing the spindle's speed with head motion. During experimental sessions, the workstation ran "standalone," isolated from the MIT network, with no time-synchronization software or other inessential processes running. Sessions lasted from 30 minutes to 8 hours. To measure disk-speed fluctuations, we repeatedly read a chosen disk block, and recorded each access-completion time. This entails so little software overhead that we could read the block on every rotation, so the completion-time differences gave a running account of the disk's period. The RT's 1024 Hz. hardware clock limited our measurement precision to $\sim$1 msec.

Our measurements were consistent with the 5 μsec variation. We considered a variety of influences whose timing effects might resemble rotational latency:

- delays within the disk controller,
- bus arbitration,
- instruction and I/O caching effects.

We believe that the RT's very simple disk controller and interrupt mechanism make these effects negligible. Our analysis of 1.7 million disk-periods showed that some noise was present in the variation, its auto-correlation fell off within 5 seconds, and its entropy amounted to about 100 bits/minute [12], enough for 2,600 highly random DES keys/day.

From 100,000 access-times, and using an FFT vector-length $T = 1000$, we gathered $\sim$50,000 complex-valued spectra, and found that

- the real and imaginary parts passed various Q-Q plot normality tests;
- the angles passed various runs tests, and their a.c.f showed no correlation;
- the angles' 64 bits were pairwise uncorrelated.

We by no means intend these tests to be definitive, because we take it for granted that *a posteriori* arguments for randomness are inconclusive.

5 Conclusion

Experiments by ourselves and others show that the disk's speed fluctuates measurably because of air turbulence and other factors. Our random number generator uses the FFT algorithm to convert the measured variations into uniformly-distributed and independent variables. In a "worst-case" experimental scenario, we have measured 100 bits/min of entropy in a quiescent disk's speed variation.

We have also sketched a mathematical justification for our claim that our generator's product is truly random. In summary,

1. Disk drives have Taylor-Couette turbulence [1];
2. The N.R.T. theorem applies to Taylor-Couette turbulence [9];
3. N.R.T. theorem $\Rightarrow$ Sensitive dependence on initial conditions [14];
4. SDIC $\Rightarrow$ cumulants decay [Section 3];
5. Cumulants decay $\Rightarrow$ independent, normal spectra [7].

Turbulence's unpredictability is formally and experimentally well-founded in nonlinear dynamics. The SDIC criterion ensures that disk access-times satisfy a statistical mixing condition, which in turn ensures that the time-series' spectra are nearly independent, nearly normal variables. Sanity check statistical analyses, of disk periods and their spectra, are consistent with our argument.

Our experimental scenario is unrealistically constrained, but yields enough random bits to meet a large installation's key-service needs. We believe we could amplify this high-quality entropy by allowing the head's motion, the disk scheduler, and spindle-speed variations to influence each other synergistically. Other hardware noise-sources offer more bandwidth, but this one costs nothing, so our "per bit" price is very competitive.

Acknowledgments We thank John Carr for his help with `microtime`(); Fred Kurzweil and Hugh Sierra (formerly of IBM), and Brian Tanner of Micropolis, for their patient explanations of disk drive design and manufacture; and Peter Constantine, John Eaton, Ciprian Foiaş, Shafi Goldwasser, Bernardo Huberman, Mark Lillibridge, Zbigniew Nitecki, Olin Sibert, Ralph Swick, Harry Swinney, Roger Temam, and Jim White, for helpful discussions, references, and advice.

References

1. S.D. Abrahamson, C. Chiang, and J.K. Eaton, "Flow structure in head-disk assemblies and implications for design," Adv. Info. Storage Syst., **1** (1991). pp. 111–132.
2. W. Alexi, B. Chor, O. Goldreich, and C.P. Schnorr, "RSA and Rabin functions: certain parts are as hard as the whole," Proc. 25th IEEE Symp. on Foundations of Computer Science, 1984, pp. 449–457; see also SIAM J. on Comput., **17**(2) (1988).
3. P. Bergé, Y. Pomeau, and C. Vidal, Order Within Chaos: Towards a Deterministic Approach to Turbulence, Wiley, New York, 1984.
4. M. Blum, "Independent unbiased coin flips from a correlated biased source: a finite state Markov chain," Proc. 25th Ann. Symp. on Foundations of Computer Science, 1984. pp. 425–33.

5. L. Blum, M. Blum, and M. Shub, "A simple unpredictable pseudo-random number generator," SIAM J. Comput., **15**(2) (1986). pp. 364–83.
6. M. Blum and S. Micali, "How to generate cryptographically strong sequences of pseudo-random bits," SIAM J. Comput., **13** (4) (Nov. 1984). pp. 850–864.
7. D. Brillinger, Time Series: Data Analysis and Theory, Holden-Day, San Francisco, 1981. Addendum.
8. B. Chor and O. Goldreich, "Unbiased bits from sources of weak randomness and probabilistic communication complexity," Proc. 26th Ann. Symp. on Foundations of Computer Science, 1985. pp. 429–42.
9. P.R. Fenstermacher, H.L. Swinney, and J.P. Gollub, "Dynamical instabilities and the transition to chaotic Taylor vortex flow," J. Fluid Mech. **94**(1) (1979). pp. 103–128.
10. T. Kenjo and S. Nagamori, Permanent-Magnet and Brushless DC Motors, Monographs in Electrical and Electronic Engineering No. 18, Clarendon Press, Oxford, UK, 1985.
11. T.G. Lewis, Distribution Sampling for Computer Simulation, Lexington Books, Lexington, Mass., 1975. p.3.
12. U. Maurer, "A universal statistical test for random bit generators," Crypto '90 Conference Proceedings, Springer-Verlag Lecture Notes in Computer Science **537**, New York, 1991. pp. 408–420.
13. S. Micali, and C.P. Schnorr, "Efficient, perfect random number generators," Crypto '88 Conference Proceedings, Springer-Verlag Lecture Notes in Computer Science **403**, New York, 1990. pp. 173–198.
14. S. Newhouse, D. Ruelle, and F. Takens, "Occurence of strange Axiom A attractors near quasi-periodic flows of T^m $m \geq 3$," Commun. Math. Phys. **64** (1978), pp. 35–40.
15. D. Ruelle, Elements of Differentiable Dynamics and Bifurcation Theory, Academic Press, San Diego, 1989.
16. M. Santha and U.V. Vazirani, "Generating quasi-random sequences from semi-random sources," J. Comput. System Sci., **33** (1986). pp. 75–87.
17. B. Schneier, Applied Cryptography: Protocols, Algorithms, and Source Code in C; Wiley, New York, 1994. p. 370.
18. H. Sierra, An Introduction to Direct-Access Storage Devices, Academic Press, Boston, Mass., 1990. pp. 100–106.
19. Brian Tanner, personal communication.
20. R. Temam, Infinite-dimensional dynamical systems in mechanics and physics, Springer-Verlag Applied Mathematical Sciences **68**, 1988, pp. 389-92.
21. D.F. Torok and R. Gronseth, "Flow and thermal fields in channels between corotating disks," IEEE Trans. on Components, Hybrids, and Manuf. Tech., **11**(4) (Dec. 1988). pp. 585–593.
22. U. Vazirani, "Towards a strong communication complexity theory, or generating quasi-random sequences from two communicating slightly-random sources," (extended abstract, undated).
23. J. von Neumann, "Various techniques used in connection with random digits," Notes by G.E. Forsythe, National Bureau of Standards, Applied Math Series, Vol. 12, pp. 36–38, Reprinted in von Neumann's Collected Works, Vol. 5, Pergamon Press (1963). pp. 768–770.
24. A.C. Yao, "Theory and applications of trapdoor functions," Proc. 23rd IEEE Symp. on Foundations of Computer Science, 1982. pp. 80–91.

Cryptanalysis of the Gemmell and Naor Multiround Authentication Protocol*

Christian Gehrmann

Dept. of Information Theory, Lund University,
Box 118, S-221 00, Lund, Sweden

Abstract. Gemmell and Naor proposed a new protocol for the authentication of long messages which was based on block codes and which used a transmission channel k times. This multiround authentication makes it possible to limit the key size independently of the message length. We propose a new attack and show that the probability analysis made by Gemmell and Naor, which was only based on the minimum distance property of the codes, does not hold for our attack. Considering also the impersonation attack we conclude that the number of rounds have to be odd.

1 Introduction

The first treatment of codes that detect deception was given by Gilbert, MacWilliams and Sloane [1]. The use of universal hashing for authentication codes (A-codes) without secrecy, so called Cartesian A-codes, was first described in [2]. The general authentication problem was formulated in information theoretic terms by Simmons [3]. Many constructions and bounds have been derived for Cartesian A-codes [4] [5], [6], [7] and it is possible to construct such codes, which are close to the theoretical bounds. However all these constructions only deal with single transmission authentication. Gemmell and Naor [9] proposed a multiple round authentication protocol. Let n denote the message length, $H(K)$ the key entropy and P_s the probability for a successful substitution attack. For single round Cartesian authentication codes it was shown that [4]

$$H(K) \approx \log(n) + 2\log(\frac{1}{P_s}) - \log\log(\frac{1}{P_s}). \tag{1}$$

The Gemmell and Naor k-round protocol obtains:

$$H(K) \approx \log^{(k-1)}(n) + 5\log(\frac{1}{P_s}) \tag{2}$$

and Gemmell and Naor proved the existence of a k-round protocol such that

$$H(K) \approx \log^{(k-1)}(n) + 2\log(\frac{1}{P_s}). \tag{3}$$

* This work was supported by the TFR grant 222 92-662.

Hence it would be possible to limit the key size independently of the message length by using multiround authentication. We will start with describing the essential properties of the protocol suggested by Gemmell and Naor. Section 3 and 4 consist of an analysis of the described scheme. We first bring to attention an impersonation attack when the number of rounds is even. Next we describe a substitution attack. Finally we give an example of the proposed substitution attack for a construction based on RS-codes.

2 The Gemmell and Naor protocol

We assume two participants A and B, who want to communicate over an insecure channel where an opponent O may introduce a new message m'_j (impersonation attack) or substitute message m_j sent by A or B for m'_j (substitution attack). Here m_j denotes the message sent by A or B in the j-th round. The first message m_0 is the information message and the rest of the messages only "check messages" in the protocol. Hence the general goal for the opponent is to send an own message m_0 or to substitute for a transmitted one. However in his attempts to succeed with this purpose he may manipulate the "check messages" as well. A and B share a secret information, i.e., the key, unknown to the opponent.

In the Gemmell and Naor protocol:

C^j: the error correcting code used in the j-th round.
$C^j(m)$: the codeword corresponding to message m when using the code C^j.
$C^j_i(m)$: the i-th code symbol of the codeword $C^j(m)$.
C^A: a Cartesian A-code.
$C^A(m)$: the authentication tag corresponding to message m when using the code C^A.
$x \circ y$: concatenation of string x and y.
p: as defined in [9].

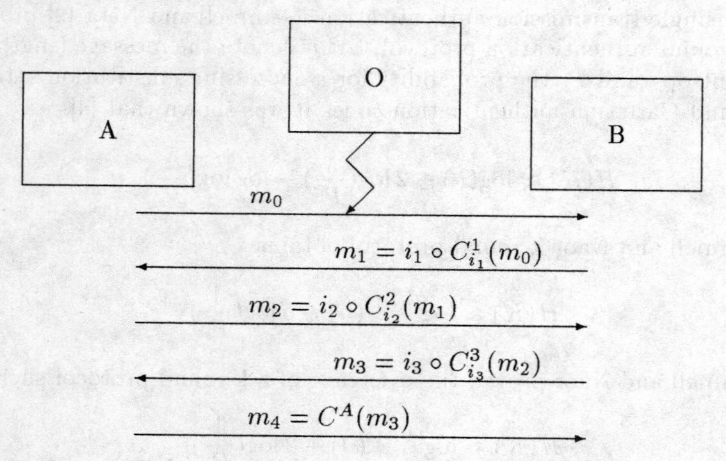

Figure 1: Multiround authentication for $k = 5$.

Assume we will use a k-round protocol and that: $C^j : \{0,1\}^n \rightarrow GF(Q_j)^{n_j}$ is a code with properties:

(i) $Q_j \geq \frac{2^{k-1-j}}{p}$
(ii) The minimum distance d_j of C^j satisfies: $d_j \geq n_j - n_j p / 2^{k-2-j}$.

The slightly modified(see below) Gemmell-Naor protocol may be described as follows:

(i) A sends message $m_0 = m, j = 0$.
(ii) $j = j+1$, B receives message m'_{j-1} and chooses a random number $i_j, 1 \leq i_j \leq n_j$. B sends message $m_j = i_j \circ C^j_{i_j}(m'_{j-1})$.
(iii) If $j = k-2$ then step vi).
(iv) $j = j+1$, A receives message m'_{j-1} and chooses a random number $i_j, 1 \leq i_j \leq n_j$. B sends message $m_j = i_j \circ C^j_{i_j}(m'_{j-1})$.
(v) if $j = k-2$ then step (vi), else back to step ii).
(vi) If k is even(odd) A(B) receives message $m'_j = m'_{k-2}$ and use a Cartesian A-code with $P_s \leq p$, to transmit $m_{k-1} = C^A(m'_{k-2})$.

We have changed the last step (vi) from the protocol in [9], by letting A(B) just send the authentication tag for the last message m'_j. This is possible because B(A) already knows the message to be authenticated. It is also important to notice that it is not possible for the opponent to freely choose the substitution message in the last step and that that this decreases the restrictions on the A-code.

For the protocol described above Gemmell and Naor stated the following:

Proposition 1 (Gemmell-Naor). *Let p be the parameter as it appears in [9]. Then*

$$\forall k \geq 2, \quad P_s \leq 2(1 - \frac{1}{2^{k-1}})p, \tag{4}$$

where P_s is the probability of a successful substitution attack.

3 An impersonation attack

First we will bring to attention an impersonation attack for the case k is even. Consider first the two round protocol($k = 2$):

(i) O sends another message $m_0 = m$.
(ii) B receives message m_0 and uses a Cartesian A-code to transmit $m_1 = C^A(m_0)$.
(iii) O absorbs the message sent by B.

This case may easy be generalised to higher order even round protocol just letting O act like A.

Proposition 2. *When the number of rounds is even, for the impersonation attack above we have*

$$P_I = 1, \tag{5}$$

where P_I is the probability of a successful impersonation attack.

Proof. A never receives the last message, i.e., the authentication tag. Hence O never will be caught.

Remark: For k odd, O succeeds in an impersonation attack if and only if he successfully authenticates the last message sent by B. But, by the definition of the A-code the probability for this event $\leq P_s$.

In the sequel we will assume k to be odd and we deal only with the substitution attacks.

4 A substitution attack

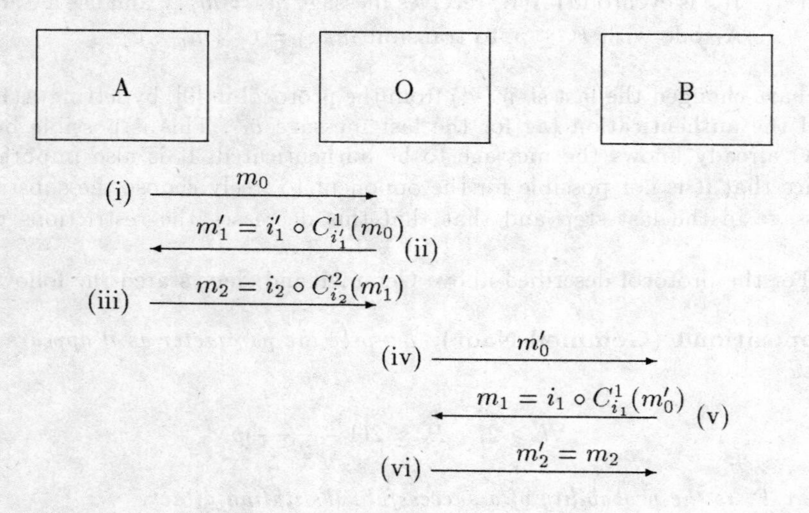

Figure 2: The attack scheme.

We will consider a specific substitution attack on the system above in the case $k \geq 5$. The attack is described by the scheme above.

(i) A sends an arbitrary message m_0 over the channel.

(ii) The opponent receives the message m_0 and chooses a random number $i'_1, 1 \leq i'_1 \leq n_1$ and sends $m'_1 = i'_1 \circ C^1_{i'_1}(m_0)$ to A.

(iii) A receives the message m'_1 and chooses a random number $i_2, 1 \leq i_2 \leq n_2$ and sends $m_2 = i_2 \circ C^2_{i_2}(m'_1)$ over the channel.

(iv) The opponent receives message m_2 from A and now substitutes the message m_0 for m'_0 and sends this to B.

(v) B receives message m'_0 and chooses a random number $i_1, 1 \leq i_1 \leq n_1$ and sends $m_1 = i_1 \circ C^1_{i_1}(m'_0)$ over the channel.

(vi) The opponent receives the message m_1 from B and just absorbs it and sends message $m_2' = m_2$ to B.

Proposition 3: For the attack scheme above

$$P_s \geq \max_{m_0' \neq m_0} \frac{|\{i_1 : C_{i_2}^2(m_1) = C_{i_2}^2(m_1')\}|}{n_1}, \tag{6}$$

where

$$m_1 = i_1 \circ C_{i_1}^1(m_0'), \quad 0 \leq i_1 \leq n_1,$$
$$m_1' = i_1' \circ C_{i_1'}^1(m_0'), \quad 0 \leq i_1' \leq n_1,$$

(See Figure 2).

Proof. Clearly if for one particular i_1 that B might choose we have

$$C_{i_2}^2(m_1) = C_{i_2}^2(m_1'),$$

then

$$m_2' = i_2 \circ C_{i_2}^2(m_1) = i_2 \circ C_{i_2}^2(m_1') = m_2$$

and adding the fact that the opponent knows i_2 before sending m_0' and that B chooses i_1 at random the result follows.

This result differs from (4) given by Gemmell and Naor and hence when $k \geq 5$ this must be taken into account when constructing the codes in the protocol. We will now construct an example that will illustrate the consequences of this result.

5 Example

Before describing an example of the G-N protocol we recall some simple facts on Reed-Solomon codes (RS-codes) [8]. We use the polynomial description of RS-codes as it appeared in the original paper by Reed and Solomon. Denote by $GF(Q)$ the Galois field with Q element. Consider the polynomial $P(x)$ of degree at most $k-1$ over $GF(Q)$, i.e.,

$$\mathcal{P} = \{P(x); m_0 + m_1 x + \cdots + m_{k-1} x^{k-1}, m_i \in GF(Q)\} \tag{7}$$

Let $\alpha \in GF(Q)$ be a primitive root. The RS-code C over $GF(Q)$ is now obtained as the set of Q-tuples

$$C = \{(P(0), P(\alpha), P(\alpha^2), ..., P(1)); P(x) \in \mathcal{P}\}$$

It is a code with k information symbols (over $GF(Q)$) and blocklength Q. We see that each codeword can be regarded as the k-tuple $m = (m_0, m_1, \cdots, m_{k-1})$ onto the Q-tuple $(P(0), P(\alpha), P(\alpha^2), \cdots, P(1))$. Let $C(m)$ denote the image of

m, i.e. $C(m) = (P(0), P(\alpha), P(\alpha^2), ..., P(1))$. As we already have seen we need for the protocol not the whole codeword but only one of it's Q coordinates. We will use the mapping

$$C_\gamma(m) = P(\gamma), \gamma \in GF(Q). \tag{8}$$

Finally we recall that the minimum distance d of the code C satisfies

$$d = Q - k.$$

We now describe an example of the G-N protocol for $k = 5$. Thus, see Figure 2, we need two codes C^1 and C^2. We will show that we can give a construction that gives a protocol for which $P_s = 1$ using the substitution attack described in Section 3.2.

We begin with setting up the protocol by chosing the codes[2] C^1 and C^2. Let q be a power of 2 and $l = q/2$. The code C^2 is chosen to be an RS-code over $GF(Q_2), Q_2 = q^2$, with $k_2 = 2l$. Hence it has block length $n_2 = q^2$ and distance $d_2 = q^2 - 2l = q^2 - q$. The code C^1 is chosen to be an RS-code over $GF(Q_1), Q_1 = Q_2^l$, with $k_1 = q^{2l-1}$. Hence it has block length $n_1 = q^{2l}$ and distance $d_1 = q^{2l} - q^{2l-1} = q^q(1 - 1/q)$. The coordinates of the codewords of C^1 are obtained by evaluating a polynomial, associated with the q^{2l-1} tuple m over $GF(Q_1)$ as specified by (8). Similar we obtain the codewords of C^2 by evaluating a polynomial associated with the $2l$ tuple over $GF(Q_2)$.

Recall that m_0 constitutes the actual information message that A wants to send to B. The second message m_1 consists of the index i_1 chosen by B (of size $\log Q_1$) and the coordinate of the corresponding codeword from C^1 selected by this index. Thus m_1 has size $2 \log Q_1 = 4l \log q$. Similary, the third message m_2 consists of the index i_2 chosen by A and the corresponding coordinate. It has size $2 \log Q_2 = 4 \log q$. Thus we see that the original message is "reduced" as illustrated below

$$\begin{array}{llll}
 & & & \text{size} \\
m_0 : & \multicolumn{2}{c}{\longleftrightarrow} & q^{2l} \log Q_1 = q^q \log q \\
m_1 : & \overset{i_1}{\longleftrightarrow} & \overset{C^1_{i_1}(m_0)}{\longleftrightarrow} & 2q \log q \\
m_2 : & \overset{i_2}{\leftrightarrow} & \overset{C^2_{i_2}(m_1)}{\leftrightarrow} & 4 \log q
\end{array}$$

Now recall from the definition of the G-N protocol property (ii) that $n_i - n_i p_i = d_i$, hence we have

$$p_1 = 1 - \frac{d_1}{n_1} = 1 - \frac{q^{2l} - q^{2l-1}}{q^{2l}} = \frac{1}{q},$$

$$p_2 = 1 - \frac{d_2}{n_2} = 1 - \frac{q^2 - q}{q^2} = \frac{1}{q},$$

[2] Actually we need also C^3 and C^A but these are irrelevant for our attack.

and hence according to proposition 1. (4) (Claim!) we would have for the probability of a successful attack in the two first rounds

$$P_s = p_1 + p_2 - p_1 p_2 < 2/q.$$

However let the received message in the first round be $m_2 = [i_2 \circ C^2_{i_2}(m'_1)]$. For simplicity define

$$\alpha = i_1, \beta = i_2, \text{ and } c = C^2_{i_2}(m'_1)$$

i.e. $m_2 = [\beta \circ c]$. The attack given in Section 4 succeeds if $C^2_\beta(m_1) = C^2_\beta(m'_1)$.

Now write

$$m_0 = m_{0,0}, m_{0,1}, ..., m_{0,q^{2l-1}-1}, \quad m_{0,j} \in GF(Q_1),$$
$$m_1 = m_{1,0}, m_{1,1}, ..., m_{1,2l-1}, \quad m_{1,j} \in GF(Q_2).$$

Thus $C^1_\alpha(m_0) = m_{0,0} + m_{0,1}\alpha + m_{0,2}\alpha^2 + \cdots + m_{0,q^{2l-1}-1}\alpha^{q^{2l-1}-1}$.

We have $\alpha \in GF(Q_1)$ and $C^1_\alpha(m_0) \in GF(Q_1)$ and since $GF(Q_1)$ is an extension field of $GF(Q_2), \alpha$ and $C^1_\alpha(m_0)$ may be represented as a l-tuple over $GF(Q_2)$,

Now recall that α consists of the first l coordinates of m_1 and $C^1_\alpha(m_0)$ of the second batch of the l coordinates, i.e.,

$$\alpha = \alpha_0, \alpha_1, ..., \alpha_{l-1} = m_{1,0}, m_{1,1}, ..., m_{1,l-1},$$

$$C^1_\alpha(m_0) = C^1_\alpha(m_0)_0, C^1_\alpha(m_0)_1, ..., C^1_\alpha(m_0)_{l-1} = m_{1,l}, m_{1,l+1}, ..., m_{1,2l-1}, m_{1,j}.$$

Thus for the whole m_1 given by
$m_1 = [\alpha \circ C^1_\alpha(m_0)] = m_{1,0}, m_{1,1}, ..., m_{1,l-1}, m_{1,l}, m_{1,l+1}, ..., m_{1,2l-1}$ we get that
$C^2_\beta(m_1) = m_{1,0} + m_{1,1}\beta + m_{1,2}\beta^2 + \cdots + m_{1,2l-1}\beta^{2l-1}, \beta \in GF(Q_2)(\subseteq GF(Q_1)!)$.

Suppose now that we choose to replace message m_0 by $m'_0 = c(\beta^l)^{-1}, -(\beta^l)^{-1}, 0, 0, ..., 0$. Note that $c(\beta^l)^{-1}, -(\beta^l)^{-1}$ are elements of the small field $GF(Q_2)$.Then

$$C^1_\alpha(m'_0) = c(\beta^l)^{-1} - (\beta^l)^{-1}\alpha.$$

Thus for m_1 we get

$$m_1 = [\alpha, C^1_\alpha(m'_0)] = \alpha_0, \alpha_1, ..., \alpha_{l-1}, (C^1_\alpha(m'_0))_0, (C^1_\alpha(m'_0))_1, ..., (C^1_\alpha(m'_0))_{l-1} =$$
$$= \alpha_0, \alpha_1, ..., \alpha_{l-1}, (c(\beta^l)^{-1} - (\beta^l)^{-1}\alpha)_0, -(\beta^l)^{-1}\alpha_1, ..., -(\beta^l)^{-1}\alpha_{l-1},$$

where we used the fact that $(c(\beta^l)^{-1} - (\beta^l)^{-1}\alpha)_i = -(\beta^l)^{-1}\alpha_i$ for $i = 1, 2, ..., l-1$. (recall that both $c(\beta^l)^{-1}$ and $-(\beta^l)^{-1}$ are elements of $GF(Q_2)$). Hence

$$C^2_\beta(m_1) = \alpha_0 + \alpha_1\beta + \alpha_2\beta^2 + \cdots + \alpha_{l-1}\beta^{l-1} +$$

$$
\begin{aligned}
&+(c(\beta^l)^{-1}-(\beta^l)^{-1}\alpha)_0\beta^l+-(\beta^l)^{-1}\alpha_1\beta^{l+1}+\cdots+-(\beta^l)^{-1}\alpha_{l-1}\beta^{2l-1}= \\
&=\alpha_0+\alpha_1\beta+\alpha_2\beta^2+\cdots+\alpha_{l-1}\beta^{l-1}+ \\
&+c-\alpha_0-\alpha_1\beta-\alpha_2\beta^2-\cdots-\alpha_{l-1}\beta^{l-1}= \\
&=c=C^2_\beta(m'_1) \quad !.
\end{aligned}
$$

and independently of the index α chosen by B, the substitution attack will succeed, i.e., $P_s = 1$.

6 Conclusion

We have analysed the Gemmell and Naor multiround protocol. We have shown that the number of rounds have to be an odd number. Furthermore, we have given a counter example to the Claim by Gemmell and Naor for the probability of a successful substitution attack.

Acknowledgements

I want to thank T. Johansson for his helpful comments and suggestions regarding this work.

References

1. E. Gilbert, F.J. MacWilliams, N. Sloane, "Codes Which Detect Deception". Bell System Technical Journal. Vol. 53. No. 3. March 1974, pp. 405-424.
2. J.L. Carter, M.N. Wegman, "New hash functions and their use in authentication and set equality", *J. Computer and System Sci.*, Vol 22, 1981, pp. 265-279.
3. G.J. Simmons, "A survey of Information Authentication", in *Contemporary Cryptology, The science for information integrity*, ed. G.J. Simmons, IEEE Press, New York, 1992.
4. T. Johansson, G. Kabatanskii, B. Smeets, "On the relation between A-codes and codes correcting independent errors", *Proceedings of Eurocrypt '93*, 1993, pp. 1-11.
5. J. Bierbrauer, T. Johansson, G. Kabatanskii, B. Smeets, "On Families of Hash Functions via Geometric Codes and Concatenation", *Proceedings of CRYPTO '93*, 1993, pp. 331-342.
6. D.R. Stinson, "Universal hashing and authentication codes", to appear in IEEE Transaction on Information Theory.
7. C. Gehrmann, "Long Message Authentication by using Pseudo-Random Functions", *Proceedings of IEEE ISIT 94*, to appear (preprint).
8. I.S. Reed, G. Solomon, "Polynomial Codes over certain Finite Fields", J. Soc. Ind. Appl. Math., vol. 8, June 1960, pp. 300-304.
9. P. Gemmell, M. Naor,"Codes for interactive authentication", *Proceedings of CRYPTO '93*, 1993, pp. 355-367.

LFSR-based Hashing and Authentication

Hugo Krawczyk

IBM T.J. Watson Research Center
Yorktown Heights, NY 10598
(hugo@watson.ibm.com)

Abstract. We present simple and efficient hash functions applicable to secure authentication of information. The constructions are mainly intended for message authentication in systems implementing stream cipher encryption and are suitable for other applications as well. The proposed hash functions are implemented through linear feedback shift registers and therefore attractive for hardware applications. As an example, a single 64 bit LFSR will be used to authenticate 1 Gbit of information with a failure probability of less than 2^{-30}. One of the constructions is the cryptographic version of the well known cyclic redundancy codes (CRC); the other is based on Toeplitz hashing where the matrix entries are generated by a LFSR. The later construction achieves essentially the same hashing and authentication strength of a completely random matrix but at a substantially lower cost in randomness, key size and implementation complexity. Of independent interest is our characterization of the properties required from a family of hash functions in order to be secure for authentication when combined with a (secure) stream cipher.

1 Introduction

In this paper we deal with the application of traditional hashing techniques (*not* one-way hashing) to cryptographic authentication of information. This investigation was initiated by Carter and Wegman [21], and further developed in subsequent works [4, 6, 19, 11, 2, 7]. We concentrate, for the sake of clarity, in the case of *message authentication*, although these techniques have broader application to different scenarios of information authentication. We assume a typical communication scenario in which two parties communicate over an unreliable link where messages can be maliciously altered. The communicating parties share a secret key unknown to the adversary.

There are two basic approaches for the application of hashing in the message authentication scenario. Both approaches use a predetermined family of hash functions from which a particular function is secretly chosen by the parties for authentication. In the first case a new hash function is selected from the family for each transmitted message. In the second, the same hash function is applied to the authentication of multiple messages, but the resultant hash values are encrypted before transmission. The advantage in the second case is that it usually requires less random (or pseudo-random) bits, and that it allows for a less frequent (possibly off-line) generation of the specific hash function.

Our work presents *very simple and practical* constructions of hashing schemes applicable to both approaches. Nevertheless, they are especially advantageous in the second case, i.e., for authentication of multiple messages, when the encryption is done via (additive) stream cipher encryption. We show these constructions to be *unconditionally secure* when used with a *perfect* one-time pad system, and therefore in most practical applications their security reduces to that of the stream cipher in use.

We start by proving a theorem on the minimal conditions required from a hash family in order to provide secure authentication when combined with one-time pad encryption, and later use this theorem to prove the security of our constructions. The characterization result is interesting independently of these specific constructions.

Our emphasis is in providing practical and secure methods for information authentication in systems that implement a stream cipher cryptosystem. We note that systems using block ciphers for secrecy often take advantage of these same ciphers for implementing a message authentication function. Additive stream ciphers, however, characterized by the use of pseudorandom generators which are essentially decoupled from the data, cannot directly be used to compute an authenticator on the data. On the other hand, the use of authentication in systems with stream cipher encryption is crucial because of the easy malleability of the plaintext through the corresponding ciphertext (e.g., flipping ciphertext bits is equivalent to flipping the same bits in the plaintext). In addition, authentication is required for validation of correct decryption and for detection of key synchronization loss. An important aspect related to stream ciphers is that these systems are often chosen for implementation (especially in hardware) because of their simplicity and efficiency. In such a case an equally simple and efficient authentication algorithm is required. *Our constructions are intended to fill this need.*

The first scheme we analyze is the cryptographic version of the well-known cyclic redundancy codes (CRC) used for non-cryptographic detection of information errors. It is based on the same operation of polynomial modular division and retains most of the simplicity of the regular CRC's except that in our case the dividing polynomial is variable. We prove the construction to be secure for authentication using the general results developed in Section 2.

The second construction is based on the well-known hashing technique that multiplies the data (seen as a binary vector) by a random matrix [5, 6]. This technique requires $m \cdot n$ random bits for specifying a hash function from m-bit messages into n-bit hash values which is prohibitive for many applications (e.g. large message and file authentication). Here, we show that essentially the same hashing and authentication effect can be achieved by choosing just n random bits and a random irreducible polynomial of degree n and then generating a Toeplitz matrix out of these initial values by a simple linear feedback shift register. In typical applications $n << m$ (n is the security parameter), and therefore the resultant savings (including number of random bits, size of hash description, key length) is enormous. It also makes possible the implementation of this technique in hardware as required in some authentication (and other hashing) applications.

Our constructions achieve up to small constants the known lower bounds on size of description and number of functions in the hash family. This is an important factor for the practicality of these constructions as they influence, for example, the amount of hardware required in their implementation. As an example, a single 64 bit LFSR will be used to authenticate 1 Gbit of information with a failure probability of less than 2^{-30}. (We stress that both of our constructions can be applied to variable length messages).

We also mention that it is possible to extend our LFSR-based Toeplitz construction to general *ε-biased sequences* [15, 1] of which LFSRs are a particular case.

RELATED WORK. Unconditionally secure authentication codes have been extensively studied in the literature (see [18] for a survey). Pioneering works were Gilbert, MacWilliams and Sloane [8], Carter and Wegman [21] and the foundational work by Simmons (see [17, 18]). Carter and Wegman were the first to interpret and construct authentication codes through hash functions. They also were first to show how one-time pad systems can be used in combination with hash functions to construct efficient authentication algorithms. This approach was further studied by Brassard [4] and Desmedt [6]. Our work follows and refines this line of research.

The hashing approach for unconditional authentication was further developed by Stinson [19] who presents improved constructions and lower bounds on the size of the required hash families. More recently, Bierbrauer, Johansson, Kabatianskii and Smeets [11, 2] and Gemmell and Naor [7] generalize and improve on the above works by noticing and exploiting the connection between hash functions and error correcting codes.

Works that directly use stream ciphers in their constructions of cryptographic checksums are Lai, Rueppel and Woollven [12] and Taylor [20]. Contrary to our approach, [12] uses for the checksum computation a number of pseudorandom bits *equal* to the number of message bits. In our construction the number of pseudorandom bits depends linearly on the security parameter (usually much smaller than the information size) and grows only *logarithmically* with the message length. The approach in [20] is essentially to generate a new member of a hash family for each message using the pseudorandom generator of the stream cipher. In our case, we reuse the same hash function for multiple messages and use the pseudorandom generator only for generating encryption pads for the hash values.

A very recent and independent paper by Johansson [10] uses a LFSR to generate an authentication matrix. However, that construction, as well as its goals, is essentially different from ours. Most notably, it requires a LFSR (and then also a key) of the length of the message itself as opposed to just the length of the security parameter as in our case.

2 Hash Functions and Message Authentication

In this section we introduce the basic concepts regarding hash functions as required in the context of message authentication, and the notion of security for the authentication functions. We also characterize the exact requirements from a hash family to be secure when combined with a one-time pad system.

2.1 Hash Functions

Definition 1. An (m, n)-*family H of hash functions* is a collection of functions that map the set of binary strings of length m into the set of binary strings of length n.

Notation: The notation $s \in_R S$ denotes that the element s is chosen with uniform probability from the set S. The expression $Pr_h(A(h))$ denotes the probability of the event $A(h)$ when $h \in_R H$, and H is a (usually implicit) set of hash functions. We will use M, M', etc., to denote arguments or inputs for the functions in the family H. When it is clear from the context, and for the sake of readability, we will omit explicit reference to the lengths of these inputs. We will usually denote the output of the hash functions by c.

A property of some hash functions that simplifies their analysis is being linear relative to the bitwise exclusive-or operation. This property is found in many natural constructions (including ours).

Definition 2. A family of functions H is $\oplus$-*linear* if for all M, M' we have $h(M \oplus M') = h(M) \oplus h(M')$.

The following property of a family of hash functions has a central role in our work, it states that elements are mapped into their images by these functions in a "balanced" way. Its importance in our context is given by Theorem 6.

Definition 3. A family of hash functions is called ε-*balanced* if

$$\forall M \neq 0,\ c,\quad Pr_h(h(M) = c) \leq \varepsilon.$$

2.2 Message Authentication

We assume a typical communication scenario in which two parties communicate over an unreliable link where messages can be maliciously altered. The communicating parties share a secret key unknown to the adversary.

For simplicity we start assuming that the parties exchange (using that secret key) only one message of length m. In that case, the secret key consists of the description of a particular hash function h drawn randomly from an (m, n)-family of hash functions and a random pad r of length n. The sender of the message M, sends M together with the "tag" $t = h(M) \oplus r$, which at reception will be recomputed and checked for consistency by the receiver.

Although in the above scheme the authentication tag looks completely random to an adversary, and therefore it learns nothing about the specific hash function, it still can use its knowledge of the hash family to try to modify consistently the message and corresponding tag such that the message alteration is not discovered. Indeed, if the family H of hash functions is not chosen carefully such an attack *is possible* (see Section 3.1). Here we derive a necessary and sufficient condition on H to make the success probability of any attack no more than a pre-specified value ε.

Let M be a message of length m authenticated with the tag $t = h(M) \oplus r$, where $h \in_R H$ and $r \in_R \{0,1\}^n$. We say that an adversary that sees M and t succeeds in breaking the authentication if it finds M' and t', where M' is different than M and $t' = h(M') \oplus r$. We assume that the adversary knows the family of hash functions, but not the particular value of h or the pad r.

Definition 4. A family H of hash functions is called *ε-otp-secure* if for any message M no adversary succeeds in the above scenario with probability larger than ε. [1]

Usually the value of ε for a given family of hash functions will depend on the parameters of this family (e.g. input and output size). The following theorem characterizes those families of hash functions that are ε-otp-secure.

Theorem 5. *A necessary and sufficient condition for a family H of hash functions to be ε-otp-secure is that*

$$\forall M_1 \neq M_2,\ c,\ \ Pr_h(h(M_1) \oplus h(M_2) = c) \leq \varepsilon.$$

Proof Sketch. The pair (M_1, t_1) is successfully replaced by the pair (M_2, t_2) only if for the secret and random h and r used by the communicating parties we have $t_1 = h(M_1) \oplus r$ and $t_2 = h(M_2) \oplus r$, or equivalently, $t_1 \oplus t_2 = h(M_1) \oplus h(M_2)$. Therefore, the success probability of the adversary is bounded by $\max_{M_1, M_2, c} Pr_h(h(M_1) \oplus h(M_2) = c)$ where c represents the difference $t_1 \oplus t_2$. Notice that this success probability is achievable whenever the transmitted message is one of the messages in which the maximum is attained (just replace (M_1, t_1) by $(M_2, t_1 \oplus c)$). □

As an immediate consequence we have the following theorem that is the main tool for proving the security of our constructions.

Theorem 6. *If H is $\oplus$-linear then H is ε-otp-secure if and only if H is ε-balanced.*

Therefore, in order to prove the security of a particular family of hash functions for implementation of a message authentication scheme of the above kind (i.e., combined with a one-time pad) it is sufficient (and necessary) to show that the family has the condition stated in Theorem 5. In case the family is also $\oplus$-linear one has to prove it to be ε-balanced.

In the typical scenario where the parties exchange multiple messages, the hash function h can be reused for the different messages, but for each new message a different random pad (each of length n) will be used for encryption of the hash value. In this case, if the hash family is ε-otp-secure then the success probability of an adversary that tries to modify a single message is still at most ε. Indeed, the fact that the adversary sees many pairs of messages and corresponding tags is useless since these tags are completely random and therefore give no information on the value of h. If the adversary can modify k of the transmitted messages its probability of success is bounded by $k\varepsilon$.

When authenticating multiple messages with the same hash function, it is desirable that this function be applicable to variable length messages (as is the case

[1] We choose the term *otp-security* to stress the essential role of the one-time pad (otp) added to the hash value for the security of the authentication scheme.

for our constructions). In this case the adversary's success probability depends on the total length of messages he or she modifies.

Remark: The above definitions of security are stated in unconditional terms (i.e., against *any adversary*). This requires the communicating parties to exchange truly random pads (of the size of the hash output) for each transmitted message. In most practical applications, however, the successive pads r will be generated using a pseudorandom generator out of a secret seed shared by the parties. In this case the security of the authentication scheme reduces to the security of the pseudorandom generator or stream cipher in use; and the computational power of the adversary is assumed to be bounded depending on the security model of the stream cipher. From a practical point of view our approach to message authentication is especially advantageous in systems implementing stream cipher encryption of the transmitted information. In these cases the hash value computed on the message is appended to the message before transmission and the combined information is then encrypted using the stream cipher.

3 Constructions

We present two simple and practical constructions of $\oplus$-linear hash functions that are ε-balanced with ε being exponentially small in the length of the hash value, and therefore suitable for information authentication in the sense described in Section 2. Both schemes can hash variable length messages.

3.1 Cryptographic CRC

The first construction is based on the operation of division modulo an irreducible polynomial over GF(2). It is a cryptographic variant of the well known Cyclic Redundancy Codes (CRC) which are commonly used as a standard error detection mechanism in data networks. CRC's are used to detect non-malicious errors and therefore there is no need for a secret key or even a family of functions; they are implemented as a fixed, public function. The simplicity of implementation and provable properties of these constructions have made them so popular; many of these advantages are inherited by the stronger cryptographic version.

To our knowledge, the first to use these functions in the cryptographic setting was Rabin [16] who proposed their use for fingerprinting information. As opposed to Rabin's application where the fingerprint value is kept secret, our setting requires the transmission of this value. Interestingly enough, even if one encrypts the fingerprint before transmissiom using a *perfect* one-time pad that scheme is insecure for message authentication. The construction presented here introduces a seemingly minor technical modification that solves that problem.

In what follows we will explicitly or implicitly associate each binary string S with the polynomial $S(x)$ over GF(2) with coefficients corresponding to the bits of S.

THE CONSTRUCTION *(Cryptographic CRC)*: We define an (m, n)-family of hash functions as follows. For each irreducible polynomial $p(x)$ of degree n over GF(2) we associate a hash function h_p such that for any message M of binary length m, $h_p(M)$ is defined as (the coefficients of) $M(x) \cdot x^n \bmod p(x)$.

Rabin's construction is essentially the same except for the multiplication by the x^n factor in the modular operation (that has the practical effect of shifting the message by n positions). Without this change the resultant hash family is only 1-balanced, and therefore breakable with probability 1. (The flipping of any bits among the n least significant bits of the message and the same bits in the encrypted authenticator will be undetected; even if the message itself is encrypted under a one-time pad!). After its modification we can prove that the proposed scheme is secure for authentication when combined with a one-time pad or secure stream cipher.

Theorem 7. *For any values of n and m the above defined family of hash functions is $\oplus$-linear and ε-balanced for $\varepsilon \leq \frac{m+n}{2^{n-1}}$, and therefore ε-otp-secure.*

Proof. The CRC family is $\oplus$-linear since division modulo a polynomial is a linear operation where addition is equivalent to a bitwise exclusive-or operation. To show that the family is also ε-balanced notice that for any polynomial $p(x)$ of degree n, any non-zero message M of length m and any string c of length n,

$$h_p(M) = c \text{ iff } M(x) \cdot x^n \bmod p(x) = c(x) \text{ iff } p(x) \text{ divides } M(x) \cdot x^n - c(x).$$

Denote $q(x) = M(x) \cdot x^n - c(x)$. Clearly, $q(x)$ is a non-zero polynomial of degree (at most) $m + n$, and $p(x)$ is an irreducible polynomial of degree n that divides $q(x)$. Because of the unique factorization property there are at most $\frac{m+n}{n}$ irreducible factors of $q(x)$ each of degree n. In other words, there are at most $\frac{m+n}{n}$ hash functions in the CRC family that map M into c. On the other hand, there are more than $\frac{2^{n-1}}{n}$ elements in this family (as the number of irreducible polynomials over GF(2) of degree n). Therefore, $Prob(h_p(M) = c) \leq \frac{(m+n)/n}{2^{n-1}/n} = \frac{m+n}{2^{n-1}}$. □

The practical consequence of this theorem and Theorem 6 is that one can safely use this very practical method of hashing for cryptographic authentication when combined with a cryptographically strong pseudorandom generator (i.e., secure stream cipher). We briefly consider some practical aspects of this construction.

VARIABLE-LENGTH MESSAGES. The hash functions in the CRC family are essentially defined by the polynomial $p(x)$ and not by the length of the messages. Therefore, they can be applied to messages of different lengths as it is desirable in practice. In this case, one has to treat the polynomial $M(x)$ corresponding to the message M as having a leading coefficient '1' (i.e., if M is of length m, then $M(x)$ is of proper degree m). This determines a 1-1 mapping between messages and polynomials and, in particular, prevents changing the message by just appending zeros to it. Also, the value of ε in Theorem 7 depends on m being the maximum size of the fake message inserted by the adversary rather than by the length of the original message.

HARDWARE IMPLEMENTATION. Implementing the above hash functions in hardware is simple and very efficient. The operation of division modulo a polynomial over GF(2) is implemented through a simple linear feedback shift register with taps or connections determined by the dividing polynomial. Since this same operation is used for standard CRC's there are plenty of references in the literature on its implementation. (Even the multiplication by x^n in our construction is implemented in many cases without penalty in hardware or performance). However, recall that in the standard CRC the dividing polynomial is fixed and known in advance, and most circuits that implement it have the particular taps hardwired into the circuit. A cryptographic CRC as proposed here needs an implementation where the connections (determined by the polynomial) are programmable. The actual value for these connections is the key for the hashing which should be changeable (and secret). We stress that CRC circuits with variable connections are already designed even for implementation of regular CRC's. One reason for that is the need to support different CRC standards (each one determines a different polynomial), and in particular different polynomial degrees. See [3] for one such example.

SOFTWARE IMPLEMENTATION. Efficient implementations of CRC's in software exist too. In these implementations significant speed up is achieved by using precomputation tables. These tables depend on the particular key polynomial. Therefore, they are computed only once per key which is affordable in many applications.

CHOOSING KEYS. The keys for the cryptographic CRC functions is a random irreducible polynomial. The time complexity of generating such a polynomial of degree n is $O(n^3)$ bit operations or, in a software implementation, $O(n^2)$ word operations (mostly XOR's and SHIFT's). Therefore, it is efficient enough for applications (as suggested here) where the key is changed only sporadically (e.g. at the beginning of a network session). Algorithms for generating random irreducible polynomials can be found in [9, 16].

A NOTE OF CARE. Efficient stream ciphers (especially in hardware) sometimes use constructions based on LFSRs. In these cases using a cryptographic CRC is especially attractive because of the similar hardware structure. However, the security of these stream ciphers (as any other encryption system) is claimed only heuristically. Therefore, special care and attention need to be devoted to the interaction between these constructions.

3.2 LFSR-based Toeplitz

Our second construction is based on the following hashing method that uses random binary matrices. Let A be an $n \times m$ Boolean matrix. Let M be a message consisting of m bits. Define $h_A(M)$ to be the Boolean multiplication of the matrix A by the column vector composed of M's bits. Carter and Wegman [5] showed that the family of functions $\{h_A : A \text{ is a } n \times m \text{ Boolean matrix}\}$, is a universal$_2$ family of hash functions. This family is according to our terminology ε-balanced for $\varepsilon = 2^{-n}$, and then its affine version (namely, $h'_{A,b}(M) = A \cdot M + b$, where b is a binary vector of length n) is strongly universal$_2$.

The description of such a hash function takes $n \cdot m$ bits (or $n(m+1)$ in the affine case). A related family with the same properties as above but with much smaller description is obtained by restricting the Boolean matrix A to be a *Toeplitz matrix*.

These are matrices where each left-to-right diagonal is fixed, i.e., if $k - i = l - j$ for any indices $1 \leq i, k \leq n$, $1 \leq j, l \leq m$, then $A_{i,j} = A_{k,l}$. (For a proof of the universality of these hash functions see e.g. [14]). Notice that by defining the first column and first row of the matrix all other entries are uniquely determined. Therefore, only $m+n-1$ bits define the whole matrix, a significant savings relative to the $n \cdot m$ bits necessary to describe the original family. However, even these functions require a description that is as long as the input (or message) to be hashed. When these inputs are significantly longer than the required output (i.e., the security parameter) then this description can be prohibitively expensive.

Our construction modifies the above Toeplitz family by restricting it even more. Indeed, we use Toeplitz matrices where consecutive columns are the consecutives states of a LFSR of length n. To see that such a construction is indeed a Toeplitz matrix just notice that Toeplitz matrices are characterized by the property that each column in the matrix is determined by shifting (down) the previous column and adding a new element to the top of the column. Therefore in our construction a function is specified by defining a particular LFSR (i.e., its connection polynomial) and its initial state, a total of $2n$ bits (recall that usually $n << m$). Interestingly enough, by limiting the connections of the LFSRs to irreducible polynomials our construction keeps most of the strength of the original Carter-Wegman family but with a much shorter description size. The price we pay is that our functions are only ε-balanced for a small ε instead of being perfectly balanced as the original. For the purpose of authentication this small ε represents no substantial loss, while the lower description size makes them significantly more practical.

THE CONSTRUCTION *(LFSR-based Toeplitz)*: Let $p(x)$ be an irreducible polynomial over GF(2) of degree n. Let $s_0, s_1, \ldots$ be the bit sequence generated by a LFSR with connections corresponding to the coefficients of $p(x)$ and initial state $s_0, s_1, \ldots, s_{n-1}$. For each such polynomial $p(x)$ and initial state $s \neq 0$ we associate a hash function $h_{p,s}$ such that for any message $M = M_0 M_1 \ldots M_{m-1}$ of binary length m, $h_{p,s}(M)$ is defined as the linear combination $\bigoplus_{j=0}^{m-1} M_j \cdot (s_j, s_{j+1} \ldots s_{j+n-1})$.

In simple words, the LFSR advances its state with each message bit. If this bit is '1' the corresponding state is accumulated into an accumulator register, if it's '0' the state is not accumulated (see Figure 1).

The main technical theorem regarding this construction is the following characterization of LFSR-based Toeplitz hashing. The proof is omitted from this abstract.

Theorem 8. *Let $p(x)$ be an irreducible polynomial of degree n over GF(2) and let $s = (s_0, \ldots, s_{n-1})^T$ be an initial state for the LFSR defined by the connection polynomial $p(x)$. Let M be an m-bit long message. Let $\lambda_1, \lambda_2, \ldots, \lambda_n$ be the n (different) roots of $p(x)$ (over GF(2^n)). Then,*

$$h_{p,s}(M) = BD_{M,p}B^{-1}s$$

where B is a non-singular $n \times n$ matrix which depends on $p(x)$ only and $D_{M,p}$ is an $n \times n$ diagonal matrix with $M(\lambda_i)$, $1 \leq i \leq n$, as its i-th diagonal entry.

From this we can derive that our construction has the required ε-balanced property.

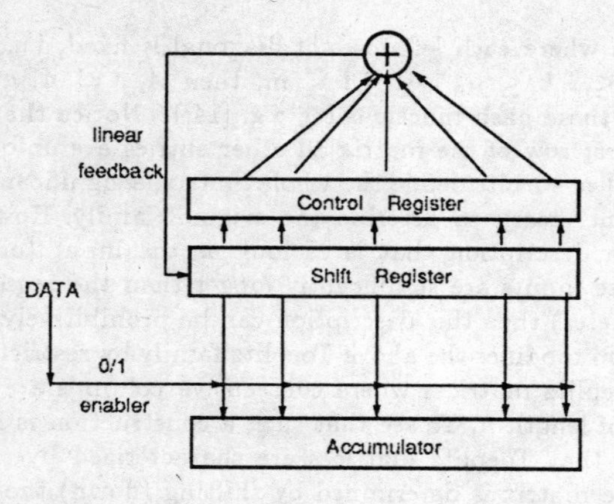

Fig. 1. A schematic implementation of the LFSR-based Toeplitz hashing

Theorem 9. *The LFSR-based Toeplitz construction defined above is ε-balanced for $\varepsilon \leq \frac{m}{2^{n-1}}$.*

Proof. Fix a message $M \neq 0$ and a hash value c. We need to bound the probability that $h_{p,s}(M) = c$ for randomly chosen irreducible polynomial $p(x)$ and initial state $s \neq 0$. We use Theorem 8 and the fact that $M(x)$ has a common root with $p(x)$ if and only if $p(x)$ divides $M(x)$. We distinguish between two cases according to the value of c.
Case I: Let $c = 0$ (i.e., c is the all-zeros vector). Since we choose $s \neq 0$ then $h_{p,s}(M) = 0$ may happen only if $D_{M,p}$ is singular (the matrices B and B^{-1} are not). This is the case only if for some i, $M(\lambda_i) = 0$, or equivalently only if $p(x)$ divides $M(x)$. The probability of such an event is at most as the number of possible irreducible factors of $M(x)$ divided by the total number of irreducible polynomials of degree n, i.e. at most $\frac{m/n}{2^{n-1}/n} = \frac{m}{2^{n-1}}$.
Case II: Let $c \neq 0$. In order for $h_{p,s}(M)$ to equal c, we need $D_{M,p}$ to be non-singular and s be the *unique* vector that is mapped by $BD_{M,p}B^{-1}$ into c. The vector s assumes this value with probability of $\frac{1}{2^n-1}$, and therefore $h_{p,s}(M) = c$ happens with at most this probability.

In either case the probability that $h_{p,s}(M) = c$ is at most $\frac{m}{2^{n-1}}$ and then our construction is $\frac{m}{2^{n-1}}$-balanced. □

Practical Considerations. Most of the remarks in Section 3.1 regarding practical implementation of CRC's hold here. We just stress that the decoupling of the LFSR from the data and having a shift register without modifications between internal stages (see Figure 1) permits significant parallelism and pipelining in the LFSR implementation which is crucial for achieving very high speeds and can be advantageous relative to the CRC construction. Notice that the LFSR-based construction can be used also for the same non-cryptographic applications where the CRC is regularly encountered.

Acknowledgment: I would like to thank Phil Rogaway, Ronny Roth and Moti Yung for their help and comments.

References

1. Noga Alon, Oded Goldreich, Johan Hastad, and Rene Peralta. Simple constructions of almost k-wise independent random variables. In 31th *Annual Symposium on Foundations of Computer Science, St. Louis, Missouri*, pages 544–553, October 1990.
2. Bierbrauer J., Johansson T., Kabatianskii G., and Smeets, B., "On Families of Hash Functions via Geometric Codes and Concatenation", *Proc. of Crypto'93*, pp. 331-342.
3. Birch, J., Christensen, L.G., and Skov, M., "A programmable 800 Mbit/s CRC check/generator unit for LANs and MANs", *Comp. Networks and ISDN Sys.*, 1992.
4. Brassard, G., "On computationally secure authentication tags requiring short secret shared keys", *Proc. of Crypto'82*, pp.79-86.
5. Carter, J.L. and Wegman, M.N., "Universal Classes of Hash Functions", *JCSS*, 18, 1979, pp. 143-154.
6. Desmedt, Y., "Unconditionally secure authentication schemes and practical and theoretical consequences", *Proc. of Crypto'85*, pp.42-55.
7. Gemmell, P., and Naor, M., "Codes for Interactive Authentication", *Proc. of Crypto'93*, pp. 355-367.
8. Gilbert, E.N., MacWilliams, F.J., and Sloane, N.J.A., "Codes which detect deception", *Bell Syst. Tech. J.*, Vol. 53, 1974, pp. 405-424.
9. John A Gordon, "Very simple method to find the minimal polynomial of an arbitrary non-zero element of a finite field", *Electronics Letters* , Vol. 12, 1976, pp. 663-664.
10. Johansson T., "A Shift Register Construction of Unconditionally Secure Authentication Codes", *Design, Codes and Cryptography*, 4, 1994, pp. 69-81.
11. Johansson T., Kabatianskii G., and Smeets, B., "On the Relation Between A-Codes and Codes Correcting Independent Errors", *Proc. of Eurocrypt'93*, pp. 1-11.
12. Lai, X., Rueppel, R.A., and Woollven, J., "A Fast Cryptographic Checksum Algorithm Based on Stream Ciphers", *Auscrypt'92*, pp. 339-348.
13. Lidl, R., and Niederreiter, H., "Finite Fields", in *Encyclopedia of Mathematics and Its Applications*, Vol 20, Reading, MA: Addison-Wesley, 1983.
14. Mansour, Y., Nisan, N., and Tiwari, P., "The Computational Complexity of Universal Hashing", *STOC'90*, pp. 235-243.
15. Joseph Naor and Moni Naor. Small bias probability spaces: efficient construction and applications. In *Proceedings of the* 22nd *Annual ACM Symposium on Theory of Computing, Baltimore, Maryland*, pages 213–223, May 1990.
16. Rabin, M.O., "Fingerprinting by Random Polynomials", Tech. Rep. TR-15-81, Center for Research in Computing Technology, Harvard Univ., Cambridge, Mass., 1981.
17. Simmons, G.J., "Authentication theory/coding theory", *Proc. of Crypto'84*, 411-431.
18. Simmons, G.J., "A Survey of Information Authentication", in Gustavos J. Simmons, editor, *Contemporary Cryptology, The Science of Information*, IEEE Press, 1992.
19. Stinson, D.R., "Universal hashing and authentication codes", *Proc. of Crypto'91*, pp. 74-85.
20. Taylor, R., "An integrity check value algorithm for stream ciphers", *Proc. of Crypto'93*, pp. 40-48.
21. Wegman, M.N., and Carter, J.L., "New Hash Functions and Their Use in Authentication and Set Equality", *JCSS*, 22, 1981, pp. 265-279.

New Bound on Authentication Code with Arbitration

Kaoru KUROSAWA

Department of Electrical and Electronic Engineering,
Faculty of Engineering,
Tokyo Institute of Technology
2–12–1 O-okayama, Meguro-ku, Tokyo 152, Japan
Tel. +81–3–5734–2577
E-mail kkurosaw@ss.titech.ac.jp

Abstract. For the authentication model with arbitration (A^2-code), Johansson showed a lower bound on the size of encoding rules. However, this bound is no longer tight if the size of source states is large. This paper presents a more tight lower bound on the size of encoding rules for large source states. An A^2-code is shown which approximately meets the proposed bound, also. Further, we show that the size of encoding rules for the transmitter can be greatly reduced if the receiver's cheating probability is slightly large.

1 Introduction

As in [8], A^2-code is described as follows. In the model for normal authentication (A-code), the transmitter and the receiver are using the same encoding rule and are thus trusting each other [1] ~ [5]. However, it is no always the case that the two communicating parties want to trust each other. Inspired by this problem Simmons has introduced an extended authentication model [6, 7], here referred to as the authentication model with arbitration (A^2-code). In this model caution is taken against deception from both outsiders (opponent) and insiders (transmitter and receiver). The model includes a fourth person, called the arbiter. The arbiter has access to all key information and is by definition not cheating. The arbiter does not take part in any communication activities on the channel but has to solve disputes between the transmitter and the receiver whenever such occur.

There are essentially five different kinds of attacks to cheat which are possible. The attacks are the following:
I, Impersonation by the opponent. The opponent sends a massage to the receiver and succeeds if the message is accepted by the receiver as authentic.
S, Substitution by the opponent. The opponent observes a message that is transmitted and substitutes this message with another. The opponent succeeds if this other message is accepted by the receiver as authentic.
T, Impersonation by the transmitter. The transmitter sends a message to the receiver and denies having sent it. The transmitter succeeds if the message is

accepted by the receiver as authentic and if the message is not one of the messages that the transmitter could have generated due to his encoding rule.
R$_0$, Impersonation by the receiver. The receiver claims to have received a message from the transmitter. The receiver succeeds if the message could have been generated by the transmitter due to his encoding rule.
R$_1$, Substitution by the receiver. The receiver receives a message from the transmitter but claims to have received another message. The receiver succeeds if this other message could have been generated by the transmitter due to his encoding rule. For each way of cheating, we denote the probability of success with P_I, P_S, P_T, P_{R_0} and P_{R_1}.

Let E_R be a set of the receiver's encoding rules and E_T be a set of the transmitter's encoding rules. Also, let S be a set of source states. Recently, Johansson showed [8] a lower bound on $|E_T|$ and $|E_R|$ to achieve $\max(P_I, P_S, P_T, P_{R_0}, P_{R_1}) = 1/q$. This bound is tight for $|S| \leq q$ because he also showed an A^2-code which meets the bound [9]. However, this bound is no longer tight if $|S| > q+1$.

On the other hand, it is known that, in A-code, the size of encoding rules is greatly reduced if P_S is slightly greater than its lower bound [11, 12, 13], most notably in [13].

This paper presents a more tight lower bound on $|E_T|$ and $|E_R|$ for $|S| > q+1$ than [8]. An A^2-code is shown which approximately meets the proposed bound, also. Further, we show that $|E_T|$ can be greatly reduced if P_{R_1} is slightly greater than $1/q$.

2 Preliminaries

2.1 Notation

- $|B|$ denotes the cardinality of a set B.
- When we write $X = \{x_{ij}\}$, X denotes a matrix whose (i,j) element is x_{ij}. We denote by x_j the j-th column vector of X. y_j denotes the j-th column vector of Y, etc.
- For a vector y, define $w(y) \triangleq$ the number of nonzero elements of y. We say that $w(y)$ is the weight of y.
- For $x_i = (x_{1i}, x_{2i}, \cdots)^T$ and $x_j = (x_{1j}, x_{2j}, \cdots)^T$, define $x_i \odot x_j = (x_{1i}x_{1j}, x_{2i}x_{2j}, \cdots)^T$.

2.2 Authentication code (A-code)

In the normal model for authentication, there are three participants, a transmitter **T**, a receiver **R** and an opponent **O**. An authentication code (A-code) is **(S,E,M)** such that $\mathbf{S} = \{s\}$ is a set of source states, $\mathbf{E} = \{e\}$ is a set of rules and $\mathbf{M} = \{m\}$ is a set of messages. T and R share e secretly. On input s, T sends m such that $m = e(s)$ to R. R accepts or rejects m based on e.

There are two kinds of attacks of the opponent O, the impersonation attack and the substitution attack. They are defined in the same way as described in

Introduction. The impersonation attack probability $\mathbf{P_I}$ is defined by

$$\mathbf{P_I} \stackrel{\triangle}{=} \max_m \Pr[\mathbf{R} \text{ accepts } m].$$

The substitution attack probability $\mathbf{P_S}$ is defined by

$$\mathbf{P_S} \stackrel{\triangle}{=} \sum_m \Pr(\mathbf{M} = m) \max_{\hat{m} \neq m} \Pr[\mathbf{R} \text{ accepts } \hat{m} \mid \mathbf{R} \text{ accepts } m]$$

Definition 1. A without secrecy A-code is such that $\forall m$ is written as (s, a), where s a a source state and a is an authenticator. For a without secrecy A-code, let $M_s \stackrel{\triangle}{=} \{m | m = (s, a)\}$.

Definition 2. An A-code is called no splitting if each e generates one message for $\forall s$.

Definition 3. A skeleton matrix for (E, M) is a $|E| \times |M|$ matrix $X = \{x_{ij}\}$ such that

$$x_{ij} = \begin{cases} 1 & \text{if } e_i \text{ accepts (or could generate) } m_j \\ 0 & \text{otherwise.} \end{cases}$$

2.3 Basic results on A-code

The following observation is a basis for the bound on P_I and P_S.

Claim 4. *Let $X = \{x_{ij}\}$ be a skeleton matrix for (E, M). Suppose that E is uniformly distributed. Then,*

$$\Pr(R \text{ accepts } m_j) = w(x_j)/|E|, \quad P_I = \max_j w(x_j)/|E|$$

$$\Pr(R \text{ accepts } m_j | R \text{ accepts } m_i) = w(x_i \odot x_j)/w(x_i)$$

Let $k \stackrel{\triangle}{=} |\, \mathbf{S} \,|$, $v \stackrel{\triangle}{=} |\, \mathbf{M} \,|$, $b \stackrel{\triangle}{=} |\, \mathbf{E} \,|$, $l \stackrel{\triangle}{=} v/k$. For simplicity, we assume that E is uniformly distributed.

Proposition 5. *[2] $P_{\mathbf{I}} \geq k/v$. For the skeleton matrix $X = \{x_{ij}\}$, $P_{\mathbf{I}} = k/v$ if and only if*

$$w(x_j)/b = k/v \text{ for } \forall j.$$

Suppose we have a without secrecy A-code with no splitting. Then,

Proposition 6. *[10] If $P_{\mathbf{I}} = k/v = 1/l$, then $P_{\mathbf{S}} \geq 1/l$. For the skeleton matrix $X = \{x_{ij}\}$, $P_I = P_S = 1/l$ if and only if, for $\forall s$, $\forall s'$ such that $s \neq s'$ and for $\forall m_i \in M_s$, $\forall m_j \in M'_s$, $w(x_i \odot w_j)/b = 1/l^2$.*

Proposition 7. *[10, 5] If $P_{\mathbf{I}} = P_{\mathbf{S}} = k/v = 1/l$, then $b \geq \max(l^2,\ k(l-1)+1)$.*

3 New bound for A^2-code

3.1 Authentication code with arbitration (A^2-code)

A^2-code is a (S, M, E_R, E_T) such that S is a set of source states, M is a set of messages, E_R is a set of the receiver's encoding rules and E_T is a set of the transmitter's encoding rules. Let $E_R \circ E_T$ denote the set of possible pairs of $E_R \times E_T$. At the preprocessing stage, the arbiter sends $f_i \in E_R$ to the receiver R and $e_i \in E_T$ to the transmitter T secretly. e_i (therefore, T) generates one message for each $s \in S$ (that is, no splitting. However, f_i would accept many messages.) R accept $m \in M$ iff $f_i(m)$ is valid. When some dispute happens between T and R, the arbiter accepts m as authentic iff e_i can generate m.

3.2 Johansson's bound

Proposition 8. *[8]. If* $\max(P_I, P_S, P_T, P_{R_0}, P_{R_1}) = 1/q$, *then*

$$|E_R| \geq q^3, \quad |E_T| \geq q^4, \quad |E_R \circ E_T| \geq q^5, \quad |M| \geq q^2|S|$$

4 Generalization of basic results on A-code

Proposition 5 can be generalized as follows.

Lemma 9. *Let* $X = \{x_{ij}\}$ *be a* $b \times v$ *binary matrix. Let the row vectors be* $g_1, g_2, \cdots,$. *Suppose that* $w(g_i) \geq k$ *for* $\forall i$. *Then,*

(1) $\max_i w(x_j)/b \geq k/v$.
(2) The equality holds if and only if $w(x_j) = kb/v$ *for* $\forall j$ *and* $w(g_i) = k$ *for* $\forall i$.

Proof. Denote by N the total number of 1s in X. Then, $N \geq kb$. Therefore, there exists x_j such that $w(x_j)/b \geq N/v \geq kb/v$. (2) is clear from the above discussion. □

Definition 10. Let $I_l(h) \triangleq \{i | (h-1)l+1 \leq i \leq hl\}$. We say that a $b \times kl$ binary matrix $X = \{x_{ij}\}$ is a (b, k, l, n_0, n_1) K- array if the following conditions are satisfied.

(1) $w(x_i) = n_0$ for $\forall i$.
(2) For $\forall h_1$, $\forall h_2$ and for $\forall x_i \in I_l(h_1)$, $\forall x_j \in I_l(h_2)$,

$$w(x_i \odot x_j) = \begin{cases} n_1 & \text{if } h_1 \neq h_2 \\ 0 & \text{if } h_1 = h_2 \text{ and } x_i \neq x_j \end{cases}$$

Then, proposition 7 can be generalized as follows.

Lemma 11. *If there exists a* (b, k, l, n_0, n_1) K- *array, then* $b \geq k(l-1)+1$.

Proof. Let $X = \{x_{ij}\}$ be a (b, k, l, n_0, n_1) K- array. Let the row vectors be $g_1, g_2, \cdots, g_b$. Define $z_h = (z_{h,1}, z_{h,2}, \cdots, z_{h,kl})$ as a row vector such that

$$z_{h,j} = \begin{cases} 1 \; j \in I_l(h) \\ 0 \text{ otherwise.} \end{cases}$$

for $1 \leq h \leq k$. Let V be a vector space over the real number field spanned by $\{g_1, \cdots, g_b, z_1, \cdots, z_{k-1}\}$. We prove that $\dim V = kl$. First, we show that $z_k \in V$. It is easy to see that $z_k = (\sum g_i)/n_0 - \sum_{i \neq k} z_i$ from the definition of K-array. Thus, $z_k \in V$. Next, define

$$u_p \triangleq (0, \cdots, \underset{(p)}{1}, \cdots, 0)$$

Let V' be the vector space spanned by $\{u_1 \cdots, u_{kl}\}$. It is clear that $V \subseteq V'$. We show the converse. For $\forall h_1$, fix $p \in I_l(h_1)$, arbitrarily. Let the row vectors of X such that $x_{ip} = 1$ be $g'_1, g'_2, \cdots, g'_{n_0}$. Then, from the definition of K-array, we have

$$\sum_{i=1}^{n_0} g'_i = n_0 u_p + n_1 \sum_{i \neq h_1} z_i$$

Hence,

$$u_p = 1/n_0 (\sum_{i=1}^{n_0} g'_i - n_1 \sum_{i \neq h_1} z_i)$$

Thus, $u_p \in V'$ for $1 \leq p \leq kl$. This means that $V' \subseteq V$. Therefore, $\dim V = \dim V' = kl$. Hence, we must have $b + k - 1 \geq kl$. Then, we have this lemma. □

Lemma 12. *Suppose we have a without secrecy A-code (E, M, S) in which $P_I = P_S = |S|/|M| = 1/l$. Then, the skeleton matrix for (E, M) is a $(|E|, |S|, l, |E|/l, |E|/l^2)$ K-array.*

Proof. Clear from proposition 5 and 6. □

4.1 New bound

We present a more tight lower bound on the size of encoding rules than proposition 8 [8] for $| S | > P_{\mathbf{I}}^{-1} + 1$. (We consider without secrecy A^2-codes with no splitting.) Let

$$E_T(f_i) \triangleq \{e_j \mid \Pr(e_j, f_i) > 0\}$$

$$E_R(e_i) \triangleq \{f_j \mid \Pr(e_i, f_j) > 0\}$$

$$M_s \triangleq \{m | m = (s, a)\}.$$

Theorem 13. *Suppose that*

(C1) $P_{\mathbf{I}} = P_{\mathbf{S}} = P_{\mathbf{R}_0} = P_{\mathbf{R}_1} = P_{\mathbf{T}} = 1/q$
(C2) $|\mathbf{M}| = q^2 |\mathbf{S}|$
(C3) $|E_{\mathbf{R}}(e_i)| = q$.
(C4) *$E_T, E_R, E_T(f_i)$ and $E_R(e_j)$ are uniformly distributed, respectively.*
(C5) *For $\forall s$, there exist $A_1, A_2, \cdots$ such that*

$$M_s = A_1 \cup A_2 \cup \cdots, \quad A_i \cap A_j = \phi, \quad |A_i| = constant.$$

$\forall f_h \in E_R$ accepts just one message in $\forall A_i$.

Then,

$$|E_R| \geq |S|q(q-1)+1, \quad |E_T \circ E_R| \geq (|S|(q-1)+1)|E_R|, \quad |E_T| = |E_T \circ E_R|/q$$

Remark. (1) Our bound is more tight than proposition 8 if $|S| > q+1$.
(2) From proposition 8, $|M| \geq q^2|S|$. (C2) requires that this equality holds.
(3) It is easy to see that $|E_R(e_i)| \geq q$ if $P_T = 1/q$. (C3) requires that this equality holds.
(4) Consider the following situation. T sends $m_i \in M_s$. The opponent changes m_i to $m_j \in M_s$ and R accepts m_j. In this case, the source state is the same. However, if some dispute happens between T and R, the arbiter does not accept m_j as authentic. This attack should also be considered as a substitution attack of the opponent. We call this attack the second type substitution attack of the opponent.

4.2 Proof

Let $X = \{x_{ij}\}$ be the skeleton matrix for (E_R, M) (see Def.3). We will show that X is a $(|E_R|, q|S|, q, |E_R|/q, |E_R|/q^2)$ K- array (see Def.10). Let

$$M_{f_i} \triangleq \{m \mid f_i \text{ accepts m }\}, \quad M_{f_i}(s) \triangleq \{m \mid m \in M_{f_i}, m = (s,a)\}.$$

Lemma 14. (1) *$w(x_i) = |E_R|/q$ for $\forall i$.*
(2) $|E_{\mathbf{T}}(f_i)| \geq |\mathbf{S}|(q-1)+1$
(3) $|M_{f_i}| = q|\mathbf{S}|$
(4) $|M_{f_i}(s)| = q$

Proof. Consider a A-code $(S, M_{f_i}, E_T(f_i))$ in which the arbiter is a receiver and the receiver is an opponent. Let $Y = \{y_{ij}\}$ be the skeleton matrix for $(E_T(f_i), M_{f_i})$. Each row vector of Y has a constant weight $|S|$ because $\forall e_j$ generates one message for each s (no splitting). Then, from lemma 9,

$$1/q = P_{R_0} = \max_i w(y_i)/|E_T(f_i)| \geq |S|/|M_{f_i}| \tag{1}$$

Therefore, $|M_{f_i}| \geq q|S|$. Now, we have shown that the weight of each row vector of X is at least $q|S|$. Then, from lemma 9 and (C2),

$$1/q = P_I = \max_i w(x_i)/|E_R| \geq q|S|/|M| = 1/q$$

This means that $\max_i w(x_i)/|E_R| = q|S|/|M|$. Then, again from lemma 9, we have

$$w(x_i) = |E_R|/q \quad \text{for } \forall i, \quad \text{and } |M_{f_i}| = q|S| \quad \text{for } \forall i$$

Then, we see that the equality of eq.(1) is also satisfied. That is,

$$1/q = P_{R_1} = P_{R_0} = |S|/|M_{f_i}|$$

Now, from lemma 11 and 12, we have (2) of this lemma. Finally, we will prove (4). Let $W = \{w_{ij}\}$ be the skeleton matrix for $(E_T(f_i), M_{f_i}(s))$. The weight of each row vector of W is 1 from the second sentence of this proof. Then, from lemma 9, we have

$$1/q = P_{R_0} = \max_i w(w_i)/|E_T(f_i)| \geq 1/|M_{f_i}(s)|$$

Therefore, $|M_{f_i}(s)| \geq q$. On the other hand, $q|S| = |M_{f_i}| = \sum_s |M_{f_i}(s)| \geq q|S|$. Hence, we must have $|M_{f_i}(s)| = q$. □

¿From lemma 14, we see that X satisfies the condition (1) of Def.10.

Lemma 15. $|M_s| = q^2$.

Proof. Let $Y = \{y_{ij}\}$ be the skeleton matrix for (E_R, M_s). Each row vector of Y has the weight $|M_{f_i}(s)| = q$ from lemma 14. Then, from lemma 9,

$$1/q = P_I = \max_i w(y_i)/|E_R| \geq q/|M_s|$$

Therefore, $|M_s| \geq q^2$. On the other hand, $q^2|S| = |M| = \sum_s |M_s| \geq q^2|S|$. Hence, it must be that $|M_s| = q^2$. □

Let $F_i \triangleq \{f_u \mid f_u \text{ accepts } m_i\}$. Note that $|F_i| = w(x_i) = |E_R|/q$ from lemma 14.

Lemma 16. *For* $\forall s, \forall s'$ *such that* $s \neq s'$, *and for* $\forall m_i \in M_s, \forall m_j \in M_{s'}$,

$$w(x_i \odot x_j) = |E_{\mathbf{R}}|/q^2.$$

Proof. Suppose that $\mathbf{T}$ sends $m_i \in M_s$. Consider a substitution attack such that $\mathbf{O}$ changes m_i to $m_j \in M_{s'}$. This attack is modeled by the skeleton matrix for $(F_i, M_{s'})$. Let the skeleton matrix be $Y = \{y_{ij}\}$. The weight of each row of Y is $|M_{f_h}(s')| = q$ from lemma 5.1(4). Then, from lemma 9 and lemma 15,

$$1/q = P_S \geq \max_j w(y_j)/|F_i| \geq |M_{f_h}(s')|/|M_{s'}| = 1/q.$$

Therefore, from lemma 9, $w(y_j) = |F_i|q^{-1} = |E_R|q^{-2}$ for $\forall j$. It is easy to see that $w(y_j) = w(x_i \odot x_j)$. □

Lemma 17. *For* $\forall s$, *there exist* $A_1 \cdots A_q$ *such that*

(1) $M_s = A_1 \cup A_2 \cup \cdots \cup A_q, |A_i| = q$

(2) *For* $\forall A_t, \forall A_u$ *and for* $\forall x_i \in A_t, \forall x_j \in A_u$,

$$w(x_i \odot x_j) = \begin{cases} |E_R|/q^2 & \text{if } A_t \neq A_u \qquad (2) \\ 0 & \text{if } A_t = A_u \text{ and } x_i \neq x_j \qquad (3) \end{cases}$$

Proof. ¿From lemma 14, $|M_{f_h}(s)| = q$ for $\forall f_h$. Then, from (C5), there exist $A_1 \cdots A_q$ such that $M_s = A_1 \cup A_2 \cup \cdots \cup A_q$ and $\forall f_h$ accepts just one message in $\forall A_t$. Because $|M_s| = q^2$ (from lemma 15), we have $|A_u| = q$.

Since $\forall f_h$ accepts just one message in $\forall A_t$, we have eq.(3). Without loss of generality, let $A_t = A_1$ and let $\hat{M} = A_2 \cup \cdots \cup A_q$. Then, $|\hat{M}| = q(q-1)$. Suppose that **T** sends $m_i \in A_1$. The second type substitution attack (see Remarks 1 (4)) is modeled by the skeleton matrix for $(F_i, \hat{M})$. Let the skeleton matrix be $Y = \{y_{ij}\}$. The weight of each row vector of Y is $|M_{f_h}(s)| - 1 = q - 1$ since m_i is excluded (see lemma 14). Then, from lemma 9,

$$1/q = P_S \geq \max_j w(y_j)/|F_i| \geq (q-1)/|\hat{M}| = 1/q.$$

Therefore, from lemma 9,

$$w(x_i \odot w_j) = w(y_j) = |F_i|q^{-1} = |E_R|q^{-2}$$

for $\forall j$. Thus, eq.(2) is proved. □

¿From lemma 5.3 and 5.4, it is easy to see that X satisfies the condition (2) of Def.4.1. ¿From this fact and from lemma 5.1(1), X is a $(|E_R|, q|S|, q, |E_R|/q, |E_R|/q^2)$ K- array. Therefore, from lemma 11,

$$|E_R| \geq q|S|(q-1) + 1$$

Then, from lemma 5.1(2), $|E_T \circ E_R| \geq (|S|(q-1)+1)|E_R|$.. ¿From (C3), $|E_T| = |E_T \circ E_R|/q$.

5 Construction of A^2-code

We show two A^2-code, α and β. A^2-code α approximately meets the bound of Theorem 13. In A^2-code β, it is shown that $|E_T|$ can be greatly reduced by letting P_{R_1} be slightly greater than $1/q$ (as long as $|S| > q + 1$).

Let q be a prime power. Let a source state s be $s = (s_1, s_2, \cdots, s_n)$, where $\forall s_i \in GF(q)$.
In the proposed codes, $f_i \in E_R$ is $f_i = (f_{i1}, f_{i2}, \cdots, f_{i,n+2})$, where $\forall f_{ij} \in GF(q)$. T sends $m = (s_1, s_2, \cdots, s_n, m_1, m_2)$. R accepts m iff

$$m_2 = s_1 f_{i,1} + s_2 f_{i,2} + \cdots + s_n f_{i,n} + f_{i,n+1} + m_1 f_{i,n+2}$$

.

(A^2-code α)

In this A^2-code, $e_i \in E_T$ is $e_i = (e_{i1}, e_{i2}, \cdots, e_{i,2n+2})$, where $\forall e_{ij} \in GF(q)$. e_i and f_i are related as follows.

$$\left.\begin{array}{rcl} e_{i1} & = & f_{i1} + e_{i,n+2} f_{n+2} \\ e_{i2} & = & f_{i2} + e_{i,n+3} f_{i,n+2} \\ & \vdots & \\ e_{in} & = & f_{in} + e_{i,2n+1} f_{i,n+2} \\ e_{i,n+1} & = & f_{i,n+1} + e_{i,2n+2} f_{i,n+2} \end{array}\right\}$$

m_1 and m_2 are computed as follows.

$$m_1 = s_{i1} e_{i,n+2} + s_{i2} e_{i,n+3} + \cdots + s_n e_{i,2n+1} + e_{i,2n+2}$$
$$m_2 = s_1 e_{i,1} + s_2 e_{i,2} + \cdots + s_n e_{i,n} + e_{i,n+1}$$

Theorem 18. *In the above scheme,* $P_{\mathbf{I}} = P_{\mathbf{S}} = P_{\mathbf{R}_0} = P_{\mathbf{R}_1} = P_{\mathbf{T}} = 1/q$, *and*

$$|E_R| = q^2|S|, \quad |E_T \circ E_R| = q^3|S|^2, \quad |E_T| = q^2|S|^2$$

(A^2-code β)

In A^2-code β, $e_i \in E_T$ is $e_i = (e_{i1}, e_{i2}, \cdots, e_{i,n+2}, e_{i,2n+2})$, where $\forall e_{ij} \in GF(q)$. This code is obtained from A^2-code α by letting

$$e_{i,n+3} = e_{i,n+2}^2, \ \cdots, \ e_{i,2n+1} = e_{i,n+2}^n$$

Theorem 19. *In the above scheme,* $P_{\mathbf{I}} = P_{\mathbf{S}} = P_{\mathbf{R}_0} = P_{\mathbf{T}} = 1/q, \quad P_{\mathbf{R}_1} \leq n/q$ *and*

$$|E_R| = q^2|S|, \quad |E_T \circ E_R| = q^4|S|, \quad |E_T| = q^3|S|$$

6 Further work

Relationships with error correcting codes and orthogonal arrays will be discussed in the final paper.

References

1. E.N.Gilbert, F.J.MacWilliams and N.J.A.Sloane, "Codes which detect deception", *Bell Syst. Tech. J.*, Vol.53, 1974, pp.405–424
2. G.J.Simmons, "Authentication theory/coding theory", in *Advances in Cryptology, Proceedings of CRYPTO 84*, G.R.Blakley and D.Chaum, Eds.Lecture notes in Computer Science, No.196. New York, NY:Springer, 1985, pp.411–431.
3. J.L.Massey, "Contemporary Cryptology, An Introduction", in *Contemporary Cryptology, The Science of Information Integrity*, G.J.Simmons, Ed., IEEE Press, 1991, pp.3–39.
4. G.J.Simmons, "A survey of Information Authentication", in *Contemporary Cryptology, The science of information integrity*, ed. G.J.Simmons, IEEE Press, New York, 1992.

5. D.R.Stinson, *Combinatorial Characterization of Authentication Codes*, Proceedings Crypto 91, Lecture Notes in Computer Science 576, Springer 1992, pp62–72.
6. G.J.Simmons, "Message authentication with arbitration of transmitter/receiver disputes", in *Proceedings of Eurocrypt '87*.
7. G.J.Simmons, "A Cartesian Product Construction for Unconditionally Secure Authentication Codes that Permit Arbitration", in *Journal of Cryptology*, Vol.2, no.2, 1990, pp.77–104.
8. Thomas Johansson,"Lower Bounds on the Probability of Deception in Authentication with Arbitation", in *Proceeding of 1993 IEEE International Symposium on Information Theory*,San Antonio,USA,January 17-22,1993,p.231
9. Thomas Johansson, "On the construction of perfect authentication codes that permit arbitration", Crypto 93
10. D.R.Stinson, "The combinatorics of authentication and secrecy codes", Journal of Cryptology, Vol.2, 1990, pp.23–49.
11. M.N.Wegman and J.L.Carter. *New hash functions and their use in authentication and set equality*, J.Comput.System Sci. 22(1981), 265–279.
12. D.R.Stinson,"Universal Hashing and authentication codes" *Proceeding of Crypto 91*,Santa,Babara,USA,1991,pp.74-85
13. Jürgen Bierbrauter, Thomas Johanssen, Gregory Kabatianskii, Ben Smeets, "On Families of Hash Functions via Geometic Codes Concatination", Crypto 93

Multi-Secret Sharing Schemes *

Extended Abstract

Carlo Blundo, Alfredo De Santis, Giovanni Di Crescenzo,
Antonio Giorgio Gaggia, and Ugo Vaccaro

Dipartimento di Informatica ed Applicazioni,
Università di Salerno, 84081 Baronissi (SA), Italy
{carblu,ads,giodic,antgio,uv}@udsab.dia.unisa.it

Abstract. A multi-secret sharing scheme is a protocol to share m arbitrarily related secrets $s_1, \ldots, s_m$ among a set of participants $\mathcal{P}$. In this paper we put forward a general theory of multi-secret sharing schemes by using an information theoretical framework. We prove lower bounds on the size of information held by each participant for various access structures. Finally, we prove the optimality of the bounds by providing protocols.

1 Introduction

A secret sharing scheme is a technique to share a secret s among a set $\mathcal{P}$ of participants in such a way that only qualified subsets, pooling together their information, can reconstruct the secret s; but subsets of participants that are not enabled to recover the secret have no information on it. Secret sharing schemes are useful in any important action that requires the concurrence of several designed people to be initiated, as launching a missile, opening a bank vault or even opening a safety deposit box. Secret sharing schemes are also used in management of cryptographic keys and multi-party secure protocols (see [10], [2]).

Secret sharing schemes were introduced by Shamir [16] and Blakley [3]. They analyzed the case when only subsets A of $\mathcal{P}$ of cardinality $|A| \geq k$, for a fixed integer k, can reconstruct the secret. These schemes are called (k, n) threshold schemes, where $n = |\mathcal{P}|$. Subsequently, Ito, Saito, and Nishizeki [11] and Benaloh and Leichter [1] described a more general method of secret sharing. They showed how to realize a secret sharing scheme for any access structure, where the access structure is the family of all subsets of participants that are able to reconstruct the secret. The recent survey by Stinson [18] contains an unified description of recent results in the area of secret sharing schemes. For different approaches to the study of secret sharing schemes, for schemes with "extended capabilities" as disenrollment, fault-tolerance, and pre-positioning and for a complete bibliography we recommend the survey article by Simmons [17].

* Partially supported by Italian Ministry of University and Research (M.U.R.S.T.) and by National Council for Research (C.N.R.).

There are several situations in which more than one secret is to be shared among participants. As an example, consider the following situation, described by Simmons [17]: there is a missile battery and not all of the missiles have the same launch enable code. The problem is to devise a scheme which will allow any one, or any selected subset, of the launch enable codes to be activated in this scheme. What is needed is an algorithm such that the same pieces of private information could be used to recover different secrets. This problem could be trivially solved by realizing different secret sharing schemes, one for each of the launch enable codes, but this solution is clearly unacceptable since each participant should remember too much information.

Another scenario in which the sharing of many secrets is important was considered by Franklin and Yung [8]. They investigated the communication complexity of unconditionally secure multi–party computation and its relations with various fault–tolerant models. They presented a general technique for parallelizing non–cryptographic computation protocols, at a small cost in fault–tolerance. Their technique replaces polynomial–based (single) secret sharing with a technique allowing multiple secrets to be hidden in a single polynomial. The technique applies to all of the protocols for secure computation which use polynomial–based threshold schemes and applies to all fault–tolerant models. Franklin and Yung [8] considered also the case of dependent secrets in which the amount of information distributed to any participant is less than the information distributed with independent schemes.

The problem of sharing more than one secret was also considered in [14].

Blundo, De Santis, and Vaccaro [5] considered the case in which m secrets are shared among participants in $\mathcal{P}$ of a single access structure $\mathcal{A}$ in such a way that: 1) any qualified subset of participants can reconstruct all the secrets, 2) any non-qualified subset has absolutely no information on any secret, and 3) any non-qualified subset knowing the value of a number of secrets might determine some (possibly no) information on other secrets. They proved lower bounds on the size of the domains from which the share given to participants are taken. Moreover, they proved that the protocol proposed by Franklin and Yung [8] is optimal with respect to the amount of information given to each participant.

Recently Jackson, Martin, and O'Keefe [12] considered the problem where participants can reconstruct more than one secret using the information that they hold. In particular, they considered the situation in which there is a secret associated with each set $K \subseteq \mathcal{P}$, where $|K| = k$. This secret can be reconstructed by any t ($t \leq k$) participants of K. They proved bounds on the size of information that participants must hold in order to ensure that up to w participants ($0 \leq w \leq n - k + t - 1$) cannot obtain any information about a secret they are not associated with. In [12] such schemes are referred as multisecret threshold schemes. Finally, in [13] the authors provide an optimal scheme, with respect to the information given to each participant, for some value of the parameters t and w.

In this paper we put forward a general theory of multi-secret sharing schemes by using an information theoretical framework. We prove lower bounds on the size of information held by each participant for various access structures. Finally, we prove the optimality of the bounds. We prove that in some cases the protocol consisting of realizing different secret sharing schemes, one for each of the secrets, is optimal with respect to the size of the share given to a single participant. In other cases the before mentioned protocol is not optimal and we exhibit schemes that give to participants shares taken from a smaller domain.

The paper is organized as follows. In Section 2 we formally define multi–secret sharing schemes by using information theoretical quantities. We consider two possible models of multi-secret sharing schemes. We model secret sharing schemes by using the entropy mainly because this leads to a compact and simple description of the scheme and because the entropy approach takes into account all probability distributions on the secret. Finally, each bound we obtain on the entropy of the share of a participant implies a bound on the amount of information held by such a participant. In Subsection 2.3 we prove that the two models proposed for multi–secret sharing are equivalent. In Section 3 we show how to construct perfect multi-secret sharing schemes for two and three secrets in which the shares distributed are taken from domains as small as possible. An important issue in the implementation of secret sharing schemes is the size of the shares given to participants, since the security of a system degrades as the amount of secret information increases. Thus, one of the basic problems in the field of secret sharing schemes is to derive bounds on the amount of information that must be kept secret. In Section 4 we prove a lower bound on the information distributed to any participant in multi–secret sharing schemes. Finally, in Section 5 we analyze the case in which all the access structures are threshold structures. We prove lower bounds on the size of information held by each participant in the scheme and provide optimal protocols for multi-secret sharing in threshold structures.
Due to the space limit on this extended abstract, some proofs are omitted. The authors will supply a complete version on request.

2 The Models

In this section we give two different definitions of multi-secret sharing schemes and show their equivalence. Let us first briefly recall the concept of secret sharing scheme.

A secret sharing scheme permits a secret to be shared among a set $\mathcal{P}$ of n participants in such a way that only qualified subsets of $\mathcal{P}$ can recover the secret, but any non-qualified subset has absolutely no information on the secret. An access structure $\mathcal{A}$ is the set of all subsets of $\mathcal{P}$ that can recover the secret.

Definition 1. Let $\mathcal{P}$ be a set of participants, a *monotone access structure* $\mathcal{A}$ on $\mathcal{P}$ is a subset $\mathcal{A} \subseteq 2^{\mathcal{P}}$, such that $A \in \mathcal{A}, A \subseteq A' \subseteq P \Rightarrow A' \in \mathcal{A}$.

Definition 2. Let $\mathcal{P}$ a set of participants and $\mathcal{A} \subseteq 2^{\mathcal{P}}$. The *closure* of $\mathcal{A}$, denoted by $\mathsf{cl}(\mathcal{A})$, is the set $\mathsf{cl}(\mathcal{A}) = \{C | B \in \mathcal{A} \text{ and } B \subseteq C \subseteq P\}$.

For a monotone access structure $\mathcal{A}$ we have $\mathcal{A} = \text{cl}(\mathcal{A})$. From now on we will consider only monotone access structures.

In multi-secret sharing schemes the problem of sharing many secrets is addressed. We consider two models of multi-secret sharing. The first model is a natural generalization of single secret sharing: we consider different access structures and in each of them we share a different secret. We will refer to this model as *Type A*. In the second model, referred to as *Type B*, each set $A \subseteq \mathcal{P}$ can recover a set S_A of secrets, where it can be $S_A = \emptyset$. This second model generalizes the one considered by Jackson, Martin, and O'Keefe [12]. Even though it could appear that the two models are different, we will show that they are indeed equivalent.

The following setting is common to both models. Let $\mathcal{P}$ be a set of participants, let S_i be the space from which the i-th secret s_i can be selected, for $i = 1, \ldots, m$, and let $\mathcal{SC}$ be the cartesian product $S_1 \times \cdots \times S_m$. Finally, let $\{p_{\mathcal{SC}}(s_1, \ldots, s_m)\}_{(s_1,\ldots,s_m)\in\mathcal{SC}}$ be a probability distribution on $\mathcal{SC}$. Let a multi-secret sharing scheme for secrets in $\mathcal{SC}$ be fixed. For any participant $P \in \mathcal{P}$, let us denote by $K(P)$ the set of all possible shares given to participant P. Given a set of participants $A = \{P_{i_1}, \ldots, P_{i_r}\} \subseteq \mathcal{P}$, where $i_1 < i_2 < \ldots < i_r$, set $K(A) = K(P_{i_1}) \times \cdots \times K(P_{i_r})$. Any multi-secret sharing scheme for secrets in $\mathcal{SC}$ and a probability distribution $\{p_{\mathcal{SC}}(s_1, \ldots, s_m)\}_{(s_1,\ldots,s_m)\in\mathcal{SC}}$ naturally induce a probability distribution on $K(A)$, for any $A \subseteq \mathcal{P}$. Denote such a probability distribution by $\{p_{K(A)}(a)\}_{a\in K(A)}$. Finally, denote by $H(S_i)$ the entropy[2] of $\{p_{S_i}(s)\}_{s\in S_i}$ and by $H(A)$ the entropy of $\{p_{K(A)}(a)\}_{a\in K(A)}$, for any $A \subseteq \mathcal{P}$. If S_A is a set of secrets $\{s_{i_1}, \ldots, s_{i_a}\}$, where $s_{i_j} \in S_{i_j}$, then denote by $H(S_A)$ the entropy of $\{p_{S_{i_1}\times\cdots\times S_{i_a}}(s_{i_1}, \ldots, s_{i_a})\}_{s_{i_j}\in S_{i_j}, j=1,\ldots,a}$. To avoid overburdening the notation, we will denote with the same symbol both a random variable and the set of its possible values. As an example, with S_i we will denote both the set in which the i-th secret is chosen and the random variable that takes values in the set S_i with probability distribution $\{p_{S_i}(s)\}_{s\in S_i}$.

We will give our two definitions of multi-secret sharing schemes first in terms of the probability distribution on the secret and on the shares given to participants, and then using the entropy function as done in [14], [15], and [6].

2.1 The First Model

In the first definition of perfect multi-secret sharing scheme, an m-tuple of secrets $(s_1, \ldots, s_m) \in S_1 \times \cdots \times S_m$ is shared in an m-tuple $(\mathcal{A}_1, \ldots, \mathcal{A}_m)$ of access structures on $\mathcal{P}$ in such a way that, for each $i = 1, \ldots, m$, the access structure $\mathcal{A}_i$ is the set of all subsets of $\mathcal{P}$ that can recover secret $s_i \in S_i$. This means that only the sets in $\mathcal{A}_i$ can recover the secret s_i, but any set $A \notin \mathcal{A}_i$ has no information on it. A multi-secret sharing scheme of Type A is defined as follows.

[2] For definition and properties of information theoretic quantities we refer to [7] and [9].

Definition 3. Let $(\mathcal{A}_1, \ldots, \mathcal{A}_m)$ be an m-tuple of monotone access structures on the set of participants $\mathcal{P}$. A *multi-secret sharing scheme of Type A* for $(\mathcal{A}_1, \ldots, \mathcal{A}_m)$ is a sharing of the secrets $(s_1, \ldots, s_m) \in S_1 \times \cdots \times S_m$ in such a way that, for $i = 1, \ldots, m$,

1. *Any subset $A \subseteq \mathcal{P}$ of participants enabled to recover s_i can compute s_i.* Formally, if $A \in \mathcal{A}_i$ then for all $a \in K(A)$ with $p_{K(A)}(a) > 0$, it holds $p(s_i|a) = 1$.
2. *Any subset $A \subseteq \mathcal{P}$ of participants not enabled to recover s_i, even knowing some of the other secrets, has no more information on s_i than that already conveyed by the known secrets.* Formally, if $A \notin \mathcal{A}_i$ then for all $a \in K(A)$ and $t \subseteq \{s_1, \ldots, s_m\} \setminus \{s_i\}$, it holds $p(s_i|at) = p(s_i|t)$.

Property 1. means that the values of the shares held by $A \in \mathcal{A}_i$ completely determine the secret s_i. Property 2. means that the probability that a secret is equal to s_i given that any subset of secrets not including s_i is equal to t and that the shares held by $A \notin \mathcal{A}_i$ are equal to a, is the same as the *a priori* probability of the secret s_i given that any subset of secrets not including s_i is equal to t. In case $t = \emptyset$, this is equivalent to say that no amount of knowledge of shares of participants not qualified to reconstruct a secret enables a Bayesian opponent to modify an *a priori* guess regarding which the secret is.

Now we can restate above conditions 1. and 2. using information theoretic tools. We model secret sharing schemes by using the entropy mainly because this leads to a compact and simple description of the scheme and because the entropy approach takes into account all probability distributions on the secret. Finally, each bound we obtain on the entropy of the share of a participant implies a bound on the amount of information held by such a participant.

Definition 4. Let $(\mathcal{A}_1, \ldots, \mathcal{A}_m)$ be an m-tuple of monotone access structures on the set of participants $\mathcal{P}$. A *multi-secret sharing scheme of Type A* for $(\mathcal{A}_1, \ldots, \mathcal{A}_m)$ is a sharing of the secrets $(s_1, \ldots, s_m) \in S_1 \times \cdots \times S_m$ in such a way that, for $i = 1, \ldots, m$,

a. *Any subset $A \subseteq \mathcal{P}$ of participants enabled to recover s_i can compute s_i.* Formally, for all $A \in \mathcal{A}_i$, it holds $H(S_i|A) = 0$.

b. *Any subset $A \subseteq \mathcal{P}$ of participants not enabled to recover s_i, even knowing some of the other secrets, has no more information on s_i than that already conveyed by the known secrets.* Formally, for all $A \notin \mathcal{A}_i$ and $T \subseteq \{S_1, \ldots, S_m\} \setminus \{S_i\}$, it holds $H(S_i|A\,T) = H(S_i|T)$.

Notice that $H(S_i|A) = 0$ means that each set of values of the shares in A corresponds to a unique value of the secret. In fact, by definition, $H(S_i|A) = 0$ is equivalent to the fact that for all $a \in K(A)$ with $p_{K(A)}(a) > 0$ it holds $p(s_i|a) = 1$. Moreover, $H(S_i|AT) = H(S_i|T)$ is equivalent to state that S_i and $K(A)$ are statistically independent, given the secrets in T; i.e., for all $a \in K(A)$ and all

$t \in T$, it holds $p(s_i|at) = p(s_i|t)$, and therefore the knowledge of a gives no information about the secret s_i that is not already given by t. Finally, notice that in the case the access structures $\mathcal{A}_1, \ldots, \mathcal{A}_m$ are all equal to the same structure $\mathcal{A}$, a multi-secret sharing scheme for secrets $s_1, \ldots, s_m$ reduces to a secret sharing scheme for the secret $s = s_1 \circ \cdots \circ s_m$ with access structure $\mathcal{A}$, where with $x \circ y$ we denote the concatenation of x and y.

2.2 The Second Model

In our second definition of perfect multi-secret sharing schemes a set $\mathcal{S} = \{s_1, \ldots, s_m\}$ of secrets, where each s_i is chosen in a set S_i, is shared among a set $\mathcal{P}$ of participants in such a way that each subset of $\mathcal{P}$ can recover a certain subset of $\mathcal{S}$, but has absolutely no information on the remaining secrets.

For each subset of participants $A \subseteq \mathcal{P}$, we denote by $S_A \subseteq \mathcal{S}$ the set of secrets that can be recovered by A, referred to as the A-secrets-set. It should be pointed out that in some cases we could have $S_A = \emptyset$. Since we only consider monotone access structures, it turns out that for any $A, B \subseteq \mathcal{P}$ if $A \subseteq B$, then $S_A \subseteq S_B$.

Definition 5. Let $\mathcal{P}$ be a set of participants, $\mathcal{S}$ be a set of secrets, and $\{S_A\}_{A \subseteq \mathcal{P}}$ be the family of A-secrets-sets. A *multi-secret sharing scheme of Type B* for $\{S_A\}_{A \subseteq \mathcal{P}}$ is a sharing of the secrets in $\mathcal{S}$ among participants in $\mathcal{P}$ in such a way that

1′. *Any subset $A \subseteq \mathcal{P}$ of participants is enabled to recover the A-secrets-set S_A.*
   Formally, for all $a \in K(A)$ with $p_{K(A)}(a) > 0$ and $s \in S_A$, it holds $p(s|a) = 1$.
2′. *Any subset $A \subseteq \mathcal{P}$ of participants has no information on any subset of secrets in $\mathcal{S} \setminus S_A$.*
   Formally, for all $A \subseteq \mathcal{P}$, for all $a \in K(A)$ and $t \subseteq \mathcal{S} \setminus S_A$, it holds $p(t|a) = p(t)$.

Property 1′. means that the value of the shares held by $A \subseteq \mathcal{P}$ completely determines the secrets in S_A. Property 2′. means that the probability that a subset of secrets is equal to t given that the shares held by A are a, is the same as the *a priori* probability of the secrets in t. Therefore, no amount of knowledge of shares of participants not qualified to reconstruct a subset of secrets enables a Bayesian opponent to modify an *a priori* guess regarding which the secrets are.

For any $A \subseteq \mathcal{P}$, if $S_A = \{s_{i_1}, \ldots, s_{i_a}\}$, then with S_A we denote the family $S_A = \{S_{i_1}, \ldots, S_{i_a}\}$. Now we can restate above conditions 1′. and 2′. using information theoretic tools.

Definition 6. Let $\mathcal{P}$ be a set of participants, $\mathcal{S}$ be a set of secrets, and $\{S_A\}_{A \subseteq \mathcal{P}}$ be the family of A-secrets-sets. A *multi-secret sharing scheme of Type B* for $\{S_A\}_{A \subseteq \mathcal{P}}$ is a sharing of the secrets in $\mathcal{S}$ among participants in $\mathcal{P}$ in such a way that

a′. *Any subset $A \subseteq \mathcal{P}$ of participants is enabled to recover the A-secrets-set S_A.*
   Formally, for all $A \subseteq \mathcal{P}$, it holds $H(S_A|A) = 0$.

b'. Any subset $A \subseteq \mathcal{P}$ of participants has no information on any subset of secrets in $\mathcal{S} \setminus S_A$.
Formally, for all $A \subseteq \mathcal{P}$ and $T \subseteq \{S_1, \ldots, S_m\} \setminus S_A$, it holds $H(T|A) = H(T)$.

Notice that $H(S_A|A) = 0$ means that each set of values of the shares of participants in A corresponds to a unique value of the secrets in S_A. Moreover, $H(T|A) = H(T)$ is equivalent to state that $T \subseteq \{S_1, \ldots, S_m\} \setminus S_A$, and $K(A)$ are statistically independent and therefore the knowledge of the shares of the participants in A gives no information about the secrets in $S \setminus S_A$.

2.3 The Equivalence of the Two Models

In this section we prove that the two definitions presented for perfect multi-secret sharing schemes are equivalent; that is, each scheme satisfying one definition satisfies also the other as stated by next theorem.

Theorem 7. *Let $\mathcal{P}$ be a set of participants and let $S_1 \times \cdots \times S_m$ be a probability space from which the secrets $(s_1, \ldots, s_m)$ are chosen. The following statements hold.*

1. *Let $\mathcal{A}_1, \ldots, \mathcal{A}_m$ be access structures on the set of participants $\mathcal{P}$. If Σ is a secret sharing scheme of Type A for $(\mathcal{A}_1, \ldots, \mathcal{A}_m)$, then Σ is a secret sharing scheme of Type B for the family $\{S_A\}_{A \subseteq \mathcal{P}}$, where $S_A = \{s_i : A \in \mathcal{A}_i,\ i \in [1, m]\}$.*
2. *Let $\{S_A\}_{A \subseteq \mathcal{P}}$ be a family of A-secret-sets. If Σ is a secret sharing scheme of Type B for $\{S_A\}_{A \subseteq \mathcal{P}}$, then Σ is a secret sharing scheme of Type A for $(\mathcal{A}_1, \ldots, \mathcal{A}_m)$, where $\mathcal{A}_i = \{A \subseteq \mathcal{P} : s_i \in S_A\}$.*

Proof: Suppose Σ is a multi-secret sharing scheme of Type A. Let $(\mathcal{A}_1, ..., \mathcal{A}_m)$ be an m-tuple of access structures on participants $\mathcal{P}$ and let $(s_1, ..., s_m) \in S_1 \times \cdots \times S_m$ be the secrets shared in $(\mathcal{A}_1, \ldots, \mathcal{A}_m)$. For any $A \subseteq \mathcal{P}$ let $S_A = \{s_i : A \in \mathcal{A}_i,\ i \in [1, m]\}$. We prove that conditions a'. and b'. of Definition 6 are satisfied. Let us prove that $H(S_A|A) = 0$. We have that

$$\begin{aligned} H(S_A|A) &= H(S_{j_1}, \ldots, S_{j_r}|A) \\ &= H(S_{j_1}|A) + \sum_{i=2}^{r} H(S_{j_i}|S_{j_1} \ldots S_{j_{i-1}} A) \\ &\leq \sum_{i=1}^{r} H(S_{j_i}|A) \\ &= 0. \end{aligned}$$

Now, we prove that for any $T \subseteq \mathcal{S} \setminus S_A$, it holds $H(T|A) = H(T)$. Suppose that $T = \{S_{j_1}, \ldots, S_{j_t}\}$. We have

$$H(T|A) = H(S_{j_1}, \ldots, S_{j_t}|A)$$

$$= H(S_{j_1}|A) + \sum_{i=2}^{t} H(S_{j_i}|S_{j_1} \dots S_{j_{i-1}} A)$$
$$= H(S_{j_1}) + \sum_{i=2}^{t} H(S_{j_i}|S_{j_1} \dots S_{j_{i-1}})$$
$$= H(S_{j_1}, \dots, S_{j_t})$$
$$= H(T).$$

Hence, if Σ is a multi-secret sharing scheme of Type A for $(\mathcal{A}_1, \dots, \mathcal{A}_m)$, then Σ is also a multi-secret sharing scheme of Type B for $\{S_A\}_{A \subseteq \mathcal{P}}$.
Now we prove that statement 2. of the theorem holds. Let $\{S_A\}_{A \subseteq \mathcal{P}}$ be a family of A-secrets-sets. Let $(\mathcal{A}_1, \dots, \mathcal{A}_m)$ be an m-tuple of access structures, where $\mathcal{A}_i = \{A \subseteq \mathcal{P} : s_i \in S_A\}$. We prove that conditions a. and b. of Definition 4 are satisfied. It is easy to prove that for all $A \in \mathcal{A}_i$ it holds $H(S_i|A) = 0$. Indeed, we get $H(S_i|A) \leq H(S_A|A)$ and since $H(S_A|A) = 0$ from a'. of Definition 6, it follows that $H(S_i|A) = 0$. Now, we prove that for all $A \notin \mathcal{A}_i$ and $T \subseteq \{S_1, \dots, S_m\} \setminus \{S_i\}$, it holds $H(S_i|A\ T) = H(S_i|T)$. Notice that if $A \notin \mathcal{A}_i$ then $s_i \notin S_A$. Let $T = T_1 \cup T_2$, where $T_1 \subseteq S_A$ and $T_2 \cap S_A = \emptyset$. We have,

$$\begin{aligned}
H(T_2) + H(S_i|T_2) &= H(S_i T_2) \\
&= H(S_i T_2|A) \\
&= H(S_i T_2|A) + H(T_1|A) \\
&= H(S_i T_1 T_2|A) \\
&= H(T_2|A) + H(T_1|A T_2) + H(S_i|A T_1 T_2) \\
&= H(T_2|A) + H(S_i|A T_1 T_2) \\
&= H(T_2) + H(S_i|A T_1 T_2)
\end{aligned}$$

From previous equalities we get $H(S_i|T_2) = H(S_i|AT)$. From well known properties of the entropy function we have $H(S_i|T_2) \geq H(S_i|T)$ and $H(S_i|AT) \leq H(S_i|T)$. Thus, the theorem holds. □

From now on, the term multi-secret sharing scheme will refer to any of the two definitions given.

3 Sharing Two and Three Secrets

In this section we describe multi-secret sharing schemes for two and three secrets. We are interested in limiting the size of the share of a fixed participant P. The scheme we propose are realized, for simplicity of the description, considering as qualified sets only pairs of participants, but they can be easily extended to handle the general case where instead of participants we have groups of them.

3.1 The Case of Two Secrets

In this section we consider the case where $\mathcal{P} = \{P, P_1, P_2\}$ and $\mathcal{S} = \{S_1, S_2\}$. Suppose that $\{P, P_1\} \in \mathcal{A}_1$, $\{P, P_2\} \in \mathcal{A}_2$, $\{P, P_1\} \notin \mathcal{A}_2$, and $\{P, P_2\} \notin \mathcal{A}_1$. If we use the single-secret sharing construction for S_1 and S_2, we obtain a perfect multi-secret sharing scheme in which the dealer gives P a share such that $H(P) \geq H(S_1 S_2)$.

Assume that $\{P_1, P_2\} \in \mathcal{A}_1 \cup \mathcal{A}_2$; we describe a scheme such that, for uniformly and independently chosen 1-bit secrets, distributes shares to participants such that $H(P) = H(S_1) = H(S_2)$. Denote by $\oplus$ the logical xor between two bits.

The dealer uniformly chooses two independent bits a, b and distributes the shares as follows:

- P gets $a \oplus b$
- P_1 gets $a \oplus s_1$, b
- P_2 gets a, $b \oplus s_2$.

In the next section (Corollary 9) we will see that in the case $\{P_1, P_2\} \notin \mathcal{A}_1 \cup \mathcal{A}_2$, all multi-secret sharing schemes must satisfy $H(P) \geq H(S_1 S_2)$.

3.2 The Case of Three Secrets

In this section we consider the case where $\mathcal{P}=\{P, P_1, P_2, P_3\}$ and $\mathcal{S}=\{S_1, S_2, S_3\}$. Assume that $\{P, P_j\} \in \mathcal{A}_j$, for each $j = 1, 2, 3$. We distinguish two cases according to which group of participants can recover a subset of the secrets and for each case we describe a multi-secret sharing scheme which gives P a share taken from a domain as small as possible.

1. $\{P_1, P_2, P_3\} \in \mathcal{A}_1 \cap \mathcal{A}_2 \cap \mathcal{A}_3$, that is, participants P_1, P_2, and P_3 together are able to recover S_1, S_2, and S_3
   (a) $\{P_1, P_2\} \in \mathcal{A}_1 \cap \mathcal{A}_2, \{P_1, P_3\} \in \mathcal{A}_1 \cap \mathcal{A}_3$, and $\{P_2, P_3\} \in \mathcal{A}_2 \cap \mathcal{A}_3$
   (b) $\{P_i, P_j\} \notin \mathcal{A}_i \cap \mathcal{A}_j$, for some $i, j \in \{1, 2, 3\}$ and $i \neq j$;
2. $\{P_1, P_2, P_3\} \notin \mathcal{A}_1 \cap \mathcal{A}_2 \cap \mathcal{A}_3$, that is, participants P_1, P_2, and P_3 together are not able to recover at least one of S_1, S_2, and S_3
   (a) $\{P_i, P_j\} \in \mathcal{A}_i \cap \mathcal{A}_j$, for some $i, j \in \{1, 2, 3\}$ and $i \neq j$;
   (b) $\{P_1, P_2\} \notin \mathcal{A}_1 \cap \mathcal{A}_2$, $\{P_1, P_3\} \notin \mathcal{A}_1 \cap \mathcal{A}_3$, and $\{P_2, P_3\} \notin \mathcal{A}_2 \cap \mathcal{A}_3$.

The above classification partitions the family of all triples of access structures we could get in four classes. We construct a multi-secret sharing scheme for each class for uniformly and independently chosen 1-bit secrets.

For all schemes the dealer uniformly chooses three independent bits a, b, and c distributing the shares as follows.

- Case 1.a:
  - P gets $a \oplus b \oplus c$

- P_1 gets $a \oplus s_1, b, c$
- P_2 gets $a, b \oplus s_2, c$
- P_3 gets $a, b, c \oplus s_3$

In this case we have that $\{P_1, P_2\} \in \mathcal{A}_1 \cap \mathcal{A}_2$, $\{P_1, P_3\} \in \mathcal{A}_1 \cap \mathcal{A}_3$, and $\{P_2, P_3\} \in \mathcal{A}_2 \cap \mathcal{A}_3$. It is easy to obtain from this scheme all possible schemes for access structures satisfying the conditions of case 1.*a* by distributing additional shares to participants P_1, P_2, P_3. For example, assume $\{P_1, P_2\} \in \mathcal{A}_1 \cap \mathcal{A}_2 \cap \mathcal{A}_3$, $\{P_1, P_3\} \in \mathcal{A}_1 \cap \mathcal{A}_3$, and $\{P_2, P_3\} \in \mathcal{A}_2 \cap \mathcal{A}_3$. Then, the dealer uniformly chooses a bit d and distributes as additional shares d to P_1 and $d \oplus s_3$ to P_2.

- Case 1.*b*:
  Assume, wlog, that $\{P_1, P_2\} \notin \mathcal{A}_1 \cap \mathcal{A}_2$.
  - P gets $a \oplus c, b$
  - P_1 gets $a \oplus s_1, b, c$
  - P_2 gets $b \oplus s_2, c$
  - P_3 gets $a, c \oplus s_3$

In this case we have that $\{P_1, P_2\} \in \mathcal{A}_2$ $\{P_2, P_3\} \in \mathcal{A}_3$, and $\{P_1, P_3\} \in \mathcal{A}_1 \cap \mathcal{A}_3$. It is easy to obtain from this scheme all possible schemes for access structures satisfying the conditions of case 1.*b* by distributing additional shares to participants P_1, P_2, P_3. For example, assume $\{P_1, P_2\} \in \mathcal{A}_2 \cap \mathcal{A}_3$, $\{P_1, P_3\} \in \mathcal{A}_1 \cap \mathcal{A}_2 \cap \mathcal{A}_3$, and $\{P_2, P_3\} \in \mathcal{A}_1 \cap \mathcal{A}_2 \cap \mathcal{A}_3$. Then, the dealer uniformly chooses three bits d, e, and f distributing as additional shares $d, f \oplus s_2$ to P_1, $d \oplus s_3, e$ to P_2, and $b, e \oplus s_1, f$ to P_3.

- Case 2.*a*:
  Assume, wlog, that $\{P_1, P_2\} \in \mathcal{A}_1 \cap \mathcal{A}_2$.
  - P gets $a \oplus b, c$
  - P_1 gets $a \oplus s_1, b$
  - P_2 gets $a, b \oplus s_2$
  - P_3 gets $a, b, c \oplus s_3$

In this case we have that $\{P_1, P_2\} \in \mathcal{A}_1 \cap \mathcal{A}_2$, $\{P_1, P_3\} \in \mathcal{A}_1$, and $\{P_2, P_3\} \in \mathcal{A}_2$. It is easy to obtain from this scheme all possible schemes for access structures satisfying conditions of case 2.*a* by distributing additional shares to participants P_1, P_2, and P_3. For example, assume $\{P_1, P_2\} \in \mathcal{A}_1 \cap \mathcal{A}_2, \{P_1, P_3\} \in \mathcal{A}_1 \cap \mathcal{A}_2$, and $\{P_2, P_3\} \in \mathcal{A}_1 \cap \mathcal{A}_2$. Then, the dealer uniformly chooses two bits d and e distributing as additional shares $d, e \oplus s_2$ to P_3 $d \oplus s_1$ to P_2, and e to P_1.

- Case 2.*b* :
  - P gets a, b, c
  - P_1 gets $a \oplus s_1$
  - P_2 gets $a, b \oplus s_2$
  - P_3 gets $a, b, c \oplus s_3$

In this case we have that $\{P_1, P_2\} \in \mathcal{A}_1$, $\{P_1, P_3\} \in \mathcal{A}_1$, and $\{P_2, P_3\} \in \mathcal{A}_2$. It is easy to obtain from this scheme all possible schemes for access structures satisfying case 2.b by distributing additional shares to participants P_1, P_2, and P_3. For example, assume $\{P_1, P_2\} \in \mathcal{A}_1, \{P_1, P_3\} \in \mathcal{A}_1 \cap \mathcal{A}_2$, $\{P_2, P_3\} \in \mathcal{A}_1 \cap \mathcal{A}_2$. Then, the dealer uniformly chooses two bits d and e distributing as additional shares $d \oplus s_2$ to P_1, e to P_2, and $d, e \oplus s_1$ to P_3.

In the next section we will prove that the above schemes are optimal with respect to the entropy of P's share.

4 Bounds on the Size of the Shares

In the previous section we have investigated the possibility of constructing perfect multi-secret sharing schemes without using necessarily different single-secret sharing schemes one for each of the secrets. We have seen that in some cases the shares given to participants are taken from smaller domains. In this section we give lower bounds on the entropy of the share of a single participant.

Theorem 8. *Let $(\mathcal{A}_1, \ldots, \mathcal{A}_m)$ be an m-tuple of access structures on the set of participants $\mathcal{P}$. Assume that for all $S_i \in \{S_1, \ldots, S_m\}$ and $T \subseteq \{S_1, \ldots, S_m\} \setminus \{S_i\}$ it holds $H(S_i|T) > 0$. If there exist a participant P and $j \leq m$ subsets of participants $X_{i_1}, \ldots, X_{i_j} \subset \mathcal{P}$, such that $\{P\} \cup X_{i_1} \cup \cdots \cup X_{i_t} \in \mathcal{A}_{i_t}$ and $X_{i_1} \cup \ldots \cup X_{i_t} \notin \mathcal{A}_{i_t}$ for $1 \leq t \leq j$, then in any multi-secret sharing scheme for $(\mathcal{A}_1, \ldots, \mathcal{A}_m)$ the entropy of the share given to P satisfies*

$$H(P) \geq H(S_{i_1}, \ldots, S_{i_j}) + H(P|X_{i_1}, \ldots, X_{i_j}, S_{i_1}, \ldots, S_{i_j}).$$

Corollary 9. *Given the secrets S_1, S_2 and the set of participants $\mathcal{P} = \{P, P_1, P_2\}$, let $(\mathcal{A}_1, \mathcal{A}_2)$ be a pair of access structures such that $\{P, P_1\} \in \mathcal{A}_1$, $\{P, P_2\} \in \mathcal{A}_2$, $\{P, P_1\} \notin \mathcal{A}_2, \{P, P_2\} \notin \mathcal{A}_1$, and $\{P_1, P_2\} \notin \mathcal{A}_1 \cap \mathcal{A}_2$. Then, in any multi-secret sharing scheme for $(\mathcal{A}_1, \mathcal{A}_2)$ the entropy of the share given to P satisfies $H(P) \geq H(S_1 S_2)$.*

Proof: Assume $\{P_1, P_2\} \in \mathcal{A}_1$. Thus, $P_1 \notin \mathcal{A}_1$ and $\{P_1, P_2\} \notin \mathcal{A}_2$. Participants P_1 and P_2 satisfy the hypothesis of Theorem 8, hence $H(P) \geq H(S_1 S_2)$. $\square$

Corollary 10. *Given the secrets S_1, S_2, and S_3, and the set of participants $\mathcal{P} = \{P, P_1, P_2, P_3\}$, let $(\mathcal{A}_1, \mathcal{A}_2, \mathcal{A}_3)$ be a triple of access structures such that $\{P, P_j\} \in \mathcal{A}_j$ and $\{P, P_j\} \notin \mathcal{A}_i$, for each $i, j \in \{1, 2, 3\}$ with $i \neq j$. Then, in any multi-secret sharing schemes for $(\mathcal{A}_1, \mathcal{A}_2, \mathcal{A}_3)$ the entropy of the share given to P satisfies*

1. *$H(P) \geq H(S_1)$ in Case 1.a of Section 3.2.*
2. *$H(P) \geq H(S_1 S_2)$ in Cases 1.b and 2.a of Section 3.2.*
3. *$H(P) \geq H(S_1 S_2 S_3)$ in Case 2.b of Section 3.2.*

The previous corollaries prove the optimality of the sharing schemes given in Sections 3.1 and 3.2 with respect to the entropy of P's share.

5 Multi-Secret Schemes for Threshold Structures

In this section we consider the problem of sharing secrets in different threshold structures. More precisely, we analyze the case in which for each secret s_i, the access structure $\mathcal{A}_i$ is the set of all subsets consisting of at least k_i participants in $\mathcal{P}_i$, and will be denoted by $\mathcal{A}_{(k_i,\mathcal{P}_i)}$. Next corollaries immediately follow from Theorem 8.

Corollary 11. *Let $(\mathcal{A}_{(k,\mathcal{P}_1)},\ldots,\mathcal{A}_{(k,\mathcal{P}_m)})$ be an m-tuple of threshold structures on a set of participants $\mathcal{P}=\cup_{i=1}^m \mathcal{P}_i$. If $\mathcal{P}_1\subseteq\mathcal{P}_2\subseteq\cdots\subseteq\mathcal{P}_m$, then in any multi-secret sharing scheme for $(\mathcal{A}_{(k,\mathcal{P}_1)},\ldots,\mathcal{A}_{(k,\mathcal{P}_m)})$ the entropy of the share given to any participant $P\in\mathcal{P}_j$ satisfies*

$$H(P)\geq H(S_jS_{j+1}\ldots S_m).$$

Proof: Let P be a participant in $\mathcal{P}_j$. Construct the sets $X_j,X_{j+1},\ldots,X_m$ as follows. Let the set X_j be equal to $X_j=\{P_{i_1},\ldots,P_{i_{k-1}}\}$, with $X_j\subseteq\mathcal{P}_j\backslash\{P\}$. For $i=j+1,\ldots,m$, let $X_i=X_j$. It is easy to see that the participant P and the sets $X_j,X_{j+1},\ldots,X_m$ satisfies the hypothesis of Theorem 8, thus the corollary is proved. □

Corollary 12. *Let $(\mathcal{A}_{(k_1,\mathcal{P}_1)},\ldots,\mathcal{A}_{(k_m,\mathcal{P}_m)})$ be an m-tuple of threshold structures on a set of participants $\mathcal{P}=\cup_{i=1}^m\mathcal{P}_i$, with $k_1\leq k_2\cdots\leq k_m$. Suppose $\cap_{i=1}^m\mathcal{P}_i\neq\emptyset$. Let $\ell<m$ be the smallest integer such that $|\cap_{i=1}^m\mathcal{P}_i|<k_\ell$. Then in any multi-secret sharing scheme for $(\mathcal{A}_{(k_1,\mathcal{P}_1)},\ldots,\mathcal{A}_{(k_m,\mathcal{P}_m)})$ the entropy of the share given to any participant $P\in\cap_{i=1}^m\mathcal{P}_i$ satisfies*

$$H(P)\geq H(S_1S_2\ldots S_{\ell+1}).$$

Remark. *If in the previous corollary an integer $\ell<m$ such that $|\cap_{i=1}^m\mathcal{P}_i|<k_\ell$ does not exist, then it can be easily proved that for any participant $P\in\cap_{i=1}^m\mathcal{P}_i$ the entropy of the share given to P satisfies $H(P)\geq H(S_1S_2\ldots S_m)$.*

Corollary 13. *Let $(\mathcal{A}_{(k_1,\mathcal{P}_1)},\mathcal{A}_{(k_2,\mathcal{P}_2)},\mathcal{A}_{(k_3,\mathcal{P}_3)})$ be an m-tuple of threshold structures on a set of participants $\mathcal{P}=\cup_{i=1}^3\mathcal{P}_i$. Suppose $\cap_{i=1}^3\mathcal{P}_i\neq\emptyset$. Then, in any multi-secret sharing scheme for $(\mathcal{A}_{(k_1,\mathcal{P}_1)},\mathcal{A}_{(k_2,\mathcal{P}_2)},\mathcal{A}_{(k_3,\mathcal{P}_3)})$ the entropy of the share given to any participant $P\in\cap_{i=1}^3\mathcal{P}_i$ satisfies*

$$H(P)\geq H(S_1S_2S_3).$$

Before to state a general theorem on a multi-threshold structure we need the following two lemmas. They hold for any multi-secret sharing scheme of Type A not just for the case of multi-threshold structures. These two lemmas are the generalization to multi-secret sharing schemes of the ones proved in [6] for the case of single secret sharing.

Lemma 14. *Let $(\mathcal{A}_1, \ldots, \mathcal{A}_m)$ be an m-tuple of access structures on the set of participants $\mathcal{P}$. Let $X, Y \subseteq \mathcal{P}$ such that $Y \notin \mathcal{A}_i$ and $X \cup Y \in \mathcal{A}_i$. Then, in any multi-secret sharing scheme, it holds $H(X|Y) = H(S_i) + H(X|YS_i)$.*

An immediate consequence of Lemma 14 is that for any $P \in \cup_{A \in \mathcal{A}_i} A$ it holds $H(P) \geq H(S_i)$. We will see that under some condition this bound can be improved when the m-tuple of access structures consists of threshold structures.

Next lemma proves that the uncertainty on shares of a non-qualified set of participants cannot be decreased by the knowledge of the secret.

Lemma 15. *Let $(\mathcal{A}_1, \ldots, \mathcal{A}_m)$ be an m-tuple of access structures on the set of participants $\mathcal{P}$. Let $X, Y \subseteq \mathcal{P}$ such that $X, Y \notin \mathcal{A}_i$ Then, in any multi-secret sharing scheme for $(\mathcal{A}_1, \ldots, \mathcal{A}_m)$, it holds $H(X|Y) = H(X|YS_i)$.*

The following theorem states a lower bound on the size of the share held by any participant in an m-tuple of threshold structures. In the following we will show that if the secrets are uniformly chosen, then the bound is tight.

Theorem 16. *Let $\mathcal{A}_{(k_1, \mathcal{P})}, \ldots, \mathcal{A}_{(k_m, \mathcal{P})}$ be threshold structures on a set of participants $\mathcal{P}$. In any multi-secret sharing scheme for $(\mathcal{A}_{(k_1, \mathcal{P})}, \ldots, \mathcal{A}_{(k_m, \mathcal{P})})$ the entropy of the share given to any participant $P \in \mathcal{P}$ satisfies*

$$H(P) \geq \sum_{i=1}^{m} H(S_i).$$

If each secret s_i is uniformly chosen in $S_i = GF(q_i)$, with q_i prime, then it is possible to realize a multi-secret sharing scheme that meets the bound of Theorem 16. To accomplish this it is enough to combine m independent threshold schemes, say Shamir's schemes [16], one for each threshold structure. In the same way we can construct an optimal multi-secret sharing sheme for the m-tuple of threshold structures $(\mathcal{A}_{(k, \mathcal{P}_1)}, \ldots, \mathcal{A}_{(k, \mathcal{P}_m)})$ considered in Corollary 11.

References

1. J. C. Benaloh and J. Leichter, *Generalized Secret Sharing and Monotone Functions*, in "Advances in Cryptology - CRYPTO '88", S. Goldwasser Ed., "Lecture Notes in Computer Science", Vol. **403**, Springer–Verlag, Berlin, pp. 27–35, 1990.
2. M. Ben-Or, S. Goldwasser, and A. Wigderson, *Completeness Theorems for Non-Cryptographic Fault–Tolerant Distributed Computation*, Proceedings of 20th Annual ACM Symposium on Theory of Computing, pp. 1–10, 1988.
3. G. R. Blakley, *Safeguarding Cryptographic Keys*, Proceedings AFIPS 1979 National Computer Conference, pp. 313–317, June 1979.
4. C. Blundo, A. De Santis, L. Gargano, and U. Vaccaro, *On the Information Rate of Secret Sharing Schemes*, in "Advances in Cryptology - CRYPTO '92", E. Brickell Ed., "Lecture Notes in Computer Science", Vol. **740**, Springer-Verlag, Berlin, pp. 149–169, 1993. To appear in Theoretical Computer Science.

5. C. Blundo, A De Santis, and U. Vaccaro, *Efficient Sharing of Many Secrets*, in "Proceedings of STACS '93 (10*th* Symp. on Theoretical Aspects of Computer Science)", P. Enjalbert, A. Finkel, K. W. Wagner Eds., "Lecture Notes in Computer Science", Vol. **665**, Springer–Verlag, Berlin, pp. 692–703, 1993.
6. R. M. Capocelli, A. De Santis, L. Gargano, and U. Vaccaro, *On the Size of Shares for Secret Sharing Schemes,* Journal of Cryptology, Vol. **6**, pp. 57–167, 1993.
7. I. Csiszár and J. Körner, *Information Theory. Coding Theorems for Discrete Memoryless Systems,* Academic Press, 1981.
8. M. Franklin and M. Yung, *Communication Complexity of Secure Computation,* Proceedings of 24th Annual ACM Symposium on Theory of Computing", pp. 699–710, 1992.
9. R. G. Gallager, *Information Theory and Reliable Communications,* John Wiley & Sons, New York, NY, 1968.
10. O. Goldreich, S. Micali, and A. Wigderson, *How to Play any Mental Game,* Proceedings of 19th ACM Symposium on Theory of Computing, pp. 218–229, 1987.
11. M. Ito, A. Saito, and T. Nishizeki, *Secret Sharing Scheme Realizing General Access Structure,* Proceedings of IEEE Global Telecommunications Conference, Globecom 87, Tokyo, Japan, pp. 99–102, 1987.
12. W.-A. Jackson, K. M. Martin, and C. M. O'Keefe, *Multisecret Threshold Schemes,* in "Advances in Cryptology - CRYPTO '93", D.R. Stinson Ed., "Lecture Notes in Computer Science", Vol. **773**, Springer-Verlag, Berlin, pp. 126–135, 1994.
13. W.-A. Jackson, K. M. Martin, and C. M. O'Keefe, *A Construction for Multisecret Threshold Schemes,* Preprint, 1994.
14. E. D. Karnin, J. W. Greene, and M. E. Hellman, *On Secret Sharing Systems,* IEEE Trans. on Inform. Theory, Vol. IT-29, no. 1, pp. 35–41, Jan. 1983.
15. S. C. Kothari, *Generalized Linear Threshold Schemes,* in "Advances in Cryptology - CRYPTO '84", G. R. Blakley, D. Chaum Eds., "Lecture Notes in Computer Science", Vol. **196**, Springer-Verlag, Berlin, pp. 231–241, 1985.
16. A. Shamir, *How to Share a Secret,* Communications of the ACM, Vol. **22**, n. 11, pp. 612–613, Nov. 1979.
17. G. J. Simmons, *An Introduction to Shared Secret and/or Shared Control Schemes and Their Application,* Contemporary Cryptology, IEEE Press, pp. 441–497, 1991.
18. D. R. Stinson, *An Explication of Secret Sharing Schemes,* Design, Codes and Cryptography, Vol. **2**, pp. 357–390, 1992.

Designing Identification Schemes with Keys of Short Size ⋆.

Jacques Stern

Laboratoire d'Informatique, École Normale Supérieure

Abstract. In the last few years, there have been several attempts to build identification protocols that do not rely on arithmetical operations with large numbers but only use simple operations (see [10, 8]). One was presented at the CRYPTO 89 rump session ([8]) and depends on the so-called Permuted Kernel problem (PKP). Another appeared in the CRYPTO 93 proceedings and is based on the syndrome decoding problem (SD) form the theory of error correcting codes ([11]). In this paper, we introduce a new scheme of the same family with the distinctive character that both the secret key and the public identification key can be taken to be of short length. By short, we basically mean the usual size of conventional symmetric cryptosystems. As is known, the possibility of using short keys has been a challenge in public key cryptography and has practical applications. Our scheme relies on a combinatorial problem which we call *Constrained Linear Equations* (CLE in short) and which consists of solving a set of linear equations modulo some small prime q, the unknowns being subject to belong to a specific subset of the integers mod q. Thus, we enlarge the set of tools that can be used in cryptography.

1 The Underlying Problem

Since the appearance of public-key cryptography, basically all practical schemes have been based on hard problems from number theory. This has remained true with zero-knowledge proofs, introduced in 1985, in a paper by Goldwasser, Micali and Rackoff ([6]) and whose practical significance was soon demonstrated in the work of Fiat and Shamir ([4]). In 1989, there were two attempts to build identification protocols that only use simple operations (see [10, 8]). One relied on the intractability of some coding problems, the other on the Permuted Kernel problem (PKP). The first of the schemes was not really practical but has been followed by a truly practical proposal based on the so-called Syndrome Decoding problem (SD). The purpose of the present paper is twofold:

- First, to introduce a new scheme based on a combinatorial problem which we call *Constrained Linear Equations* (CLE in short) and which consists of solving a set of linear equations modulo some small prime q, the unknowns being subject to belong to a specific subset of the integers mod q.

⋆ PATENT CAUTION: This document may reveal patentable subject matter

– Second, to demonstrate in this setting, the possibility to have an identification scheme where both the secret key and the public identification key can be taken to be of short length. By short, we basically mean the usual size of conventional symmetric cryptosystems, i.e. 64 or 80 bits.

We briefly comment on the second point. Besides having been a long time challenge in public key cryptography, the question of short keys may be of practical importance. As is known, identification schemes avoiding large integers such as PKP or SD, are not identity based. This means that public keys should be related to the user's identity by a signature of some authority or by a directory, which the verifier has access to. If the second option is taken, then the key length becomes an important issue. It is even more important for the prover, a smart card in many practical applications: short keys may save space in the physically protected area of the card where they are stored and thus may allow the use of relatively low cost cards.

We now turn to our basic problem:

Constrained Linear Equations (CLE)
instance: A (small) prime number q, a system S of r homogeneous linear equations with k unknowns and whose coefficients are integers mod q, a subset X of the integers mod q.
Question Is there a solution of S consisting of k elements of the given set X?

It is easily seen that the problem is $\mathcal{NP}$-complete. Our further assumption is that it is intractable in the following sense:

Intractability of CLE Assumption: No probabilistic polynomial time algorithm can take as its input the values of q, S, X and output, with non negligible probability, a solution of S consisting of k elements of the given set X.
As usual, non negligible stands for bounded from below by the inverse of some power of the size of the input.

From the practical point of view, we mention, as a minimal choice, the case where $q = 257$, $k = 40$, $r = 20$, $|X| = 16$. We do not really advocate these figures for highly secure applications but we use them as a convenient benchmark in order to establish comparisons with the minimal sizes provided for PKP or SD. The minimal size suggested for the SD identification scheme has been carefully analyzed in [3], where it is shown that the workload of the best possible known attacks is about 2^{68}. The minimal size of the parameters in the original PKP proposal (see [8]) has been extensively discussed in [1, 7]. Attacks based on intelligent gaussian elimination and a space-time trade off yield a workload of 2^{52}. Similar attacks can be carried against CLE and it can be seen that the figures chosen above yield a similar 2^{52} workload. Thus, the comparisons in terms of key size will be significant. Whether or not this is enough for applications is open for discussion. We feel that, for secure applications, it is safer to recommend the following parameters for CLE: $q = 257$, $k = 48$, $r = 24$, $|X| = 16$.

2 Key Generation

The key generation algorithm is based on a trick which combines a linear operation and a highly non linear one. We feel that this trick might find further applications in other areas of cryptography. The prime number q is such that $q-1$ is the product of two almost equal integers. Thus, 251 is $15*16+1$ and 257 is 16^2+1. We write $q-1=c*d$. We next consider the multiplicative group of non-zero integers mod q. This is a cyclic group and we can easily build a subgroup G of order c. Picking one element in each class mod G, we get a set X, consisting of d elements such that any integer between 1 and $q-1$ can be uniquely written as the product mod q of an element of G and an element of X. We let $g(u)$ the element of G appearing in the unique decomposition of u and, similarly, we let $k(u)$ be the corresponding element of X. If U is a vector whose coordinates are non-zero integers mod q, we let $g(U)$ be the vector obtained by applying g coordinatewise. $k(U)$ is defined accordingly.

Besides a fixed prime number q, a subgroup G and a fixed subset X as above, the proposed scheme uses a fixed $(n \times m)$-matrix M whose coefficients are randomly chosen integers mod q. This matrix is common to all users and is originally built randomly. Each user receives a secret key S which is a vector with m coordinates, each a member of X. The public identification is computed as

$$P = g(M(S))$$

Note that there is a (slight) chance that the computation of $g(M(S))$ cannot be carried through if $M(S))$ has some zero coordinate.A heuristic analysis shows that this happens with probability close to $1-(1-\frac{1}{q})^n$. With the figures of the numerical example provided above, this is 0.07 and therefore, after a few trials, one can reach the desired value of P.

Clearly, recovering the secret key S from the public data P amounts to solving the equation

$$P \otimes T = M.S$$

where $\otimes$ denotes coordinatewise multiplication mod q and where S, T are two unknown vectors, having respectively m and n coordinates, subject to the condition that these coordinates are members of X . Thus, one has to solve an instance of the CLE problem with n equations and $m+n$ unknowns.

If we turn to our minimal size numerical example, we see that one can take $q = 257$ and $n = m = 20$. Furthermore, X has 16 elements as well as G. Thus both S and P can be coded on 80 bits. Without apparent consequence on the intractability of the combinatorial problem to solve, S can be generated deterministically from a (say) 64-bit seed. It is also possible to fix the first four coordinates of the public key P or (better) to derive them from the public identity of the user. The resulting public key is stored on 64 bits. The key generation uses, as above, the idea of multiple trials. The expected number of trials until an acceptable key is found is about $7*10^4$, which is still reasonable. A few more bits could even be saved by analogous tricks. We feel that these manipulations do not affect the security of the scheme that we will present, but this opinion should be further investigated. If we turn to alternative numerical example mentioned above, we find the following figures: $q = 257$, $n = m = 24$, $|X| = |G| = 16$. Using the tricks just described, this is still compatible with secret and public keys of 80 bits.

We will now describe, in the style of [8, 11], two interactive identification protocols by which a prover demonstrates possession of the secret key S corresponding to the public key P.

3 A Three Pass Identification Protocol

As is the case for the PKP and SD schemes, we will need some cryptographic hash function H. This hash function should be collision free as will be discussed further. For practical implementations, a standard hash function such as Rivest's MD5 ([9]) can be used. The protocol includes several rounds, each of these being performed as follows:

1. The prover picks at random two vectors U, V having respectively m and n coordinates, each an integer mod q. He also chooses two random permutations σ and τ. σ operates on the integers $\{1 \cdots m\}$ and τ on the integers $\{1 \cdots n\}$. Then he sends commitments h_1, h_2, h_3 respectively computed as
$$h_1 = H(\sigma, \tau, M.U + P \otimes V)$$
$$h_2 = H(U.\sigma, V.\tau)$$
$$h_3 = H((U + S).\sigma, (V - T).\tau)$$
In the above, H denotes the cryptographic hash function, $+$, $-$ and $\otimes$ denote coordinatewise operations mod q and $U.\sigma$ stands for the action of σ on U, that is to say the vector $U_{\sigma(i)}$, $1 \leq i \leq m$. Also, T is the vector with n coordinates defined by $T = k(M(S))$, with the notations of section 2.
2. The verifier sends a random element b of $\{0, 1, 2\}$.
3. If b is 0, the prover reveals σ, τ U and V. If b is 1, the prover reveals σ, τ and the two vectors $U' = (U + S)$ and $V' = (V - T)$. Finally, if b equals 2, the prover discloses vectors $U.\sigma$, $V.\tau$ together with vectors $U" = (U + S).\sigma$ and $V" = (V - T).\tau$.
4. If b equals 0, the verifier checks that commitments h_1 and h_2 have been computed honestly. This is possible since, using the values of σ, τ, U and V disclosed at step 2, he can compute the respective values of $M.U + P \otimes V$, $U.\sigma$, $V.\tau$. From these values he checks that $h_1 = H(\sigma, \tau, M.U + P \otimes V)$ and that $h_2 = H(U.\sigma, V.\tau)$.
If b equals 1, the verifier checks that commitments h_1 and h_3, were correct: note that σ, τ are known from step 3 and that
$$M.U + P \otimes V = M.(U + S) + P \otimes (V - T) - M.S + P \otimes T = M.U' + P \otimes V'$$
This allows the verifier to check the equality $h_1 = H(\sigma, \tau, M.U' + P \otimes V')$. He can also check that $h_3 = H(U'.\sigma, V'.\tau)$.

Now, if b is 2, the verifier checks commitments $h_2 = H(U.\sigma, V.\tau)$ and $h_3 = H(U", V")$. Furthermore, the verifier computes the two vectors $U'' - U.\sigma$ and $V.\tau - V''$ and verifies that all of their coordinates are members of X.

The number r of consecutive rounds depends on the required level of security and will be discussed further on.

4 A Five Pass Identification Protocol

We new describe an alternative protocol allowing identification. Again, this protocol includes several rounds, each of these being performed as follows:

1. The prover picks at random two vectors U, V having respectively m and n coordinates, each an integer mod q. He also chooses two random permutations σ and τ. σ operates on the integers $\{1 \cdots m\}$ and τ on the integers $\{1 \cdots n\}$. Then he sends commitments h_1, h_2 respectively computed as
$$h_1 = H(\sigma, \tau, M.U + P \otimes V)$$
$$h_2 = H(S.\sigma, T.\tau, U.\sigma, V.\tau)$$
In the above, all notations are as in the previous section.
2. The verifier sends a random element a of between 0 and $q - 1$.
3. The prover computes the pair $Y = (aS + U).\sigma$, $Z = (aT - V).\tau$ and sends back these two vectors to the verifier.
4. The verifier sends a random bit b of $b = 0$ or 1.
5. If b is 0, the prover reveals σ and τ. If b is 1, the prover discloses vectors $S.\sigma$ and $T.\tau$
6. If b equals 0, the verifier checks commitments h_1. This is possible since, using the values of σ and τ disclosed at step 2, he can compute successively the respective values of $Y' = Y.\sigma^{-1}$ and $Z' = Z.\tau^{-1}$, and then $M(Y' - P \otimes Z')$. Provided the answer is correct this last vector equals $M.U + P \otimes V$, From these values he checks that $h_1 = H(\sigma, \tau, M.U + P \otimes V)$.
If b equals 1, the verifier checks that commitment h_2 was correct: note that if the correct values of $S.\sigma$ and $T.\tau$ have been received, the verifier can compute vectors $Y - aS.\sigma$ and $aT.\tau - Z$. These are respectively equal to $U.\sigma$ and $V.\tau$, so that one should have $h_2 = H(S.\sigma, T.\tau, Y - aS.\sigma, aT.\tau - Z)$. Having checked this equality, the verifier also tests that the vectors received at step 5 are such that all of their coordinates are members of X.

5 Security of the Scheme

It is apparent that the security of the scheme relies on the difficulty of solving the equation
$$P \otimes T = M.S$$

where P is the public key of a specific user and S, T are two unknown vectors, having respectively m and n coordinates, subject to the condition that these coordinates are members of X .

In order to perform the first protocol without knowing the secret key, various strategies can be used.

- Having only U, V, σ and τ ready for the verifier's query and replacing the unknown S, T by arbitrary vectors with coordinates in X. In this case, the false prover hopes that b is 0 or 2 and the probability of success is 2/3 for a single round and $(2/3)^r$ in general, where r is the number of rounds. A similar strategy can be defined with $U + S$, $V - T$ in place of U, V.
- Having simultaneously U, V, $U + S'$ and $V - T'$ ready where S', T' is a regular solution of

$$P \otimes T = M.S$$

(i.e. without the constraint about X). This yields the same probability of success.

It is fairly clear that shifting beetween one strategy to another has also the same probability of success. Similar strategies can be designed for the second protocol with probability of success $\frac{q+1}{2q}$ and $\left(\frac{q+1}{2q}\right)^r$ if the protocol is repeated r times. In the reverse direction, we have:

Theorem 1. *Assume that some probabilistic polynomial-time adversary is accepted with probability* $\geq (2/3)^r + \epsilon$ *after playing a constant number r of rounds of the first protocol with a fair verifier , then there exists a polynomial-time probabilistic machine which extracts a secret pair S, T from the public data or output collisons for the hash function, with overwhelming probability.*

remark There is an analogous result for the second protocol with $(2/3)^r$ replaced by $\left(\frac{q+1}{2q}\right)^r$.

proof: Consider the tree $T(\omega)$ of all 3^r executions corresponding to all possible questions of the verifier when the adversary has a fixed random tape ω. Let

$$\alpha = Pr(T(\omega) \quad \text{has a vertex with 3 sons} \quad)$$

If α is $< \epsilon$, then, it is easily seen that the probability of succes of the adversary is bounded by $(2/3)^r + \epsilon$: $(2/3)^r$ comes from the case where $T(\omega)$ has no vertex with 3 sons and ϵ from the other case. Thus α is at least ϵ and by resetting the adversary $1/\epsilon$ times, one finds, with constant probability an execution tree with a vertex having 3 sons. Repeating again, the probability can be made very close to one. Now a vertex with 3 sons corresponds to a situation where 3 commitments h_1, h_2, h_3 have been made and where the adversary can provide answers to the 3 possible queries of the verifier. Consider the answer σ, τ, U, V to the question $b = 0$ and the answer σ', τ', U', V' to the question $b = 1$. Since

$$H(\sigma, \tau, M.U + P \otimes V) = h_1 = H(\sigma', \tau', M.U' + P \otimes V')$$

we conclude that either a collision for the hash function H has been found or else $\sigma = \sigma'$, $\tau = \tau'$ and $M.U + P \otimes V = M.U' + P \otimes V'$. Similar arguments show that, unless an H-collision has been found, the answer to $b = 2$ consists of $U.\sigma$, $V.\tau$, $U'.\sigma$, $V'.\tau$. We note that, since the last answer is accepted, both $(U' - U).\sigma$ and $(V - V').\tau$ have all coordinates in X. Also $M.(U' - U) = P \otimes (V - V')$, as observed above. It follows that the underlying system of constrained linear equations has been solved.

Following the techniques in [5], it is possible to prove a more foundational result, which shows that repetition of either protol is a *proof of knowledge* of a solution of the constrained system

$$P \otimes T = M.S$$

We state such a result for our first protocol. We let N denote the size of the public data.

Theorem 2. *Assume that some probabilistic polynomial-time adversary is accepted with non negligible probability after playing with a fair verifier a number of rounds of the first protocol that is* $\Theta(\log N)$, *then there exists a polynomial-time probabilistic machine which extracts a secret pair* S, T *from the public data or outputs collisons for the hash function, with overwhelming probability.*

Before we turn to zero-knowledge, let us observe that, at step 3 of the first protocol, the prover eventually discloses the image of the secret pair S, T, under two random permutations σ and τ. A similar remark applies to the other protocol. Thus, the exact repartition of the values of the unknowns in S and separately in T have to be considered as public data. This information makes the computation of the solutions of

$$P \otimes T = M.S$$

a bit easier. We have taken this into account when analyzing the security of the CLE problem. Still, it is advisable to avoid irregular distributions (e.g. where an element of S appears many times).

It can be proved formally that both schemes are zero-knowledge. We will only give a brief hint for the first protocol. As we observed above, anyone can be ready to answer two queries among the three possible ones at each round. Hence, by using the standard idea of resettable simulation (see [6]), one can devise a polynomial-time simulation algorithm that mimics the fair communication between the prover and the verifier in expected time $O(2/3.r)$. Some remarks are in order here:

1. As was just observed, the exact repartition of the values of the unknowns in S and T are basically public data. This does not contradict zero-knowledge as they leak equally from the actual executions and the simulated ones.
2. Hash values make the simulation a bit harder: a convenient setting is the so-called *random oracle model* (see [2]). Alternatively, one has to assume specific statistical independance properties for the hash function.

6 Performances of the Scheme.

The performances of our scheme are very comparable to those of [8, 11] and we will restrict ourselves to various remarks.

1. As for previous schemes of the same family, the memory needed to implement the scheme is not large: especially, it is not necessary to store all of M. One can only store words corresponding to some chosen locations and extend these by a fixed software random number generator.
2. The operations to perform are very simple and well suited to the environment of 8-bit microprocessors.
3. The communication complexity of the protocol is quite acceptable: if we assume that hash values are 128 bits long, we obtain an average number of bits per round which is close to 840 bits for the first protocol and 725 for the second. This is for the minimal suggested size of parameters. For the alternate choice, these figures go up to 940 and 824. There is a trick that can save one hash value, at least for the first protocol. It consists of replacing h_1, h_2, h_3 by $H(h_1, h_2, h_3)$ and providing the missing hash value at step 3 (for example transmitting h_3 if $b = 0$). This yields similar communication complexities for both schemes.
4. In order to achieve a level of security of 10^{-6}, the first protocol has to be repeated 35 times and the second one only 20 times. Whether or not this is a serious drawback should be discussed with practical implementations in mind. We simply note that the number of interactions is almost the same in both case, because the second protocol needs more passes.
5. As is the case for PKP and SD, our scheme is not identity based. This means that public keys have to be certified by the issuing authority or that the verifier needs to access a directory. As emphasized in the introduction, the distinctive character of the scheme, namely the short key length, is a definite advantage in the latter case.

7 An Additive Variant

Before concluding the paper, we briefly mention an alternative approach for key generation: let X and Y be two subsets of the set of integers mod q, such that any integer can be written (non necessarily in a unique way) as the sum mod q of an element of X and an element of Y (it is not difficult to construct such subsets). From a random vector S xith m coordinates, all in X, one can compute $M.S$ as $T + P$, where T, P are vectors with n coordinates respectively in X and Y. One can take P as a public key and keep S, T secret. The resulting CLE problem is written

$$M.S = T + P$$

where S, T are unknown vectors with coordinates in X. Protocols to prove knowledge of the secret data are simple variants of those described above. We do not know whether this alternative key generation method is weaker than the original one. We suspect it might be the case if X and Y are chosen in a simple way (e.g. by specifying that elements of X are those with prescribed bits equal to zero).

8 Conclusion

We have defined a new practical identification scheme based on a combinatorial problem which we call CLE (Constrained Linear Equations). This scheme allows the use of keys of short length (64 or 80 bits). We have proposed two protocols using CLE: both only use very simple operations and thus widen the range of techniques that can be applied in cryptography. We welcome attacks from readers and, as is customary when introducing a new cryptographic tool, we suggest that the scheme should not be adopted prematurely for actual use.

Acknowledgements

I wish to thank A. Shamir for his comments on an earlier presentation of this work and for suggestions which led to the alternative approach mentioned in section 7. I also want to thank my student J.B. Fischer for help in the evaluation of the security of CLE.

References

1. T. Baritaud, M. Campana, P. Chauvaud and H. Gilbert: On the security of the permuted kernel identification scheme. In: Proceedings of Crypto 92. Lecture Notes in Computer Science 740. Berlin: Springer 1993, pp. 305-311.
2. M. Bellare and P. Rogaway: Random oracles are practical: a paradigm for designing efficient protocols. In: Proceedings of the 1st ACM Conference on Computer and Communications Security, 1993, pp. 62-73.
3. F. Chabaud: On the security of some cryptosystems based on error-correcting codes. In: Proceedings of Eurocrypt 94. Lecture Notes in Computer Science, to appear.
4. A. Fiat and A. Shamir: How to prove yourself: Practical solutions to identification and signature problems. In: Proceedings of Crypto 86. Lecture Notes in Computer Science 263. Berlin: Springer 1987, pp. 181-187.
5. U. Feige, A. Fiat and A. Shamir: Zero-knowledge proofs of identity. In: Proc. 19th ACM Symp. Theory of Computing, 1987, pp. 210-217 and *J. Cryptology* 1, 77-95 (1988).
6. S. Goldwasser, S. Micali and C. Rackoff: The knowledge complexity of interactive proof systems. In: Proc. 17th ACM Symp. Theory of Computing, 1995, pp.291-304.
7. J. Patarin and P. Chauvaud: Improved algorithms for the permuted kernel prolem. In: Proceedings of Crypto 93, Lecture Notes in Computer Science 773. Berlin: Springer 1994, pp. 391-402.
8. A. Shamir: An efficient identification scheme based on permuted kernels. In: Proceedings of Crypto 89. Lecture Notes in Computer Science 435. Berlin: Springer 1990, pp. 606-609.

9. R. L. Rivest: The MD5 Message Digest Algorithm. In: Proceedings of Crypto 90. Lecture Notes in Computer Science 537. Berlin: Springer 1991, pp. 303-311.
10. J. Stern: An alternative to the Fiat-Shamir protocol. In: Proceedings of Eurocrypt 89. Lecture Notes in Computer Science 434. Berlin: Springer 1990, pp. 173-180.
11. J. Stern: A new identification scheme based on syndrome decoding. In: Proceedings of Crypto 93. Lecture Notes in Computer Science 773. Berlin: Springer 1994 , pp. 13-21.

Proofs of Partial Knowledge and Simplified Design of Witness Hiding Protocols

Ronald Cramer, CWI
Ivan Damgård, Aarhus University, Denmark
Berry Schoenmakers, CWI

Abstract. Suppose we are given a proof of knowledge $\mathcal{P}$ in which a prover demonstrates that he knows a solution to a given problem instance. Suppose also that we have a secret sharing scheme $\mathcal{S}$ on n participants. Then under certain assumptions on $\mathcal{P}$ and $\mathcal{S}$, we show how to transform $\mathcal{P}$ into a witness indistinguishable protocol, in which the prover demonstrates knowledge of the solution to some subset of n problem instances out of a collection of subsets defined by $\mathcal{S}$. For example, using a threshold scheme, the prover can show that he knows at least d out of n solutions without revealing which d instances are involved. If the instances are independently generated, we get a witness hiding protocol, even if $\mathcal{P}$ did not have this property. Our results can be used to efficiently implement general forms of group oriented identification and signatures. Our transformation produces a protocol with the same number of rounds as $\mathcal{P}$ and communication complexity n times that of $\mathcal{P}$. Our results use no unproven complexity assumptions.

1 Introduction

In this work[1], we assume that we are given an interactive proof where the prover P convinces the verifier V that P knows some secret. Typically, the secret is the preimage under some one-way function of a publicly known piece of information. Thus the secret could be for example a discrete log or an RSA root. Such a proof is called a proof of knowledge [5], and can be used in practice to design identification schemes or signature systems.

We assume in the following that the proof of knowledge has a special form in that the verifier only sends uniformly chosen bits. This is also known as a *public coin protocol*. For simplicity, we restrict ourselves to 3-round protocols, where the prover speaks first (generalization of our results to any number of rounds is possible). We also assume that the protocol is honest verifier zero-knowledge (HVZK), i.e. the protocol does not reveal anything (for example about the prover's secret) to the honest verifier, but it is not necessarily secure against a cheating verifier.

Numerous protocols are known to satisfy the conditions described above. Concrete examples are Schnorr's discrete log protocol [13] and Guillou-Quisquater's RSA root protocol [8]. None of these protocols are known to be zero-knowledge or even witness hiding. In general, a parallelization of a sequential zero-knowledge (ZK) proof [7] will often satisfy the conditions.

---
[1] Partly done while visiting Aarhus University.

The second ingredient we need is a secret sharing scheme, i.e. a scheme for distributing a secret among a set of participants such that some subsets of them are qualified to reconstruct the secret while other subsets have no information about it. The collection of qualified subsets is called the access structure. The secret sharing scheme has to satisfy some properties which will be made more precise below. Shamir's secret sharing scheme [14] has the properties we need.

Our main result uses a proof of knowledge $\mathcal{P}$, an access structure Γ for n participants, and a secret sharing scheme $\mathcal{S}$ for the access structure dual to Γ to build a new protocol, in which the prover shows that he knows solutions to a subset of n problem instances corresponding to a qualified set in the access structure of Γ (see Section 3 for details on access structures). The protocol is witness indistinguishable, i.e. the prover reveals no Shannon information about which qualified subset of solutions he knows. The new protocol has the same number of rounds as $\mathcal{P}$ and communication complexity roughly n times that of $\mathcal{P}$. We also show that for some access structures, the new protocol is in fact witness hiding (WH), i.e. even even a cheating verifier will not learn enough to be able to compute the prover's secret. Although WH is a weaker property than general ZK, it can replace ZK in many protocol constructions, including identification schemes.

Since a simple 1 out of 2 structure is enough for our result to produce a WH protocol, we obtain as a corollary a general method simplifying the design of WH protocols: first build a protocol $\mathcal{P}$ with properties as described above - for security against the verifier only the weak and therefore easy to obtain property of HVZK is needed. Then apply our result using a 1 out of 2 structure to get a WH protocol. This new protocol will have complexity equivalent to running $\mathcal{P}$ twice in parallel.

After surveying related work, we give in the following two sections more details on the protocols and the secret sharing schemes we consider. Section 4 then contains the main result and corollaries, and Section 5 describes a nice application of our results to group oriented identification and signatures.

1.1 Related Work

Our techniques are to some extent related to those of De Santis et al. [11]. The models are quite different, however: [11] considers non-interactive proofs of membership, while we consider interactive proofs of knowledge. Also, [11] considers variants of the quadratic residuosity problem, while we consider any problem that affords a protocol of the right form.

In some independent work, De Santis et al. [12] apply techniques similar to ours to proofs of membership in random self-reducible languages. This leads to perfect ZK proofs for monotone Boolean operations over such languages.

In [4], Feige and Shamir introduce the concepts of witness indistinguishable (WI) and witness hiding (WH) protocols and prove the existence of WH protocols for a large class of problems, including the ones we consider (Corollary 4.4). This was done using general zero-knowledge techniques and the assumption that one-way functions exist. Compared to [4], our result shows that if we start from a proof of knowledge with properties as described above, WH protocols can be constructed much more efficiently and without using computational assumptions.

In [3], a transformation from HVZK proofs was given for protocols including the type we consider. That transformation produced ZK protocols, but on the other hand greatly increased the communication and round complexity so that, contrary to ours, the practical value of that transformation is quite limited. If the target is ZK, however, the increased round complexity seems to be unavoidable.

2 Proofs of Knowledge

Let a binary relation $R = \{(x, w)\}$ be given, for which membership can be tested in polynomial time. For any x, its *witness set* $w(x)$ is the set of w's, such that $(x, w) \in R$.

In the following, we assume that we are given a protocol $\mathcal{P}$, which is a *proof of knowledge* for R, i.e. there is a common input x (of length k bits) to prover P and verifier V and a private input w to P. The prover tries to convince the verifier that $w \in w(x)$. Refer to [5] or [4] for a formal definition.

In order for the constructions in the following to work, $\mathcal{P}$ needs to satisfy a few special properties.

First, we will assume that $\mathcal{P}$ is a three round public coin protocol (although the three round restriction can be removed). Conversations in the protocol will be ordered triples of the form

$$m_1, c, m_2$$

The second message in the protocol is a random bit string c chosen by the verifier. We refer to this as a challenge, and to the prover's final message as the answer.

We also assume that completeness holds for $\mathcal{P}$ with probability 1, i.e. if indeed $w \in w(x)$, then the verifier always accepts.

We assume that $\mathcal{P}$ satisfies knowledge soundness in the following sense: the length of c is such that the number of possible c-values is super-polynomial in k, and for any prover P^*, given two conversations between P^* and V, (m_1, c, m_2) and (m_1, c', m_2'), where $c \neq c'$, an element of $w(x)$ can be computed in polynomial time. We call this the *special soundness property.* It is easily seen to imply the standard soundness definition, which calls for the existence of a knowledge extractor, which can extract a witness in polynomial time from any prover that is successful with non-negligible probability.

Although special soundness is less general than the standard definition, all known proofs of knowledge have this property, or at least a variant where computation of the witness follows from some small number of correct answers. Assuming special soundness is therefore not a serious restriction.

Finally, we assume that $\mathcal{P}$ is *honest verifier zero-knowledge*: there is a simulator S that on input x produces conversations that are indistinguishable from real conversations with input x between the honest prover and the honest verifier. For simplicity we assume perfect indistinguishability in the following; generalization to other flavors of indistinguishability is easy. Most known honest verifier zero-knowledge protocols in fact satisfy something stronger, viz. that there is a procedure that can take any c as input and produce a conversation indistinguishable from the space of all conversations between the honest prover and verifier in which c is the challenge. We call this *special honest verifier zero-knowledge.*

We will later need the concepts of *witness indistinguishable* (WI) and *witness hiding* (WH) protocols, which were introduced in [4]. Informally, a protocol is witness indistinguishable if conversations generated with the same x but different elements from $w(x)$ have indistinguishable distributions, i.e. even a cheating verifier cannot tell which witness the prover is using. If the problem instance x is generated with a certain probability distribution by a generator G which outputs pairs (x, w) with $w \in w(x)$, we can define the concept of *witness hiding.* A protocol is witness hiding over G, if it does not help even a cheating verifier to compute a witness for x with non-negligible probability when the x is generated by G. We refer to [4] for details.

With respect to the witness indistinguishable property, we can already now note the following:

Proposition 1. *Let $\mathcal{P}$ be a three round public coin proof of knowledge for relation R. If $\mathcal{P}$ is honest verifier zero-knowledge, then $\mathcal{P}$ is witness indistinguishable.*

Proof. We trivially have WI for conversations with the honest verifier: The use of any witness w leads to the distribution produced by the simulator. This implies that the distribution of m_2, given any fixed m_1 and c, is independent of w. The proposition then follows from noting that in conversations with a general verifier, the distribution of m_1, and hence of c, is independent of w.

In many concrete cases, this proposition is not interesting because there is only one witness, in which case WI is trivial and cannot imply anything. Nevertheless, Proposition 1 will be needed in the following for technical reasons.

2.1 An Example

As a concrete example of a protocol with the properties we need, we present Schnorr's protocol from [13] for proving knowledge of a discrete log in a group G of prime order q. Let $g \neq 1$, and let $x = g^w$ be the common input. P is given w as private input. In the language of the above section, the protocol is a proof of knowledge for the relation that consists of pairs $(\,(x, g, G),\; w)$ such that $x = g^w$ in G. Then the protocol works as follows:

1. The prover chooses z at random in $[0..q)$, and sends $a = g^z$ to V.
2. The verifier chooses c at random in $[0..q)$, and sends it to P.
3. P sends $r = (z + cw) \bmod q$ to V, and V checks that $g^r = a\, x^c$.

Completeness trivially holds with probability 1. Correct answers to two different c-values give two equations $r_1 = z + wc_1 \bmod q$ and $r_2 = z + wc_2 \bmod q$ so we find that $w = (r_1 - r_2)/(c_1 - c_2) \bmod q$. So special soundness holds also. Finally, note that by choosing c and r at random, we can make a simulated conversation $(g^r x^{-c}, c, r)$ between the honest verifier and prover. Since c can be chosen freely, we even get special honest verifier zero-knowledge.

3 Secret Sharing

A secret sharing scheme is a method by which a secret s can be distributed among n participants, by giving a *share* to each participant. The shares are computed in such a way that some subsets of participants can, by pooling their shares, reconstruct s. These subsets are called *qualified* sets. Participants forming a non-qualified set should be able to obtain no information whatsoever about s. Such a secret sharing scheme is called *perfect.*

The collection of qualified sets is called the *access structure* for the secret sharing scheme. Clearly if participants in some set can reconstruct s, so can any superset, and therefore in order for the scheme to make sense, it must be the case that if A is a qualified set, then any set containing A is also qualified. An access structure with this property is called *monotone.*

A special case of monotone access structures is structures containing all subsets larger than some threshold value. Such structures are called *threshold structures.*

Any monotone access structure has a natural dual structure. This concept was first defined in [15].

Definition 2. Let Γ be an access structure containing subsets of a set M. If $A \subseteq M$, then $\bar{A}$ denotes the complement of A in M. Now Γ^*, *the dual access structure* is defined as follows:

$$A \in \Gamma^* \Leftrightarrow \bar{A} \notin \Gamma.$$

The next propositions follow directly from the definition.

Proposition 3. *The dual Γ^* of a monotone access structure is monotone as well, and satisfies*

$$(\Gamma^*)^* = \Gamma.$$

Furthermore, if Γ is a threshold structure, then so is Γ^*.

Proposition 4. *Let Γ be monotone. A set is qualified in Γ exactly when it has a non-empty intersection with every qualified set in Γ^*.*

In the next section, we will assume we are given a protocol of the form described in Section 2. For each input length k we will assume we are given a monotone access structure $\Gamma(k)$ on n participants, where $n = n(k)$ is a polynomially bounded function of k. Thus we have a *family of access structures*

$\{\Gamma(k) | \ k = 1, 2, \ldots\}$ We can then build a new protocol for proving statements on n problem instances provided we have a perfect secret sharing scheme $\mathcal{S}(k)$ for $\Gamma(k)^*$ satisfying certain requirements to be defined below.

Let $D(s)$ denote the joint probability distribution of all shares resulting from distributing the secret s. For any set A of participants, $D_A(s)$ denotes the restriction of $D(s)$ to shares in A. As $\mathcal{S}(k)$ is perfect, $D_A(s)$ is independent from s for any non-qualified set A. So we will write D_A instead of $D_A(s)$, whenever A is non-qualified. The requirements then are:

1. All shares generated in $\mathcal{S}(k)$ have length polynomially related to k.
2. Distribution and reconstruction of a secret can be done in time polynomial in k.
3. Given secret s and a full set of n shares, one can test in time polynomial in k that the shares are all consistent with s, i.e. that all qualified sets of shares determine s as the secret.
4. Given any secret s, a set of shares for participants in a non-qualified set A (distributed according to D_A) can always be completed to a full set of shares distributed according to $D(s)$ and consistent with s. This completion process can be done in time polynomial in k.
5. For any non-qualified set A, the probability distribution D_A is such that shares for the participants in A are independent and uniformly chosen.

Definition 5. A perfect secret sharing scheme satisfying requirements 1–4 is called *semi-mooth*. If, in addition, requirement 5 is satisfied it is called *smooth*.

It is natural to ask if for any family of monotone access structures there is a family of smooth secret sharing schemes. This question is easy to answer in case of threshold structures. In that case it is clear that Shamir's secret sharing scheme [14] can be used. This scheme is even *ideal*, i.e. the shares are of the same length as the secret. Given d or more shares, the secret s can be found, whereas with $d-1$ or fewer shares, s is completely unknown.

The following alternative to Shamir's scheme (which is also ideal) can lead to more efficient protocols than Shamir's when used in our construction (Theorem 8) with a threshold structure when $d < n/2$.

Again $s \in GF(q)$ is the secret, but the i-th share now is a number $c_i \in GF(q)$, $1 \le i \le n$, such that $Bc = se_1$. Here, B is a $n-d+1$ by n matrix over $GF(q)$, $c = (c_1, \ldots, c_n)$, and $e_1 = (1, 0, \ldots, 0)$ is a vector of length $n-d+1$. Matrix B should be such that any $n-d+1$ columns are linearly independent (which implies that the rank of B is equal to $n-d+1$). An appropriate choice for B is therefore the first $n-d+1$ rows of a Vandermonde matrix over $GF(q)$, say:

$$B = \begin{pmatrix} 1 & 1 & \cdots & 1 \\ 1 & 2 & \cdots & n \\ \vdots & \vdots & \ddots & \vdots \\ 1 & 2^{n-d} & \cdots & n^{n-d} \end{pmatrix}.$$

The secret s can be recovered from any d shares as follows. Since $Bc = se_1$, it follows that $s = \sum_{i=1}^{n} c_i$. Furthermore, when d entries of c are known, the remaining $n-d$ entries follow uniquely from the equation $B'\mathbf{c}=\mathbf{o}$, where B' is the matrix B with the first row removed and $\mathbf{o}$ denotes a vector of $n-d$ zeros. This is true because B' is a $n-d$ by n matrix for which any $n-d$ columns are linearly independent. In case less than d shares are known, the remaining shares can be chosen such that any secret is matched.

For families of access structures other than threshold ones, the answer to the question on existence of smooth secret sharing schemes depends on whether the parameter n is a constant, or is allowed to increase polynomially as a function of k.

In case n is a constant, there exists a smooth secret sharing scheme for any monotone access structure. For any minimal qualified set A, we do the following: choose $s_1, \ldots, s_{|A|}$ at random under the condition that $s_1 \oplus \cdots \oplus s_{|A|} = s$, and give one s_i to each participant in A. This scheme was first proposed in [9].

It is easy to check that this scheme is smooth. In particular, the size of shares and the work needed in this scheme is linear in k, but the constant involved depends of course on n and on the access structure. However, the number of possible subsets is exponential in n, so for non-constant n this scheme will not necessarily be smooth.

For non-constant n, it is an open question whether there are secret sharing schemes of the kind we need for any sequence of access structures. Benaloh and Leichter [1] have proposed secret sharing schemes for more general access structures defined by monotone formulae, i.e. Boolean formulae containing only AND and OR operators.

Consider a monotone formula F with n variables. Any subset A of n participants corresponds in a natural way to a set of values of the n variables by assigning a variable to each participant and let each variable be 1 if the corresponding participant is in A and 0 otherwise. We let $F(A)$ be the bit resulting from evaluating F on inputs corresponding to A. Then we can define an access structure Γ_F by

$$A \in \Gamma_F \Leftrightarrow F(A) = 1$$

We let F^* denote the *dual formula*, which results from replacing in F all AND operators by OR's and vice versa. It is not hard to show the following proposition.

Proposition 6. *If F is monotone then Γ_F is also monotone. Conversely, for any monotone access structure Γ, there is a monotone formula F, such that $\Gamma = \Gamma_F$. We have that $(\Gamma_F)^* = \Gamma_{F^*}$.*

In [1], a generic method is given that, based on any monotone formula F, builds a perfect secret sharing scheme for the access structure Γ_F. The formula F may contain general threshold operators, in addition to simple AND and OR operators. For a polynomial size formula, it can be shown that the secret sharing scheme from [1] satisfies all of the above requirements except possibly requirement 5. This leads to:

Proposition 7. *Let $\{\Gamma(k)\}$ be a family of access structures such that $\Gamma(k) = \Gamma_{F_k}$ for a family of polynomial size monotone formula $\{F_k\}$. Then there exists a family of semi-smooth secret sharing schemes for $\{\Gamma(k)\}$.*

A final comment before we go on to the main result is that we will need to distribute secrets of length $t = t(k)$ bits, where t is polynomially bounded in k. This does not impose any restrictions on $\mathcal{S}(k)$ because any secret sharing scheme can distribute secrets of any length by running an appropriate number of copies of the scheme in parallel. We therefore assume that $\mathcal{S}(k)$ always distributes secrets of length t. Note that, if n is constant as a function of k, only one access structure and secret sharing scheme are involved.

4 Main Result

The next theorem describes the construction of a proof of knowledge from a basic proof of knowledge $\mathcal{P}$ for a relation R and a family of secret sharing schemes. In the constructed proof of knowledge both prover and verifier are probabilistic polynomial time machines, using the prover and verifier of $\mathcal{P}$, respectively, as subroutines.

For the statement of the result we need some notation. Let $\Gamma = \{\Gamma(k)\}$ be a family of access structures on $n(k)$ participants. Then R_Γ is a relation defined by the following condition: $((x_1, ..., x_m), (w_1, ..., w_m)) \in R_\Gamma$ iff all x_i's are of the same length, say, k bits, $m = n(k)$, and the set of indices i for which $(x_i, w_i) \in R$ corresponds to a qualified set in $\Gamma(k)$. In a proof of knowledge for relation R_Γ the prover thus proves to know witnesses to a set of the x_i's corresponding to a qualified set in $\Gamma(k)$.

Theorem 8. *Let $\mathcal{P}$ be a three round public coin, honest verifier zero-knowledge proof of knowledge for relation R, which satisfies the special soundness property. Let $\Gamma = \{\Gamma(k)\}$ be a family of monotone access structures and let $\{\mathcal{S}(k)\}$ be a family of smooth secret sharing schemes such that the access structure of $\mathcal{S}(k)$ is $\Gamma(k)^*$. Then there exists a three round public coin, witness indistinghuisable proof of knowledge for relation R_Γ.*

Proof. To improve readability we drop in the following the dependency on k from the notation, and write $\mathcal{S} = \mathcal{S}(k)$, $\Gamma = \Gamma(k)$ and $n = n(k)$. We will distribute secrets of length t in $\mathcal{S}$. If the length of any share resulting from this is larger than t, we will replace $\mathcal{P}$ by a number of parallel executions of $\mathcal{P}$ to make sure that a challenge is at least as long as any share.[2] Note that this does not violate the honest verifier zero-knowledge nor the special soundness property. A basic idea in the following will be to interpret a challenge as a share. If challenges are longer than shares, we will simply take the first appropriate number of bits of the challenge to be the corresponding share. If c is a challenge, *share*(c) will denote the corresponding share.

The following now describes the new protocol, in which $A \in \Gamma$ denotes the set of indices i for which P knows a witness for x_i:

[2] For some secret sharing schemes, there is a lower bound on the length of shares in terms of n. For Shamir's scheme, the length of shares is at least $\log_2(n+1)$. If t is smaller than this bound, we can again replace $\mathcal{P}$ by a number of parallel executions.

1. For each $i \in \overline{A}$, P runs simulator S on input x_i to produce conversations (m_1^i, c_i, m_2^i). For each $i \in A$, P determines m_1^i as what the prover in $\mathcal{P}$ would send as m_1 given a witness for input x_i. P then sends the values m_1^i, $i = 1, \ldots, n$ to V.
2. V chooses a t-bit string s at random and sends it to P.
3. Consider the set of shares $\{share(c_i) | i \in \overline{A}\}$ that correspond to the c_i from the simulation in Step 1. As $\overline{A}$ is non-qualified in Γ^*, requirement 4 guarantees that P can complete these shares to a full set of shares consistent with s. P then forms challenges c_i for indices $i \in A$, such that $share(c_i)$ equals the share produced in the completion process. This is done by simply copying the bits of the shares and padding with random bits if necessary. In Step 1, S has produced a final message m_2^i in $\mathcal{P}$ for $i \in \overline{A}$. For $i \in A$, P knows a witness for x_i, and can therefore find a valid m_2^i for m_1^i and c_i by running the prover's algorithm from $\mathcal{P}$. Finally, P sends the set of messages c_i, m_2^i, $i = 1, \ldots, n$ to V.
4. V checks that all conversations (m_1^i, c_i, m_2^i) now produced would lead to acceptance by the verifier in $\mathcal{P}$, and that the shares $share(c_i)$ are consistent with secret s. He accepts if and only if these checks are satisfied.

It is clear from the assumptions on $\mathcal{S}$ that P and V need only poly-time and access to the prover and verifier of $\mathcal{P}$. It therefore remains to be seen that the protocol is a proof of knowledge and that it is witness indistinguishable.

Completeness is trivially seen to hold by inspection of the protocol. For *soundness*, assume that some prover P^* for a given first message $\{m_1^i | \ i = 1, \ldots, n\}$ can answer correctly a non-negligible fraction of the possible choices of s. This means that by rewinding P^*, we can efficiently get correct answers to two different values, say s and s'.[3] Let the shares of s and s' sent in the protocol be $share(c_i)$ and $share(c_i')$, $i = 1, \ldots, n$, respectively. Then for every qualified set $B \in \Gamma^*$, there must be an $i \in B$, such that $share(c_i) \neq share(c_i')$ since otherwise it would follow that $s = s'$. But then we also have that $c_i \neq c_i'$ and so by assumption on $\mathcal{P}$, we can compute a witness for x_i. So P^* knows a witness in every qualified set of Γ^*. On account of Proposition 4 the set of witnesses we thus extract is a qualified set in the access structure Γ.

As for *witness indistinguishability*, we have to show that the distribution of the conversation is independent of which qualified set $A \in \Gamma$ the prover uses. First observe that the distribution of each m_1^i depends only on x_i and equals the distribution of the prover's first message in an execution of $\mathcal{P}$ with x_i as input. This follows from Proposition 1, using that $\mathcal{P}$ is honest verifier zero-knowledge. In particular, the joint distribution of the m_1^i's, and hence the verifier's choice of s, is independent of A.

Since the set $\{share(c_i)\}$ is constructed by completing a set of uniformly distributed shares in a non-qualified set of $\mathcal{S}$, the joint distribution of the $share(c_i)$'s is simply $D(s)$. Since the c_i's are constructed from the shares by possibly padding with random bits, the joint distribution of the c_i's is independent of A. Fi-

[3] There are 2^t possible s-values which is super-polynomial in k, whence any polynomial fraction of these contain at least 2 values for all large enough k.

nally, Proposition 1 implies that the distribution of each m_2^i depends only on x_i, m_1^i and c_i, and is therefore also independent of A.

Remark. If the secret sharing schemes are ideal, the communication complexity of the protocol in Theorem 8 is at most t bits plus n times that of $\mathcal{P}$. Note that instead of taking several instances of the same proof of knowledge, it is also possible to combine different proofs of knowledge. In this way, one may for instance prove knowledge of either a discrete log or an RSA root without revealing which.

Theorem 9. *As Theorem 8, but with $\mathcal{P}$* special *honest verifier zero-knowledge and $\mathcal{S}(k)$* semi-*smooth.*

Proof. In this case the protocol from Theorem 8 is changed as follows. In Step 1, the prover uses $\mathcal{S}$ to distribute an arbitrary secret, and discards all shares in A. The remaining shares are distributed according to $D_{\overline{A}}$. He then runs the special simulator on the corresponding challenges. Note that the completion process can still be performed on account of requirement 4, and as before, the honest prover can counter any challenge s by the verifier. Soundness is proven in the same way as before. Therefore, the modified scheme still constitutes a proof of knowledge for relation R_Γ.

As for witness indistinguishability, we only have to note that the distribution of any m_1^i generated by the (special) simulator is the same for any particular challenge value c_i used, because m_1^i in a real execution of $\mathcal{P}$ is independent of the challenge. Therefore the joint distribution of the m_1^i's is the same as in the case of Theorem 8. The rest of the proof is therefore the same as for Theorem 8.

The witness indistinguishable property of the protocol from Theorem 8 leads us to a generalization of Theorem 4.3 of [4]. To state the result, we need to introduce the concept of an *invulnerable generator* G for a relation R. Such generators were first introduced in [6] and later used in slightly modified form in [4]. Such a generator is a probabilistic polynomial time algorithm which outputs a pair $(x, w) \in R$. The generator is invulnerable if no probabilistic polynomial time enemy given only x can compute an element in $w(x)$ with non-negligible probability, taken over the coin flips of both G and the enemy.

Thus, asserting the existence of an invulnerable generator for a relation is a way of stating that it is feasible to generate hard, solved instances of the underlying computational problem.

For any generator G, we let G^n denote the generator that produces an n-tuple of pairs in R by running G independently n times in parallel. We will also need some notation for access structures: for a monotone access structure Γ, we let the sets in Γ correspond to subsets of the index set $N = \{1, ..., n\}$. Now let the set $I_\Gamma \subseteq N$ be defined by: $i \in I_\Gamma$ iff i is contained in every qualified set in Γ. It is easy to see by monotonicity of Γ that $i \in I_\Gamma$ precisely if $N \setminus \{i\}$ is not qualified (using Proposition 4).

Theorem 10. *Let $\mathcal{P}$ be a witness indistinguishable proof of knowledge for the relation R_Γ, where $\Gamma = \{\Gamma(k)\}$ is a family of monotone access structures on $n(k)$ participants, and R is a binary relation. If for all k, $\Gamma(k)$ contains at least two different minimal qualified sets, and there is an invulnerable generator G for R, then $\mathcal{P}$ is witness hiding over $G^{n(k)}$.*

Proof. We follow the line of reasoning from Thm. 4.3 of [4]. Suppose we are given an probabilistic polynomial time enemy $\mathcal{A}$ that has non-negligible probability of computing a witness, using the honest prover in the scheme from Theorem 8 as a subroutine. We show that $\mathcal{A}$ can be compiled into an algorithm that solves with non-negligible probability random instances x generated by G, thus contradicting the invulnerability of the generator (see [4]).

From the assumption on $\Gamma(k) = \Gamma$ (at least two minimal qualified sets) it follows that $N \setminus I_\Gamma$ must contain at least two elements, and that I_Γ is not qualified.

Our compilation now works as follows:

1. Recall that our input is a problem instance x generated by G. We now form an n tuple of instances $(x_1, ..., x_n)$ as follows: choose at random $j \in N$, and let $x_j = x$. For all other indices i, run G to produce a solved instance x_i and save the witness w_i.
2. Give $x_1, ..., x_n$ as input to $\mathcal{A}$. When $\mathcal{A}$ needs to interact with the prover, we simply simulate the prover's algorithm in $\mathcal{P}$. If $j \notin I_\Gamma$, this can be done, since then $N \setminus \{j\}$ is qualified and we know witnesses of all instances except x_j. If $j \in I_\Gamma$, we fail and stop.
3. If $\mathcal{A}$ is successful, it outputs a witness for the relation R_Γ which by definition is a set of witnesses $\{w_i\}$ corresponding to a qualified set A in Γ. If $j \in A$, we have success and can output w_j. Else output something random.

We now show that this compilation finds a witness for x with non-negligible probability. It is sufficient to show that we find a witness with non-negligible probability given that $j \notin I_\Gamma$ since this happens with probability at least $2/n$. Now note that the joint distribution of the x_i's we give to $\mathcal{A}$ is the same as in an ordinary interaction with the prover. Therefore $\mathcal{A}$ is successful with non-negligible probability by assumption. We therefore only have to bound the probability that j is in A, the set of witnesses we get from $\mathcal{A}$. Since I_Γ is not qualified, A must contain at least one index not in I_Γ. By witness indistinguishability, $\mathcal{A}$ has no information about which j in $N \setminus I_\Gamma$ we have chosen, and so the probability that $j \in A$ is at least $1/|N \setminus I_\Gamma|$. Hence if $\mathcal{A}$ has success probability ϵ, our success probability given that $j \notin I_\Gamma$ is at least ϵ/n, which is non-negligible.

Note that an access structure has at least two minimal qualified sets exactly when the corresponding minimal CNF-formula contains at least one OR-operator.

Note also that this result only shows that an enemy cannot compute a complete qualified set of witnesses. It does not rule out that the protocol could help him to compute a small, non-qualified set. Ideally, we would like to prove that the enemy cannot compute even a single witness. With a stronger assumption on the access structure, this can be done:

Corollary 11. *Let $\mathcal{P}$ be a witness indistinguishable proof of knowledge for the relation R_Γ, where $\Gamma = \{\Gamma(k)\}$ is a family of monotone access structures on $n(k)$ participants, and R is a binary relation. Suppose that for all k the set $I_{\Gamma(k)}$ is empty. Suppose finally that there is an invulnerable generator G for R, and that inputs for $\mathcal{P}$ are generated by $G^{n(k)}$. Then no probabilistic polynomial time enemy interacting with the honest prover can with non-negligible probability compute a witness for any of the x_i in the input to the protocol.*

Proof. Since $I_{\Gamma(k)}$ is non qualified, there are at least two minimal sets, and therefore the proof is the same as for Theorem 10, except that it follows from the assumption that the index j is always chosen among all indices. Hence if the enemy outputs at least one correct witness, there is a non-negligible probability of at least $1/n$ that this is the witness we are looking for.

A certain special case of Theorem 8 is interesting in its own right:

Corollary 12. *Let $\mathcal{P}$ be a three round public coin, honest verifier zero-knowledge proof of knowledge for relation R, which satisfies the special soundness property. Then for any n, d there is a protocol with the same round complexity as $\mathcal{P}$ in which the prover shows that he knows d out of n witnesses without revealing which d witnesses are known.*

Proof. Use Theorem 8 with, for example, Shamir's secret sharing scheme for $\mathcal{S}$ and a threshold value of $n - d + 1$.

Corollary 13. *Consider the protocol guaranteed by Corollary 12, let $n = 2$ and $d = 1$, i.e. the prover proves that he knows at least 1 out of 2 solutions. For any generator G generating pairs in R, this protocol is witness hiding over G^2.*

Proof. Since protocols constructed from Theorem 8 are always witness indistinguishable, we can use Theorem 4.2 of Feige and Shamir[4].

Note that for this corollary, we do not need the assumption that G is invulnerable, as in Theorem 10.

To build the protocol of Corollary 13, we need a 2 out of 2 threshold scheme. Such a scheme can be implemented by choosing random shares c_1, c_2 such that $c_1 \oplus c_2$ equals the secret. Therefore, in the simple case of Corollary 13, the protocol constructed by Theorem 8 simply becomes a game where the verifier chooses a random s, and the prover shows that he can answer correctly a pair of challenges c_1, c_2, such that $s = c_1 \oplus c_2$. In the prover's final message, he only has to send c_1 because the verifier can then compute c_2 himself. Hence the communication complexity of the new protocol is exactly twice that of $\mathcal{P}$, whence the new protocol is just as practical as $\mathcal{P}$.

Corollary 14. *Let* $\{\Gamma(k) = \Gamma_{F_k}\}$ *be a family of monotone access structure on* $n(k)$ *participants defined by a polynomial size family of formulas* $\{F_k\}$*, and let* $\mathcal{P}$ *be a three round public coin,* special *honest verifier zero-knowledge proof of knowledge for relation* R*, which satisfies the special soundness property. Then there exists a witness-indistinguishable proof of knowledge for relation* $R_{\Gamma(k)}$*. Let* $M(k)$ *be the maximal number of occurrences of a variable in* $F(k)$*. Then the communication complexity of the new protocol is at most* $nM(k)$ *times that of* $\mathcal{P}$ *plus t bits.*

Proof. By Proposition 3, $\Gamma(k)^* = \Gamma_{F_k^*}$, and since the size of F_k^* is the same as that of F_k, we can use the secret sharing scheme guaranteed by Proposition 7 when we do the construction of Theorem 8. The statement on the communication complexity follows from the fact that the shares of the secret sharing scheme constructed in [1] from $F(k)$ have maximal size $tM(k)$ bits, so that we have to use $M(k)$ parallel executions of $\mathcal{P}$ in the construction of Theorem 8.

5 Application to Identification and Signatures

Suppose we have n users, for example employees of a company, such that the i-th user has a public key x_i and secret key $w_i \in w(x_i)$. Suppose also that certain subsets of users are qualified in the sense that they are allowed to initiate certain actions, sign letters on behalf of the company, etc. This defines an access structure on the set of users. Theorem 8 now gives a way in which a subset of users can collaborate to identify themselves as a qualified subset, without revealing anything else about their identities. This makes good sense, if they are to assume responsibility on behalf of the company, rather than personally.

This also extends to digital signatures, since by using a hash function, any three round proof of knowledge as the one produced by Theorem 8 can be turned into a signature scheme by computing the challenge as a hash value of the message to be signed and the prover's first message (this technique was introduced in [5]). By this method, a signature can be computed which will show that a qualified subset was present, without revealing which subset was involved. This is a generalization of the results from e.g. [10] and also of the group signature concept, introduced by Chaum and Van Heyst [2]. One aspect of group signatures which is missing here, however, is that it is not possible later to "open" signatures to discover the identities of users involved.

Note also that our method allows participants to form groups completely freely, using the same keys in all groups. For example, two participants who normally use Schnorr signatures individually can go together and form a "1 out of 2" signature without changing their keys or the basic algorithms in which they are used.

6 Open Problems

Two obvious open problems remain. First, can Theorem 8 be proved assuming ordinary soundness of $\mathcal{P}$, and not special soundness? And secondly, can it be generalized to other types of protocols than public coin protocols?

Acknowledgement We thank Douglas Stinson for helping us with information about results on secret sharing schemes, and Matthew Franklin for useful discussions and comments on the presentation.

References

1. J. Benaloh and J. Leichter: *Generalized Secret Sharing and Monotone Functions*, Proc. of Crypto 88, Springer Verlag LNCS series, 25–35.
2. D. Chaum and E. van Heyst: *Group Signatures*, Proc. of EuroCrypt 91, Springer Verlag LNCS series.
3. I. Damgård: *Interactive Hashing can Simplify Zero-Knowledge Protocol Design Without Complexity Assumptions*, Proc. of Crypto 93, Springer Verlag LNCS series.
4. U. Feige and A. Shamir: *Witness Indistinguishable and Witness Hiding Protocols*, Proc. of STOC 90.
5. U. Feige, A. Fiat and A. Shamir: *Zero-Knowledge Proofs of Identity*, Journal of Cryptology 1 (1988) 77–94.
6. M. Abadi, E. Allender, A. Broder, J. Feigenbaum and L. Hemachandra: *On Generating Solved Instances of Computational Problems*, Proc. of Crypto 88, Springer Verlag LNCS series.
7. S. Goldwasser, S. Micali and C. Rackoff: *The Knowledge Complexity of Interactive Proof Systems*, SIAM Journal on Computing 18 (1989) 186–208.
8. L. Guillou and J.-J. Quisquater: *A Practical Zero-Knowledge Protocol fitted to Security Microprocessor Minimizing both Transmission and Memory*, Proc. of EuroCrypt 88, Springer Verlag LNCS series.
9. M. Ito, A. Saito, and T. Nishizeki: *Secret Sharing Scheme Realizing any Access Structure*, Proc. Glob.Com. (1987).
10. T.Pedersen: *A Threshold Cryptosystem without a Trusted Third Party*, Proc. of EuroCrypt 91.
11. A. De Santis, G. Di Crescenzo and G. Persiano: *Secret Sharing and Perfect Zero-Knowledge*, Proc. of Crypto 93, Springer Verlag LNCS series.
12. A. De Santis, G. Persiano, M. Yung: *Formulae over Random Self-Reducible Languages: The Extended Power of Perfect Zero-Knowledge*, manuscript.
13. C.P. Schnorr: *Efficient Signature Generation by Smart Cards*, Journal of Cryptology 4 (1991) 161–174.
14. A. Shamir: *How to Share a Secret*, Communications of the ACM 22 (1979) 612–613.
15. G.J. Simmons, W.A. Jackson and K. Martin: *The Geometry of Shared Secret Schemes*, Bulletin of the Institute of Combinatorics and its Applications 1 (1991) 71–88.

Language Dependent Secure Bit Commitment

Toshiya Itoh[1] Yuji Ohta[1] Hiroki Shizuya[2]

[1] Department of Information Processing,
Interdisciplinary Graduate School of Science and Engineering,
Tokyo Institute of Technology,
4259 Nagatsuta, Midori-ku, Yokohama 227, Japan.
[2] Education Center for Information Processing, Tohoku University,
Kawauchi, Aoba-ku, Sendai 980, Japan.

Abstract. In this paper, we define two classes of languages, one induces opaque/transparent bit commitments and the other induces transparent/opaque bit commitments. As an application of opaque/transparent and transparent/opaque properties, we first show that if a language L induces an opaque/transparent bit commitment, then there exists a prover-practical perfect zero-knowledge proof for L, and we then show that if a language L induces a transparent/opaque bit commitment, then there exists a bounded round perfect zero-knowledge proof for L.

1 Introduction

A bit commitment is a two party (interactive) protocol between a sender S and a receiver R in which after the sender S commits to a bit $b \in \{0,1\}$ at hand, (1) the sender S cannot change his mind in a computational or an information-theoretic sense; and (2) the receiver R learns nothing about the bit $b \in \{0,1\}$ in a computational or an information-theoretic sense. Bit commitments have diverse applications to cryptographic protocols, especially to zero-knowledge proofs (see, e.g., [6], [1], [11], [9], [4], etc). For simplicity, we assume that a bit commitment f is noninteractive, i.e., the sender S sends to the receiver R only a single message C. According to computational power of senders and receivers, bit commitments can be classified into the following four possible types (see, e.g., [12]).

	Power of Sender S	Power of Receiver R
Type A	poly-time bounded	poly-time bounded
Type B	poly-time bounded	unbounded
Type C	unbounded	poly-time bounded
Type D	unbounded	unbounded

Feige and Shamir [6] used a bit commitment of Type A to show that any language $L \in \mathcal{NP}$ has a two round perfect zero-knowledge proof of knowledge. Brassard, Chaum, and Crépeau [1] and Naor et al [11] showed that any language $L \in \mathcal{NP}$ has a perfect zero-knowledge argument assuming the existence of a bit commitment of Type B and Bellare, Micali, and Ostrovsky [4] showed that any honest verifier statistical zero-knowledge proof for a language L can be transformed to a statistical zero-knowledge proof for the language L assuming the existence of a bit commitment of Type B. In addition, Goldreich, Micali, and Wigderson [9] used a bit commitment of Type C to show that any language $L \in \mathcal{NP}$ has a computational zero-knowledge proof. Now we look at the properties required to bit commitments for each possible type above.

Assume that the sender S is computationally unbounded. If there exist $r, s \in \{0,1\}^k$ such that $f(0,r) = f(1,s)$, then a cheating sender S^* chooses $r \in \{0,1\}^k$ to compute $C = f(0,r)$ and reveals 1 and $s \in \{0,1\}^k$ to change his mind. Thus any $r, s \in \{0,1\}^k$ must satisfy that $f(0,r) \neq f(1,s)$. Here we refer to such a bit commitment f as *transparent*. Assume that the receiver R is computationally unbounded. If the distribution of $f(0,r)$ is apart from that of $f(1,r)$, then a cheating receiver R^* might learn something about the value of the bit $b \in \{0,1\}$ only looking at $C = f(b,r)$. Thus the distributions of $f(0,r)$ and $f(1,s)$ must be almost identical. Here we refer to such a bit commitment f as *opaque*.

If both the sender S and the receiver R are computationally unbounded, then any bit commitment f must be transparent and opaque, however it is impossible to algorithmically implement such a bit commitment. This implies that there exists inherently no way of designing bit commitments of Type D. Thus only possible way of doing this is to physically implement such a bit commitment. This is referred to as an *envelope*. Assuming the existence of the envelope, Goldreich, Micali, and Wigderson [9] showed that any language $L \in \mathcal{NP}$ has a perfect zero-knowledge proof and then Ben-Or et al [2] showed that any language $L \in \mathcal{IP}$ has a perfect zero-knowledge proof. The goal of this paper is to algorithmically construct a bit commitment of Type D in a somewhat different setting.

In this paper, we consider the following framework: Our bit commitment f is allowed to have an additional input $x \in \{0,1\}^*$ and its property heavily depends on the additional input $x \in \{0,1\}^*$. In this setting, we define two classes of languages, one induces opaque/transparent bit commitments and the other induces transparent/opaque bit commitments. Informally, a language L induces an opaque/transparent bit commitment f_L if (1) for every $x \in L$, the distribution of $f_L(x,0,r)$ is *identical* to that of $f_L(x,1,r)$; and (2) for every $x \notin L$, the distribution of $f_L(x,0,r)$ is *completely different* from that of $f_L(x,1,r)$, and L induces a transparent/opaque bit commitment f_L if $\overline{L}$ induces an opaque/transparent bit commitment $f_{\overline{L}}$. Then we can show the following theorems:

Theorem 18: If a language L induces an opaque/transparent bit commitment, then there exists a prover-practical perfect zero-knowledge proof for L.

Theorem 21: If a language L induces a transparent/opaque bit commitment, then there exists a bounded round perfect zero-knowledge proof for L.

2 Preliminaries

Here we present several definitions necessary to the subsequent discussions.

Definition 1 [8]. Let $L \subseteq \{0,1\}^*$. A probability ensemble $\{U(x)\}_{x\in L}$ is said to be identical to a probability ensemble $\{V(x)\}_{x\in L}$ on L if for every $x \in L$,

$$\sum_{\alpha\in\{0,1\}^*} |\mathrm{Prob}\,\{U(x)=\alpha\} - \mathrm{Prob}\,\{V(x)=\alpha\}| = 0.$$

Let k be a security parameter. Let $g(b,r)$ be a polynomial (in k) time computable function. A function g is a noninteractive bit commitment if after the sender S sends $C = g(b,r)$ to the receiver R, (1) any cheating sender S^* cannot change his mind, i.e., S^* cannot reveal $r, s \in \{0,1\}^k$ such that $C = g(0,r) = g(1,s)$; and (2) any cheating receiver R^* learns nothing about the bit $b \in \{0,1\}$ only looking at $C = g(b,r)$. As a modification, let us consider bit commitments in the following setting: Let L be a language and let k be a polynomial. Assume that $f_L(x,b,r)$ is a polynomial (in $|x|$) time computable function for any $b \in \{0,1\}$ and any $r \in \{0,1\}^{k(|x|)}$.

Definition 2. A language L is said to induce an opaque/transparent (O/T for short) bit commitment f_L if

- opaque: for every $x \in L$, the distribution of $f_L(x,0,r)$ is identical to that of $f_L(x,1,r)$;
- transparent: for every $x \notin L$, there do not exist $r \in \{0,1\}^{k(|x|)}$ and $s \in \{0,1\}^{k(|x|)}$ such that $f_L(x,0,r) = f_L(x,1,s)$,

where k is a polynomial that guarantees the security of f_L.

The opaque/transparent property guarantees that for every $x \in L$, any all powerful cheating receiver R^* cannot guess better at random the value of the bit $b \in \{0,1\}$ after receiving $f_L(x,b,r)$ from the sender S and for every $x \notin L$, any all powerful cheating sender S^* cannot change his mind after sending $f_L(x,b,r)$ to the receiver R. Let $\mathcal{OT}$ be the class of languages that induce O/T bit commitments. From Definition 2, it is clear that $\mathcal{OT} \subseteq \mathcal{NP}$.

Definition 3. A language L is said to induce a transparent/opaque (T/O for short) bit commitment f_L if $\overline{L}$ induces an O/T bit commitment $f_{\overline{L}}$.

Contrary to the opaque/transparent property, the transparent/opaque property guarantees that for every $x \in L$, any all powerful cheating sender S^* cannot change his mind after sending $f_L(x,b,r)$ to the receiver R and for every $x \notin L$, any all powerful cheating receiver R^* cannot guess better at random the value of the bit $b \in \{0,1\}$ after receiving $f_L(x,b,r)$ from the sender S. Let $\mathcal{TO}$ be the class of languages that induce T/O bit commitments. From Definitions 2 and 3, it is obvious that co-$\mathcal{TO} = \mathcal{OT} \subseteq \mathcal{NP}$.

Definition 4 [8]. An interactive protocol $\langle P,V\rangle$ is an interactive proof system for a language L if there exists an honest verifier V that satisfies the following:

- completeness: there exists an honest prover P such that for every $k > 0$ and for sufficiently large $x \in L$, $\langle P,V\rangle$ halts and accepts $x \in L$ with probability at least $1 - |x|^{-k}$, where the probabilities are taken over the coin tosses of P and V.

- soundness: for every $k > 0$, for sufficiently large $x \notin L$, and for any cheating prover P^*, $\langle P^*, V\rangle$ halts and accepts $x \notin L$ with probability at most $|x|^{-k}$, where the probabilities are taken over the coin tosses of P^* and V.

It should be noted that the resource of P is computationally unbounded while the resource of V is bounded by probabilistic polynomial (in $|x|$) time.

In the remainder of this paper, we assume that a term "zero-knowledge" implies "blackbox simulation" zero-knowledge.

Definition 5 [10]. An interactive proof system $\langle P, V\rangle$ for a language L is said to be (blackbox simulation) perfect zero-knowledge if there exists a probabilistic polynomial time Turing machine M_U such that for any (cheating) verifier V^* and for sufficiently large $x \in L$, the probability ensemble $\{M_U(x; V^*)\}_{x\in L}$ is identical to the probability ensemble $\{\langle P, V^*\rangle(x)\}_{x\in L}$ on L, where $M(\cdot; A)$ denotes a Turing machine with blackbox access to a Turing machine A.

From a practical purpose, Boyar, Friedl, and Lund [3] defined a notion of *prover-practical* (zero-knowledge) interactive proof systems.

Definition 6 [3]. An interactive proof system $\langle P, V\rangle$ for a language $L \in \mathcal{NP}$ is said to be prover-practical if the honest prover P runs in probabilistic polynomial time and some trapdoor information on input $x \in L$ is initially written on the private auxiliary tape of P.

Let $A, B \in \mathcal{NP}$ and let g be a reduction from A to B, i.e., g is a polynomial time computable function and for any $x \in \{0,1\}^*$, $x \in A$ iff $g(x) \in B$.

Definition 7 [6]. Let $A, B \in \mathcal{NP}$. A reduction g from A to B is said to be witness-preserving if there exists a polynomial time computable function h that given a witness w for any $x \in A$, $h(x, w)$ is a witness for $g(x) \in B$.

Definition 8 [6]. Let $A, B \in \mathcal{NP}$. A reduction g from A to B is said to be polynomial time invertible if there exists a polynomial time computable function γ that given a witness w' for $g(x) \in B$, $\gamma(g(x), w')$ is a witness for $x \in A$.

3 Examples

It is obvious from the Definitions 2 and 3 that $L \in \mathcal{OT}$ iff $\overline{L} \in \mathcal{TO}$. Thus we only exemplify several languages that induce O/T bit commitments.

For graphs G and H, we use $G \simeq H$ to imply that G is isomorphic to H and use $G \not\simeq H$ to imply that G is not isomorphic to H.

Definition 9. For an integer $h > 0$, Universal Graph Isomorphism Tuple UGIT is defined to be UGIT $= \{\langle h, \langle G_1^0, G_1^1\rangle, \langle G_2^0, G_2^1\rangle, \ldots, \langle G_h^0, G_h^1\rangle\rangle \mid G_i^0 \simeq G_i^1$ for each i $(1 \le i \le h)\}$.

Definition 10. For an integer $h > 0$, Existential Graph Isomorphism Tuple EGIT is defined to be EGIT $= \{\langle h, \langle G_1^0, G_1^1\rangle, \langle G_2^0, G_2^1\rangle, \ldots, \langle G_h^0, G_h^1\rangle\rangle \mid G_i^0 \simeq G_i^1$ for some i $(1 \le i \le h)\}$.

Definition 11. Let $N = p_1^{e_1} p_2^{e_2} \cdots p_h^{e_h}$ be the prime factorization of N. Define cMODd to be $N \in c\mathrm{MOD}d$ if and only if $p_i \equiv c \pmod{d}$ for each i $(1 \leq i \leq h)$.

In the following, we show that the languages UGIT, EGIT, and 1MOD4 induce O/T bit commitments f_{UGIT}, f_{EGIT}, and f_{1MOD4}, respectively.

Lemma 12. *The language* UGIT *induces an* O/T *bit commitment* f_{UGIT}.

Proof: For $x = \langle h, \langle G_1^0, G_1^1\rangle, \langle G_2^0, G_2^1\rangle, \ldots, \langle G_h^0, G_h^1\rangle\rangle$, let V_i $(1 \leq i \leq h)$ be a set of vertices for G_i^0 and G_i^1, and let $b \in \{0,1\}$ be a bit that a sender S wishes to send to a receiver R. Here we define a bit commitment f_{UGIT} for UGIT as follows: For each i $(1 \leq i \leq h)$, S chooses $\pi_i \in_{\mathrm{R}} \mathrm{Sym}(V_i)$. Then S computes a graph $H_i = \pi_i(G_i^b)$ and sends $\langle H_1, H_2, \ldots, H_h\rangle$ to R.

Assume that $x \in$ UGIT. It follows from Definition 9 that $G_i^0 \simeq G_i^1$ for each i $(1 \leq i \leq h)$. Then the distribution of $\langle H_1, H_2, \ldots, H_h\rangle$ for $b = 0$ is *identical* to that of $\langle H_1, H_2, \ldots, H_h\rangle$ for $b = 1$. Assume that $x \notin$ UGIT. It follows from Definition 9 that there exists at least an i_0 $(1 \leq i_0 \leq h)$ such that $G_{i_0}^0 \not\simeq G_{i_0}^1$. This implies that $\pi_{i_0}(G_{i_0}^0) \neq \varphi_{i_0}(G_{i_0}^1)$ for any $\pi_{i_0}, \varphi_{i_0} \in \mathrm{Sym}(V_{i_0})$. Then for any $\pi_i, \varphi_i \in \mathrm{Sym}(V_i)$ $(1 \leq i \leq h)$,

$$f_{\mathrm{UGIT}}(x, 0, \langle \pi_1, \pi_2, \ldots, \pi_h\rangle) \neq f_{\mathrm{UGIT}}(x, 1, \langle \varphi_1, \varphi_2, \ldots, \varphi_h\rangle).$$

Thus the language UGIT induces an O/T bit commitment f_{UGIT}. ■

For an integer $h > 0$, define Universal Quadratic Residuosity Tuple UQRT to be UQRT $= \{\langle h, \langle x_1, N_1\rangle, \ldots, \langle x_h, N_h\rangle\rangle \mid x_i$ is a square modulo N_i for each i $(1 \leq i \leq h)\}$. Then in a way similar to Lemma 12, we can show the following:

Lemma 13. *The language* UQRT *induces an* O/T *bit commitment* f_{UQRT}.

Let us proceed to show the other examples.

Lemma 14. *The language* EGIT *induces an* O/T *bit commitment* f_{EGIT}.

Proof: Let $x = \langle h, \langle G_1^0, G_1^1\rangle, \langle G_2^0, G_2^1\rangle, \ldots, \langle G_h^0, G_h^1\rangle\rangle$ and let V_i $(1 \leq i \leq h)$ be a set of vertices for G_i^0 and G_i^1. Let $b \in \{0,1\}$ be a bit that a sender S wishes to send to a receiver R. Here we define a bit commitment f_{EGIT} for EGIT as follows: For each i $(1 \leq i \leq h)$, S first chooses $e_i \in_{\mathrm{R}} \{0,1\}$ and $\pi_i \in_{\mathrm{R}} \mathrm{Sym}(V_i)$. Then S computes $c \equiv e_1 + e_2 + \cdots + e_h + b \pmod{2}$ and a graph $H_i = \pi_i(G_i^{e_i})$ $(1 \leq i \leq h)$ and sends $\langle c, H_1, H_2, \ldots, H_h\rangle$ to R.

Assume that $x \in$ EGIT. It follows from Definition 10 that there exists at least an i_0 $(1 \leq i_0 \leq h)$ such that $G_{i_0}^0 \simeq G_{i_0}^1$. Then on that position i_0 $(1 \leq i_0 \leq h)$, the distribution of $\pi_{i_0}(G_{i_0}^0)$ is *identical* to that of $\pi_{i_0}(G_{i_0}^1)$. This implies that the distribution of $\langle c, H_1, H_2, \ldots, H_h\rangle$ for $b = 0$ is *identical* to that of $\langle c, H_1, H_2, \ldots, H_h\rangle$ for $b = 1$. Assume that $x \notin$ EGIT. It follows from Definition 10 that for every i $(1 \leq i \leq h)$, $G_i^0 \not\simeq G_i^1$. Then for any $e_i, d_i \in \{0,1\}$ and $\pi_i, \varphi_i \in \mathrm{Sym}(V_i)$ $(1 \leq i \leq h)$,

$$f_{\mathrm{EGIT}}(x, 0, \langle e_1, \ldots, e_h\rangle, \langle \pi_1, \ldots, \pi_h\rangle) \neq f_{\mathrm{EGIT}}(x, 1, \langle d_1, \ldots, d_h\rangle, \langle \varphi_1, \ldots, \varphi_h\rangle).$$

Thus the language EGIT induces an O/T bit commitment f_{EGIT}. ■

For an integer $h > 0$, define Existential Quadratic Residuosity Tuple EQRT to be EQRT $= \{\langle h, \langle x_1, N_1\rangle, \ldots, \langle x_h, N_h\rangle\rangle \mid x_i$ is a square modulo N_i for some i $(1 \le i \le h)\}$. Then in a way similar to Lemma 14, we can show the following:

Lemma 15. *The language* EQRT *induces an* O/T *bit commitment* f_{EQRT}.

The final example has different flavor from those of the examples above.

Lemma 16. *The language* 1MOD4 *induces an* O/T *bit commitment* f_{1MOD4}.

Proof: Let $x = p_1^{e_1} p_2^{e_2} \cdots p_h^{e_h}$ be the prime factorization of x. Let $b \in \{0,1\}$ be a bit that a sender S wishes to send to a receiver R. Define a bit commitment f_{1MOD4} for 1MOD4 as follows: First S chooses $r \in_{\mathrm{R}} Z_x^*$. Then S computes $c \equiv (-1)^b r^2 \pmod{x}$ and sends $c \in Z_x^*$ to R. It should be noted that -1 is a square modulo x if and only if $x \in$ 1MOD4.

Assume that $x \in$ 1MOD4. From Definition 11 and the fact that -1 is a square modulo x, it follows that $c \in Z_x^*$ is always a square modulo x regardless of the value of $b \in \{0,1\}$. This implies that the distribution of $c \in Z_x^*$ for $b = 0$ is *identical* to that of $c \in Z_x^*$ for $b = 1$. Assume that $x \notin$ 1MOD4. From Definition 11 and the fact that -1 is not a square modulo x, it follows that for any $r \in Z_x^*$, $c \equiv (-1)^b r^2 \pmod{x}$ is a square modulo x if and only if $b = 0$. Then for any $r, s \in Z_x^*$, $f_{\mathrm{1MOD4}}(x, 0, r) \neq f_{\mathrm{1MOD4}}(x, 1, s)$. Thus the language 1MOD4 induces an O/T bit commitment f_{1MOD4}. ■

It is easy to show that (1) $2 \in Z_N^*$ is a square modulo N if and only if $N \in \pm$1MOD8; (2) $3 \in Z_N^*$ is a square modulo N if and only if $N \in \pm$1MOD12; and (3) $5 \in Z_N^*$ is a square modulo N if and only if $N \in \pm$1MOD5. Then in a way similar to Lemma 16, we can show the following:

Lemma 17. *The languages* ±1MOD8, ±1MOD12, *and* ±1MOD5 *induce* O/T *bit commitments* $f_{\pm\mathrm{1MOD8}}$, $f_{\pm\mathrm{1MOD12}}$, *and* $f_{\pm\mathrm{1MOD5}}$, *respectively.*

4 Opaque/Transparent Bit Commitments

Assume that a language L induces an O/T bit commitment f_L. Now let us consider the interactive protocol $\langle A, B\rangle$ on input $x \in \{0,1\}^*$: (A1) A chooses $b \in_{\mathrm{R}} \{0,1\}$ and $r \in_{\mathrm{R}} \{0,1\}^{k(|x|)}$ and sends $a = f_L(x, b, r)$ to B; (B1) B chooses $e \in_{\mathrm{R}} \{0,1\}$ and sends $e \in \{0,1\}$ to A; (A2) A sends to B $\sigma \in \{0,1\}^{k(|x|)}$ such that $a = f_L(x, e, \sigma)$; and (B2) B checks that $a = f_L(x, e, \sigma)$. After $n = |x|$ independent invocations from step A1 to step B2, V accepts $x \in \{0,1\}^*$ if and only if every check in step B2 is successful.

By the opaque/transparent property of f_L, we can show in almost the same way as the case of random self-reducible languages [13] that L has a perfect zero-knowledge proof. In the protocol $\langle A, B\rangle$, however, A needs to evaluate $\sigma \in \{0,1\}^{k(|x|)}$ such that $a = f_L(x, e, \sigma)$ for each iteration. Thus in general, $\langle A, B\rangle$ could not be prover-practical. In this section, we show a stronger result, i.e., L has a prover-practical perfect zero-knowledge proof.

Theorem 18. *If a language L induces an* O/T *bit commitment, then there exists a* prover-practical *perfect zero-knowledge proof for the language L.*

Proof: Let f_L be an O/T bit commitment induced by a language L. From Definition 2, we have an $\mathcal{NP}$-statement below:

$$x \in L \Longleftrightarrow \exists r, s \in \{0,1\}^{k(|x|)} \text{ s.t. } f_L(x,0,r) = f_L(x,1,s). \tag{1}$$

Let us consider the following interactive protocol $\langle P, V\rangle$ for L.

Interactive Protocol $\langle P, V\rangle$ for L

common input: $x \in \{0,1\}^*$.

P0-1: P reduces an $\mathcal{NP}$-statement of Eq.(1) to a directed Hamiltonian graph $G = (V, E)$, where $|V| = n = |x|^c$ for some constant $c > 0$.
P0-2: P defines an adjacency matrix $A_G = (a_{ij})$ of $G = (V, E)$.
V0-1: V reduces an $\mathcal{NP}$-statement of Eq.(1) to a directed Hamiltonian graph $G = (V, E)$, where $|V| = n = |x|^c$ for some constant $c > 0$.
V0-2: V defines an adjacency matrix $A_G = (a_{ij})$ of $G = (V, E)$.
P1-1: P chooses $\pi \in_R \mathrm{Sym}(V)$ and $s_{ij} \in_R \{0,1\}^{k(|x|)}$ $(1 \le i, j \le n)$.
P1-2: P computes $c_{ij} = f_L(x, a_{\pi(i)\pi(j)}, s_{ij})$.
$P \to V$: $C = (c_{ij})$ $(1 \le i, j \le n)$.
V1: V chooses $e \in_R \{0,1\}$.
$V \to P$: $e \in \{0,1\}$.
P2-1: For $e = 0$, P assigns $\langle \pi, s_{11}, \ldots, s_{nn}\rangle$ to w.
P2-2: For $e = 1$, P assigns $\langle\langle i_1, j_1\rangle, \ldots, \langle i_n, j_n\rangle, s_{i_1 j_1}, \ldots, s_{i_n j_n}\rangle$ to w such that $\langle i_1, j_1\rangle, \ldots, \langle i_n, j_n\rangle$ is a single cycle.
$P \to V$: w.
V2-1: For $e = 0$, V checks that $c_{ij} = f_L(x, a_{\pi(i)\pi(j)}, s_{ij})$ for each i, j $(1 \le i, j \le n)$.
V2-2: For $e = 1$, V checks that $\langle i_1, j_1\rangle, \ldots, \langle i_n, j_n\rangle$ is indeed a single cycle and that $c_{i_m j_m} = f_L(x, 1, s_{i_m j_m})$ for each m $(1 \le m \le n)$.

After $n = |V|$ independent invocations from step P1-1 to step V2-2, V accepts $x \in \{0,1\}^*$ if and only if every check in step V2-1 and step V2-2 is successful.

We show that the protocol $\langle P, V\rangle$ is a prover-practical perfect zero-knowledge proof for the language L if L induces an O/T bit commitment f_L.

Completeness: If L induces an O/T bit commitment f_L, then $L \in \mathcal{NP}$, i.e.,

$$x \in L \Longleftrightarrow \exists r, s \in \{0,1\}^{k(|x|)} \text{ s.t. } f_L(x,0,r) = f_L(x,1,s).$$

Assume that for the common input $x \in L$ to $\langle P, V\rangle$, the honest prover P has $r, s \in \{0,1\}^{k(|x|)}$ such that $f_L(x,0,r) = f_L(x,1,s)$. Since the reduction from any $L \in \mathcal{NP}$ to a directed Hamiltonian graph (DHAM) is known to be witness-preserving, P can compute in polynomial (in $|x|$) time a Hamiltonian cycle H of $G = (V, E)$ in step P0-1. Then P can execute in polynomial (in $|x|$) time every process of $\langle P, V\rangle$. It is obvious that P always causes V to accept $x \in L$.

Soundness: From Eq.(1), it follows that for any $x \notin L$, there does not exist $r, s \in \{0,1\}^{k(|x|)}$ such that $f_L(x,0,r) = f_L(x,1,s)$. This implies that $G = (V,E)$ generated in step V0-1 is not a Hamiltonian graph. We show the soundness condition of $\langle P, V\rangle$ by contradiction. Assume that for some $k_0 > 0$ and infinitely many $x \notin L$, there exists a cheating prover P^* that causes V to accept $x \notin L$ with probability at least $|x|^{-k_0}$. Let $L' \subseteq \overline{L}$ be an infinite set of such $x \notin L$. Then from a standard analysis (see, e.g., [5]), it follows that there must exist $C = (c_{ij})$ that passes both tests in steps V2-1 and V2-2. We note that for any $x \in L'$, there do not exist $r, s \in \{0,1\}^{k(|x|)}$ such that $f_L(x,0,r) = f_L(x,1,s)$. This implies that P^* cannot change his mind after step P1-2 even if P^* is infinitely powerful. To pass the test in step V2-1, $C = (c_{ij})$ must be an encoding of a non-Hamiltonian graph $G = (V,E)$ generated in step V0-1, while to pass the test in step V2-2, $C = (c_{ij})$ must be an encoding of a Hamiltonian graph $\tilde{G} = (\tilde{V}, \tilde{E})$. This contradicts the assumption that $G = (V,E)$ generated in step V0-1 is not a Hamiltonian graph. Then for each $k > 0$ and sufficiently large $x \notin L$, any cheating prover P^* causes V to accept $x \notin L$ with probability at most $|x|^{-k}$.

Perfect Zero-Knowledgeness: This can be shown in a way similar to the case of random self-reducible languages [13]. The construction of M_U for any cheating verifier V^* is as follows:

Construction of M_U

common input: $x \in L$.

M0-1: **count** := 0; and **conv** := ε, where ε is a null string.
M0-2: M_U provides V^* with r_{V^*} as random coin tosses for V^*.
M0-3: M_U simulates steps P0-1 and P0-2.
M1-1: M_U chooses $\alpha \in_R \{0,1\}$.
M1-2: M_U chooses an n vertex random cycle of which adjacency matrix is $H = (h_{ij})$.
M2-1: If $\alpha = 0$, then M_U simulates steps P1-1 and P1-2.
M2-2: If $\alpha = 1$, then M_U chooses $s_{ij} \in_R \{0,1\}^{k(|x|)}$ and computes $c_{ij} = f(x, h_{ij}, s_{ij})$.
M3: M_U runs V^* on input $\langle x, r_{V^*}, \mathbf{conv}, C\rangle$ to generate e.
M4-1: If $e \notin \{0,1\}$, then M_U halts and outputs $\langle x, r_{V^*}, \mathbf{conv}\|\langle C, e\rangle\rangle$, where $x\|y$ denotes the concatenation of strings $x, y \in \{0,1\}^*$.
M4-2: If $e \neq \alpha$, then go to step M1-1.
M4-3: If $e = \alpha$, then M_U simulates steps P2-1 and P2-2 depending on $\alpha \in \{0,1\}$.
M5-1: M_U sets **conv** := **conv**$\|\langle C, e, w\rangle$ and **count** := **count** + 1.
M5-2: If **count** $< n$, then go to step M1-1; otherwise M_U halts and outputs $\langle x, r_{V^*}, \mathbf{conv}\rangle$.

Note that for any $x \in L$, the distribution of $f_L(x,0,r)$ is identical to that of $f_L(x,1,r)$. This implies that the distribution of $f_L(x, a_{\pi(i)\pi(j)}, s_{ij})$ is identical to that of $f_L(x, h_{ij}, s_{ij})$ for every $x \in L$. Then the probability that $e = \alpha$ in step M4-3 is *exactly* 1/2. Since M_U iterates $n = |x|^c$ times the procedure from step M1-1 to step M5-2, M_U runs in expected polynomial (in $|x|$) time. Note again that for every $x \in L$, the distribution of $f_L(x, 0, r)$ is identical to that of $f_L(x, 1, r)$. Then the probability ensemble $\{\langle P, V^*\rangle(x)\}_{x \in L}$ is identical to the probability ensemble $\{M_U(x; V^*)\}_{x \in L}$ on L.

Thus the protocol $\langle P, V\rangle$ is a prover-practical perfect zero-knowledge proof for L if L induces an O/T bit commitment f_L. ■

For a language $L \in \mathcal{NP}$, define a polynomial time computable relation R_L to be $\langle x, y\rangle \in R_L$ if and only if $\rho(x, y) = \mathtt{true}$, where ρ is a polynomial (in $|x|$) time computable predicate that witnesses the language $L \in \mathcal{NP}$. As immediate corollaries to Theorem 18, we can show the following:

Corollary 19 (to Theorem 18). *Let L be $\mathcal{NP}$-complete. If the language L induces an* O/T *bit commitment, then the polynomial time hierarchy collapses.*

Corollary 20 (to Theorem 18). *If a language L induces an* O/T *bit commitment, then there exists a perfect zero-knowledge proof of knowledge for R_L.*

5 Transparent/Opaque Bit Commitments

Here we consider the case that L induces a T/O bit commitment (see Definition 3), and show that if a language L induces a T/O bit commitment, then there exists a bounded round perfect zero-knowledge proof for L.

Theorem 21. *If a language L induces a* T/O *bit commitment, then there exists a two round prefect zero-knowledge proof for the language L.*

Proof: Let L be a language that induces a T/O bit commitment f_L. Here we overview the outline of the protocol $\langle P, V\rangle$ for L. Let $x \in \{0,1\}^*$ be a common input to $\langle P, V\rangle$. For each i ($1 \leq i \leq |x|$), V chooses $e_i \in_{\mathrm{R}} \{0,1\}$, $r_i \in_{\mathrm{R}} \{0,1\}^{k(|x|)}$ and computes $\alpha_i = f_L(x, e_i, r_i)$. Then V reduces the following $\mathcal{NP}$-statement,

$$\exists e_1, e_2, \ldots, e_{|x|} \exists r_1, r_2, \ldots, r_{|x|} \text{ s.t. } \bigwedge_{i=1}^{|x|} \alpha_i = f_L(x, e_i, r_i), \tag{2}$$

to a directed Hamiltonian graph $G = (V, E)$, where $|V| = |x|^d$ for some constant $d > 0$. Let H be a Hamiltonian cycle of G. From the witness-preserving property of the reduction from any $L \in \mathcal{NP}$ to DHAM, there exist polynomial time computable functions g and h that satisfy

$$G = g(\alpha_1, \alpha_2, \ldots, \alpha_{|x|});$$
$$H = h(\langle \alpha_1, \alpha_2, \ldots, \alpha_{|x|}\rangle, \langle e_1, e_2, \ldots, e_{|x|}; r_1, r_2, \ldots, r_{|x|}\rangle).$$

Here V generates many random copies of G and commits to them with the T/O bit commitment f_L. After these preliminary steps, V shows to P that V knows the Hamiltonian cycle H of G. If V succeeds to convince P, then P shows to V that P knows $e_1, e_2, \ldots, e_{|x|}$.

Interactive Protocol $\langle P, V\rangle$ for L

common input: $x \in \{0,1\}^*$.

V1-1: V chooses $e_i \in_R \{0,1\}$ and $r_i \in_R \{0,1\}^{k(|x|)}$ for each i $(1 \le i \le |x|)$.
V1-2: V computes $\alpha_i = f_L(x, e_i, r_i)$.
V1-3: V computes $G = g(\alpha_1, \ldots, \alpha_{|x|})$, i.e., V reduces the $\mathcal{NP}$-statement of Eq.(2) to a directed Hamiltonian graph $G = (V, E)$, where $|V| = n = |x|^d$ for some $d > 0$.
V1-4: V defines an adjacency matrix $A_G = (a_{ij})$ of $G = (V, E)$.
V1-5: V computes $H = h(\langle \alpha_1, \ldots, \alpha_{|x|}\rangle, \langle e_1, \ldots, e_{|x|}; r_1, \ldots, r_{|x|}\rangle)$, where H is one of Hamiltonian cycles of $G = (V, E)$.
V1-6: V chooses $\pi_\ell \in_R \mathrm{Sym}(V)$ $(1 \le \ell \le n^2)$ and $s_{ij}^\ell \in_R \{0,1\}^{k(|x|)}$ $(1 \le i, j \le n)$.
V1-7: V computes $c_{ij}^\ell = f_L(x, a_{\pi_\ell(i)\pi_\ell(j)}, s_{ij}^\ell)$.
$V \to P$: $\langle \alpha_1, \alpha_2, \ldots, \alpha_{|x|}\rangle$, $\langle (c_{ij}^1), (c_{ij}^2), \ldots (c_{ij}^{n^2})\rangle$ $(1 \le i, j \le n)$.
P1: P chooses $b_\ell \in_R \{0,1\}$ for each ℓ $(1 \le \ell \le n^2)$.
$P \to V$: $\langle b_1, b_2, \ldots, b_{n^2}\rangle \in \{0,1\}^{n^2}$.
V2-1: If $b_\ell = 0$ $(1 \le \ell \le n^2)$, V assigns $\langle \pi_\ell, s_{11}^\ell, s_{12}^\ell, \ldots, s_{nn}^\ell\rangle$ to w_ℓ.
V2-2: If $b_\ell = 1$ $(1 \le \ell \le n^2)$, V assigns

$$\langle\langle i_1^\ell, j_1^\ell\rangle, \langle i_2^\ell, j_2^\ell\rangle, \ldots, \langle i_n^\ell, j_n^\ell\rangle, s_{i_1^\ell j_1^\ell}^\ell, s_{i_2^\ell j_2^\ell}^\ell, \ldots, s_{i_n^\ell j_n^\ell}^\ell\rangle$$

to w_ℓ such that $\langle i_1^\ell, j_1^\ell\rangle, \langle i_2^\ell, j_2^\ell\rangle, \ldots, \langle i_n^\ell, j_n^\ell\rangle$ is a single cycle.
$V \to P$: $\langle w_1, w_2, \ldots, w_{n^2}\rangle$.
P2-1: P computes $G = g(\alpha_1, \alpha_2, \ldots, \alpha_{|x|})$ and an adjacency matrix $A_G = (a_{ij})$ of G.
P2-2: For each $b_\ell = 0$ $(1 \le \ell \le n^2)$, if $c_{ij}^\ell = f_L(x, a_{\pi_\ell(i)\pi_\ell(j)}, s_{ij}^\ell)$ for each i, j $(1 \le i, j \le n)$, then P continues; otherwise P halts and rejects $x \in \{0,1\}^*$.
P2-3: For each $b_\ell = 1$ $(1 \le \ell \le n^2)$, if $\langle i_1^\ell, j_1^\ell\rangle, \langle i_2^\ell, j_2^\ell\rangle, \ldots, \langle i_n^\ell, j_n^\ell\rangle$ is indeed a single cycle and $c_{i_m^\ell j_m^\ell}^\ell = f_L(x, 1, s_{i_m^\ell j_m^\ell}^\ell)$ for each m $(1 \le m \le n)$, then P continues; otherwise P halts and rejects $x \in \{0,1\}^*$.
P2-4: If there exist $\beta_i \in \{0,1\}$, $t_i \in \{0,1\}^{k(|x|)}$ such that $\alpha_i = f_L(x, \beta_i, t_i)$ for every i $(1 \le i \le |x|)$, then P continues; otherwise P halts and rejects $x \in \{0,1\}^*$.
$P \to V$: $\langle \beta_1, \beta_2, \ldots, \beta_{|x|}\rangle$.
V3: If $\beta_i = e_i$ for every i $(1 \le i \le |x|)$, then V halts and accepts $x \in \{0,1\}^*$; otherwise V halts and rejects $x \in \{0,1\}^*$.

Now we turn to show that if L induces a T/O bit commitment f_L, then the protocol $\langle P, V\rangle$ for L is a two round perfect zero-knowledge proof for L.

Completeness: Assume here that $x \in L$. If V follows the protocol above, then $G = (V, E)$ is always a Hamiltonian graph. From the T/O property of f_L, it follows that for every $x \in L$, there does not exist $r, s \in \{0,1\}^{k(|x|)}$ such that $f_L(x, 0, r) = f_L(x, 1, s)$. Thus for each i $(1 \le i \le |x|)$, P can find in step P2-4 a *unique* $\beta_i \in \{0,1\}$ such that $\alpha_i = f_L(x, e_i, t_i)$ for some $t_i \in \{0,1\}^{k(|x|)}$. Then V always halts and accepts $x \in L$ in step V3.

Soundness: Assume that $x \notin L$. Define an interactive protocol $\langle A, B\rangle$ for $\overline{L} \in \mathcal{OT}$ to be on input $x \in \{0,1\}^*$ (1) A (resp. B) plays the role of V (resp. P); and (2) $\langle A, B\rangle$ simulate $\langle P, V\rangle$ except that the process from step V1-6 to step P2-3 in $\langle P, V\rangle$ is executed in serial.

From the T/O property of f_L, it follows that for every $x \notin L$, the distribution of $f_L(x, 0, r)$ is identical to that of $f_L(x, 1, s)$. Then the protocol $\langle A, B\rangle$ can be simulated in a perfect zero-knowledge manner for every $x \notin L$ by using the resettable simulation technique [9]. It turns out that the subprotocol of $\langle P, V\rangle$, from step V1-6 to step P2-3, is *perfectly witness indistinguishable* [6], because it can be regarded as the parallel composition of the protocol $\langle A, B\rangle$ by exchanging the roles of A and B. Then in the protocol $\langle P, V\rangle$, any cheating P^* cannot guess better at random the value of $e_i \in \{0,1\}$ for each i $(1 \le i \le |x|)$. Thus for each $k > 0$ and sufficiently large $x \notin L$, any cheating prover P^* causes V to accept $x \notin L$ with probability at most $|x|^{-k}$.

Perfect Zero-Knowledgeness: This can be shown in almost the same way as the case of graph nonisomorphism [9]. From the polynomial time invertible property of the reduction from any $L \in \mathcal{NP}$ to DHAM, there exist polynomial time computable functions g and γ that satisfy

$$g(\alpha_1, \ldots, \alpha_{|x|}) = G; \quad \gamma(G, H) = \langle \beta_1, \ldots, \beta_{|x|}; t_1, \ldots, t_{|x|}\rangle,$$

where H is one of Hamiltonian cycles of G and $\alpha_i = f_L(x, \beta_i, t_i)$ for each i $(1 \le i \le |x|)$. Here we use H_t to denote the t-*th* (n-vertex) single cycle for each t $(1 \le t \le n!)$ in the lexicographic order. Then the construction of M_U for any cheating verifier V^* is as follows:

Construction of M_U

common input: $x \in L$.

M0-1: **count** := 1; and **conv** := ε, where ε is a null string.
M0-2: M_U provides V^* with r_{V^*} as random coin tosses for V^*.
M1-1: M_U runs V^* on input x, r_{V^*} to generate $\langle \alpha_1, \ldots, \alpha_{|x|}\rangle, \langle (c_{ij}^1), \ldots (c_{ij}^{n^2})\rangle$.
M1-2: **conv** := **conv**$\|\langle\langle \alpha_1, \ldots, \alpha_{|x|}\rangle, \langle (c_{ij}^1), \ldots (c_{ij}^{n^2})\rangle\rangle$.
M2: M_U chooses $b_\ell \in_R \{0,1\}$ for each ℓ $(1 \le \ell \le n^2)$.
M3-1: M_U runs V^* on input $x, r_{V^*}, \langle b_1, b_2, \ldots, b_{n^2}\rangle$ to generate $\langle w_1, \ldots, w_{n^2}\rangle$.
M3-2: **conv** := **conv**$\|\langle\langle b_1, \ldots, b_{n^2}\rangle, \langle w_1, \ldots, w_{n^2}\rangle\rangle$.
M4-1: M_U computes $G = g(\alpha_1, \ldots, \alpha_{|x|})$ and an adjacency matrix $A_G = (a_{ij})$ of G.
M4-2: For each $b_\ell = 0$ $(1 \le \ell \le n^2)$, if $c_{ij}^\ell = f_L(x, a_{\pi_\ell(i)\pi_\ell(j)}, s_{ij}^\ell)$ for each i, j $(1 \le i, j \le n)$, then M_U continues; otherwise M_U halts and outputs $\langle x, r_{V^*}, \mathbf{conv}\rangle$.
M4-3: For each $b_\ell = 1$ $(1 \le \ell \le n^2)$, if $\langle i_1^\ell, j_1^\ell\rangle, \langle i_2^\ell, j_2^\ell\rangle, \ldots, \langle i_n^\ell, j_n^\ell\rangle$ is indeed a single cycle and $c_{i_m^\ell j_m^\ell}^\ell = f_L(x, 1, s_{i_m^\ell j_m^\ell}^\ell)$ for each m $(1 \le m \le n)$, then M_U continues; otherwise M_U halts and outputs $\langle x, r_{V^*}, \mathbf{conv}\rangle$.

M5-1: M_U resets V^* to the state of step M1-2.
M5-2: If count $> n!$, then M_U halts and outputs $\langle x, r_{V^*}, \mathtt{conv}\rangle$.
M5-3: If $H_{\mathtt{count}}$ is a Hamiltonian cycle of G, then $H := H_{\mathtt{count}}$ and go to step M7-2.
M5-4: M_U chooses $\tilde{b}_\ell \in_R \{0,1\}$ for each ℓ $(1 \le \ell \le n^2)$.
M6-1: M_U runs V^* on input $x, r_{V^*}, \langle \tilde{b}_1, \ldots, \tilde{b}_{n^2}\rangle$ to generate $\langle \tilde{w}_1, \ldots, \tilde{w}_{n^2}\rangle$.
M6-2: For each $\tilde{b}_\ell = 0$ $(1 \le \ell \le n^2)$, if $c^\ell_{ij} = f_L(x, a_{\tilde{\pi}_\ell(i)\tilde{\pi}_\ell(j)}, \tilde{s}^\ell_{ij})$ for each i, j $(1 \le i, j \le n)$, then M_U continues; otherwise count := count + 1 and go to step M5-1.
M6-3: For each $\tilde{b}_\ell = 1$ $(1 \le \ell \le n^2)$, if $\langle \tilde{i}^\ell_1, \tilde{j}^\ell_1\rangle, \langle \tilde{i}^\ell_2, \tilde{j}^\ell_2\rangle, \ldots, \langle \tilde{i}^\ell_n, \tilde{j}^\ell_n\rangle$ is a single cycle and $c^\ell_{\tilde{i}^\ell_m \tilde{j}^\ell_m} = f_L(x, 1, \tilde{s}^\ell_{\tilde{i}^\ell_m \tilde{j}^\ell_m})$ for each m $(1 \le m \le n)$, then M_U continues; otherwise count := count + 1 and go to step M5-1.
M7-1: If $b_\ell \neq \tilde{b}_\ell$ for some ℓ $(1 \le \ell \le n^2)$, then M_U computes a Hamiltonian cycle H of $G = (V, E)$ from w_ℓ and $\tilde{w}_\ell$; otherwise count := count + 1 and go to step M5-1.
M7-2: M_U computes $\langle \beta_1, \beta_2, \ldots, \beta_{|x|}; t_1, t_2, \ldots, t_{|x|}\rangle = \gamma(G, H)$.
M7-3: If $\alpha_i = f_L(x, \beta_i, t_i)$ for every i $(1 \le i \le |x|)$, then set conv := conv$\|\langle \beta_1, \beta_2, \ldots, \beta_{|x|}\rangle$; otherwise M_U halts and outputs $\langle x, r_{V^*}, \mathtt{conv}\rangle$.
M7-4: M_U halts and outputs $\langle x, r_{V^*}, \mathtt{conv}\rangle$.

We first show that M_U terminates in expected polynomial (in $|x|$) time for any cheating verifier V^*. Define $K \subseteq \{0,1\}^{n^2}$ to be a subset of $\langle b_1, b_2, \ldots, b_{n^2}\rangle \in \{0,1\}^{n^2}$ for which V^* passes the tests in steps M4-2 and M4-3. Then the following three cases are possible: (C1) $\|K\| \ge 2$; (C2) $\|K\| = 1$; and (C3) $\|K\| = 0$, where $\|A\|$ denotes the cardinality of a finite set A.

In the case of (C1), the expected number I_{C1} of invocations of V^* satisfies

$$I_{C1} \le 1 + \frac{\|K\|}{2^{n^2}} \cdot \left(\frac{\|K\| - 1}{2^{n^2}}\right)^{-1} = 1 + \frac{\|K\|}{\|K\| - 1} \le 3.$$

In the case of (C2), the probability that V^* passes the tests in steps M4-2 and M4-3 is exactly 2^{-n^2}. Then M_U halts and outputs $\langle x, r_{V^*}, \mathtt{conv}\rangle$ in step M4-2 or M4-3 with probability $1 - 2^{-n^2}$. If V^* passes the tests in steps M4-2 and M4-3, then M_U must exhaustively searches a Hamiltonian cycle H of G at most in $n!$ steps. Thus it turns out that the expected number I_{C2} of invocations is bounded by $I_{C2} = 1 + 2^{-n^2} \cdot n! < 2$. In the case of (C3), M_U always halts and outputs $\langle x, r_{V^*}, \mathtt{conv}\rangle$ with a single invocation of V^*. Thus M_U terminates in expected polynomial (in $|x|$) time for any cheating verifier V^*.

We then show that for any verifier V^*, M_U on any input $x \in L$ simulates the real interactions between P and V^* in a perfect zero-knowledge manner.

In the case of (C3), M_U always halts in step M4-2 or step M4-3 and outputs $\langle x, r_{V^*}, \mathtt{conv}\rangle$ with the distribution identical to one in $\langle P^*, V\rangle$.

In the case of (C1), the following three cases are possible: (C1-1) M_U halts in step M4-2 or step M4-3 and outputs $\langle x, r_{V^*}, \mathtt{conv}\rangle$; (C1-2) M_U halts in step M5-2 or step M7-3 and outputs $\langle x, r_{V^*}, \mathtt{conv}\rangle$; and (C1-3) M_U halts in step M7-4 and outputs $\langle x, r_{V^*}, \mathtt{conv}\rangle$. In the case of (C1-1), it is obvious that the distribution of

$\langle x, r_{V^*}, \mathbf{conv}\rangle$ is identical to one in $\langle P, V^*\rangle$. Note that P returns $\langle \beta_1, \beta_2, \ldots, \beta_{|x|}\rangle$ iff every α_i $(1 \leq i \leq |x|)$ is properly generated. From the polynomial time invertible property of the reduction from any $L \in \mathcal{NP}$ to DHAM, it follows that every α_i $(1 \leq i \leq |x|)$ is properly generated iff $G = g(\alpha_1, \alpha_2, \ldots, \alpha_{|x|})$ is a Hamiltonian graph. Then in the case of (C1-2), the distribution of $\langle x, r_{V^*}, \mathbf{conv}\rangle$ is identical to one in $\langle P, V^*\rangle$. Let us consider the case that M_U in step M7-1 finds $b_\ell \neq \tilde{b}_\ell$ for some ℓ $(1 \leq \ell \leq n^2)$. We assume without loss of generality that $b_\ell = 0$ and $\tilde{b}_\ell = 1$. Then

$$w_\ell = \langle \pi_\ell, s_{11}^\ell, s_{12}^\ell, \ldots, s_{nn}^\ell\rangle;$$
$$\tilde{w}_\ell = \langle \langle \tilde{i}_1^\ell, \tilde{j}_1^\ell\rangle, \langle \tilde{i}_2^\ell, \tilde{j}_2^\ell\rangle, \ldots, \langle \tilde{i}_n^\ell, \tilde{j}_n^\ell\rangle, \tilde{s}_{\tilde{i}_1\tilde{j}_1}^\ell, \tilde{s}_{\tilde{i}_2\tilde{j}_2}^\ell, \ldots, \tilde{s}_{\tilde{i}_n\tilde{j}_n}^\ell\rangle.$$

From the assumption that $b_\ell = 0$ and $\tilde{b}_\ell = 1$, it follows that w_ℓ passes the test in step M4-2 and $\tilde{w}_\ell$ passes the test in step M6-3. Thus the Hamiltonian cycle H of G is given by

$$H = \langle \langle \pi_\ell^{-1}(\tilde{i}_1^\ell), \pi_\ell^{-1}(\tilde{j}_1^\ell)\rangle, \langle \pi_\ell^{-1}(\tilde{i}_2^\ell), \pi_\ell^{-1}(\tilde{j}_2^\ell)\rangle, \ldots, \langle \pi_\ell^{-1}(\tilde{i}_n^\ell), \pi_\ell^{-1}(\tilde{j}_n^\ell)\rangle\rangle.$$

From the polynomial time invertible property of the reduction from any $L \in \mathcal{NP}$ to DHAM, it follows that $\gamma(G, H) = \langle \beta_1, \beta_2, \ldots, \beta_{|x|}; t_1, t_2, \ldots, t_{|x|}\rangle$ and $\alpha_i = f_L(x, \beta_i, t_i)$ $(1 \leq i \leq |x|)$. The T/O property of f_L guarantees that for every $x \in L$, there does not exist $r, s \in \{0,1\}^{k(|x|)}$ such that $f_L(x, 0, r) = f_L(x, 1, s)$. Then $\beta_i = e_i$ for each i $(1 \leq i \leq |x|)$ and thus in the case of (C1-3), the distribution of $\langle x, r_{V^*}, \mathbf{conv}\rangle$ is identical to one in $\langle P, V^*\rangle$.

In the case of (C2), the following three cases are possible: (C2-1) M_U halts in step M4-2 or step M4-3 and outputs $\langle x, r_{V^*}, \mathbf{conv}\rangle$; (C2-2) M_U halts in step M5-2 or step M7-3 and outputs $\langle x, r_{V^*}, \mathbf{conv}\rangle$; and (C2-3) M_U halts in step M7-4 and outputs $\langle x, r_{V^*}, \mathbf{conv}\rangle$. In a way similar to the case of (C1), we can show that in the cases of (C2-1), (C2-2), and (C2-3), the distribution of $\langle x, r_{V^*}, \mathbf{conv}\rangle$ is identical to one in $\langle P, V^*\rangle$. Then for any cheating verifier V^*, M_U on input $x \in L$ simulates $\langle P, V^*\rangle$ in a perfect zero-knowledge manner.

Thus the interactive protocol $\langle P, V\rangle$ is a two round perfect zero-knowledge proof for L if L induces a T/O bit commitment f_L. ∎

6 Concluding Remarks

From Theorem 18, it follows that any language $L \in \mathcal{OT}$ has an unbounded round perfect zero-knowledge Arthur-Merlin proof. This however could be improved, because any language $L \in \mathcal{OT}$ has an $\mathcal{NP}$-proof [8]. Then

1. If a language L induces an O/T bit commitment, then does there exist a bounded round perfect zero-knowledge proof for the language L?

To affirmatively solve this, a verifier will have to flip private coins, because Goldreich and Krawczyk [7] showed that there exists a bounded round (blackbox simulation) zero-knowledge Arthur-Merlin proof for L, then $L \in \mathcal{BPP}$.

Languages that induce O/T or T/O bit commitments might have diverse applications to many cryptographic protocols. Then

2. What is the other application of languages that induce O/T or T/O bit commitments?

Every known random self-reducible language [13], e.g., graph isomorphism, quadratic residuosity, etc., induces an O/T bit commitment. Then finally

3. For any language L, if L is random self-reducible, then does L induce an O/T bit commitment?

References

1. Brassard, G., Chaum, D., and Crépeau, C., "Minimum Disclosure Proofs of Knowledge," *J. Comput. System Sci.*, Vol.37, No.2, pp.156-189 (1988).
2. Ben-Or, M., Goldreich, O., Goldwasser, S., Håstad, J., Kilian, J., Micali, S., and Rogaway, P., "Everything Provable is Provable in Zero-Knowledge," *Proceedings of Crypto'88*, Lecture Notes in Computer Science 403, pp.37-56 (1990).
3. Boyar, J., Friedl, K., and Lund, C., "Practical Zero-Knowledge Proof: Giving Hints and Using Deficiencies," *J. Cryptology*, Vol.4, No.3, pp.185-206 (1991).
4. Bellare, M., Micali, S., and Ostrovsky, R., "The (True) Complexity of Statistical Zero-Knowledge," *Proceedings of the 22nd Annual ACM Symposium on Theory of Computing*, pp.494-502 (1990).
5. Feige, U., Fiat, A., and Shamir, A., "Zero-Knowledge Proofs of Identity," *J. Cryptology*, Vol.1, No.2, pp.77-94 (1988).
6. Feige, U. and Shamir, A., "Zero-Knowledge Proofs of Knowledge in Two Rounds," *Proceedings of Crypto'89*, Lecture Notes in Computer Science 435, pp.526-544 (1990).
7. Goldreich, O. and Krawczyk, H., "On the Composition of Zero-Knowledge Proof Systems," *Proceedings of ICALP'90*, Lecture Notes in Computer Science 443, pp.268-282 (1990).
8. Goldwasser, S., Micali, S., and Rackoff, C., "The Knowledge Complexity of Interactive Proof Systems," *SIAM J. Comput.*, Vol.18, No.1, pp.186-208 (1989).
9. Goldreich, O., Micali, S., and Wigderson, A., "Proofs That Yield Nothing But Their Validity or All Languages in $\mathcal{NP}$ Have Zero-Knowledge Proof Systems," *J. Assoc. Comput. Mach.*, Vol.38, No.1, pp.691-729 (1991).
10. Goldreich, O. and Oren, Y., "Definitions and Properties of Zero-Knowledge Proof Systems," *J. Cryptology*, Vol.7, No.1, pp.1-32 ((1994).
11. Naor, M., Ostrovsky, R., Venkatesan, R., and Yung, M., "Perfect Zero-Knowledge Arguments for $\mathcal{NP}$ Can Be Based on General Complexity Assumptions," *Proceedings of Crypto'92*, Lecture Notes in Computer Science 740, pp.196-214 (1993).
12. Ostrovsky, R., "Comparison of Bit-Commitment and Oblivious Transfer Protocols when Players have Different Computing Power," *DIMACS Technical Report* #90-41, pp.27-29 (1990).
13. Tompa, M. and Woll, H., "Random Self-Reducibility and Zero-Knowledge Interactive Proofs of Possession of Information," *Proceedings of the 28th Annual IEEE Symposium on Foundations of Computer Science*, pp.472-482 (1987).

On the length of cryptographic hash-values used in identification schemes

Marc Girault

SEPT
42 rue des Coutures, BP 6243
14066 Caen, France.
e-mail : marc.girault@sept.fr

Jacques Stern

Laboratoire d'Informatique
Ecole Normale Supérieure
45 rue d'Ulm, 75230 Paris, France.
e-mail : jacques.stern@ens.fr

Abstract. Many interactive identification schemes based on the zero-knowledge concept use cryptographic hash-values, either in their basic design or in specific variants. In this paper, we first show that 64-bit hash-values, a length often suggested, definitely decrease the level of the security of all these schemes. (Of course, this does not compromise the security of the schemes by themselves). Then we prove that collision-resistance is a sufficient condition to achieve the claimed level of security. Finally, by using a weaker notion of collision-resistance, we present interesting variants of some of these schemes (in particular the Schnorr and the Guillou-Quisquater schemes) which minimize the number of communication bits for a given level of security.

1 Introduction

In recent years, several interactive identification schemes have been proposed based on the zero-knowledge concept [GMR85]. In all these schemes, the prover starts by committing himself to some secret values he picks at random. To compute this commitment, hash-functions are often used, either in the basic design of the scheme or in specific variants.

The first scheme of this type was the one by Fiat and Shamir, presented at CRYPTO'86 conference [FS86]. This scheme is based on the modular square root extraction problem. The basic protocol, to be repeated several times, has three passes. In a variant whose the goal is to minimize the communication bits, the prover sends in the first pass a cryptographic hash-value of some elements selected by him. The length suggested by the authors for this hash-value was 128 bits.

At CRYPTO'89 conference, Schnorr presented an identification scheme based on the discrete logarithm problem [Sc89]. The protocol of this scheme has three passes. In a variant whose the goal is to minimize the communication bits, the prover sends in the first pass a cryptographic hash-value of some elements selected by him. The suggested length for this hash-value was at least k bits, where $1-2^{-k}$ is the level of security (impostor detection probability) to be achieved.

At the rump session of CRYPTO'89 conference, Shamir presented an identification scheme based on an NP-complete problem, the so-called permuted kernel problem [Sh89]. The basic protocol of this scheme, to be repeated several times, has five passes and, in the first pass, the prover sends two cryptographic hash-values of some elements selected by him. Hash-values of 64 bits were specifically mentioned in the paper.

Finally, at CRYPTO'93 conference, Stern presented an identification scheme based on an NP-complete problem, the so-called syndrome decoding problem [St93] (a first tentative had already been presented at EUROCRYPT'89 [St89]). The basic protocol of this scheme, to be repeated several times, has three passes and, in the first pass, the prover sends three cryptographic hash-values of some elements selected by him. Hash-values of 64 bits were specifically mentioned in the paper.

The goal of this paper is to discuss the appropriate length of these hash-values. First, we show that, if 2^{32} operations are deemed to be computationally feasible (we will make this assumption all along the paper), then 64-bit hash-values definitely decrease the level of security of all these schemes. This is shown by exhibiting, for each these schemes, specific attacks based on the birthday paradox. Of course, these attacks do not compromise the security of the schemes by themselves, but only suggest to use longer hash-values.

Second, we formally prove that collision-resistance is a sufficient condition to achieve the level of security which is claimed by the authors. As a consequence, a length of 128 bits (if 2^{64} operations are deemed to be computationally infeasible) or more for the hash-values seems to be convenient.

Third, by using a weaker notion of collision-resistance (the so-called *r*-collision resistance), we present interesting variants of some of these schemes (in particular the Schnorr and the Guillou-Quisquater scheme) which minimize the number of communication bits for a given level of security.

2 The Fiat-Shamir scheme

2.1 Description

The identification scheme presented by Fiat and Shamir at CRYPTO'86 conference [FS86] is based on the difficulty of extracting square roots modulo a composite number whose factors are unknown. A trusted center is used to compute the users' secret keys. The universal parameters, i.e. those shared by all the users, are :

- a large composite modulus n (whose factors are only known to the center)
- two small integers k and t
- a pseudo-random function f.

The recommended size for n was (at least) 512 bits in 1986 (today, a larger size would probably be recommended). Values for k and t are closely related to the level of security of the protocol (see further). Typical values are 6 and 5 (or 9 and 8 for the related signature scheme).

The prover's public key is his "identity" (i.e. a string I which contains relevant informations about him and/or his device). His secret key is composed of k values s_j computed as follows : Let $v_j = f(I,j)$ for k small values of j such that v_j is a quadratic residue modulo n (for convenience, we assume that $j = 1...k$). Then $s_j^2 v_j = 1 \pmod n$ for each value of j.

The basic (3-pass) protocol is the following.

1. The prover randomly selects an integer r in $\{0...n\}$, computes $x = r^2 \pmod n$ and sends x to the verifier.
2. The verifier randomly selects an element $e = (e_1, e_2, ..., e_k)$ of $\{0,1\}^k$ and sends e to the prover.
3. The prover computes $y = r\prod_{e_j=1} s_j \pmod n$ and sends y to the verifier.
4. The verifier computes all the v_j and checks : $x = y^2 \prod_{e_j=1} v_j \pmod n$.

Note that an impostor (who ignores s) can easily deceive the verifier with probability 2^{-k}, by selecting an integer y, "guessing" an element e and computing x as in step 4. As he is provably unable, if factoring is difficult, to deceive the verifier with probability essentially greater than 2^{-k}, then it suffices to repeat t times the basic protocol to obtain a level of security (i.e. the impostor detection probability) greater than or equal to $1-2^{-kt}$.

In order to decrease the number of communication bits, Fiat and Shamir have suggested to send to the verifier at step 1 the first 128 bits of $f(x)$ instead of x. Let us call c the result and h the function which maps x to c. The check equation of step 4 then becomes :

$$c = h(y^2 \prod_{e_j=1} v_j \pmod n)$$

We will call the new scheme the h-variant of the Fiat-Shamir scheme. In the following section, we show that some bad choices for h reduce the level of the security of the scheme. To be clear, these "bad" choices are *not* mentioned in the paper by Fiat and Shamir.

2.2 Too short (or ill-chosen) hash-values decrease the level of security

Case : c is 64-bit long. If c is too short (say 64 bits), then there is an easy (at least to design) attack using the birthday paradox : first, the impostor selects a set E of 2^{32} integers y and two distincts elements $e^{(1)}$ and $e^{(2)}$ of $\{0,1\}^k$. For each element y of E, he computes :

$$h(y^2 \prod_{e_j^{(1)}=1} v_j \pmod n) \text{ and } h(y^2 \prod_{e_j^{(2)}=1} v_j \pmod n).$$

Due to the birthday paradox, there exist with high probability two integers y_1 and y_2 such that :

$$h(y_1^2 \prod_{e_j^{(1)}=1} v_j \pmod n) = h(y_2^2 \prod_{e_j^{(2)}=1} v_j \pmod n)$$

Let us call c this common value (the "collision"). If no collision occurred, the impostor has to increase slightly the size of E.

Now, at each execution of the protocol, the impostor does the following. He sends c to the verifier, who sends back e to him. Then, if $e = e^{(1)}$, he replies with y_1. If $e = e^{(2)}$, he replies with y_2. So, in two cases, the verifier will be satisfied with the reply and the impostor acceptance probability is 2^{-k+1} instead of 2^{-k}. In particular, if $k = 1$, then the impostor is always accepted.

Case : c is a truncation of x. If h is only a truncation of x, for instance if c is composed of the 128 rightmost bits of x, then another type of attack allows the impostor to achieve the same probability of acceptance as above. The impostor selects an integer z less than n whose 128 rightmost bits are zero, and two distinct elements $e^{(1)}$ and $e^{(2)}$ of $\{0..1\}^k$. Let m be equal to $z \prod_{e_j^{(1)}=1} v_j^{-1} \pmod n$ and k be equal to $-\prod_{e_j^{(2)}=1} v_j \prod_{e_j^{(1)}=1} v_j^{-1} \pmod n$.

By using the Pollard-Schnorr attack of the Ong-Schnorr-Shamir signature scheme [PS87], it is possible to find in polynomial time two integers y_1 and y_2 such that :

$$y_1^2 + k y_2^2 = m \pmod n$$

which implies, by multiplying with $\prod_{e_j^{(1)}=1} v_j \pmod n$:

$$y_1^2 \prod_{e_j^{(1)}=1} v_j - y_2^2 \prod_{e_j^{(2)}=1} v_j = z \pmod n$$

With probability about 1/2, $y_1^2 \prod_{e_j^{(1)}=1} v_j \pmod n$ is greater than $y_2^2 \prod_{e_j^{(2)}=1} v_j \pmod n$. In such a case, the above equation implies that the 128 rightmost bits of $y_1^2 \prod_{e_j^{(1)}=1} v_j \pmod n$ are equal to the 128 rightmost bits of $y_2^2 \prod_{e_j^{(2)}=1} v_j \pmod n$. (In the other case, try with another value of z). Let us call c this common value. Then the rest is as above. Note that this attack can be adapted to the case where c is composed of the 128 leftmost bits (or even any 128 consecutive bits) of x.

2.3 The level of security of the h-variant

We now show that the h-variant does achieve the security level which is claimed, if h is collision-free. In fact, we prove a more general theorem, for which we need the following definitions :

Definition 1. A r-collision for a function h is a r-tuple $(x_1, x_2, \ldots, x_r)$ of r pairwise distinct values such that $h(x_1) = h(x_2) = \ldots = h(x_r)$.

Definition 2. (*informal*) A function h is r-collision-free (or r-collision-resistant) if it is computationally infeasible to find a r-collision for h.

We now establish a precise connection between r-collision resistance and the level of security of the h-variant.

Theorem 1. *If there exists a PPTM (Probabilistic Polynomial Turing Machine) M such that the probability that M be accepted by an honest verifier is greater than* $(r-1)2^{-k}+\varepsilon$, *with* $\varepsilon>0$, *then there exists a PPTM* $\tilde{M}$ *which with overwhelming probability either computes the square root of one product of the form :*

$$\prod_{j=1}^{k} v_j^{c_j} \pmod n.$$

where $c_j = -1, 0$ *or* $+1$ *(not all of them zero) or finds a r-collision for h.*

Remark: As observed in [FFS88], the first conclusion contradicts the intractability of factoring assumption, as a coalition of the legitimate user and of the potential attacker could factor n. The second conclusion implies that h is not r-collision-free.

Proof. Let Ω be the set of m elements in which M picks its random values and E be the set $\{0,1\}^k$, both of them with the uniform distribution. For each value (ω,e) of $\Omega\times E$, M passes the protocol (we say it is a success) or not. Let S be the subset of $\Omega\times E$ composed of all the successes. Our assumption is that :

$$\frac{\mathrm{Card}(S)}{\mathrm{Card}(\Omega\times E)} > (r-1)2^{-k}+\varepsilon$$

with $\varepsilon>0$ and $\mathrm{Card}(\Omega\times E) = m.2^k$.

Let Ω_r be the section $\{\omega\in\Omega : \mathrm{Card}\{e\in E : (\omega,e) \text{ is a success}\} \geq r\}$. We have :

$$\mathrm{Card}(S) \leq \mathrm{Card}(\Omega_r).2^k + (r-1)(m-\mathrm{Card}(\Omega_r))$$

Then :

$$\frac{\mathrm{Card}(S)}{\mathrm{Card}(\Omega\times E)} \leq \left[\frac{\mathrm{Card}(\Omega_r)}{\mathrm{Card}(\Omega)} + (r-1)(2^{-k} - \frac{\mathrm{Card}(\Omega_r)}{\mathrm{Card}(\Omega\times E)})\right] \leq \left[\frac{\mathrm{Card}(\Omega_r)}{\mathrm{Card}(\Omega)} + (r-1)2^{-k}\right]$$

which implies :

$$\frac{\mathrm{Card}(\Omega_r)}{\mathrm{Card}(\Omega)} \geq \varepsilon.$$

Let $\tilde{M}$ be the PPTM obtained by resetting M ε^{-1} times. With constant probability, $\tilde{M}$ picks ω in Ω_r and the probability can be made close to 1 by repeating the execution of $\tilde{M}$. At the end, r values $y_1, y_2..., y_r$ are found such that, for distinct $e^{(1)}, e^{(2)},..., e^{(r)}$:

$$h(y_1^2 \prod_{e_j^{(1)}=1} v_j \pmod n) = h(y_2^2 \prod_{e_j^{(2)}=1} v_j \pmod n) = ... = h(y_r^2 \prod_{e_j^{(r)}=1} v_j \pmod n)$$

Now, there are two possibilities :

a) either two of the values (say $y_1^{\,2}\prod_{e_j^{(1)}=1} v_j \pmod n$ and $y_2^{\,2}\prod_{e_j^{(2)}=1} v_j \pmod n$) are equal before hashing. In that case, y_1/y_2 is a square root of a product of the form $\prod_{j=1}^{k} v_j^{\,c_j} \pmod n$, where $c_j = -1$, 0 or $+1$ (not all of them zero) ;

b) or all these values are pairwise distinct and a r-collision for h has been found. □

This result suggests to use hash-functions which are only resistant to r-collisions (with $r > 2$), so that the hash-values computed in the first pass can be made much shorter. Indeed the decrease of the level of security can be balanced by sending a slightly larger value of e in the second pass (more precisely, if $r = 2^u$, $e \in \{0,1\}^{k+u}$ instead of $e \in \{0,1\}^k$). But this is not so interesting in the Fiat-Shamir scheme, as it would also imply a larger number of secrets s_j, a very undesirable feature. On the contrary, this idea is particularly attractive in the Schnorr scheme, as shown in the following section.

3 The Schnorr scheme

3.1 Description

The identification scheme presented by Schnorr at CRYPTO'89 conference [Sc89] is based on the difficulty of computing a discrete logarithm. The universal parameters are :

- a large prime p
- a prime q such that $q \mid p-1$
- an integer α (the "base") such that $\alpha^q = 1 \pmod p$
- a small integer k.

The recommended sizes for n and q were respectively (at least) 512 bits and 140 bits in 1989. The value of k is closely related to the level of security of the protocol (see further). A typical value is 40 (or 72 for the related signature scheme).

The prover's secret key is an integer s in $\{1...q\}$. His public key is $v = \alpha^{-s} \pmod p$. The basic (3-pass) protocol is the following.

1.	The prover randomly selects an integer r in $\{1...q\}$, computes $x = \alpha^r \pmod p$ and sends x to the verifier.
2.	The verifier randomly selects an element e of $\{0...2^k - 1\}$ and sends e to the prover.
3.	The prover computes $y = r + se \pmod q$ and sends y to the verifier.
4.	The verifier checks : $x = \alpha^y v^e \pmod p$.

Note that an impostor (who ignores s) can easily deceive the verifier with probability 2^{-k}, by selecting an integer y, "guessing" an element e and computing x as in step 4. As he is provably unable, if computing discrete logarithms is difficult, to deceive the verifier with probability essentially greater than 2^{-k}, then the level of security is equal to $1-2^{-k}$.

In order to decrease the number of communication bits, Schnorr has suggested to send to the verifier at step 1 $c=h(x)$ where h is a k-bit hash-function. The check equation of step 4 then becomes :

$$c=h(\alpha^{y}v^{e}(\bmod p))$$

We will call the new scheme the h-variant of the Schnorr scheme. In the following section, we show that some bad choices for h reduce the level of security of the scheme.

3.2 Too short hash-values decrease the level of security

The first observation is the same as in the Fiat-Shamir scheme : if c is too short, then the security level may be lower than expected. For example, if c is only 64-bit long, then a birthday attack quite similar to the one described in subsection 2.2 can be designed (no matter how the function h is defined). As a consequence, c should be at least 128-bit long if we want the security level be equal to 2^{-k}.

The second observation differs from Fiat-Shamir case : as far as we are aware, the level of security does not decrease if c is only a truncation of x, provided the number of bits is large enough (say 128 bits, because of the first observation). This shows that, in this scheme, one-wayness (and a fortiori collision-resistance) does not seem to be a necessary condition for h, in order to achieve a security level be equal to 2^{-k}. Nevertheless, collision-resistance remains a sufficient condition to achieve this security level, as shown now.

3.3 The level of security of the h-variant

We can state a similar theorem to the one of section 2 :

Theorem 2. *If there exists a PPTM M such that the probability that M be accepted by an honest verifier is greater than* $(r-1)2^{-k}+\varepsilon$, *with* $\varepsilon>0$, *then there exists a PPTM* $\tilde{M}$ *which with overwhelming probability either computes the discrete logarithm of* v *(modulo* q *in base* α*) or finds a r-collision for h.*

Proof. The proof is quite similar to the theorem of section 2. At the end, r values y_1, y_2,..., y_r are found such that, for pairwise distinct e_1, e_2, ..., e_r :

$$h(\alpha^{y_1}v^{e_1}(\bmod p))=h(\alpha^{y_2}v^{e_2}(\bmod p))=..........=h(\alpha^{y_r}v^{e_r}(\bmod p))$$

Now, there are two possibilities :

a) either two of the values are equal before hashing, say : $\alpha^{y_1}v^{e_1} \pmod p$ and $\alpha^{y_2}v^{e_2} \pmod p$. In that case, $\frac{y_1 - y_2}{e_2 - e_1} \pmod q$ is the discrete logarithm of v modulo q in base α.

b) or all these values are pairwise distinct and a r-collision for h has been found. □

3.4 An interesting optimization

The preceding theorem leads to an interesting optimization of the Schnorr scheme. The idea is to use r-collision hash-functions with $r > 2$, so that the hash-values computed in the first pass can be made much shorter. The decrease of the level of security is compensated by sending a slightly larger value of e in the second pass. Contrary to the Fiat-Shamir scheme, this does not have any undesirable consequence.

In order to make a precise statement, we need a result related to the birthday paradox :

Lemma. *Let E be a set of cardinality n, F_m a sample of size m drawn from E with replacements and r an integer. Let us call r-coincidence an element of E to which exactly r elements of F_m are equal to. For n a sufficiently large integer, $m = (\lambda r!)^{1/r} n^{r-1/r}$, with $\lambda \leq 1$ and $m/n \leq 1/128$, the probability that there is no s-coincidence for $s \geq r$ is very close to $e^{-\lambda}$ (hence greater than $1/e$).*

Proof (sketch). By a classical result from von Mises (see e.g. [Fe68] page 106), the probability $p(i,r)$ that there are exactly i r-coincidences is :

$$p(i,r) \approx e^{-\Lambda} \frac{\Lambda^i}{i!}$$

with :

$$\Lambda = n \frac{e^{-m/n}}{r!} \left(\frac{m}{n} \right)^r$$

If $m = (\lambda r!)^{1/r} n^{r-1/r}$ with $m/n \leq 1/128$, then $\Lambda = \lambda e^{-m/n} \approx \lambda$. Hence the probability $p(0,r)$ that there is no r-coincidence at all is very close to $e^{-\lambda}$, and so is the probability that there is no s-coincidence for $s \geq r$ if $\lambda \leq 1$. (Intuitively, the reason why is the following : if the probability that there is a s-coincidence for $s \geq r+1$ were not nearly equal to zero, then the probability $1 - p(0,r)$ that there is a r-coincidence would be nearly equal to 1 ; this is not the case since $1 - p(0,r) \approx 1 - e^{-\lambda} \leq 1 - 1/e \approx 0.632$). □

This result allows to specify a version of the Schnorr scheme which minimizes the number of communication bits. Let m be an integer greater than or equal to the number of operations deemed to be computationally infeasible (typically $m = 2^{64}$), and h be a pseudo-random hash-function whose the values range in the interval $\{0...n-1\}$ with $m = (\lambda r!)^{1/r} n^{r-1/r}$, $m/n \leq 1/128$ and $\lambda \leq 1$. Then, using a traditional argument, h can be considered as r-collision-free, as the probability that a r-collision is found with a number of computations supposed to be infeasible is substantially less than 1 (to be precise, less than 0.632 and even less than $1-e^{-\lambda}$). For instance, if $r = 2$, the usual choice : $m = n^{1/2}$ gives $\lambda = 1/2$. If $n \geq 2^{14}$, the assumptions of the lemma are satisfied, and the function can be considered as (2-) collision-resistant.

If r > 2, according to the theorem of subsection 3.2.1, the level of security has decreased to $1-(r-1)2^{-k}$. But we can compensate this decrease by choosing e in the range $\{0 \ldots 2^{k+\log_2(r-1)}\}$. For that reason, it may be convenient that r-1 is a power of 2.

As a consequence, some practical values for r are 5 and 9. Fixing m to 2^{64} and λ to 1/2 as above, we have $r = 5 \Rightarrow n \geq 2^{80}/2.27$ and $r = 9 \Rightarrow n \geq 2^{72}/3.84 \Rightarrow n \geq 2^{71}$ (since we must have $m/n \leq 1/128$).

In the first case, e must be k+2-bit long and the hash-values 79-bit long, if we want to achieve a level of security equal to $1-2^{-k}$.

In the second case, e must be k+3-bit long and the hash-values 71-bit long. We therefore have saved about 57 bits in the first pass (compared to a 128-bit hash-value computed from a (2-)collision-resistant hash-function) and, in the second pass, we have only three bits more to transport. Globally, we have saved 54 bits, for a total of 71 + (k + 3) + 140 = 214 + k, instead of 268 + k. If k = 40, then we have a total of 254 instead of 308, i.e. the number of communication bits has decreased by about 18%.

The same optimization applies to the Guillou-Quisquater scheme [GQ88]. In that case, the number of communication bits can be typically decreased from 680 to 626 bits, i.e. by about 8%.

4 The Stern identification scheme

4.1 Description

The identification scheme presented by Stern at CRYPTO'93 conference [St93] is based on the difficulty of the syndrome decoding problem, that is the (NP-complete) problem of finding a word of given syndrome and of given weight. The universal parameters are :

- a random binary $(k \times n)$ matrix A $(k < n)$
- an integer p
- a hash-function h.

Typical values for (k,n,p) are (256,512,56) or, still better, (512,1024,110). The prover's secret key is a random n-bit word s of weight (i.e. the number of '1' bits of s, denoted by $|s|$) equal to p. His public key is $v = A(s)$. The basic (3-pass) protocol is as follows.

1. The prover randomly selects an n-bit word y and a permutation σ of the integers $\{1...n\}$, computes $c_0 = h(\sigma, A(y))$, $c_1 = h(y.\sigma)$, $c_2 = h(y'.\sigma)$, with $y' = y \oplus s$ (where $\oplus$ stands for bitwise addition modulo 2 and $y.\sigma$ refers to the image of y under permutation σ), and sends c_0, c_1 and c_2 to the verifier.

2. The verifier randomly selects an element b of $\{0,1,2\}$ and sends b to the prover.

3. The prover sends to the verifier :

- if $b=0$: (y,σ)
- if $b=1$: (y',σ)
- if $b=2$: $(y.\sigma, y'.\sigma)$

4. The verifier checks :

$h(\sigma, A(y)) = c_0$ and $h(y.\sigma) = c_1$
$h(\sigma, A(y') \oplus v) = c_0$ and $h(y'.\sigma) = c_2$
$h(y.\sigma) = c_1$, $h(y'.\sigma) = c_2$ and $|y.\sigma \oplus y'.\sigma| = p$

Note that an impostor (who ignores s) can deceive the verifier with probability 2/3, by using one of the three following strategies :

a) he selects a permutation σ and two words y and y' such that $A(y') = A(y) \oplus v$, and computes c_0, c_1 and c_2 as above. Then he is able to answer correctly to $b=0$ and $b=1$, but not to $b=2$.

b) he selects a permutation σ and two words y and y' such that $|y \oplus y'| = p$, and computes c_0, c_1 and c_2 as above. Then he is able to answer correctly to $b=0$ and $b=2$, but not to $b=1$.

c) he selects a permutation σ and two words y and y' such that $|y \oplus y'| = p$, computes $c_0 = h(\sigma, A(y') \oplus v)$ and computes c_1 and c_2 as above. Then he is able to answer correctly to $b=1$ and $b=2$, but not to $b=0$.

Provided an impostor cannot deceive the verifier with probability greater than 2/3, then it suffices to repeat t times the basic protocol in order to obtain an impostor detection probability greater than or equal to $1-(2/3)^t$.

4.2 Too short hash-values decrease the level of security

We now show that an impostor can deceive the verifier with probability 1 if hash-values c_0, c_1 and c_2 are 64-bit long. More precisely, we show that, if only one of these hash-values is 64-bit long, then the impostor has a strategy, based on the birthday paradox, which allows him to deceive the verifier with probability equal to 1.

Case : c_0 is 64-bit long. The attack is as follows. First, the impostor selects two words z and z' such that $|z \oplus z'| = p$. Then he prepares a set Σ of 2^{32} permutations σ of $\{1...n\}$. For each permutation σ of this set, he computes $h(\sigma, A(z.\sigma^{-1}))$ and $h(\sigma, A(z'.\sigma^{-1}) \oplus v)$. Due to the birthday paradox, there exist with high probability two permutations σ and σ' in Σ such that $h(\sigma, A(z.\sigma^{-1})) = h(\sigma', A(z'.\sigma'^{-1}) \oplus v)$. Let us call c_0 this common value (the "collision"), and set $y = z.\sigma^{-1}$ and $y' = z'.\sigma'^{-1}$. If no collision occurred, the impostor has to increase slightly the size of Σ.

Now, at each execution of the basic protocol, the impostor does the following. He sends c_0, $c_1 = h(z)$ and $c_2 = h(z')$ to the verifier. If the verifier sends $b=0$, he replies with (y,σ). If $b=1$, he replies with (y',σ'). If $b=2$, he replies with (z,z'). In each case, the verifier will be satisfied with the reply.

Case : c_1 is 64-bit long. The attack is as follows. The impostor selects a permutation σ and a word y'. Then he prepares a set E_1 of 2^{32} words y_1 of syndrome equal to $A(y') \oplus v$, i.e. words y_1 such that $A(y_1) = A(y') \oplus v$. (Note that there are about 2^{n-k} such words, hence more than 2^{256}, and that they can be computed in a straightforward manner). He also prepares a set E_2 of 2^{32} words y'_2 such that $|y' \oplus y_2| = p$. With high probability, there exist y_1 in E_1 and y_2 in E_2 such that $h(y_1.\sigma) = h(y_2.\sigma)$. Let us call c_1 this common value.

Now, at each execution of the basic protocol, the impostor does the following. He sends $c_0 = h(\sigma, A(y_1))$, c_1 and $c_2 = h(y'.\sigma)$ to the verifier. If the verifier sends $b=0$, he replies with (y_1,σ). If $b=1$, he replies with (y',σ). If $b=2$, he replies with $(y_2.\sigma, y'.\sigma)$. In each case, the verifier will be satisfied with the reply.

Case : c_2 is 64-bit long. The attack is essentially the same as the previous one. First, the impostor selects a permutation σ and a word y. Then he prepares a set E_1 of 2^{32} words y'_1 of syndrome equal to $A(y) \oplus v$ and a set E_2 of 2^{32} words y'_2 such that $|y \oplus y'_2| = p$. With high probability there exist a word y'_1 in E_1 and a word y'_2 in E_2 such that $h(y'_1.\sigma) = h(y'_2.\sigma)$. Let us call c_2 this common value.

Now, at each execution of the basic protocol, the impostor does the following. He sends $c_0 = h(\sigma, A(y))$, $c_1 = h(y.\sigma)$ and c_2 to the verifier. Then, if $b=0$, he replies with(y,σ). If $b=1$, he replies with (y'_1,σ). If $b=2$, he replies with $(y.\sigma, y'_2.\sigma)$. In each case, the verifier will be satisfied with the reply.

4.3 The level of security

The level of security results from the following theorem, implicitly contained in [St93] :

Theorem 3. *If there exists a PPTM M such that the probability that M be accepted by an honest verifier is greater than 2/3 + ε, with ε>0, then there exists a PPTM $\tilde{M}$ which, with overwhelming probability, either computes a word of weight p and of syndrome v or finds a collision for h.*

5 The Shamir identification scheme

5.1 Description

The identification scheme presented by Shamir at CRYPTO'89 conference [Sh89] is based on the permuted kernel problem, that is the (NP-complete) problem of finding a permutation which puts a given vector into the kernel of a given matrix. The universal parameters are :

- a (small) prime p
- a random p-ary $(k \times n)$ matrix A $(k < n)$
- a hash-function h.

Typical values for (p,k,n) are $(251,16,32)$ or, still better, $(251,37,64)$. All calculations are done modulo p. The prover's secret key is a random permutation π of the integers $\{1...n\}$. His public key is an n-vector v such that $v.\pi \in \mathrm{Ker}A$ (where, as above, $v.\pi$ refers to the image of v under permutation π). The basic (5-pass) protocol is as follows.

1. The prover randomly selects an n-vector y and a permutation σ of the integers $\{1...n\}$, computes $c_0 = h(\sigma, A(y))$ and $c_1 = h(\sigma', y.\sigma)$, with $\sigma' = \pi\sigma$, and sends c_0 and c_1 to the verifier.

2. The verifier randomly selects an integer d in $\{0...p-1\}$ and sends d to the prover.

3. The prover computes $w = y.\sigma + dv.\sigma'$ and sends w to the verifier.

4. The verifier randomly selects a bit b and sends b to the prover.

5. The prover sends to the verifier :	6. The verifier checks :
- if $b=0$: σ	$h(\sigma, A(w.\sigma^{-1})) = c_0$
- if $b=1$: σ'	$h(\sigma', w-dv.\sigma') = c_1$

Note that an impostor (who ignores π) can deceive the verifier with probability $(p+1)/2p$, by using one of the two following strategies :

a) he selects an n-vector y, two permutations σ and σ' and an integer d_1 in $\{0...p-1\}$, then computes c_0 as above and $c_1 = h(\sigma', y.\sigma - d_1 v\sigma')$. At step 3, he sends $w = y.\sigma$ to the verifier, whatever d is. Then he is able to answer correctly either to $b=0$ and any d, or to $b=1$ and $d=d_1$, but not to $b=1$ and $d \neq d_1$.

b) he selects an n-vector y, two permutations σ and σ' and an integer d_1 in $\{0...p-1\}$, then computes $c_0 = h(\sigma, A(y + d_1 v.\sigma'\sigma^{-1}))$. At step 3, he sends $w = y.\sigma + dv.\sigma'$ to the verifier. Then he is able to answer correctly either to $b=1$ and any d, or to $b=0$ and $d=d_1$, but not to $b=0$ and $d \neq d_1$.

Provided an impostor cannot deceive the verifier with probability greater than $(p+1)/2p$, then it suffices to repeat t times the basic protocol in order to obtain an impostor detection probability greater than or equal to $1-(p+1/2p)^t$ ($\approx 1-2^{-t}$ if p is not too small).

5.2 Too short hash-values decrease the level of security

We now show that an impostor can deceive the verifier with probability $(p+2)/2p$ if hash-values c_0 and c_1 are 64-bit long. More precisely, we show that, if only one of these hash-values is 64-bit long, then the impostor has a strategy, based on the birthday paradox, which allows him to deceive the verifier with probability equal to $(p+2)/2p$. This is of particular significance if the value sent by the verifier in the second pass is restricted to 0 or 1 (a possibility mentioned in [Sh89]), since the impostor acceptance probability then grows up from 3/4 to 1.

Case : c_0 is 64-bit long. The attack is as follows. First, the impostor selects a permutation σ', an n-vector z and two distinct elements d_1 and d_2. Then he prepares a set Σ of 2^{32} permutations σ of $\{1...n\}$. (Note that there are $n! \geq 32! \geq 2^{117}$ such permutations). For each permutation σ of this set, he computes $h(\sigma, A(z.\sigma^{-1} + d_1 v.\sigma'\sigma^{-1}))$ and $h(\sigma, A(z.\sigma^{-1} + d_2 v.\sigma'\sigma^{-1}))$. Due to the birthday paradox, there exist with high probability two permutations σ_1 and σ_2 in Σ such that $h(\sigma_1, A(z.\sigma_1^{-1} + d_1 v.\sigma'\sigma_1^{-1})) = h(\sigma_2, A(z.\sigma_2^{-1} + d_2 v.\sigma'\sigma_2^{-1}))$. Let us call c_0 this common value.

Now, at each execution of the basic protocol, the impostor does the following : he sends c_0 and $c_1 = h(\sigma', z)$ to the verifier, who sends back d to him. He sends $w = z + dv.\sigma'$ to the verifier, who sends back b to him. Then, if $b = 1$, he replies with σ'. If $b = 0$ and $d = d_1$ (resp. $d = d_2$), he replies with σ_1 (resp. σ_2). In other cases, he sends anything. So, in $p+2$ cases, the verifier will be satisfied with the reply.

Case : c_1 is 64-bit long. First, the impostor selects a permutation σ, an n-vector y and two distinct elements d_1 and d_2. Then he prepares a set Σ' of 2^{32} permutations σ' of $\{1...n\}$. For each permutation σ' of this set, he computes $h(\sigma', y.\sigma - d_1 v.\sigma')$ and $h(\sigma', y.\sigma - d_2 v.\sigma')$. With high probability, there exist two permutations σ'_1 and σ'_2 in Σ' such that $h(\sigma'_1, y.\sigma - d_1 v.\sigma'_1) = h(\sigma'_2, y.\sigma - d_2 v.\sigma'_2)$ and we call c_1 this common value.

Now, at each execution of the basic protocol, the impostor does the following : he sends $c_0 = h(\sigma, A(y))$ and c_1 to the verifier, who sends back d to him. He sends $w = y.\sigma$ to the verifier, who sends back b to him. Then, if $b = 0$, he replies with σ. If $b = 1$ and $d = d_1$ (resp. $d = d_2$), he replies with σ'_1 (resp. σ'_2). In other cases, he sends anything. So, in $p+2$ cases, the verifier will be satisfied with the reply.

5.3 The level of security

The level of security results from the following theorem, implicitly contained in [Sh89] :

Theorem 4. *If there exists a PPTM M such that the probability that M be accepted by an honest verifier is greater than* $(p+1)/2p + \varepsilon$, *with* $\varepsilon > 0$, *then there exists a PPTM* $\tilde{M}$ *which with overwhelming probability either computes a permutation* π *such that* $v.\pi \in \mathrm{Ker}A$ *or finds a collision for h.*

6 Conclusion

First, we have considered several identification schemes using hash-values in the first pass (either in their basic design or in specific variants) and given evidence that 64-bit hash-values were too short to achieve the level of the security claimed by their authors, by exhibiting for each of them one or more specific attacks. Second we have proved that collision-resistance was a sufficient condition to achieve this level. Third we have shown that the number of communication bits could be minimized in schemes based on modular arithmetic by using *r*-collision resistant hash-functions instead of collision-resistant hash-functions. As an example, the number of bits transported in the first pass of the Schnorr scheme can be decreased from 128 to 71, and the total number of bits transported from 308 to 254, i.e. by 18%, for a level of security equal to $1-2^{-40}$.

References

[Fe68] W. Feller, An introduction to probability theory and its applications, J. Wiley & Sons, 1968.

[FFS88] U. Feige, A. Fiat and A. Shamir, Zero-knowledge proofs of identity, *Journal of Cryptology*, Vol. 1, N°2, 1988, pp. 77-94.

[FS86] A. Fiat and A. Shamir, How to prove yourself : Practical solutions to identification and signature problems, *Advances of Cryptology : Proceedings of CRYPTO'86*, Lecture Notes in Computer Science, Vol. 263, Springer-Verlag, Berlin, 1987, pp. 186-194.

[GMR85] S. Goldwasser, S. Micali and C. Rackoff, The knowledge of interactive proof-systems, *Proceedings of 17th ACM Symposium on Theory of Computing,* 1985, pp. 291-304.

[GQ88] L.C. Guillou and J.J. Quisquater, A practical zero-knowledge protocol fitted to security microprocessors minimizing both transmission and memory, *Advances of Cryptology : Proceedings of EUROCRYPT'88*, Lecture Notes in Computer Science, Vol. 330, Springer-Verlag, Berlin, 1988, pp. 123-128.

[PS87] J.M. Pollard and C.P. Schnorr, An efficient solution of the congruence $x^2 + ky^2 = m \pmod{n}$, *IEEE Transactions on Information Theory*, Vol. IT-33, N°5, 1987, pp. 702-709.

[Sc89] C.P. Schnorr, Efficient identification and signature for smart cards, *Advances of Cryptology : Proceedings of CRYPTO'89*, Lecture Notes in Computer Science, Vol. 435, Springer-Verlag, Berlin, 1987, pp. 239-252.

[Sh89] A. Shamir, An efficient identification scheme based on permuted kernels, *Advances of Cryptology : Proceedings of CRYPTO'89*, Lecture Notes in Computer Science, Vol. 435, Springer-Verlag, Berlin, 1987, pp. 606-609.

[St89] J. Stern, An alternative to the Fiat-Shamir protocol, *Advances of Cryptology : Proceedings of EUROCRYPT'89,* Lecture Notes in Computer Science, Vol. 434, Springer-Verlag, Berlin, 1990, pp. 173-180.

[St93] J. Stern, A new identification scheme based on syndrome decoding, *Advances of Cryptology : Proceedings of CRYPTO'93,* to appear.

Incremental Cryptography: The Case of Hashing and Signing

Mihir Bellare[1] and Oded Goldreich[2] and Shafi Goldwasser[3]

[1] Advanced Networking Laboratory, IBM T.J. Watson Research Center, P.O. Box 704, Yorktown Heights, NY 10598, USA. e-mail: mihir@watson.ibm.com.

[2] Department of Applied Mathematics and Computer Science, Weizmann Institute of Science, Rehovot, Israel. e-mail: oded@wisdom.weizmann.ac.il.

[3] Department of Applied Mathematics and Computer Science, Weizmann Institute of Science, Rehovot, Israel, and MIT Laboratory for Computer Science, 545 Technology Square, Cambridge MA 02139, USA. e-mail: shafi@wisdom.weizmann.ac.il.

Abstract. We initiate the investigation of a new kind of efficiency for cryptographic transformations. The idea is that having once applied the transformation to some document M, the time to update the result upon modification of M should be "proportional" to the "amount of modification" done to M. Thereby one obtains much faster cryptographic primitives for environments where closely related documents are undergoing the same cryptographic transformations.

We provide some basic definitions enabling treatment of the new notion. We then exemplify our approach by suggesting incremental schemes for hashing and signing which are efficient according to our new measure.

1 Introduction

We initiate an investigation of *incremental* algorithms for cryptographic functions. The idea, taking digital signatures as an example, is to have a signature which is easy to update upon modification of the underlying message. Thus, suppose you have signed message M to obtain signature σ. Now you make some change to M. For example, you might replace one block by another, insert a new block, or delete an old one. Let M' be the modified message. You want to update the signature σ into a signature σ' of M'. The time to update should be somehow proportional to the "amount of change" you have made in M to get M', and not the same as simply signing M' anew.

1.1 A wide range of applications

Incrementality is suitable for an environment in which the documents undergoing cryptographic transformations are altered versions of documents which have already undergone the same cryptographic transformations.

For example suppose you are sending the same message to many different users so that the text is essentially the same except for header information, and you want to sign each copy.

A second example is video traffic. Here one can take advantage of the well-known fact that successive video frames usually differ only slightly.

A third example is the use of authentication tags for virus protection. We imagine a user who has a PC and stores his files on a remote, possibly insecure host which could be attacked by a virus. The user authenticates the files so that he can detect their modification by a virus. When a user modifies his files, he must re-authenticate them, and an incremental scheme might be useful. (The type of authentication used for virus protection can vary. The most likely choice may be a private key authentication scheme such as a finger-print. But one can envisage applications where the user's signature has to be checked also by others and a full fledged digital signature is desirable).

In general it seems clear that incrementality is a nice property to have for any cryptographic primitive (eg. finger-printing, message authentication, digital signatures, hashing, encryption, etc.).

1.2 Problems considered in this abstract

It quickly becomes apparent that incrementality presents a large area of research. The goal of this (preliminary) abstract is to draw attention to this area, lay some basic definitions, and provide some examples. For simplicity we restrict the scope of the work presented here in two ways. First, we focus on just one update operation on the underlying message, namely the replacement of one block by another. Second, we limit the primitives we consider to digital signatures and the collision-free hashing primitive via which we approach it.

We view a message $M = M[1] \ldots M[n]$ as a sequence of b-bit blocks. Let $M\langle j, m\rangle$ denote M with $M[j]$ replaced by the b-bit string m. The problem, for collision-free hashing, is to design a scheme for which there exists an efficient "update" algorithm: this algorithm is given the hash function H, the hash $h = H(M)$ of M and the "replacement request" (j, m), and outputs the hash $H(M\langle j, m\rangle)$ of the modified message. Similarly, for signing, the update algorithm is given the signing key Sk, a message M, its signature σ, and the replacement request (j, m), and must produce the signature σ' of $M\langle j, m\rangle$. Ideally, in either case, the update time should depend only on the block size b and the underlying security parameter k, and not on the length of the message.[4] A scheme is said to be ideally incremental if it possesses an ideal update algorithm.

In work in progress we address other operations on messages, like insertion or deletion of blocks, and we also expand the scope to consider more primitives, namely finger-printing and message authentication. See Section 5 for more information.

[4] Some care must be taken in formalizing this since the update algorithm receives the entire nb-bit message as input and in a standard Turing machine model could not even read the relevant part in poly(k, b) time. Thus we assume a RAM type computation model and in the formal definition allow the update time to depend on $\log n$.

1.3 Incremental collision-free hashing

We pin-point an ideal incremental scheme for collision-free hashing which is based on the hardness of the discrete logarithm in groups of prime order. Hashing an n-block message (each block k bits long) to a k bit string takes n exponentiations, while updating takes two exponentiations; these exponentiations are modulo a $O(k)$-bit prime.

The special case of this hash function in which the number of blocks n is a constant was presented and analyzed by Chaum, Heijst and Pfitzmann [7]. Brands [5] provided a proof of security for $n = \text{poly}(k)$. The observation that this hash function possesses an (ideal) update algorithm identifies for the first time a crucial use for it: we know no other collision-free hashing scheme that is ideally incremental.

We make an additional contribution by considering the "exact security" of the reduction via which security is proved, and presenting a new reduction which is essentially optimal. The motivation and impact of this result, as explained in Section 3.4, is practical: it enables a user to get the same level of security for a smaller value of the security parameter, leading to greater efficiency.

Note the hash functions we discuss here are ones of public description. That is, the description of the function is provided to the adversary trying to find collisions. This is unlike the hash functions used in applications like finger-printing, where the description of the function is not available to the collision-finder!

1.4 Incremental digital signatures

With an (ideal) incremental hash function available, an (ideal) incremental signature scheme is easily derived by a slight variation of the standard hash-and-sign construction. Namely, hash the message M with the incremental hash function to get h; sign h with some standard scheme to get σ^*; and use (h, σ^*) as the signature. (The variation is that the hash value must be included in the signature). To update, update the hash (fast by assumption) and then sign the new hash value from scratch (fast because we are signing a k-bit string).

When evaluating the security of this scheme (or any other) scheme one must be careful: the presence of the update algorithm entails new security considerations. In keeping with the philosophy of an adaptive chosen message attack [13], we must allow the adversary to obtain examples of signatures under the update algorithm. In general these could be differently distributed from signatures created by the signing algorithm, and the extra power could help the adversary.

We will show that the above scheme achieves what we call *basic* security. Here, in addition to being able to get examples of signatures from the signing algorithm, the adversary can point to any past message/signature pair and obtain the result of an arbitrary update on this pair.

1.5 Practical issues

Incrementality is fundamentally a practical concern because it is a measure of efficiency.

Clearly, an (ideal) incremental scheme is a win over a standard one as message sizes get larger. The practical concern is what is the cross-over point: if incrementality only helps for messages longer than I am ever likely to get, I am not inclined to use the incremental scheme. The cross over point for our schemes is low enough to make them interesting (cf. Section 3.4).

We prefer schemes with low memory requirements. Signatures of size proportional to the message, as permitted in theoretical definitions, are not really acceptable. Thus we want schemes with $\text{poly}(k, b)$ size signatures independent of the number n of message blocks. (This consideration eliminates some trivial incremental schemes like the tree hash signature scheme. See Section 4.4). This is achieved in our constructions.

Finally, we analyze and state all our security results exactly (as opposed to asymptotically) and strive for the best possible reductions.[5]

1.6 An interesting open question

The notion of basic security makes an assumption. Namely, that the signer is in a setting where the integrity of messages and signatures which he is updating is assured. That is, when a signer applies the update algorithm to update M and its signature σ, he is confident that this data has not been tampered with since he created it. This is reflected in the fact that adversary's attack on the update algorithm consists of pointing to a past (authentic) message/signature pair.

This is the right assumption in the majority of applications of digital signatures. For example, in the case where I am sending the same message to many parties except with different headers, I sign one copy and update to obtain the rest. But I keep the original copy and its signature on my machine— when I update I know the original is authentic.

But there are some situations in which one might want an even stronger form of security. For example, suppose you are remote editing a file residing on an insecure machine, and at any time the machine could be hit by a virus which would tamper with the data. For efficiency you are incrementally signing the file every time you make a change to it. But when you run the update algorithm, you can't be sure the data is still authentic. (It is impractical to verify authenticity before updating because verification takes time depending on n and the whole point of incrementality is to update the signature quick).

We formalize a new notion of security under *substitution attacks* appropriate to the above setting. We then show that substitution attacks can be used to break the above hash-and-sign scheme when the hash function is our discrete log based one. This is interesting in two ways— it illustrates the strength of the new attacks, and it shows that a "standard" construction (namely hash-and-sign) can become insecure in a new setting!

---

[5] Exact security is not new. Although the majority of theoretical works only make asymptotic statements, the exact tradeoffs can be derived from the proofs. (However these tradeoffs are sometimes quite bad). Moreover several previous works explicitly address exact security with concern for tradeoff quality, eg. [12, 14, 17, 11, 1].

We leave it as an open problem to find ideal incremental signature schemes secure against substitution attack.

2 Preliminaries

We follow the notation for algorithms and probabilistic experiments that originates in [13] and refer the reader there for a detailed exposition. Let's briefly recall that $z \stackrel{R}{\leftarrow} A(x, y, \cdots)$ is the experiment of running probabilistic algorithm A and letting z be its output, and $[A(x, y, \cdots)]$ is the set of all strings output by $A(x, y, \cdots)$ with positive probability. PPT denotes "probabilistic, polynomial time."

Our results require us to be careful about the model of computation. Rather than the traditional Turing machine model, we use the (in any case more realistic) RAM model. In this model any algorithm $A(x, y, \cdots)$ has random access to each of of its inputs $x, y, \cdots$.

A message is viewed as a sequence of blocks. The block size is denoted b and we let $B_b = \{0,1\}^b$ be the domain within which blocks range. We let n denote the number of blocks and B_b^n the space of n-block messages. With b understood, $M[i]$ is the i-th block of $M \in B_b^n$.

A *replacement request* has the form (j, m) with $1 \leq j \leq n$ and $m \in B_b$. We let $M\langle j, m\rangle$ denote the message consisting of M with block j replaced by m. We'll say M has been *updated* or *incremented* by (j, m).

3 Incremental collision-free hashing

3.1 Families of hash functions

We need to extend usual definitions of hash families to allow independent consideration of the security parameter, the block size and the number of blocks. These parameters are denoted k, b, n, respectively. Below the string H is (the description of) a particular hash function.

Definition 1. A family of hash functions is specified by a pair $\mathcal{H} = (\mathsf{HGen}, \mathsf{HEval})$ of algorithms.

- The PPT generator HGen takes as input $1^k, 1^b, 1^n$ and outputs a string H
- The polynomial time hash evaluation algorithm HEval takes H and a message $M \in B_b^n$ and outputs a k bit string called the hash of M under H.

When the family (HGen, HEval) is clear from the context, we will identify H with $\mathsf{HEval}(H, \cdot)$ and regard it as a map of B_b^n to $\{0,1\}^k$. In particular we will write $H(M)$ for $\mathsf{HEval}(H, M)$.

3.2 Incrementality

The following definition says that an update algorithm IncH is one that can turn the hash of M into the hash of $M\langle j, m\rangle$.

Definition 2. Let $\mathcal{H}$ = (HGen, HEval) specify a family of hash functions. We say that IncH is an update algorithm for $\mathcal{H}$ with running time $T(\cdot,\cdot,\cdot)$ if

$$\forall\, k, b, n,\ \ \forall\, H \in [\mathsf{HGen}(1^k, 1^b, 1^n)]\ \ \forall\, j \in \{1, \ldots, n\},\ \ \forall\, m \in B_b,$$

if $h = \mathsf{HEval}(H, M)$ then it is the case that $\mathsf{IncH}(H, M, h, (j, m))$ halts in $T(k, b, n)$ steps with output equal to $\mathsf{HEval}(H, M\langle j, m\rangle)$.

The IncH-*augmentation* of $\mathcal{H}$ = (HGen, HEval) is the triple $\mathcal{H}^+$ = (HGen, HEval, IncH).

Notice this definition makes no requirement on the running time T of IncH. So, in particular, an update algorithm can just run $\mathsf{HEval}(H, M\langle j, m\rangle)$ to compute its output. We don't wish to exclude this— it is a legitimate update algorithm. But of course an update algorithm will be interesting only when it runs faster than HEval. The term "incremental hash family" will be loosely used to refer to a hash family possessing some "non-trivial" update algorithm.

We would like to say that an "ideal" update algorithm is one whose running time does not depend on n. Such an algorithm would random access a small number of relevant memory blocks (this is where we need the RAM model as opposed to the Turing machine model) and do some quick computation before writing the output back to memory. The formal definition that follows, however, allows a dependence of the time on $\log n$ (because otherwise the blocks cannot be accessed even in a RAM) but this quantity will be much smaller than, say, $k + b$, in any realistic situation, and thus we view the running time as being independent of n.

Definition 3. An update algorithm IncH for $\mathcal{H}$ is ideal if its running time $T(k, b, n)$ is polynomial in k, b and $\log n$.

We'll say that $\mathcal{H}$ is an *ideal incremental* scheme if it possesses an ideal update algorithm.

3.3 Collision-freeness

Incrementality does not necessitate any additions to the usual notions of attacks on the hash family. (This is in contrast to the situation for signatures, where the presence of incrementality will introduce new security issues). Thus we just have the usual notion of collision-freeness. A *collision-finder* for $\mathcal{H}^+$ is a probabilistic algorithm A. We discuss security via the following experiment.

Experiment describing A's attack on $\mathcal{H}^+(k, b, n)$:–

(1) Run $\mathsf{HGen}(1^k, 1^b, 1^n)$ to get H.

(2) Run A on input H. We ask that at the end of her execution, A outputs a pair of *distinct* messages $M_1, M_2 \in B_b^n$. We say A is *successful* if $H(M_1) = H(M_2)$.

We say that A *succeeds in (t,ϵ)-breaking* $\mathcal{H}^+(k,b,n)$ if, in the above experiment, A runs for time t and is successful with probability at least ϵ. We don't say what it means for a scheme to be "secure:" there is no need, because we will make stronger statements on the exact security (cf. Theorem 4) which imply the usual asymptotic notion.

The fact that the hash function is "public" is captured in the fact that the adversary is given its descripton when trying to find collisions.

3.4 An incremental hash family

Discrete log in groups of prime order

We fix a PPT algorithm PrimeGen which on input 1^k outputs a $k+1$ bit prime p identifying a group G_p of (prime) order p. We let $\mathcal{G} = \bigcup_k \mathcal{G}(k)$, where $\mathcal{G}(k) = \{ G_p \ :\ p \in [\mathsf{PrimeGen}(1^k)] \}$, be the set of all these groups, and we assume the discrete log problem for $\mathcal{G}$ is hard (when the prime is chosen according to PrimeGen). Such groups have been used for cryptography by Croft and Harris [8], Schnorr [17], Chaum and Van Antwerpen [6], and others, and we refer the reader to these works for how to choose such groups. In particular, with appropriate assumptions on the distribution of primes if necessary, it can be done so that G_p is a subgroup of Z_q^* for some q of size $O(k)$, so that we may assume efficient group operations. In particular exponentiation takes $O(k^3)$ time.

A key fact is that since G_p has prime order, every non-trivial element is a generator. We let $\mathsf{index}_g^{G_p}(x) \in \{0,1,\ldots,p-1\}$ denote the discrete logarithm of x to (non-trivial) base g in the group G_p. A *discrete log finder* is a probabilistic algorithm B.

Experiment describing B's attack on $\mathcal{G}(k)$:–

(1) Let $p \xleftarrow{R} \mathsf{PrimeGen}(1^k)$; $g \xleftarrow{R} G_p - \{1\}$; $x \xleftarrow{R} G_p - \{1\}$.

(2) Run B on input p, g, x. We say that she is successful if her output is $\mathsf{index}_g^{G_p}(x)$.

We way that B *succeeds in (t,ϵ)-breaking* $\mathcal{G}(k)$ if in the above experiment she halts in t steps and is successful with probability at least ϵ.

We denote by $\langle\cdot\rangle$: $B_b \to \{1,\ldots,2^b\}$ an encoding of message blocks into non-zero integers. To be specific, we set $\langle m \rangle$ to 1 plus the number whose binary expansion is m. Thus for any prime p of length at least $b+1$ and any $g \in G_p$ we can compute $g^{\langle m \rangle}$, and if g is non-trivial so is $g^{\langle m \rangle}$.

The hash family and the update algorithm

The block size will be set equal to the security parameter, $b = k$. (Formally have the following algorithms output junk when $b \neq k$. Theorem 4 only addresses the case $b = k$).

The hash family $\mathcal{H} = (\mathsf{HGen}, \mathsf{HEval})$ is specified as follows. On input $1^k, 1^k, 1^n$ the generator HGen runs PrimeGen(1^k) to get a $k+1$ bit prime p. It then selects $g_1, \ldots, g_n$ at random from $G_p - \{1\}$ and outputs $(p; g_1, \ldots, g_n)$ as the description of the hash function H. The value of the hash function $H = (p; g_1 \ldots g_n)$ on a given message $M = M[1] \ldots M[n] \in B_k^n$ is given by

$$H(M[1] \ldots M[n]) \stackrel{\text{def}}{=} \mathsf{HEval}(H, M[1] \ldots M[n]) = \prod_{i=1}^n g_i^{\langle M[i] \rangle} ,$$

the operations being of course in the group G_p.

The interest of [7] in this family seemed to stem from its efficiency as compared, for example, to that of the (discrete log based) hash family of [9]. Brands [5] mentions the family in the context of a general exposition of the "representation" problem. The (seemingly rare) incremental property that we next observe it possesses seems for the first time to pinpoint a crucial use of this family, and in some sense answers a question of [5] who asked for interesting uses of the representation problem when n was more than a constant.

Define the algorithm IncH by

$$\mathsf{IncH}(H, M, h, (j, m)) = h \cdot g_j^{-\langle M[j] \rangle} \cdot g_j^{\langle m \rangle} .$$

It is easy to see that if $h = H(M)$ then $\mathsf{IncH}(H, M, h, (j, m)) = H(M\langle j, m \rangle)$. Now note IncH can random access the $O(nk)$ bit description of H to get g_j in poly($k, \log n$) time, and similarly for the other inputs. Its output is then given by a polynomial time computation on $O(k)$ bit inputs and hence the algorithm runs in poly(k) time. Thus it is an ideal update algorithm for $\mathcal{H}$.

Security

A proof of the security of $\mathcal{H}^+$ takes the following form. Given a collision finder A for $\mathcal{H}^+(k, k, n)$ we construct a discrete log finder B for $\mathcal{G}(k)$. Now suppose A succeeds in (t, ϵ)-breaking $\mathcal{H}^+(k, k, n)$. The question we consider is for what values of t', ϵ' the constructed algorithm B succeeds in (t', ϵ')-breaking $\mathcal{G}(k)$.

Previous works [7, 5] have only discussed asymptotic security, where one sets $n = n(k)$ to some fixed polynomial in k, regards $t, \epsilon, t', \epsilon'$ as functions of k, assumes t, ϵ are polynomial and non-negligible, respectively, and then shows that t', ϵ' are also polynomial, and non-negligible, respectively. But for practice it is important to know *exactly* how the resources and achievements of B compare to those of A, so that we may know what size to choose for the prime p and what adversaries we can tolerate with a specific security parameter. Moreover, it is important to strive for the tightest possible reduction, because this means that the same "security" can be obtained with a smaller value of the security parameter, meaning greater efficiency. Thus we want the effort and success of B should be as close to those of A as possible.

In this light let's look at the existing reductions to see what they achieve. The proof of [7] only applies to the case of $n = O(1)$ block messages, and in fact t' seems to grow exponentially with n, so that this reduction is not suitable for our purposes. Brands [5] proposes a reduction which removes the restriction on n and achieves $t' = t + O(nk^3)$ and $\epsilon' = \epsilon/n$. The running time of B here is essentially optimal: we must think of t as much larger than n or k, and additive terms like the $O(nk^3)$ correspond to overhead of B coming from simple and unavoidable arithmetic operations. The loss in the success probability is more serious. Note that (particularly in our case) n may be very large. Thus even if A is successful with high probability, the above may only let us conclude that B is successful with low probability.

We improve the reduction to be essentially optimal. We preserve the current quality of the running time, and achieve for B a success probability within a small constant factor of that of A.

The big-oh notation, both in the time as given above and in the following theorem, hides a constant which depends only on the underlying machine model and can be taken as small in a reasonable setting. U denotes some oracle machine which depends only on our proof and the given family $\mathcal{H}$. Although the statement of the theorem does not say anything about the "size" of U, the proof shows that it is "small," and this is important in practice. $\mathcal{H}^+$ is the IncH-augmentation of $\mathcal{H}$.

Theorem 4. There is an oracle machine U such that the following is true. Suppose collision-finder A succeeds in (t, ϵ)-breaking $\mathcal{H}^+(k, k, n)$. Then discrete log finder $B \stackrel{\text{def}}{=} U^A$ succeeds in (t', ϵ')-breaking $\mathcal{G}(k)$ where $t' = t + O(nk^3)$ and $\epsilon' = \epsilon/2$.

The proof is in Appendix A.

Efficiency

Hashing an n-block message takes n exponentiations (equivalently, one multiplication per message bit) modulo a $O(k)$-bit prime. This is quite good for a number-theory based scheme.

How does it compare with standard hash functions like MD5 or SHA? Let's fix $k = 512$. In hashing from scratch there is no comparison— MD5 on $512n$ bits is far better than n exponentiations. But assume we are in a setting with frequent updates. With MD5 we have no choice but to hash from scratch, while in our scheme we can use the update algorithm to update the hash in two exponentiations. Thus to compare the efficiency we should ask how large is n before the time to do two exponentiations of 512 bit numbers is less than the time to evaluate MD5 on a $512n$ bit string. A computation yields a reasonable value.

Note there are heuristics (based on vector-chain addition) to compute $\prod_{i=1}^{n} g_i^{\langle M[i] \rangle}$ faster than doing n modular exponentiations [4].

A practical version with small description size

The size of (the description of) the hash function in the above is $O(nk)$ so that it depends on the message size, which we assume large. In practice this is too much. Here we suggest a way to reduce the size to $O(k)$. We let $f\colon \{0,1\}^k \to \{0,1\}^{O(k)}$ be the restriction of some "standard" hash function, such as MD5, to inputs of length k. We now set $g_i = f(i)$ to be the result of evaluating f at i. Now the description of the hash function is just the prime p and anyone can quickly compute $g_1, \ldots, g_n$ for themselves. The loss in efficiency is negligible since the time for the arithmetic operations dwarfs the MD5 computation time.

Although such a construction must ultimately be viewed as heuristic, its security can be discussed by assuming f is a random function. Extending our proof of security to this setting is not difficult and we can conclude (the following statement is informal) that the scheme just described satisfies Theorem 4 in the random oracle model. As discussed by [2], although this approach (namely prove security in a random oracle model and then instantiate the random oracle with a standard hash function) does not yield provable security, it provides a better guarantee than purely heuristic design, and protocols designed in this manner seem to be secure in practice. We refer the reader to this paper also for more suggestions on functions with which to "instantiate" f.

4 Incremental Signing

4.1 Signature schemes

Definition 5. A signature scheme is a triple $\mathcal{S} = (\mathsf{KGen}, \mathsf{Sig}, \mathsf{Vf})$ of algorithms. There is a polynomial $s(\cdot,\cdot,\cdot)$ called the signature size such that

- The PPT key generator KGen takes as input $1^k, 1^b, 1^n$ and outputs a pair (Sk, Vk) of strings called, respectively, the signing and (corresponding) verifying keys.
- The PPT signing algorithm Sig takes as input Sk and $M \in B_b^n$ and outputs a $s(k, b, n)$-bit string called the signature of M.
- The polynomial time verifying algorithm Vf outputs a bit and satisfies $\mathsf{Vf}(Vk, M, \sigma) = 1$ for every $M \in B_b^n$ and every $\sigma \in [\mathsf{Sig}(Sk, M)]$.

The assumption that Vf is deterministic is for simplicity only: in general one can consider probabilistic verifiers.

We'll say that a signature scheme has *short signatures* if the signature size depends only on k. In such a case we abuse notation and write the signature size as $s(k)$.

4.2 Incrementality

An update algorithm is one that can turn a signature of M into *some* signature of $M\langle j, m\rangle$.

Definition 6. Let $\mathcal{S} = (\mathsf{KGen}, \mathsf{Sig}, \mathsf{Vf})$ be a signature scheme. We say that IncSig is an update algorithm for $\mathcal{S}$ with running time $T(\cdot,\cdot,\cdot)$ if

$$\forall\, k, b, n, \;\; \forall\, (Sk, Vk) \in [\mathsf{KGen}(1^k, 1^b, 1^n)] \;\; \forall\, j \in \{1, \ldots, n\}, \;\; \forall\, m \in B_b,$$

if $\mathsf{Vf}(Vk, M, \sigma) = 1$ then it is the case that $\mathsf{IncSig}(Sk, M, \sigma, (j, m))$ halts in $T(k, b, n)$ steps with output σ' satisfying $\mathsf{Vf}(Vk, M\langle j, m\rangle, \sigma') = 1$.

The IncSig-*augmentation* of $\mathcal{S} = (\mathsf{KGen}, \mathsf{Sig}, \mathsf{Vf})$ is $\mathcal{S}^+ = (\mathsf{KGen}, \mathsf{Sig}, \mathsf{Vf}, \mathsf{IncSig})$.

Note that the output of $\mathsf{IncSig}(Sk, M, \sigma, (j, m))$ is not required to be distributed in the same way as that of $\mathsf{Sig}(Sk, M\langle j, m\rangle)$— IncSig just has to return something that Vf would accept.

The term "incremental signature scheme" will be loosely used to refer to a signature scheme possessing some "non-trivial" update algorithm. Ideality of an update algorithm is defined in analogy to Definition 3, and an ideal incremental signature scheme is one that possesses an ideal update algorithm.

The schemes we prefer have short signatures, but it is possible to discuss update algorithms (even ideal ones) even if the signatures are long. In such a case IncSig will not be able to output the entire signature— one imagines that it modifies σ in a few chosen places and the result is what we view as σ'.

Analogously one can define the notion of incremental verification. We leave it to the reader.

4.3 Basic security

We recall that we will be evaluating the security of signature schemes at two levels, motivated by differing security demands of applications. The basic level we present here is suitable for settings in which a signer updating signature σ of message M is guaranteed that these quantities are authentic. In the majority of applications of digital signatures this assumption is valid.

The definition extends the notion of existential forgery under adaptive chosen message attack to allow the adversary access to $\mathsf{IncSig}(Sk, \cdots)$. (This is necessary because signatures produced by IncSig might be from a different distribution than those produced by Sig and perhaps the adversary can gain an advantage by seeing examples from this new distribution). The restriction that updates only be made on authentic data is captured below in the fact that the incremental signing requests simply point to a message and signature from the past.

Experiment describing F's attack on $\mathcal{S}^+(k, b, n)$:–

(1) Run $\mathsf{KGen}(1^k, 1^b, 1^n)$ to get keys Sk, Vk.

(2) Initialize: Set $\alpha = 0$.

(3) Run the adversary F on input Vk. Her oracle queries are answered as follows.

(3.1) Suppose F makes a simple signing request— this has the form of a message $M \in B_b^n$. Let $\sigma \stackrel{R}{\leftarrow} \mathsf{Sig}(Sk, M)$ and return σ to F. Let $\alpha \leftarrow \alpha + 1$. Let $M_\alpha \leftarrow M$ and $\sigma_\alpha \leftarrow \sigma$.

(3.2) Suppose F makes an incremental signing request— this has the form $((j,m),\beta)$ with $\beta \in \{1,\ldots,\alpha\}$. Let $\sigma' \stackrel{R}{\leftarrow} \mathsf{IncSig}(Sk, M_\beta, \sigma_\beta, (j,m))$ and return σ' to F. Let $\alpha \leftarrow \alpha+1$. Let $M_\alpha \leftarrow M_\beta\langle j,m\rangle$ and $\sigma_\alpha \leftarrow \sigma'$.

(4) We ask that at the end of her execution, F output a pair (M,σ) such that $M \notin$ Legal, where Legal $= \{ M_1,\ldots,M_\alpha \}$. We say that F is successful if $\mathsf{Vf}(Vk, M, \sigma) = 1$.

We say that F *succeeds in* $(t, q_{\mathrm{sig}}, q_{\mathrm{inc}}, \epsilon)$*-breaking* $\mathcal{S}^+(k,b,n)$ *with a basic attack* if, in the above experiment, she runs for t steps, makes $q_{q_{\mathrm{sig}}}$ simple signing requests, makes $q_{q_{\mathrm{inc}}}$ incremental signing requests, and succeeds with probability at least ϵ.

4.4 Incremental signature schemes achieving basic security

In what follows $\mathcal{S}^* = (\mathsf{KGen}^*, \mathsf{Sig}^*, \mathsf{Vf}^*)$ denotes a standard (ie. not necessarily incremental) signature scheme as per Definition 5, assumed secure against existential forgery under adaptive chosen message attack in the standard sense of [13]. Exact security is discussed by saying that an adversary F^* *succeeds in* (t,q,ϵ)*-breaking* $\mathcal{S}^*(k,b,n)$ *with an adaptive chosen message attack* if in this attack she runs in time t, makes q signing queries, and succeeds in existential forgery with probability at least ϵ. We consider two standard transformations.

Incremental hash-and-sign

Given an incremental hash function, a slight variation of the standard hash-and-sign method yields an incremental signature scheme. Security must however be reconsidered, in light of the fact that our basic attacks allow attacks on the update algorithm. Luckily they do not cause any damage. For completeness we provide details below.

Let $\mathcal{H}^+ = (\mathsf{HGen}, \mathsf{HEval}, \mathsf{IncH})$ be a family of hash functions together with an update algorithm. We specify $\mathcal{S}^+ = (\mathsf{KGen}, \mathsf{Sig}, \mathsf{Vf}, \mathsf{IncSig})$ as follows.

On input $1^k, 1^b, 1^n$ algorithm KGen runs $H \stackrel{R}{\leftarrow} \mathsf{HGen}(1^k, 1^b, 1^n)$; $(Sk^*, Vk^*) \stackrel{R}{\leftarrow} \mathsf{KGen}^*(1^k, 1^k, 1^1)$. It outputs the signing key $Sk = (Sk^*, H)$ and the verifying key $Vk = (Vk^*, H)$. Note the keys of the original signature scheme are chosen to sign messages consisting of one k-bit block only.

The signature of $M \in B_b^n$ given the above keys is $\mathsf{Sig}(Sk, M) = (H(M), \mathsf{Sig}^*(Sk^*, H(M)))$. Namely the hash of the message, together with its signature under the original scheme. (Including the hash $h = H(M)$ in the signature is the slight variation. It may seem redundant since anyone can compute it given M, H, but it is important for incrementality). Note Sig^* is being applied only to a k-bit string.

Given the verification key $Vk = (Vk^*, H)$ and a string $\sigma = (h, \sigma^*)$ the algorithm Vf outputs 1 iff $h = H(M)$ and $\mathsf{Vf}^*(Vk^*, \sigma^*) = 1$.

Given $Sk, M, \sigma, (j,m)$ (with $Sk = (H, Sk^*)$ and $\sigma = (h, \sigma^*)$) the update algorithm IncSig first updates the hash by $h' = \mathsf{IncH}(H, M, h, (j,m))$. Then it computes from scratch a signature $\sigma' = \mathsf{Sig}^*(Sk^*, h')$ of (the k-bit string) h' under the original scheme. It outputs (h', σ').

Note that the signatures in this scheme are short, namely poly(k) bits.

The following theorem says that if $\mathcal{S}^+$ can be broken then either $\mathcal{S}^*$ or $\mathcal{H}^+$ can be broken, and specifies the exact security corresponding to this statement. The function $\tau(k, b, n)$ represents time depending only on the algorithms defining the schemes. It should be viewed as much smaller than t and its exact value can be derived from the proof.

Theorem 7. There are oracle machines U_1, U_2 and a function $\tau(k, b, n)$ such that the following is true. Suppose F succeeds in $(t, q_{\rm sig}, q_{\rm inc}, \epsilon)$-breaking $\mathcal{S}^+(k, b, n)$ with a basic attack, and let $q = q_{\rm sig} + q_{\rm inc}$. Then one of the following is true:

(1) Either $F^* \stackrel{\rm def}{=} U_1^F$ succeeds in $(t + q \cdot \tau(k, b, n),\, q,\, \epsilon/2)$-breaking $\mathcal{S}^*(k, k, 1)$ with an adaptive chosen-message attack, or

(2) $A \stackrel{\rm def}{=} U_2^F$ succeeds in $(t + q \cdot \tau(k, b, n),\, \epsilon/2)$-breaking $\mathcal{H}^+(k, b, n)$.

The proof of Theorem 7 is simple and is omitted.

The hardness of discrete log implies, via [15], the existence of standard (ie. non-incremental) signature schemes which can play the role of $\mathcal{S}^*$ in the above. Combining this with the results of Section 3.4 we have established the existence of an incremental signature scheme with short signatures given the hardness of the discrete log in groups of prime order. This construction however is not too practical because of the use of the result of [15]. For a practical version we could use El Gamal's scheme [10] or RSA in the role of $\mathcal{S}^*$ and the practical version of our hash function (cf. Section 3.4) in the role of $\mathcal{H}$.

The public file is large because the hash function has poly(n, k) size. But it isn't necessary that each user publish a hash function. Rather, some (trusted) center can publish a single hash function for use by all users. Now, a user's public file is just that of the original non-incremental scheme, and this is poly(k).

The tree hash scheme uses too much memory

The tree-hash scheme is probably the first thing that comes to mind when asked to find an incremental signature scheme.

Assuming for simplicity that $b = k$ we recall that the scheme makes use of a standard (ie. not necessarily incremental) collision-free hash function $H\colon \{0,1\}^{2k} \to \{0,1\}^k$. The message is hashed by the binary tree construction. That is, in each stage, adjacent blocks are hashed together to yield a single block, halving the number of blocks per stage. In $\lg(n)$ stages we have the final hash value. This can be signed under the standard scheme.

Now suppose we store all the internal nodes of the tree: formally, include them in the signature. Now the hash can he incremented by just recomputing the tree nodes indicated by the path from the updated block to the root of the tree.

The security needs again to be reconsidered because we allow the adversary to attack the update algorithm (cf. Section 4.3) but some thought shows that the scheme satisfies our basic security requirement.

But the signature is long— incrementality is at the cost of storing about twice as many bits as in the message. Thus while this scheme may be incremental under our formal definition, it is too memory inefficient to be interesting in most applications. We want schemes with short signatures and hence prefer the method of Section 4.4.

4.5 Security against substitution attacks

We provide here a stronger notion of security for incremental signature schemes, suitable for applications like remote editing a file on an insecure machine. We let $\mathcal{S}^+ = (\mathsf{KGen}, \mathsf{Sig}, \mathsf{Vf}, \mathsf{IncSig})$ be an augmented signature scheme. The adversary's incremental signing requests now have a new form: she supplies $M, \sigma, (j, m), \beta$. We first describe the experiment then provide explanation and discussion.

Experiment describing F's attack on $\mathcal{S}^+(k, b, n)$:–

(1) Run $\mathsf{KGen}(1^k, 1^b, 1^n)$ to get keys Sk, Vk.

(2) Initialize: Set $\alpha = 0$.

(3) Run the adversary F on input Vk. Her oracle queries are answered as follows.

  (3.1) Suppose F makes a simple signing request— this has the form of a message $M \in B_b^n$. Let $\sigma \stackrel{R}{\leftarrow} \mathsf{Sig}(Sk, M)$ and return σ to F. Let $\alpha \leftarrow \alpha + 1$ and let $M_\alpha \leftarrow M$.

  (3.2) Suppose F makes an incremental signing request— this has the form $(M, \sigma, (j, m), \beta)$ with $\beta \in \{1, \ldots, \alpha\}$. Let $\sigma' \stackrel{R}{\leftarrow} \mathsf{IncSig}(Sk, M, \sigma, (j, m))$ and return σ' to F. Let $\alpha \leftarrow \alpha + 1$ and let $M_\alpha \leftarrow M_\beta\langle j, m\rangle$.

(4) We ask that at the end of her execution, F output a pair (M, σ) such that $M \notin \text{Legal}$, where $\text{Legal} = \{ M_1, \ldots, M_\alpha \}$. We say that F is successful if $\mathsf{Vf}(Vk, M, \sigma) = 1$.

We say that F *succeeds in* $(t, q_{\text{sig}}, q_{\text{inc}}, \epsilon)$*-breaking* $\mathcal{S}^+(k, b, n)$ *with a substitution attack* if, in the above experiment, she runs for t steps, makes $q_{q_{\text{sig}}}$ simple signing requests, makes $q_{q_{\text{inc}}}$ incremental signing requests, and succeeds with probability at least ϵ.

Recall that the assumption in basic security was that when the signer applies the update algorithm it is to "authentic" data. We are assuming we are in a situation where this assumption is not realistic; for example, the data is on an insecure medium and when the signer accesses it to update a message and signature, he cannot be sure it has not been tampered with. In the worst case, he must assume it has been adversarially tampered with.

To model this the adversary is asked, as before, to point, via β, to that message out of the past on which she is requesting an update, and to supply the update request (j, m). The novel element is that she will additionally supply

M, σ, to be taken to mean that she has substituted these for M_β, σ_β. That is, she has tampered with the data.

The index β is not reflected in the way her query is answered— the answer is obtained by applying $\mathsf{IncSig}(Sk, \cdots)$ to the message M and accompanying σ, (j, m) that F provides. But β is used to update the signer's own "view" of what is happening. The idea is that that signer has "accepted" to update M_β according to (j, m), and thus has, from his point of view, willingly signed $M_\beta\langle j, m\rangle$. In other words, we can view the set Legal, at the end of the experiment, as being all those messages which the signer believes he has signed.

The notion of existential forgery says F is successful if she outputs a message M not previously queried of $\mathsf{Sig}(Sk, \cdot)$, and passing verification. We recall that the intuition is that "legitimately signed" messages are excluded. Thus according to the above discussion, we should declare F successful if she forges the signature of a message not in Legal.

Why would such an attack help the adversary? The reason is that IncSig was designed to be used on inputs $Sk, M, \sigma, (j, m)$ for which $\mathsf{Vf}(Vk, M, \sigma) = 1$, and we don't know what happens when this algorithm is run on strange inputs. One might ask why IncSig doesn't simply check that $\mathsf{Vf}(Vk, M, \sigma) = 1$. The reason is that in general this could defeat the efficiency we are trying to gain. For example, if IncSig is ideal it has only $\text{poly}(k, b, \log n)$ time and verification takes $\text{poly}(k, b, n)$ time.

It is important to note that we do not view the adversary as having legitimately obtained the signature of $M\langle j, m\rangle$— what the signer believes he has signed is $M_\beta\langle j, m\rangle$.

4.6 A successful substitution attack

We illustrate the strength of substitution attacks by showing how the scheme of Section 4.4 can be broken, in this setting, when we use, as the hash family, the one of Section 3.4. (In particular this means the scheme in question should not be used in applications like remote editing a file on a machine which could be unexpectedly hit by a virus).

The attack is interesting in illustrating how substitution attacks work. It is also interesting in illustrating how a "standard" construction like hash-and-sign which is secure in the usual sense fails to be secure in a new setting.

For simplicity assume the messages consist of just one block ($n = 1$): the attack easily generalizes to arbitrary n. The hash function is described by $(p; g)$ and reduces simply to $H(M) = g^{\langle M\rangle} = g^{1+M}$, the operations being in G_p. We let Sk^* be the signing key under the standard scheme, so that the signature of M is $\sigma = (g^{1+M}, \sigma^*)$ where $\sigma^* \stackrel{R}{\leftarrow} \mathsf{Sig}^*(Sk^*, g^{1+M})$.

The adversary F begins with the simple signing request A. The reply she obtains has the form $\sigma_A = (h_A, \sigma_A^*)$ where $h_A = g^{1+A}$. Think of it as the signer having signed A and stored A, σ_A on the insecure medium. We set $M_1 = A$.

Now, F make the incremental signing request $(B, \sigma_A, (1, C), 1)$. That is, on the insecure medium, she changes A to B, and asks the signer to substitute C for the first (and only) block of this message. According to our scheme, the signer first applies the hash update algorithm to update the hash: $h_F = h_A \cdot g^{-(1+B)} \cdot g^{1+C} = g^{1+A-B+C}$. Then he re-signs via $\sigma_F^* \stackrel{R}{\leftarrow} \mathsf{Sig}^*(Sk^*, h_F)$. The reply to F is $\sigma_F = (h_F, \sigma_F^*)$.

What is important to note at this point is that what the signer really believes himself to have signed is C. That is, in terms of the experiment of Section 4.5, we have $M_2 = C$. Thus, the adversary can simply output $(A - B + C, \sigma_F)$ as a forgery. The verification algorithm will accept σ_F as the signature of $A - B + C$. But at this point the set of messages whose signatures have been legally obtained is $\text{Legal} = \{A, C\}$. For appropriate choices of B, C (it suffices that $B \notin \{A, C\}$) it is the case that $A - B + C \notin \text{Legal}$. Thus the adversary is successful, and the scheme is broken with probability one.

Notice that the attack did not find collisions in H, nor did it forge signatures under Sk^*.

We don't know whether the attack applies to *any* instance of the hash-and-sign paradigm, but the above is sufficient to show hash-and-sign is not in general secure against substitution attack.

We leave as an open problem to design an incremental signature scheme secure against substitution attack, under the restrictions that the signature be short and the update algorithm be ideal. Some progress towards this question is described below.

5 Work in progress

In [3] we expand the scope of this research in the following directions. First, we consider more complex update operations on messages such as insertion (of a new block into the message) or deletion (of an existing block). These are clearly important in applications. Second, we consider other primitives such as finger-printing and message authentication. We appropriately extend the notion of subsitution attack to these contexts. Our main result is a finger-printing scheme which permits insertion and deletion and is secure against substitution attack.

Acknowledgments

We thank Hugo Krawczyk for many informative discussions on this materiel.

The research of the second author was partially supported by grant No. 92-00226 from the US-Israel Binational Science Foundation (BSF), Jerusalem, Israel. The research of the third author was partially supported by NSF FAW grant CCR-9023313 and DARPA grant N00014-92-J-1799.

References

1. M. Bellare, J. Kilian and P. Rogaway. The security of cipher block chaining. *Advances in Cryptology – Crypto 94 Proceedings.*

2. M. Bellare and P. Rogaway. Random oracles are practical: A paradigm for designing efficient protocols. *Proceedings of the First Annual Conference on Computer and Communications Security*, ACM, 1993.
3. M. Bellare, O. Goldreich and S. Goldwasser. Work in progress.
4. J. Bos and M. Coster. Addition chain heuristics. *Advances in Cryptology – Crypto 89 Proceedings,* Lecture Notes in Computer Science Vol. 435, Springer-Verlag, G. Brassard, ed., 1989.
5. S. Brands. An efficient off-line electronic cash system based on the representation problem. CWI Technical Report CS-R9323.
6. D. Chaum and H. Van Antwerpen. Undeniable signatures. *Advances in Cryptology – Crypto 89 Proceedings,* Lecture Notes in Computer Science Vol. 435, Springer-Verlag, G. Brassard, ed., 1989.
7. D. Chaum, E. Heijst and B. Pfitzmann. Cryptographically strong undeniable signatures, unconditionally secure for the signer. *Advances in Cryptology – Crypto 91 Proceedings,* Lecture Notes in Computer Science Vol. 576, Springer-Verlag, J. Feigenbaum, ed., 1991.
8. Croft and Harris. Public key cryptography and re-usable shared secrets. In *Cryptography and Coding*, Clarendon Press, 1989.
9. I. Damgård. Collision free hash functions and public key signature schemes. *Advances in Cryptology – Eurocrypt 87 Proceedings,* Lecture Notes in Computer Science Vol. 304, Springer-Verlag, D. Chaum, ed., 1987.
10. T. El Gamal. A public key cryptosystem and a signature scheme based on discrete logarithms. *IEEE Trans. Info. Theory,* Vol. IT 31, 1985.
11. S. Even, O. Goldreich and S. Micali. On-line/Off line digital signatures. Manuscript. Preliminary version in Crypto 89.
12. O. Goldreich and L. Levin. A hard predicate for all one-way functions. *Proceedings of the Twenty First Annual Symposium on the Theory of Computing,* ACM, 1989.
13. S. Goldwasser, S. Micali and R. Rivest. A digital signature scheme secure against adaptive chosen-message attacks. *SIAM Journal of Computing,* 17(2):281–308, April 1988.
14. R. Impagliazzo, L. Levin and M. Luby. Pseudo-random generation from one-way functions. *Proceedings of the Twenty First Annual Symposium on the Theory of Computing,* ACM, 1989.
15. M. Naor and M. Yung. Universal One-Way Hash Functions and their Cryptographic Applications. *Proceedings of the Twenty First Annual Symposium on the Theory of Computing,* ACM, 1989.
16. R. Rivest. The MD5 message-digest algorithm. *IETF Network Working Group,* RFC 1321, April 1992.
17. C. Schnorr. Efficient identification and signatures for smart cards. *Advances in Cryptology – Crypto 89 Proceedings,* Lecture Notes in Computer Science Vol. 435, Springer-Verlag, G. Brassard, ed., 1989.

A Proof of Theorem 4

We first describe the algorithm $B = U^A$. Then we argue that its running time is as claimed and finally that its success probability is as claimed.

On inputs p, g, x algorithm B selects $r_1, \ldots, r_n \in \{0,1\}$ at random and $u_1, \ldots, u_n \in \{0, 1 \ldots, p-1\}$ at random. For $i = 1, \ldots, n$ it sets

$$g_i = \begin{cases} g^{u_i} \text{ if } r_i = 0 \\ x^{u_i} \text{ if } r_i = 1 . \end{cases}$$

It sets $H = (p\,;\, g_1, \ldots, g_n)$. Now it invokes $A(H)$ and obtains distinct messages

$$M_1 = M_1[1] \ldots M_1[n] \text{ and } M_2 = M_2[1] \ldots M_2[n] . \tag{1}$$

For $j = 1, 2$ it is now convenient to set $t_{j,i} = \langle M_j[i] \rangle$. Algorithm B sets $a = \sum_{r_i=1} u_i(t_{1,i} - t_{2,i})$, the arithmetic here being modulo p. If this quantity is 0 then B has failed, and it halts with no output. So assume it is non-zero. Now compute an inverse b of a mod p. (That is, $ba \equiv 1$ mod p. Such an inverse always exists since p is prime, and it can be found via Euclid's algorithm). B outputs $\alpha = b \cdot \sum_{r_i=0} u_i(t_{2,i} - t_{1,i})$ mod p and halts.

B invokes A once. In addition it performs some arithmetic modulo p of which the dominant part is $O(n)$ exponentiations. This accounts for the claimed running time. We now turn to justifying the claimed success probability.

Note that the distribution of $g_1, \ldots, g_n$ is uniform and independent and is the same as the distribution over these quantities that HGen would generate. So the messages found by B in Equation 1 are a collision —ie. $H(M_1) = H(M_2)$— with probability at least ϵ. Now assuming they are a collision we have

$$\prod_{i=1}^{n} g_i^{t_{1,i}} = \prod_{i=1}^{n} g_i^{t_{2,i}} .$$

Using the definition of $g_1, \ldots, g_n$ and re-arranging terms in the above we get

$$\prod_{r_i=1} x^{u_i(t_{1,i}-t_{2,i})} = \prod_{r_i=0} g^{u_i(t_{2,i}-t_{1,i})} .$$

Note that the left hand side is x^a. We now claim that with probability at least 1/2 we have $a \neq 0$. Given this, raise both sides of the above equation to the power b to get

$$x = x^{ab} = \prod_{r_i=0} g^{bu_i(t_{2,i}-t_{1,i})} = g^{\alpha} ,$$

showing that α is indeed $\mathsf{index}_g^{G_p}(x)$. It remains to justify the claim. We will argue this informally. We will use the following technical fact.

Technical Fact. Let $a_1, \ldots, a_n$ be numbers with the property that $\sum_{i=1}^{n} a_i \neq 0$. Let $X_1, \ldots, X_n$ be independent random variables defined by $\Pr[\, X_i = a_i \,] = \Pr[\, X_i = 0 \,] = 1/2$ for each $i = 1, \ldots, n$. Let $X = \sum_{i=1}^{n} X_i$. Then $\Pr[\, X \neq 0 \,] \geq 1/2$.

We note that the distribution on $g_1, \ldots, g_n$ is independent of $r_1, \ldots, r_n$. Thus we may think of the experiment as the following game. We choose $g_1, \ldots, g_n$ at random and obtain the collision from A. We let $a_i = u_i(t_{1,i} - t_{2,i})$ for $i = 1, \ldots, n$. Now we choose $r_1, \ldots, r_n$ at random and compute $\sum_{r_i=1} a_i$. Viewed this way we can see it is the same as the technical fact stated above.

An Efficient Existentially Unforgeable Signature Scheme and its Applications

Cynthia Dwork[1] and Moni Naor[2] *

[1] IBM Almaden Research Center, USA. e-mail:dwork@almaden.ibm.com
[2] Weizmann Institute, Israel. e-mail: naor@wisdom.weizmann.ac.il

Abstract. We present a practical existentially unforgeable signature scheme and point out applications where its application is desirable. A signature scheme is *existentially unforgeable* if, given any polynomial (in the security parameter) number of pairs

$$(m_1, S(m_1)), (m_2, S(m_2)), \ldots (m_k, S(m_k))$$

where $S(m)$ denotes the signature on the message m, it is computationally infeasible to generate a pair $(m_{k+1}, S(m_{k+1}))$ for any message $m_{k+1} \notin \{m_1, \ldots m_k\}$. We have developed a signature scheme that requires at most 6 times the amount of time needed to generate a signature using RSA (which is *not* existentially unforgeable).

1 Introduction

Consider the problem of providing a "receipt" for data stored in a *document repository*, where the data can be of arbitrary form, much as one is provided with a claim check at a left luggage counter. In the most simple implementation, the receipt should just be a pair consisting of an identifier and a signature on this identifier. If the signature scheme is existentially *forgeable*, then anyone can produce what appears to be a valid receipt, or claim check; an existentially unforgeable signature scheme prevents this.

In order to ensure that the retrieved document is authentic, it should be signed by the owner. Suppose, for example, that the document is a will. If the signature is computed using an existentially forgeable scheme, then the will can be challenged by anyone producing a possibly nonsensical "will" and a signature on this "will." However, with an existentially unforgeable signature scheme any signed document, nonsensical or otherwise, has necessarily been signed by the claimed signer.

Our interest in finding efficient existentially unforgeable signature schemes comes from the problem of signing FAXed documents. Since FAXed documents have received legal standing in court, it is essential to find a signature scheme

* Research Performed when this author was with the IBM Almaden Research Center.

appropriate to this environment. Let D be a document, let h be a collision-intractable hash function, and let S be a signing function. Assume the sender sends to the receiver $D, h(D), S(h(D))$. Even if the receiver can check that the document has been correctly hashed and that the signature on the hash is valid, once D is printed out on a FAX machine there is no way to re-capture D from the printed image. For example, scanning the printed image optically will probably produce some $D' \neq D$, since the scanned image may be slightly tilted, or dirty, etc. Either the receiver must have a small disk drive, or even a tape drive on which to store D, or some other party must store the data and issue a receipt to the receiver. In particular, the (not necessarily trustworthy) sender could store the data; the receipt could be the pair $(h(D), S(h(D)))$. Since $h(D)$ looks "random," the signature scheme must be existentially unforgeable, otherwise, as in the claim check example, anyone could generate what appears to be a valid receipt[3].

In this paper we present an efficient existentially unforgeable signature scheme which we believe is the first practical one. The security of the scheme relies only on the RSA assumption: *it is computationally infeasible to extract pth roots mod N, where p is a prime and N is a product of two large primes.* The cost of implementing our scheme is close to that of RSA: for all reasonable parameters, the cost of signing and verifying is only six times that of RSA. Thus in almost every scenario where it is feasible to apply RSA it should be feasible to use our scheme.

The paper is organized as follows: in the next section we describe the history of digital signatures, emphasizing work relevant to our scheme. In Section 3 the scheme is presented. Section 4 provides the outline of the proof of security. In Section 5 we describe how to use the proposed scheme in the context of signing FAXes.

2 Related Work

Since the introduction of the concept of digital signatures by Diffie and Hellman [3] and the first proposals of candidates for implementation [13, 17] the subject has been widely studied. We briefly outline the major developments (not necessarily in chronological order), especially those pertaining to the scheme proposed in this paper.

Goldwasser, Micali, and Rivest [10] formalized the notion of security of a signature scheme. The highest form of security they proposed was called *existentially unforgeable under an adaptive chosen plaintext attack*. In this attack the adversary (or forger) gets to see a signature on any message of its choice, in an adaptive manner. The forger has then to produce, *without* the cooperation of the signature algorithm, a signature on *one* message that was not previously signed. (See exact definition in [10]; we use their definitions in this paper.) This

[3] Actually, at the cost of complicating the protocol, we can assume h is chosen, by the receiver, from a family of universal one-way hash functions [15]. This is discussed further in Section 5.

notion was considered too strong for practical purposes, but as the application to signing FAXed documents briefly sketched in the Introduction shows, it does appear in the "real-world". The RSA [17] and Rabin's [16] schemes are known to be *not* existentially unforgeable.

The implementation of an existentially unforgeable signature scheme suggested in [10] was based on the hardness of factoring. Constructions based on weaker assumptions (trapdoor permutation, 1-1 one-way functions and one-way functions) were given in [1, 15, 18]. These schemes are all rather inefficient. For instance, the scheme in [10] employs a tree whose height is proportional to the length of the message (or some digest of it). [15, 18], who use Merkle's authentication tree method [12], need a tree whose height is a least the logarithm of the total number of messages signed by the system. Signing and verifying both involve tracing a path from the root to a leaf, where moving from node to node is quite expensive (e.g. an RSA computation in [10], as optimized in [9]).

Our scheme employs a tree as well. However it is very shallow one; in a lifetime of a system it is very unlikely to need more than three levels.

There are several constructions of one-time or fixed-time signature schemes that are existentially unforgeable. (One-time means that the public key is good for one signature only; fixed-time means that there is an *a priori* upper bound on the number of messages the scheme can sign. The size of the public key is usually related to this number.) These are the original Lamport scheme (described in [3] and used in [12, 15, 18]). This requires as many invocations of a one-way function as there are bits to be signed (Some improvements are known). The scheme of Bos and Chaum [2] can be considered as a fixed-time signature scheme. The size of the public information needed grows at least as fast as the square root of the number of messages the scheme should be able to sign. See more about it below.

Even, Goldreich, and Micali [5] tried to combat the computational cost of signature schemes by distinguishing between on-line and off-line computation. Their scheme requires extensive pre-computation, "between" signing of different documents, but the on-line computation required for signing is very efficient. The size of a signature is rather large.

The El-Gamal scheme [4] relies on no cleanly specified function; moreover, given a legitimately signed document in that scheme, it is possible to generate other legitimate signatures and messages and thus is not existentially unforgeable. The scheme of Fiat and Shamir [7], and its descendants [8, 14, 19], are very efficient, since, unlike RSA and related schemes, they do not require modular exponentiation. However, they do require that the "one-way hash" function actually be something stronger, more like a black-box random function (no precise definition of the assumptions needed appeared). None of them is known to be existentially unforgeable.

Most of the ingredients of our scheme have appeared before; the contribution of this work is merely in finding the right mixture that makes the full scheme efficient. The idea of using exponentiation to hide information appears in the original RSA signature scheme [17]. Fiat and Shamir employ the subset product technique for signing [7]. Merkle [12] suggested the tree authentication scheme, but in his scheme the tree cannot be shallow. The scheme in [2] is similar in spirit to the one-time version of our scheme used in every node. Bellare and Micali [1] suggested a tree based scheme where nodes are "revived" by choosing a new trapdoor permeation which, in turn, is authenticated by the parent of the current node. Our scheme can be seen as an efficient way of performing this, by replacing the trapdoors of [1] with "masks" from the Fiat-Shamir scheme [7].

3 The Scheme

3.1 Outline

In rough outline, the scheme works as follows (details are given in the next section). Every signer s has, as in any signature scheme, a pair of keys. The *public* key is used to verify the signature on messages supposedly signed by i. The *private* key contains information known only to s, and is used by s to compute signatures of given messages.

The signer maintains a short, very bushy tree. Every message is associated with a leaf of the tree. Thus, the tree will be quite large; however, as we shall see, at any time the signer need only store information associated with a single root-leaf path in the tree. For concreteness, we take the outdegree of the tree to be $l = 1000$. If the tree is of height 3 then it has a billion leaves, (which should be sufficient to sign all the message's s should desire). Associated with every node is a 1000-bit number. This number is random for all internal nodes and a message for the the leaves. Assume all messages have length at most one thousand bits (larger messages may be hashed down to this length or broken into pieces, with each piece signed individually). A signer's public key is an integer N, the product of two large primes, and y_{root}, the random number associated with the root of the tree. (y_{root} can be the same for all signers, as long as it chosen as a random l-bits number initially.) At first all leaves are unused. To sign a message m, the signer associates m with the leftmost unused leaf, say w, and authenticates the path from the root to w. That is, starting with the root, for each step along the (length 3) path from root to leaf, the signer authenticates the string associated with the child, using the string associated with the parent. This is done via a basic authentication step which involves using subset product [7] and extracting pth roots [17].

Intuitively, the reason the scheme is practical is that we have found a way to "re-use" an internal node many times (once for each leaf in the subtree rooted at this node).

Common to all users (signers and verifiers) are two lists: P contains one thousand primes, and X contains one thousand random strings. Let v be an internal node in the tree, with associated string y_v, and let w be the jth child of v. In the basic authentication step we use the bits of the string associated with w to select elements of X. We compute the product of y_v and the selected elements. Finally, we compute the p_jth root of this product. Note that since the values associated with the internal nodes are all random, there is virtually no chance that any two such strings are identical. Thus, we never use the same y_v, p_j pair twice in a basic authentication step. This is critical in obtaining existential unforgeability.

3.2 Detailed Description

Let l be the security parameter of the scheme (l should be such that it is infeasible to factor l-bit numbers). In practice we would take l to be about 1000, though of course we can take it to be larger. There are two sets $X = \{x_1, x_2, \ldots, x_l\}$ and $P = \{p_1, p_2, \ldots, p_l\}$ of integers. The x_i's are random integer of length at most l. The p_i's are primes, they can be either the l smallest primes or l random primes of length l (This depends on the RSA assumption we are making, whether it is " it is hard to extract pth roots mod$N$ for small p" or "it is hard to extract pth roots mod$N$ for random p"). These lists are fixed and the same for all signers. All signers and verifiers should have access to them.

Each signer maintains a tree of height d and outdegree l. As mentioned above, d need not be larger than 3, since l^d is a lower bound on the number of messages it is possible to sign. Associated with each internal node v is an l-bit number y_v. Associated with each leaf is an l-bit string representing a message of up to l bits. The value associated with the root, y_{root}, is random and is part of the public information, like P and X. The value associated with a non-root internal node is chosen at random and on the fly, and is authenticated in a basic authentication step that uses the (authenticated) value associated with the parent node. The y's within any given tree must be distinct; there is virtually no chance that the same y will be chosen twice. Similarly, a message, which is conceptually just a value associated with a leaf, is authenticated using the the value stored at the parent node. The full signature of a message is the path from the root to the leaf, together with the basic authentication steps for all values associated with the nodes along the path.

The public key: the public key of each signer s is a number N_s, where N_s is an l bit number which is a product of two primes. It is important to chose the primes at random (and of course independent of the list X).

The basic authentication step: For the basic authentication step at a non-root node v, let y_v be the value associated with v (a message, if v is a leaf), and let z_v be the value associated with the parent of v. We use the bits of y_v as selectors of the elements of X. Let y_{vi} denote the ith bit of y_v, for $1 \leq i \leq l$. Let v be the jth child of its parent. Then $auth(y_v)$ is given by

$$\left(z_v \prod_{y_{vi}=1} x_i \right)^{1/p_j} \mod N_s.$$

The authentication of y_v can be verified as follows. Given a string α purported to be $auth(y_v)$, and given also z_v, y_v, j, and the public lists P and X, one computes

$$z_v \prod_{y_{vi}=1} x_i \mod N_s$$

and checks that it is equal to $\alpha^{p_j} \bmod N_s$.

A signature of a message m associated with leaf w is composed of m, the path from the root to w (i.e. the indices of the children), and for each node v along the path (excluding the root and including w), y_v (the value associated with v) and its authentication, as computed in the basic authentication step just described. The signature is verified by verifying the basic authentication steps for all the nodes on the path.

3.3 Computational Requirements:

The complexity time of signing

Signing a message involves

- d RSA computations (i.e. exponentiations mod N_s)
- d subset multiplications, i.e. multiplying a random subset of l numbers. This is roughly equivalent to modular exponentiation.

Since we can assume that $d = 3$, we say that the complexity of the scheme is six times that of RSA. The complexity of verifying a signature is similar.

Size of signature

A valid signature consists of d numbers, each l bits long, plus d numbers to describe the path from the root to the leaf which are $\log l$ bits each. Therefore the size of a signature is roughly 6 times that of RSA.

Size of public key and storage requirements

The size of public key is an l-bits number (similar to RSA). A cost that RSA does not have is the storage of the lists X and P, common to all users, which require $2l^2$ bits altogether. This is roughly 1,000,000 bits which should be accessible to every user. This is feasible if both signer and receiver are a "full" computer, but may be an obstacle in using the scheme in a smartcard environment. However, as smartcards are getting more powerful, storing 128K bytes in ROM on a smartcard is not impossible.

Apart from the lists X and P, the memory needed to run the signature scheme is not large, one should essentially maintain a path from the root to a leaf, i.e $d \log l$ bits.

3.4 Remarks on Implementation

Consider a particular path in the tree. Since the tree is of height 3, the path has 3 internal nodes: the root, and two others, say, v and w. Let y_{root}, y_v, and y_w be the values associated with these internal nodes. Then, although the first time a message associated with a child of w is signed, the signer must perform the computation needed to authenticate y_w using y_v, this information can be saved and used for the remaining $l - 1$ (about 1000) messages associated with children of w. Similarly, the authentication of y_w using y_{root} can be re-used l^2 times (about 1,000,000 times).

The p_i's may be chosen to be small in order to speed up verification of a signature. Furthermore, Fiat [6] suggested a way of amortizing RSA computation. His method fits very well with our scheme, since we use different roots anyway.

Other suggestions to speed up RSA and the Fiat-Shamir signature scheme are applicable to our scheme as well. The signer can do its computation modulo each of the factors of N_s separately and combine them using Chinese remaindering. To speed up the subset product one can pre-process the list X and partition X to small sets and for every set compute all products of its subsets. This can decrease significantly the time spent on the subset multiplication.

4 Security of the Scheme

Our definition of security is that of Goldwasser, Micali and Rivest [10]. We only sketch why the proposed scheme obeys this definition. The proof of security is by showing that the ability to generate a single $(m, S(m))$ pair, even for a nonsensical m, for an m on which the signer was not explicitly asked to sign, violates the RSA assumption: it is computationally infeasible to extract pth roots mod N, where p is a (random) prime of length l and N is a product of two large primes chosen at random, also of length l.

Intuitively, the security of the scheme rests on the important observations made in [20] and [7]:

- Having a black box that computes x^{1/p_1} mod N for random x does not help in evaluating x^{1/p_2} mod N, if p_1 and p_2 are relatively prime.
- For numbers $x_1, x_2, \ldots x_l$, for arbitrary subset $S \subset \{1, \ldots, l\}$ and random $y \in Z_N$ the value of $\left(y \prod_{i \in S} x_i\right)^{1/p}$ mod N yields no information about any of the $x_i^{1/p}$.

Suppose that the scheme can be broken, i.e. there is a algorithm $\mathcal{A}$ that operates in time T and has probability ρ of breaking the scheme. We show that there is an algorithm $\mathcal{B}$ that works in expected time $O(T)$ and can extract pth roots with probability at least ρ/l^2.

The input to $\mathcal{B}$ is x, N and p. The algorithm consists of two phases, a preprocessing phase where the public key and public information are generated, and a simulation phase where the algorithm $\mathcal{A}$ is simulated on the public key generated in the previous phase.

If a tree based scheme (i.e a system where parents vouch for the authenticity of there children) is broken, then there must be the first time an illegitimate value (i.e. a value not authenticated by the signer) is authenticated at some node w. we can guess with probability $1/l$ at which child $1 \leq j \leq l$ of w this will occur. Furthermore, we can guess with probability at least $1/l$ an index $1 \leq i \leq l$ where the legitimate value authenticated with w and j and the forged value differ.
Preprocessing phase

1. Choose random $1 \leq i \leq l$ and $1 \leq j \leq l$.
2. Set $N_s = N$.

3. Choose random l-bit primes $p_1, p_2, \ldots p_{j-1}, p_{j+1}, \ldots p_l$. Set $p_j = p$ and let $P = \{p_1, p_2, \ldots, p_l\}$.
4. To choose $x_1, x_2, \ldots x_{i-1}, x_{i+1}, \ldots x_l$, generate $l-1$ random values in Z_{N_s} and raise them to the power $p_1 \cdot p_2 \cdots p_l$ modulo N_s. Thus all the p_kth roots of the elements $X \setminus \{x_i\}$ are known and can be computed efficiently. Choose x_i to be $(r \cdot x)^{p_1 \cdots p_{j-1} \cdot p_{j+1} \cdots p_l} \bmod N_s$, for a random $r \in Z_{N_s}$. The important thing to note is that given x_i^{1/p_j} it is possible to extract $(r \cdot x)^{1/p_j}$ from it (see [20]) and hence x.
5. We now determine y_v for all internal nodes v that are ancestors of the first T (the upper bound on the number of steps of $\mathcal{A}$) leaves. The tree is constructed bottom up. Our intuition is: given that v is the jth child of its parent, then
   - if $y_{vi} = 1$ then z_v (the value associated with the parent of v) should be chosen as $\beta_v^{p_1 \cdot p_2 \cdots p_l} / x_i$ for random $\beta_v \in Z_{N_s}$.
   - if $y_{vi} = 0$, then z_v should be chosen as $\beta_v^{p_1 \cdot p_2 \cdots p_l}$ for random $\beta_v \in Z_{N_s}$.

   Though the above procedure suggests a bottom-up method for choosing the y_v, we prefer the following two steps description: start with assigning random bits $b_u \in \{0, 1\}$ as the values of the y_{ui}'s (i.e. the ith bit of y_u) to all the nodes u of the tree. Note that for a leaf u, b_u is necessarily a guess, and is correct with probability 1/2. However, all the other b_u's will indeed be the values of y_{ui}.

   Given an assignment b_u to y_{ui} for all node u, it is possible to quickly find a solution to the y_v's together with the appropriate β_v's satisfying the requirement that $y_{vi} = b_v$'s for all *internal* nodes v: choose $\beta_v \in Z_{N_s}$ at random and set y_v by the rule above. With probability 1/2 $y_{vi} = b_v$ (i.e. it has the right y_{vi}). If they are equal, then accept y_v and continue with the preprocessing; otherwise, repeat until successful.
6. Fix y_{root} as the value the above procedure gave to the root.

Before we proceed in describing the simulation we should note that in the description above there is an inaccuracy: we choose values $z \in Z_N$ at random (where z stands for either the x_k's or the y_v's), whereas in a regular execution of $\mathcal{A}$ it should be that z is a random l-bit number. This can be corrected by replacing z by a random z' of length l such that $z = z' \bmod N$. However, this gives a certain advantage to the z's such that $z > (2^l \bmod N)$ (since there are fewer z''s that $z = Z' \bmod N$). Hence we reject such z's (and repeat the process we used to choose it) with probability proportional to their bias - $\frac{\lfloor 2^l/N \rfloor}{\lceil 2^l/N \rceil}$. This cannot increase the expected work by more the a factor of 2.

The simulation

1. present $\mathcal{A}$ with (X, P, N_s, y_{root}) as determined in the preprocessing phase.
2. Start the simulation of $\mathcal{A}$; at every step t, $\mathcal{A}$ provides the signer with a message m_t and requests a signature. The algorithm $\mathcal{A}$ receives as requested a signature on m_t using the path to the tth leftmost leaf of the tree chosen in the preprocessing. The signature is generated according to the following:
   - To handle a node which is not the jth child of its parent is not a problem, since we can extract all p_kth roots when $k \neq j$.

- To handle a node which is a jth child but is an *internal* node can also be done easily because of the way y_v was chosen in step 5 in the preprocessing phase.
- When handling a leaf u that is the jth child of its parent v, then with probability 1/2 the incorrect b_u was chosen (i.e. not fitting the ith bit of the message m_t for which $\mathcal{A}$ requests a signature). In this case
  - Rewind $\mathcal{A}$ for j steps back, to a stage just before parent v is used for the first time.
  - Choose a new value to b_u at random.
  - Choose a new value y_v, where v is the parent of u; the value of y_v should obey the restriction b_v on y_{vi} imposed in the preprocessing step, thus preventing further propagation of the rollback.

  As discussed below, the rewinding may increases the *expected* run-time by a factor of at most two.

Claim 4.1 *The distribution the $\mathcal{A}$ in the simulation witnesses (i.e. the signatures) is the same distribution $\mathcal{A}$ sees in a regular execution.*

Proof. The lists X is of l numbers uniformly chosen from the l-bits numbers. The list P is a list of l primes of length l, given that the input p was chosen at random (which is our assumption). All the y_v are uniformly distributed l-bits numbers. Hence, by induction on the number of steps the distribution $\mathcal{A}$ witnesses in the simulation is the same as in a regular execution.

Claim 4.2 *The expected time to run the simulation is $O(T)$.*

Proof. There are two possibilities that may force us to spend more time than T: one is Step 5 of the preprocessing phase where we may fail in choosing y_v that satisfies the requirement on y_{vi}. However this happens with probability 1/2 and hence doubles the expected amount of work. The second possibility is in Step 2 of the simulation, where we may have to rewind $\mathcal{A}$ for j steps. However, as before, this happens with probability 1/2 and forces an expected increase of $2j/l \leq 2$ of the work.

Claim 4.3 *The probability of success is at least ρ/l^2.*

Proof. Since the $\mathcal{A}$ in our simulation sees the same distribution as in a regular execution, the probability it breaks the system is ρ. furthermore, the distribution $\mathcal{A}$ in the simulation sees is independent of i and j chosen by $\mathcal{B}$. If a $\mathcal{A}$ breaks the signatures scheme there is the first time an illegitimate value is authenticated by some node w. By illegitimate we mean that the signer in the simulation authenticated a different value at w. Hence, with probability at least $1/l^2$ we have that the forgery occurred at a jth child and the value authenticated by the signer and the value forged differ at the ith bit. If this is the case, then we can extract $x^{1/p} \bmod N$ from the two authenticated values.

We therefore can conclude

Theorem 1. *Any algorithm for breaking the provided signature scheme can be used at a similar cost and probability of success to extract modular roots.*

5 Application to Signing Faxed Documents

We assume that at least one of the sender or the receiver is using a computer-FAX, or *CFAX*, system. There are three scenarios to consider: CFAX to CFAX (straightforward) CFAX to FAX (interesting), and FAX to CFAX (same as CFAX to CFAX). As mentioned in the Introduction, the CFAX to FAX scenario is interesting because the receiver cannot store the bit-stream image of the FAXed document, and yet this is the data that has been signed. Thus, the sender must store the data, and issue to the receiver some type of receipt.

We next discuss the different scenarios in more detail. We do not assume the existence of a collision-intractable hash function. Rather, we rely only on the existence of a family of universal one-way hash functions [15]. These functions have the following property: *Fix a document D. Let h be chosen at random from a family of universal one-way hash functions. Then it is computationally infeasible to find a document D' such that $h(D) = h(D')$*[4].

Impagliazzo and Naor have described a one-way hash function based on subset sum [11]; in particular, breaking the assumed universal one-way hash property of this family of functions is as hard as solving subset sum. They also suggested less efficient schemes based on factoring.

Throughout this section, we assume the existence of an existentially unforgeable signature scheme secure against chosen plaintext attack, such as the one described in Section 3.

5.1 When the Receiver is a CFAX

For documents prepared on-line this is straightforward. The receiver doesn't really care who is actually sending the transmission, since when it verifies the signature it knows that the claimed sender actually signed the document originally. The procedure for sending a document D is as follows. The sender chooses at random a universal one-way hash function h hashing documents down to *fingerprints* of, say, 1000 bits. Letting K_S (respectively, L_S) be the public (resp. private) signature key of the sender, the sender sends to the verifier S (the sender's name), D (the document), and $L_S(h(D))$ (a signed fingerprint of D).

The properties of h make it virtually impossible for the receiver to find a document D' with the same fingerprint as D, *i.e.* , satisfying $h(D') = h(D)$[5]. Since the signature scheme is existentially unforgeable, it is virtually impossible to produce $L_S(m')$ for any m', despite seeing any (polynomial) number of other signed messages (or hashes of messages) $L_S(m_1), L_S(m_2), \ldots$ without knowing the private key.

---

[4] This is weaker than collision-intractability, which allows D to be chosen after h is known.

[5] Note that, if the sender did not choose h at random, then the properties of UOWHFs do not prevent finding such a D'. However, the sender is protected by choosing h at random, and the receiver is protected because it is the responsibility of the receiver to store and later to produce D together with $h(D)$.

If the document is sent as a series of blocks, or pages, then each page must have an identifying header and a page number, as in all legal documents (e.g., "Page 2 of 5"). The identifying nonce can include things like time of day.

Physical documents can be scanned into the sender's machine and treated from then on as if prepared on-line.

5.2 The Procedure for CFAX to FAX

Here we have a CFAX machine C sending a document to an ordinary FAX F. In this scenario the receiver cannot alone convince a third party that the document was signed by the claimed sender, but it can show to a third party a signed statement promising cooperation in the judgment process. This weakness can be circumvented by adding a tape drive to the box. The idea is that the tapes are used only if the sender refuses to cooperate. As pointed out below, refusal on the part of the sender to cooperate may actually permit the receiver to forge, so it is definitely in the sender's interest to cooperate. In the following, we do not assume that the receiver has a long-term storage device.

To avoid having to maintain and access a directory of public keys, there can be a central agency with whom public keys are registered. The central agency has its own pair of keys, K_{Center}, L_{Center}, where all users know K_{Center} (rather than having to know all public keys).

The CFAX sender, C, first forwards to the recipient the statement L_{Center} ("C's public signature key is K_C"), where L_{Center} is the private key of the trusted center. The recipient knows K_{Center} and can therefore be certain of using the correct public key for F. The agents proceed as follows.

1. F chooses at random a UOWHF h. It *does not* reveal this to C.
2. C sends to F the document D; F hashes D on-line, computing $h(D)$ and temporarily saving this; it also prints D.
3. F sends h to C.
4. C computes $h(D)$ and sends $L_C(h), L_C(h(D))$ to F.
5. Let the pair received by F be α, β. F checks that h is indeed the function it sent to C above. It verifies that $\alpha = L_C(h)$. It verifies that $\beta = L_C(h(D))$, using $h(D)$ that it computed and stored above. F then prints $h, L_C(h), h(D), L_C(h(D))$ in hex or a bar code.

For particularly important transactions C may store D on tape. This is discussed next.

Handling Disputes The tape should be used only as a last resort. Since the signature scheme is existentially unforgeable, the tuple $h, L_C(h), h(D), L_C(h(D))$ constitutes a promise by C to produce a document that hashes to $h(D)$, since F cannot generate the first two components of the triple because h is chosen at random. Moreover, D was fixed by C without knowledge of h, so the only satisfactory document that C can produce is D. We are relying here on the properties of universal one-way hash functions: If D is fixed and h is chosen at random, then it is "impossible" to find a D' such that $h(D') = h(D)$.

If the sender C refuses to produce D, then if D has been stored on tape F can produce the tape. However, since F is the one that chose h this exposes C to "forgery," in that if F had been dishonest and chosen h dependent on D (after Step 2, rather than before Step 2), then F may be able to produce $D' \neq D$ such that $h(D') = h(D)$.

References

1. M. Bellare and S. Micali, *How to Sign Given Any Trapdoor Function*, Proc. 20th ACM Annual Symposium on the Theory of Computing, 1988, pp.32-42.
2. J. Bos and D. Chaum, *Provably Unforgeable Signatures*, Proc. Advances in Cryptology - Crypto'92 Proceedings, Springer Verlag, 1993, pp. 1–14.
3. W. Diffie and M. Hellman, *New Directions in Cryptography* , IEEE Trans. on Information Theory 22(6), 1976, pp. 644-654.
4. T. El Gamal, *A Public Key Cryptosystem and a Signature Scheme Based on Discrete Logarithms*, IEEE Trans. Inform. Theory, IT-31(4), 1985, pp. 469–472
5. S. Even, O. Goldreich, and S. Micali, *On-line/Off-line Digital Signatures*, Proc. Advances in Cryptology – Crypto '89, Springer Verlag, pp. 263–275, 1990.
6. A. Fiat, *Batch RSA*, Proc. Advances in Cryptology – Crypto '89, Springer Verlag, 1990.
7. A. Fiat and A. Shamir, *How to Prove Yourself*, Proc. of Advances in Cryptology - Crypto '86, Springer Verlag, 1987, pp. 641–654.
8. A. Fiat and A. Shamir, *Method, Apparatus, and Article for Identification and Signature*, United States Patent 4,748,668 (5/31/88)
9. O. Goldreich, *Two Remarks Concerning the Goldwasser-Micali-Rivest Signature Scheme*, Proc. Advances in Cryptology – Crypto '86, Springer Verlag, 1987.
10. S. Goldwasser, S. Micali, and R. Rivest, *A Digital Signature Scheme Secure Against Adaptive Chosen-Message Attacks*, SIAM J. Computing 17(2), pp. 281–301, 1988.
11. R. Impagliazzo and M. Naor, *Efficient Cryptographic Schemes Provably as Secure as Subset Sum*, Proc. of the 30th Symp. on Foundations of Computer Science, 1989, pp. 236–241. Full version: Technical Report CS93-12, Weizmann Institute.
12. R. Merkle, *A Digital Signature Based on a Conventional Encryption Function*, Proc. Advances in Cryptology – Crypto '87, Springer Verlag, 1988, pp. 369–378.
13. R. C. Merkle and M. Hellman, *Hiding information and Signature in Trapdoor Knapsack*, IEEE Transaction on Information Theory, Vol 24, 1978, pp. 525–530.
14. S. Micali and A. Shamir, *An Improvement of the Fiat-Shamir Identification and Signature Scheme*, Proc. Advances in Cryptology – Crypto '88, *LNCS 403*, Springer-Verlag, pp. 244–247, 1990
15. M. Naor and M. Yung, *Universal One Way Hash Functions and Their Cryptographic Applications*, Proc. 21st ACM Annual Symposium on the Theory of Computing, 1989, pp. 33–43.
16. M. O. Rabin *Digital Signatures and Public Key Functions as Intractable as Factoring*, Technical Memo TM-212, Lab. for Computer Science, MIT, 1979.
17. R. Rivest, A. Shamir, and L. Adelman, *A Method for Obtaining Digital Signature and Public Key Cryptosystems*, Comm. of ACM, 21 (1978), pp. 120-126.

18. J. Rompel, *One-way Function are Necessary and Sufficient for Signatures*, Proc. 22nd ACM Annual Symposium on the Theory of Computing, 1990, pp. 387–394.
19. C. P. Schnorr, *Efficient Signature Generation by Smart Cards*, J. Cryptology 4, pp. 161–174, 1991.
20. A. Shamir, *On the Generation of Cryptographically Strong Pseudo-Random Number Sequences*, ACM Trans. Comput. Sys., 1 (1983), pp. 38–44.

Bounds for Resilient Functions and Orthogonal Arrays

Extended Abstract

Jürgen Bierbrauer[1], K. Gopalakrishnan[2] and D. R. Stinson[2,3]

[1] Mathematisches Institut der Universität
Im Neuenheimer Feld 288, 69120 Heidelberg, Germany
[2] Department of Computer Science and Engineering
University of Nebraska - Lincoln, Lincoln, NE 68588
[3] Center for Communication and Information Science
University of Nebraska - Lincoln, Lincoln, NE 68588

Abstract. Orthogonal arrays (OAs) are basic combinatorial structures, which appear under various disguises in cryptology and the theory of algorithms. Among their applications are universal hashing, authentication codes, resilient and correlation-immune functions, derandomization of algorithms, and perfect local randomizers. In this paper, we give new bounds on the size of orthogonal arrays using Delsarte's linear programming method. Then we derive bounds on resilient functions and discuss when these bounds can be met.

1 Introduction

Orthogonal arrays (OAs) are basic combinatorial structures. They and some natural generalizations appear under various disguises in cryptology and the theory of algorithms. Among the applications we mention *universal hashing* and *authentication codes*, *resilient* and *correlation-immune functions*, *derandomization of algorithms* and *perfect local randomizers*.

Here, we concentrate on resilient functions, two possible applications of which are mentioned in [2] and [6]. The first application concerns the generation of shared random strings in the presence of faulty processors. The second involves renewing a partially leaked cryptographic key (one setting in which this would be relevant is quantum cryptography [1]). Correlation-immune functions are used in stream ciphers as combining functions for running-key generators that are resistant to a correlation attack (see, for example, Rueppel [13]).

In this abstract, we give new bounds on the size of orthogonal arrays using Delsarte's linear programming method. Then we derive bounds on resilient functions and discuss when these bounds can be met. Complete proofs of all results, as well as further related results, can be found in the full paper [10].

The concept of binary resilient functions was introduced and studied in the papers Chor *et al* [6], Bennett *et al* [2], Stinson [14] and Friedman [8]. An (n, m, t)-resilient function is a function $f : \{0,1\}^n \longrightarrow \{0,1\}^m$ such that every possible output m-tuple is equally likely to occur when the values of t arbitrary inputs are fixed by an opponent and the remaining $n-t$ input bits are chosen independently at random. More formally, the property can be stated as follows: For every t-subset $\{i_1, \ldots, i_t\} \subseteq \{1, \ldots, n\}$, for every choice of $z_j \in \{0,1\}(1 \leq j \leq t)$, and for every $(y_1, \ldots, y_m) \in \{0,1\}^m$, we have

$$Pr(f(x_1, \ldots, x_n) = (y_1, \ldots, y_m)|x_{i_j} = z_j, 1 \leq j \leq t) = \frac{1}{2^m}.$$

A basic problem is to maximize t given m and n, i.e., to determine the largest value of t such that an (n, m, t)-resilient function exists. It was proved by Chor et al in [6] that an $(n, 2, t)$ - resilient function exists if and only if $t < \lfloor \frac{2n}{3} \rfloor$. The corresponding question for $m = 3$ or higher remained open. As a consequence of our new bounds, we have completed the determination of the optimal resiliency of resilient functions with $m = 3$, and we have also done most of the cases for $m = 4$.

Here are some examples of resilient functions from [6] (all addition is modulo 2):

(1) $m = 1, t = n - 1$. Define $f(x_1, \ldots, x_n) = x_1 + \ldots + x_n$.
(2) $m = n - 1, t = 1$. Define $f(x_1, \ldots, x_n) = (x_1 + x_2, x_2 + x_3, \ldots, x_{n-1} + x_n)$.
(3) $m = 2$, $n = 3h$, $t = 2h - 1$. Define

$$f(x_1, \ldots, x_{3h}) = (x_1 + \ldots + x_{2h}, x_{h+1} + \ldots + x_{3h}).$$

2 Orthogonal Arrays

An *orthogonal array* $OA_\lambda(t, k, v)$ is a $\lambda v^t \times k$ array of v symbols, such that in any t columns of the array every one of the possible v^t ordered pairs of symbols occurs in exactly λ rows. Usually t is referred to as the *strength* of the orthogonal array, k is called the number of *factors*, v is called the number of *levels* and λ is called the *index* of the orthogonal array. An orthogonal array is said to be *simple* if no two rows are identical. Of course, an array with $\lambda = 1$ is simple. In this paper, we consider only simple arrays.

A *large set of orthogonal arrays* $LOA_\lambda(t, k, v)$ is a set of v^{k-t}/λ simple arrays $OA_\lambda(t, k, v)$ such that every possible k-tuple of symbols occurs in exactly one of the orthogonal arrays in the set.

The proof of the following theorem, which elucidates the connection between resilient functions and orthogonal arrays, can be found in [14].

Theorem 1. *An (n, m, t) - resilient function is equivalent to a large set of orthogonal arrays $LOA_{2^{n-m-t}}(t, n, 2)$.*

The correspondence between resilient functions and orthogonal arrays is as follows. For any m-tuple $(y_1, \ldots, y_m)$, the inverse image $f^{-1}(y_1, \ldots, y_m)$ of an (n, m, t) - resilient function, say f, is an orthogonal array $OA_{2^{n-m-t}}(t, n, 2)$; and the 2^m OA's thus obtained comprise a large set.

In view of Theorem 1, any necessary condition for the existence of an orthogonal array $OA_{2^{n-m-t}}(t, n, 2)$ is also a necessary condition for the existence of an (n, m, t) - resilient function. Classical bounds for orthogonal arrays include the Rao and Bush bounds; see [14] for their application to resilient functions.

3 Bounds Based on Linear Programming

The most powerful bounds on orthogonal arrays are obtained from Delsarte's linear programming bound; this is the main theme of this section. While developing the bounds based on linear programming techniques, we will be using several standard results from coding theory without proof; the reader is referred to MacWilliams and Sloane [11] for background information on error-correcting codes.

An (n, M, d) binary error correcting code $\mathcal{C}$ is a set of M vectors of length n such that the Hamming distance between any two vectors in $\mathcal{C}$ is at least d. The *distance distribution* of the code is defined to be the sequence $(B_0, B_1, \ldots, B_n)$, where

$$B_i = \frac{1}{M}|\{(u, v) : u, v \in \mathcal{C}, d(u, v) = i\}|.$$

Note that $B_0 = 1$ and $B_0 + B_1 + \ldots + B_n = M$.

The *Krawtchouck transform* of the distance distribution is another sequence, denoted $(B_0^{'}, B_1^{'}, \ldots, B_n^{'})$, where

$$B_k^{'} = \frac{1}{M}\sum_{i=0}^{n} B_i P_k(i).$$

$P_k(i)$ is the value of the *Krawtchouk polynomial* $P_k(x)$ at the integer i and can be defined explicitly as

$$P_k(i) = \sum_{j=0}^{k}(-1)^j \binom{i}{j}\binom{n-i}{k-j}.$$

If $B_i^{'} = 0$ for $1 \leq i \leq d^{'} - 1$ and $B_{d^{'}}^{'} \neq 0$, then $d^{'}$ is called the *dual distance* of the code $\mathcal{C}$. This concept was defined by Delsarte [7]. It is a well known theorem that if we write the vectors in $\mathcal{C}$ as rows of an $M \times n$ array, then any set of $r \leq d^{'} - 1$ columns contains each r-tuple exactly $M/2^r$ times, and $d^{'}$ is the largest number with this property (see for example [11, p. 139], [7]). In other words $\mathcal{C}$ is an orthogonal array $OA_\lambda(d^{'} - 1, n, 2)$ where $\lambda = M/2^{d^{'}-1}$.

It is clear that for any code $\mathcal{C}$, we have $B_i \geq 0$ for $i = 0, 1, \ldots n$. On the other hand, it is a non-trivial theorem (see for example [11, p. 139],[7]) that $B_i^{'} \geq 0$ for $i = 0, 1, \ldots n$.

Suppose an $OA_\lambda(t, n, 2)$ exists. Let M be the number of rows in this orthogonal array. A lower bound on M can be obtained by solving a suitable linear programming problem (see [11, §4 of Ch.17], [7] for similar approaches to a different problem), if we view the rows of this orthogonal array as codewords of a code

$\mathcal{C}$. Let $(B_0, B_1, \ldots, B_n)$ be the distance distribution of $\mathcal{C}$ and $(B_0', B_1', \ldots, B_n')$ be its transform. Then

$$M = B_0 + B_1 + \ldots + B_n.$$

Also

$$\begin{aligned} B_k' &= \frac{1}{M} \sum_{i=0}^{n} B_i P_k(i) \\ &= \frac{1}{M} \left(\sum_{i=1}^{n} B_i P_k(i) + \binom{n}{k} \right). \end{aligned}$$

As $B_k' \geq 0$, we get

$$\sum_{i=1}^{n} B_i P_k(i) \geq -\binom{n}{k} \text{ for } 1 \leq k \leq n.$$

Note further that the dual distance of this code $\mathcal{C}$ is at least $t+1$ (as it is an orthogonal array of strength t). So we can formulate the following linear programming problem which we will refer to as *LP1*:

Minimize $B_1 + B_2 + \cdots + B_n$ subject to
$\sum_{i=1}^{n} B_i P_k(i) = -\binom{n}{k}$ for $1 \leq k \leq t$
$\sum_{i=1}^{n} B_i P_k(i) \geq -\binom{n}{k}$ for $t+1 \leq k \leq n$
$B_i \geq 0$ for $1 \leq i \leq n$

Let $B = B(n,t)$ be the optimal solution to the above linear programming problem. Then we have

$$M = \sum_{i=0}^{n} B_i \geq 1 + B \tag{1}$$

Now let us return to our original problem of establishing stronger upper bounds for the optimal value of t. In view of Theorem 1, an (n, m, t) resilient function exists if only if $LOA_{2^{n-m-t}}(t, n, 2)$ exists. Clearly $LOA_{2^{n-m-t}}(t, n, 2)$ exists only if an $OA_{2^{n-m-t}}(t, n, 2)$ exists. The number of rows in this orthogonal array is given by

$$M = 2^{n-m-t} 2^t = 2^{n-m}.$$

In view of the bound of inequality (1), this immediately implies that

$$m \leq n - \log_2(1 + B).$$

Thus we have an upper bound for the optimal value of m, given n and t. The upper bound for the optimal value of t for a given n and m can be trivially computed once we have a table of upper bounds for the optimal value of m.

4 Explicit Bounds

The disadvantage of the linear programming bound is that one needs to solve a different LP for every parameter situation. Thus it is of interest to try to derive explicit bounds as corollaries of the LP bound. We will pursue this idea in this section.

Let us first form the dual of the linear programming problem *LP1*. We will refer to the dual as *LP2*.

$$\text{Maximize} \sum_{i=1}^{n} x_i \binom{n}{i}$$

subject to

$$\sum_{i=1}^{n} x_i P_i(k) \geq -1 \text{ for } k = 1 \text{ to } n$$

$x_1, x_2, \ldots x_t$ unrestricted
$x_{t+1}, x_{t+2}, \ldots x_n \leq 0$

It is a standard theorem in the theory of linear programming that in a pair of primal-dual linear programs the optimal value of the minimization problem will always be greater than or equal to the value attained by the objective function of the maximization problem at any feasible solution vector. So any feasible solution to the dual linear program *LP2* yields a lower bound on the size of the orthogonal array of strength t and length n and consequently an upper bound on m of t-resilient functions on n-tuples.

Consider the solution vector

$$x_i = 1 - \frac{i}{t+1}.$$

We will first show that this is indeed a feasible solution. In the proof, we will make use of the following standard properties (see problem 45 in page 153 of [11]) of the Krawtchouk polynomials.

$$\sum_{i=0}^{n} P_i(k) = 2^n \delta_{k,0}, \tag{2}$$

$$\sum_{i=0}^{n} i P_i(k) = 2^{n-1} \left(n\delta_{k,0} - \delta_{k,1} \right), \tag{3}$$

where $\delta_{r,s}$ is the Kronecker symbol defined by $\delta_{r,s} = 1$ if $r = s$ and $\delta_{r,s} = 0$ if $r \neq s$.

$$\begin{aligned}
\sum_{i=1}^{n} x_i P_i(k) &= \sum_{i=1}^{n} \left(1 - \frac{i}{t+1}\right) P_i(k) \\
&= \sum_{i=1}^{n} P_i(k) - \frac{1}{t+1} \sum_{i=1}^{n} i P_i(k) \\
&= 2^n \delta_{k,0} - 1 - \frac{2^{n-1}}{t+1} (n\delta_{k,0} - \delta_{k,1}) \\
&= -1 + \frac{2^{n-1}}{t+1} \delta_{k,1} \text{ as } k \neq 0 \text{ in our case} \\
&\geq -1.
\end{aligned}$$

Clearly $x_{t+1} = 0$. As the x_i's form a decreasing sequence, it follows that the condition $x_{t+1}, \ldots, x_n \leq 0$ is also satisfied by the above solution vector. Now that we have shown that the x_i's form a feasible solution, let us compute the value of the objective function at this solution vector.

$$\begin{aligned}
\sum_{i=1}^{n} x_i \binom{n}{i} &= \sum_{i=1}^{n} \left(1 - \frac{i}{t+1}\right) \binom{n}{i} \\
&= \sum_{i=1}^{n} \binom{n}{i} - \frac{1}{t+1} \sum_{i=1}^{n} i \binom{n}{i} \\
&= 2^n - 1 - \frac{1}{t+1} \sum_{i=1}^{n} n \binom{n-1}{i-1} \\
&= 2^n - 1 - \frac{n}{t+1} \sum_{j=0}^{n-1} \binom{n-1}{j} \\
&= 2^n - 1 - \frac{n 2^{n-1}}{t+1}
\end{aligned}$$

Hence it follows that

$$M \geq 1 + B \geq 2^n - \frac{n 2^{n-1}}{t+1}. \tag{4}$$

As a consequence, we get the following Theorem, which was first proved by Friedman [8] using very different methods.

Theorem 2. *If an (n, m, t)-resilient function exists, then*

$$t \leq \left\lfloor \frac{2^{m-1} n}{2^m - 1} \right\rfloor - 1.$$

Similarly, by proving that the vector

$$x_i = 1 - \frac{\lceil \frac{i}{2} \rceil}{\lceil \frac{t+1}{2} \rceil}.$$

is a feasible solution to the dual linear program *LP2*, we can show that

$$M \geq 1 + B \geq 2^n - \frac{2^{n-2}(n+1)}{\lceil \frac{t+1}{2} \rceil}. \tag{5}$$

As a consequence, we get the following new bound, which complements the bound of Theorem 2.

Theorem 3. *If an (n, m, t)-resilient function exists, then*

$$t \leq 2 \left\lfloor \frac{2^{m-2}(n+1)}{2^m - 1} \right\rfloor - 1.$$

The bounds of inequalities (4) and (5) are easily computed. Also for many parametric situations, these two bounds are as powerful as the linear programming bound itself. We record this fact as the following Theorem; however, the proof is omitted due to space limitations.

Theorem 4. *When t is odd and $t+1 \leq n \leq 2t+1$, the bound of equation (4) is as powerful as the linear programming bound. When t is even and $t+1 \leq n \leq 2t+2$, the bound of equation (5) is as powerful as the linear programming bound.*

The result of Theorem 2 was conjectured in [6], where it was shown to be true for $m = 1, 2$ (and in these cases, the bound is tight). Theorem 2 establishes that the conjecture is true for arbitrary m.

By considering a slightly different (but equivalent) version of the linear programming problem, we can exhibit a duality between codes and orthogonal arrays in the sense that each bound on block codes which can be derived from Delsarte's LP-bound yields a bound on orthogonal arrays as well. In this sense the trivial bound $M \geq 2^t$ is the dual of the Singleton bound for codes, the Rao bound is the dual of the sphere packing bound for codes, and the bounds of inequalities (4) and (5) are the dual of the Plotkin bounds for codes.

The MRRW-bound on codes (see [12]), or at least the easier part of it, also follows from the LP-bound. So this bound also carries over to orthogonal arrays. We state an asymptotic version, as applied to resilient functions:

Theorem 5. *If there is an infinite series of resilient functions with parameters (n, m, t), where $n \longrightarrow \infty, t/n \longrightarrow \delta$, then m/n is asymptotically bounded by $H_2(\frac{1}{2} - \sqrt{\delta(1-\delta)})$, where H_2 is the binary entropy function. In particular, positive rates m/n can be attained asymptotically only if $\delta < 1/2$.*

5 Optimal Resilient Functions

Any method of construction of resilient functions yields a lower bound on the optimal value of t. The most important construction method for resilient functions uses linear error correcting codes. A resilient function constructed in this way is said to be a *linear* resilient function.

An (n, m, d) *linear code* is an m-dimensional subspace C of $[GF(2)]^n$ such that the Hamming distance between any two vectors in C is at least d. Let G be an $m \times n$ matrix whose rows form a basis for C; G is called a *generating matrix* for C. The following construction for resilient functions was given in [6, 2]:

Theorem 6. *Let G be a generating matrix for an (n, m, d) linear code C. Define a function $f : [GF(2)]^n \longrightarrow [GF(2)]^m$ by the rule $f(x) = xG^T$. Then f is an $(n, m, d-1)$ - resilient function.*

The *simplex code* $\mathcal{S}_m$ is the dual of the Hamming code $\mathcal{H}_m$. The generator matrix of $\mathcal{S}_m$ is of size $m \times 2^m - 1$ and has all non-zero binary m-tuples as its columns. It is not too difficult to see that each non-zero codeword of the simplex code has weight 2^{m-1}. Thus $\mathcal{S}_m$ is a linear constant-weight code with parameters $(2^m - 1, m, 2^{m-1})$. We shall make use of the simplex code in the proof of the following theorem.

Theorem 7. *Suppose there exists an (n, m, t) linear resilient function which is optimal by virtue of meeting one of the two bounds of Theorem 2 and Theorem 3. Then there exists an $(n + 2^m - 1, m, t + 2^{m-1})$ linear resilient function which is also optimal by virtue of meeting the same bound.*

Proof. Since the (n, m, t) resilient function is linear, it must have been constructed from an $(n, m, t+1)$ linear code $\mathcal{C}$. Now consider the linear code $\mathcal{C}'$ whose generating matrix is obtained by pasting the generating matrices of the code $\mathcal{C}$ and the simplex code $\mathcal{S}_m$, i.e.,

$$G_{\mathcal{C}'} = [G_{\mathcal{C}} \mid G_{\mathcal{S}_m}]$$

Clearly $\mathcal{C}'$ is a linear $(n+2^m-1, m, t+1+2^{m-1})$ code and from this code we can construct a linear resilient function with parameters $(n + 2^m - 1, m, t + 2^{m-1})$.

By assumption the (n, m, t) resilient function is optimal as the parameters meet one of the two bounds derived in the last section. If we increase n by $2^m - 1$, both the bounds go up by 2^{m-1} and so it immediately follows that the linear resilient function with parameters $(n + 2^m - 1, m, t + 2^{m-1})$ is also optimal by virtue of meeting the same bound. □

The simplex code itself yields a $(2^m - 1, m, 2^{m-1} - 1)$ linear resilient function, which meets the bound of Theorem 2 and hence is optimal. Applying Theorem 7, we see that optimal linear resilient functions exist whenever $n \equiv 0 \bmod 2^m - 1$, a result first shown by Friedman [8].

In view of Theorem 6, whenever a linear (n, m, d) code exists, so does an $(n, m, d-1)$ resilient function. Brouwer and Verhoeff [4] provides a compilation of best known linear binary codes. It so happens that when $m = 3$ and $4 \leq n \leq 10$, the linear resilient functions constructed from the best known linear codes are optimal by virtue of meeting one of the two bounds of Theorem 2 and Theorem 3. In fact the bound of Theorem 2 is always met except when $n = 9$, in which case the bound of Theorem 3 is met. This fact coupled with Theorem 7 completely determines the optimal resiliency of resilient functions with $m = 3$. We record the result as the following theorem.

Theorem 8. *The optimal resiliency t of resilient functions with $m = 3$ is*

$$\lfloor \tfrac{4n}{7} \rfloor - 1 \text{ if } n \not\equiv 2 \bmod 7 \text{ and}$$

$$\lfloor \tfrac{4n}{7} \rfloor - 2 \text{ if } n \equiv 2 \bmod 7.$$

Similar analysis settles the question for $m = 4$ in all cases except for two congruence classes modulo 15, as described in the following theorem.

Theorem 9. *The optimal resiliency t of resilient functions with $m = 4$ is*

$$\lfloor \tfrac{8n}{15} \rfloor - 1 \text{ if } n \not\equiv 2,3,4,6 \text{ or } 10 \bmod 15 \text{ and}$$

$$\lfloor \tfrac{8n}{15} \rfloor - 2 \text{ if } n \equiv 2,6 \text{ or } 10 \bmod 15.$$

Here, it turns out that the bound of Theorem 2 is met in those cases when $n \not\equiv 2,3,4,6,10 \bmod 15$, and the bound of Theorem 3 is relevant for $n \equiv 2,6,10 \bmod 15$. The two congruence classes $n \equiv 3,4 \bmod 15$ are unsolved at present.

It is also possible to determine other classes of parameters for which the optimal resiliency can be computed, by using the method of *anticodes* [11, Ch. 17, §6]. For each fixed m, we get many infinite classes of optimal resilient functions. The following is one theorem that can be proved by these methods.

Theorem 10. *Let m be a fixed integer. Let $m > u_1 \geq u_2 \geq 1$. Then for the following values of n, the upper bound on resiliency given in Theorem 2 is tight and it can be attained by linear resilient functions.*

$$n = s(2^m - 1) - 2^{u_1} + 1 \text{ for all } s \geq 1$$
$$n = s(2^m - 1) - 2^{u_1} - 2^{u_2} + 2$$
$$\text{for } s = 1 \text{ if } u_1 + u_2 \leq m, \text{ and for all } s \geq 2.$$

6 Related Results

Most of the techniques used in this paper are applicable mutatis mutandis to non-binary orthogonal arrays and resilient functions. Some results on non-binary resilient functions are proved in [9]. As well, the following bound on non-binary orthogonal arrays is proved in [3]:

Theorem 11. *If there exists an $OA_\lambda(t, n, v)$, then*

$$M \geq v^n \left(1 - \frac{(v-1)n}{v(t+1)}\right).$$

Acknowledgements

Research of the second and third authors is supported by NSF grant CCR-9121051.

References

1. C. H. Bennett, G. Brassard, A. K. Ekert: Quantum cryptography. Scientific American **267(4)** (1992), 26–33
2. C. H. Bennett, G. Brassard, J. M. Robert: Privacy amplification by public discussion. SIAM J. Computing **17** (1988), 210–229
3. J. Bierbrauer: Bounds on orthogonal arrays and codes. Preprint
4. A. E. Brouwer, T. Verhoeff: An updated table of minimum-distance bounds for binary linear codes. IEEE Transactions on Information Theory **39** (1993), 662–677
5. P. Camion, C. Carlet, P. Charpin, N. Sendrier: On correlation-immune functions. In: Advances in Cryptology – CRYPTO '91. Lecture Notes in Computer Science **576** (1992), 86–100
6. B. Chor, O.Goldreich, J. Håstad, J. Friedman, S. Rudich, R. Smolensky: The bit extraction problem or t–resilient functions. In: 26th IEEE Symposium on Foundations of Computer Science, 1985, pp. 396–407
7. P. Delsarte: Four fundamental parameters of a code and their combinatorial significance. Information and Control **23** (1973), 407–438
8. J. Friedman: On the bit extraction problem. In: 33rd IEEE Symposium on Foundations of Computer Science, 1992, pp. 314–319
9. K. Gopalakrishnan, D. R. Stinson: Characterizations of non-binary correlation-immune and resilient functions. Technical Report UNL-CSE-93-010, University of Nebraska, March 1993. Submitted to: Designs, Codes and Cryptography
10. K. Gopalakrishnan, D. R. Stinson, J. Bierbrauer: Orthogonal arrays, resilient functions, error correcting codes and linear programming bounds. Preprint
11. F. J. MacWilliams, N. J. A. Sloane: The Theory of Error-Correcting Codes, North-Holland, 1977
12. R. J. McEliece, E. R. Rodemich, H. Rumsey, L. R. Welch: New upper bounds on the rate of a code via the Delsarte-McWilliams inequalities. IEEE Trans. on Information Theory **23** (1977), 157–166
13. R. A. Rueppel: Analysis and Design of Stream Ciphers, Springer-Verlag, 1986
14. D. R. Stinson: Resilient functions and large sets of orthogonal arrays. Congressus Numerantium **92** (1993), 105–110

Tracing Traitors

Benny Chor[1] and Amos Fiat[2] and Moni Naor[3]

[1] Dept. of Computer Science, Technion, Haifa 32000, Israel.
[2] Dept. of Computer Science, School of Mathematics, Tel Aviv University, Tel Aviv, Israel, and Algorithmic Research Ltd.
[3] Dept. of Computer Science and Applied Math, Weizmann Institute, Rehovot, Israel.

Abstract. We give cryptographic schemes that help trace the source of leaks when sensitive or proprietary data is made available to a large set of parties. This is particularly important for broadcast and database access systems, where the data should be accessible only to authorized users. Such schemes are very relevant in the context of pay television, and easily combine with and complement the Broadcast Encryption schemes of [6].

1 Introduction

If only one person is told about some secret, and this next appears on the evening news, then the guilty party is evident. A more complex situation arises if the set of people that have access to the secret is large. The problem of determining guilt or innocence is (mathematically) insurmountable if all people get the exact same data and one of them behaves treacherously and reveals the secret.

Any data that is to be available to some while it should not be available to others can obviously be protected by encryption. The *data supplier* may give authorized parties cryptographic keys allowing them to decrypt the data. This does not solve the problem above because it does not prevent one of those authorized to view the message (say, Alice) from transferring the *cleartext* message to some unauthorized party (say, Bob). Once this is done then there is no (cryptographic) means to trace the source of the leak. We call all such unauthorized access to data *piracy*. The *traitor* or *traitors* is the (set of) authorized user(s) who allow other, non-authorized parties, to obtain the data. These non-authorized parties are called *pirate users*.

In many interesting cases it is somewhat ineffective piracy if the relevant *cleartext* messages must be transmitted by the "traitor" to the "enemy". Typical cases where this is so include

- Pay-per-view or subscription television broadcasts. It is simply too expensive and risky to start a pirate broadcast station.
- CD ROM distribution of data where a surcharge is charged for different parts of the data. The cleartext data can only be distributed on a similar storage device.
- Online databases, freely accessible (say on the internet) where a charge may be levied for access to all or certain records.

In all these cases, transmitting the cleartext from a traitor, Alice, to an pirate-user, Bob, is either irrelevant or rather expensive. As piracy in all these cases is a criminal commercial enterprise the risk/benefit ratio becomes very unattractive. These three examples can be considered generic examples covering a wide range of data services offered.

Our contribution in this paper may be viewed in the following manner: Consider a ciphertext that may be decrypted by a large set of parties, but each and every party is assigned a different *personal key* used for decrypting the ciphertext. (We use the term personal key rather than private key to avoid confusion with public key terminology). Should the personal key be discovered (by taking apart a television pirate decoder or by counter-espionage), the traitor will be identified.

We note that in fact, our schemes have the very desirable property that the identity of the traitor can be established by considering the pirate decryption process as a black box. It suffices to capture one pirate decoder and it's behavior will identify the traitor, there is no need to "break it open" or read any data stored inside. We use the term pirate decoder to represent the pirate decryption process, this may or may not be a physical box, this may simply be be some code on a computer.

Clearly, a possible solution is to encrypt the data separately under different personal keys. This means that the total length of the ciphertext is at least n times the length of the cleartext, where n is the number of authorized parties. This is clearly impossible in any broadcast environment. This is also very problematic in the context of CD ROM distributed databases because this means that every CD ROM must be different. An encrypted online database, freely accessible as above, must store an individually encrypted copy of the database for each and every authorized user.

The underlying security assumption of our schemes is either information theoretic security (where the length of the personal keys grows with the length of the messages to be transmitted) or it may be based on the security of any symmetric scheme of your choice. In both cases, security depends on a scheme parameter k, the largest group of colliding traitors.

In practice today it is often considered sufficient to prevent piracy by supplying the authorized parties with so-called secure hardware solutions that are designed to prevent interference and access to enclosed cryptographic keys (smart-cards and their like). Our schemes do not require any such assumption, they obtain their claimed security without any secure hardware requirements. Should such devices be used to store the keys they will undoubtedly make the attack more expensive, but this is not a requirement.

Fighting piracy in general has the following components:

1. Identify that piracy is going on and prevent the transmittal of information to pirate users, while harming no legitimate users.
2. Take legal measures against the source of such piracy, supply legal evidence of the pirate identity.

Any solution to fighting piracy must be considered in light of the following performance parameters:

(a) What are the memory and computation requirements per authorized user?
(b) What are the memory and computation requirements for the data supplier?
(c) What is the data redundancy overhead? This is measured in multiples of the the cryptographic security parameter and refers to the communications overhead (in broadcast or online systems) or the additional "wasted" storage in CD ROM type systems.

Consider a pirate user who has already obtained all keys required to read a CD ROM in it's entirety. Clearly, there is little one can do technically to prevent her from continuing to use the CD ROM. The situation is somewhat different if the system requires some action on behalf of the data supplier, *e.g.*, television broadcast or online database.

The broadcast encryption scheme of Fiat and Naor [6] deals with disabling active pirate users very efficiently. These schemes allow one to broadcast messages to any dynamic subset of the user set, this is specifically suitable for pay-per-view TV applications but also implies the piracy protection above. These schemes require a single short transmission to disable all pirate decoders if they were manufactured via a collaborative effort of no more than k traitors.

The number of traitors above, k, is a parameter of the broadcast encryption schemes. While this may not be evident at first, the same scheme could be used by any online database supplier to kill off illegitimate access simply by telling users who log on what users are currently blacklisted.

The goal of this paper is to deal *traitor tracing* (item 2 above), *i.e.*, to identify the source of the problem and to deal with it via legal or extra-legal means. Our solution, called traitor tracing, is valid for all examples cited above, broadcast, online, and CD ROM type systems.

We devise *k-resilient traceability* schemes with the following properties:

1. Either the cleartext information itself is continuously transmitted to the enemy by a traitor, or
2. Any captured pirate decoder will correctly identify a traitor and will protect the innocent even if up to k traitors combine and collude.

It would make sense to have both broadcast encryption and traitor tracing schemes available, at different security levels. The costs of such schemes are measured in the memory requirements at the user end and in the total transmission length required. In practice one would want a broadcast encryption scheme with a different security level (measured in the numbers of traitors required to disable the scheme). Fortunately, both types of scheme, at arbitrary security levels, can be trivially combined simply by XOR'ing the results.

We deal with schemes of the following general form: The data supplier generates a base set R of r random keys and assigns subsets of these keys to users, m keys per user (these parameters will be specified later). These m keys jointly form the user personal key. Note that different personal keys may have a nonempty intersection. We denote the personal key for user u by $P(u)$, this is a set of keys over the base set R.

A traitor tracing message consists of many pairs of (*enabling block*, *cipher block*). The cipher block is the symmetric encryption of the actual data (say a few seconds of a video clip), under some secret random key S. Alternately, it could simply be the XOR of the message with S and we would get an information theoretic secure version of the scheme. The enabling block allows authorized users to obtain S. The enabling block consists of encrypted values under some or all of the r keys at the data supplier. Every user will be able to compute S by decrypting the values for which he has keys and then computing the actual key from these values. The computation on the user end, for all schemes we present, is simply the exclusive or of all values the user has been able to decrypt.

Figures 1 and 2 describe the general nature of our traitor tracing schemes.

Traitors may conspire and give an unauthorized user (or users) a subset of their keys so that the unauthorized user will also be able to compute the real message key from the values he has been able to decrypt. The goal of the system designer is to assign keys to the users such that when a pirate decoder is captured and the keys it possesses are examined, it should be possible to detect at least one traitor, subject to the limitation that the number of traitors of is at most k. (We cannot hope to detect all traitors as one traitor may simply provide his personal key and others may provide nothing).

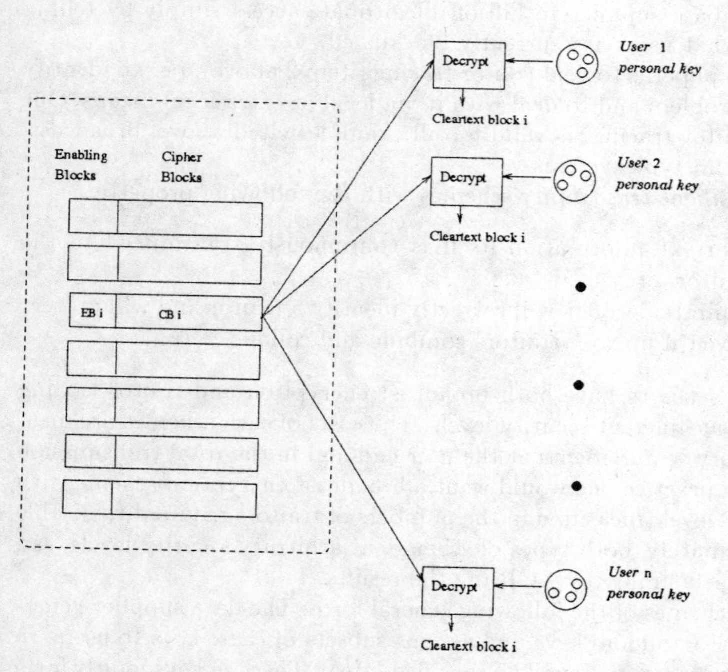

Fig. 1.

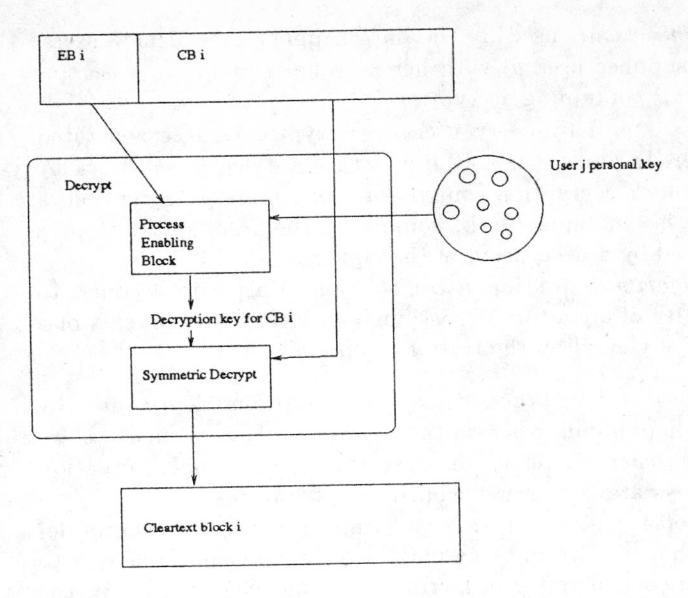

Fig. 2.

We remark that in many cases it is preferable to predetermine a fixed number of users n, and to assign them personal keys, even if the actual number of users is smaller. Later users who join the system by purchasing a subscription to a television station, online database, or CD ROM access privilege are assigned personal keys from those preinstalled. This is especially important in the case of data distributed on CD ROM.

1.1 An Example

Using the 1-level secret scheme described hereinafter, allocating 5% of a compressed MPEG II digital video channel to the traitor tracing scheme allows us to change keys every minute or so (a new enabling block every minute).

The traitor tracing scheme is resilient to $k = 32$ traitors, with probability $1 - 2^{-10}$, and can accommodate up to 1,000,000,000 authorized users. The total number of keys stored by the data supplier (the television broadcaster) is 2^{19}, the personal key of every user consists of 2^{13} keys. These parameters are overly pessimistic because they are derived from the general theorem concerning the scheme using the Chernoff bound.

In practice, there is no real need to change keys every minute, even changing keys once every hour will make any pirate broadcaster give up in despair.

2 Definitions

For messages generated by a data supplier for a set of n users, we define three elements that jointly constitute a *traceability scheme*:

- A *user initialization scheme*, used by the data supplier to add new users. The data supplier supplies user u_i with her personal key, in our case this consists of a set $P(u_i)$ containing decryption keys.
- A *decryption scheme*, used by every user to decrypt messages generated by the data supplier. In our schemes, the messages are decrypted block by block where every block decryption consists of a preliminary decryption of encrypted keys in the enabling block, combining the results to obtain a common key, followed by a decryption of the cipher block.
- A *traitor tracing algorithm*, used upon confiscation of a pirate decoder, to determine the identity of a traitor. We assume below that the contents of a pirate decoder can be viewed by the traitor tracing algorithm.

We distinguish between circumstances where the decryption schemes used by all users are in the public domain, whereas the decryption keys themselves are kept secret, called *open schemes*, versus the case where the actual decryption scheme as well as the keys are kept secret, called *secret schemes*.

The goal of an adversary is to construct a pirate decoder that allows decryption and prevents the guilty from being identified. In particular, one way to ensure that the guilty are safe is to try to incriminate someone else. Clearly, the adversaries task is no harder with an open scheme compared to a secret scheme. On the other hand, secret schemes pose additional security requirements at the data supplier cite and the correctness of the traitor identity may be based on probabilistic arguments, which may be somewhat less convincing in a court of law.

We present efficient schemes of both types, and our constructions give better results for secret schemes. It is clearly advantageous to use secret schemes in practice, and any real implementation will do so.

To simplify the definitions below we will assume that it is impossible to guess a secret key. The probability of guessing a secret key is exponentially small in the length of the key, and thus we will ignore this question in the definitions below. An alternative would be to talk about probability differences rather than absolute probabilities, this is done in [6] and analogous definitions could be used here.

Definition 1. An n user open traceability scheme is called k*–resilient* if for every coalition of at most k traitors the following holds: Suppose the coalition uses the information its members got in the initialization phase to construct a pirate decoder. If this decoder is capable of applying the decryption scheme, then the traitor tracing algorithm will correctly identify one of the coalition members.

Definition 2. An n user secret traceability scheme is called (p,k)*–resilient* if for all but at most p of the $\binom{n}{k}$ coalition of k traitors the following holds: Suppose the coalition uses the information its members got in the initialization phase to construct a pirate decoder. If this decoder is capable of applying the decryption scheme, then the traitor tracing algorithm will identify one of the coalition members with probability at least $1-p$.

3 Construction of Traceability Schemes

In this section we describe three constructions of k-resilient traceability schemes. All these schemes are based on the use of hash functions, combined with *any* private key cryptosystem. (For more information on hash functions and their applications, see [7, 3, 9, 5].) The basic use of hash functions is to assign decryption keys to authorized users in a manner which prevents any coalition of traitors from combining keys taken from the personal keys of its members into a set of keys that allows decryption yet is "close" to the personal key of any innocent user.

The first scheme is the simplest one. It is an open scheme, based on "one level" hash functions. Each hash function maps the n users into a set of $2k^2$ decryption keys. The keys themselves are kept secret, but the mapping (which user is mapped to what key) is publicly known. This is a simple scheme, but its performance can be improved upon: Every user personal key consists of $O(k^2 \log n)$ decryption keys, and the enabling block consists of $O(k^4 \log n)$ encrypted keys.

The second scheme is an open "two level" scheme. Here, a set of first level hash functions map the n users into a set of size k. each function thereby induces a partition of the n users to k subsets. Each of these subsets is mapped separately by "second level" hash functions into $\log^2 k$ decryption keys. This scheme requires $O(k^2 \log^2 k \log n)$ keys per user, and an enabling block of $O(k^3 \log^4 k \log n)$ encrypted keys.

The third scheme is a "one level" secret scheme. Here, we assume that the hash functions, as well as the decryption keys, are kept secret. There is a positive probability p $(0 < p < 1)$ that the adversary will be able to produce pirate decoders which prevent the identification of any traitor.

However, even if the keys known to the k collaborators enable the construction of such "wrongly incriminating" pirate decoders, choosing such set is highly improbable. Even if this unlikely event occurs, the adversary will not know that this is the case.

Being a secret scheme implies that the adversary does not know what keys corresponds to any specific user. The personal key consists of $O(k \log(n/p))$ decryption keys, and has $O(k^2 \log(n/p))$ encrypted keys per enabling block.

All schemes are constructed by choosing hash values at random, and using probabilistic arguments to assert that the desired properties hold with overwhelming probability. Therefore, these schemes are not constructive. However, the properties of the simplest scheme can be verified.

3.1 A Simple Scheme

Let k be an upper bound on the number of traitors. Every enabling block consists of r encryptions, and m denotes the number of keys comprising a user personal key.

We first deal with the case $k = 1$. The data supplier generates $r = 2 \log n$ keys $s_1^0, s_1^1, s_2^0, s_2^1, \ldots, s_{\log n}^0, s_{\log n}^1$. The personal key for user i is the set of $m = \log n$ keys $s_1^{b_1}, s_2^{b_2}, \ldots, s_{\log n}^{b_{\log n}}$, where b_i is the ith bit in the ID of u.

To encrypt a secret s, the data supplier splits it into $\log n$ secrets $s_1, s_2, \ldots, s_{\log n}$, i.e., the data supplier chooses random $s_1, s_2, \ldots, s_{\log n}$ such that q is the bitwise XOR of the s_i's (its j-th bit equals $s^{(j)} = XOR_{i=1}^{\ell} s_i^{(j)}$). The value s_i is encrypted using keys s_i^0 and s_i^1 and both encryptions are added to the enabling block. Every user u can reconstruct all the s_i's and hence can decrypt s. On the other hand, any pirate decoder must contain a key for every $1 \leq i \leq \log n$ (otherwise s_i would remain unknown and consequently s could not be obtained).

Thus, given that at most one traitor is involved, the keys stored in the pirate decoder uniquely identify the traitor.

When dealing with larger coalition, the idea is to generalize the above scheme. Instead of one bit per index we will have larger domains (and have a key for every element in the domain). We will also split s into more than $\log n$ parts and have appropriately more indices or hash functions. The major difficulty we encounter is in the procedure for detecting traitors. Since, unlike the case $k = 1$, keys may be mixed from several members of the coalition, we must make sure that the two users are not only different on some indices, but are different in almost all indices. A detailed description of the scheme is given below.

Initialization: A set of ℓ "first level" hash functions $h_1, h_2, \ldots, h_\ell$ is chosen at random by the data supplier. Each hash function h_i maps $\{1, \ldots, n\}$ into an independent set of $2k^2$ random keys, $S_i = \{s_{i,1}, s_{i,2}, \ldots, s_{i,2k^2}\}$. The personal key for user u is the set $P(u) = \ell$ keys $\{h_1(u), h_2(u), \ldots, h_\ell(u)\}$.

Distributing a Key: For each i ($i = 1, 2, \ldots, \ell$) the data supplier encrypts a key s_i under each of the $2k^2$ keys in S_i. The final key s is the bitwise XOR of the s_i's (its j-th bit equals $s^{(j)} = XOR_{i=1}^{\ell} s_i^{(j)}$). Each authorized user has one key from S_i, so he can decrypt every s_i, and thus compute s.

Parameters: The memory required per user is $m = \ell$ keys. An enabling block to encode a secret value s consists of $= 2k^2\ell$ key encryptions.

Fraud: The k traitors can get together and combine their personal keys. They may choose one key from every set S_i ($i = 1, 2, \ldots, \ell$). These ℓ keys are put together in a pirate decoder. This set of keys F, enables the purchaser of such a decoder to decrypt every s_i, and thus compute s.

Detection of Traitors: Upon confiscation of a pirate decoder, the set of keys in it, F, is exposed. The set F must contain at least ℓ keys (at least one key per set S_i). Denote the key with minimum index in S_i by $f_i \in S_i$. For each i, the users in $h_i^{-1}(f_i)$ are identified and marked. The user with largest number of marks is exposed.

Goal: We want to show that for all coalitions of size k, the probability of exposing a user who is not a traitor is negligible.

Clearly, at least one of the traitors contributes at least ℓ/k of the keys to the pirate decoder (we ignore duplicate keys from the same S_i). We want to show that the probability (over all choices of hash functions) that a good user is marked ℓ/k times is negligible. Consider a specific user, say 1, and a specific coalition T of k traitors (which does not include 1). As hash functions are chosen at random, the value $a_i = f_i(1)$ is uniformly distributed in S_i. The coalition gets at most k keys in S_i. The probability that a_i is among these keys is at most $1/2k$.

Let X_i be a zero-one random variable, where $X_i = 1$ if $a_i \in h_i(T)$. The mean value of $\sum_{i=1}^{\ell} X_i$ is $\ell/2k$. We use the following version of Chernoff bound (see [2], Theorem A.12) to bound the probability that $\sum_{i=1}^{\ell} X_i \geq \ell/k$. Let $X_1, \ldots, X_\ell$ be mutually independent random variables, with

$$Pr(X_j = 1) = p$$
$$Pr(X_j = 0) = 1 - p$$

Then, for all $\beta \geq 1$

$$Pr\left(\frac{1}{\ell}\sum_{j=1}^{\ell} X_j \geq \beta p\right) < \left(\frac{e^{\beta-1}}{\beta^\beta}\right)^{p\ell}.$$

In our case, substituting $p = 1/2k$ and $\beta = 2$, we have

$$Pr\left(\frac{1}{\ell}\sum_{i=1}^{\ell} X_i \geq \frac{1}{k}\right) < \left(\frac{e}{4}\right)^{\ell/2k} < 2^{-\ell/4k}.$$

In order to overcome all $\binom{n}{k}$ coalitions and all n choices of users, we choose ℓ satisfying

$$n \cdot \binom{n}{k} \cdot 2^{-\ell/4k} < 1,$$

namely $\ell > 4k^2 \log n$. With this parameter, there is a choice of ℓ hash functions such that for every coalition and every authorized user not in the coalition, the user is not incriminated by the tracing algorithm. We summarize these results in the next theorem:

Theorem 3. *There is an open k-resilient traceability scheme, where a user personal key consists of $m = 4k^2 \log n$ decryption keys, and an enabling block consists of $r = 8k^4 \log n$ key encryptions.*

Explicit Constructions: The discussion above shows the existence of open k resilient traceability schemes, and does provide us with a randomized method for constructing the scheme that works with high probability. It does not, however, suggests an explicit construction. Note however that a given construction can be verified quite efficiently. The idea is to examine all the pairs of elements u, v and check the number of function h_i such that $h_i(v) = h_j(u)$. If this number is smaller than ℓ/k^2 than we can conclude that no coalition T of at most k elements "covers" more than a $1/k$ fraction of the keys of u and hence cannot incriminate u.

By considering pairwise differences, we can phrase the construction problem as a problem in coding theory (see [8] for more information): construct a code with n codewords over an alphabet of size $2k^2$ of length ℓ such that the distance between every two codewords is at least $\ell - \ell/k^2$. The goal is construct such a code with as small ℓ as possible. There are no known explicit construction that match the probabilistic bound. For the best known construction see [1] and references therein. For small k the constructions there yields a scheme with $m \in O(k^6 \log n)$ and $r \in O(k^8 \log n)$.

3.2 An Open Two Level Scheme

The "two level" traceability scheme, described in this subsection, more complicated than the simple scheme, but it saves about a factor of k in the broadcast overhead.

Theorem 4. *There is an open k-resilient traceability scheme, where a user personal key consists of $m = 2k^2 \log^2 k \log n$ decryption keys, and an enabling block consists of $r = 4k^3 \log^4 k \log n$ key encryptions.*

Proof. We describe the system, step by step. As in the one-level scheme, the proof is existential. We do not know however how to verify efficiently that a given scheme is "good".

Initialization: A set of ℓ "first level" hash functions $h_1, h_2, \ldots, h_\ell$, each mapping $\{1, \ldots, n\}$ to $\{1, \ldots, k\}$, is chosen at random. For each i $(i = 1, 2, \ldots, \ell)$ and each element a in $\{1, \ldots, k\}$, a set of d "second level" hash functions $g_{i,a,1}, \ldots, g_{i,a,d}$ is chosen at random. Each second level function $g_{i,a,j}$ maps the users in $h_i^{-1}(a) \subset \{1, \ldots, n\}$ into a set of $4 \log^2 k$ random independent keys (the ranges of different functions are independent).

Each user $u \in \{1, \ldots, n\}$ receives $\ell \cdot d$ keys

$$\begin{array}{ccc} g_{1,h_1(u),1}(u), & \ldots & , g_{1,h_1(u),d}(u) \\ \vdots & & \vdots \\ g_{\ell,h_\ell(u),1}(u) & , \ldots, & g_{\ell,h_\ell(u),d}(u) \end{array}$$

Distributing a Key: The data supplier chooses at random ℓ independent keys $s_1, \ldots, s_\ell$. The final key is $s = BITWISE - XOR_{i=1}^{\ell}(s_i)$.

For each i $(i = 1, 2, \ldots, \ell)$, a $(a = 1, \ldots, k)$, and j $(j = 1, \ldots, d)$, let $s_{i,a,j}$ be an independent random key, satisfying

$$\begin{aligned} s_i &= BITWISE - XOR(s_{i,1,1}, \ldots, s_{i,1,d}) \\ &= BITWISE - XOR(s_{i,2,1}, \ldots, s_{i,2,d}) \\ &\vdots \\ &= BITWISE - XOR(s_{i,k,1}, \ldots, s_{i,k,d}) \end{aligned}$$

The key $s_{i,a,j}$, encrypted under each of the $4 \log^2 k$ keys in the range of the function $g_{i,a,j}$, is added to the enabling block.

User u possesses the d keys $g_{i,h_i(u),1}(u), \ldots, g_{i,h_i(u),d}(u)$ and so he is capable of decoding $s_{i,h_i(u),1}, \ldots, s_{i,h_i(u),d}$, allowing him to reconstruct s_i and then compute the final key s.

Parameters: The personal key consists of $m = \ell d$ keys. The total number of key encryptions in an enabling block encoding s is $4k\ell d \log^2 k$.

Fraud: The k traitors can get together and expose their own keys in order to construct a pirate decoder. By the bit sensitivity of XOR, the box must be able to decrypt every s_i $(i = 1, 2, \ldots, \ell)$. To do this, the decoder must be able to decrypt a complete row $s_{i,a,1},\ldots,s_{i,a,d}$ for some a, $1 \le a \le k$. So, for each i ($i = 1,2,\ldots,\ell$) the traitors choose $a = h_i(u)$ for some $u \in T$, and d keys $g_{i,a,1}(u_1),\ldots,g_{i,a,d}(u_d)$, where $u_1,\ldots,u_d \in T$ and $h_i(u_1) = h_i(u_2) = \ldots = h_i(u_d) = a$. For every i, these d keys are placed in the pirate decoder.

Detection of Traitors: Upon confiscation of a pirate decoder, the set of keys in it, F, is exposed. As argued above, the decoder must contain a block of d keys of the form $k_{i,a,1} = g_{i,a,1}(u_1),\ldots,k_{i,a,d} = g_{i,a,d}(u_d)$ corresponding to each i ($i = 1,2,\ldots,\ell$). (If more than one row is in the decoder, only the one with minimum a is used by the detection algorithm.) For each $j = 1,\ldots,d$, the detective identifies the users in $g^{-1}_{i,a,j}(k_{i,a,j})$. Each of these users is called *marked.* All users who are marked at least $d/\log k$ times, are suspects for s_i. The user who is a suspect for the largest number of s_i's is identified as a traitor.

Goal: We want to show that there is a choice of hash functions such that for all coalitions, a good user is never identified as a traitor.

Consider a specific user, say 1, and a specific coalition T of k traitors (which does not include 1). We first bound the probability that user 1 will be a suspect for s_i. The first level hash function h_i partitions the users to k subsets $\{h_i^{-1}(1),\ldots,h^{-1}(k)\}$. The expected maximum number of traitors in these k subsets is $\log k/\log\log k$. The probability that user 1 is hashed to a subset together more than $\log k$ traitors is at most $1/16k$ [2]. Denote $h_i(1) = a$. Consider the conditional probability space where $T \cap h_i^{-1}(a)$ contains indeed at most $\log k$ traitors. In this conditional space, the d keys $k_{i,a,1},\ldots,k_{i,a,d}$ in the pirate decoder come from the personal keys of $T \cap h_i^{-1}(a)$. As this set contains fewer than $\log k$ members, there must be at least one member in $T \cap h_i^{-1}(a)$ who is marked at least $d/\log k$ times. Therefore at least one member of T is a suspect for s_i.

Returning to our innocent user 1, the detective marks user 1 with respect to $g_{i,a,j}$ if there is some $u \in T \cap h_i^{-1}(a)$ such that $g_{i,a,j}(1) = g_{i,a,j}(u)$. The range of $g_{i,a,j}$ contains $4\log^2 k$ keys. At most $\log k$ of these are in $g_{i,a,j}\left(T \cap h_i^{-1}(a)\right)$. So the probability that user 1 is marked with respect to $g_{i,a,j}$ is at most $1/(4\log k)$. The expected number of times user 1 will be marked, with respect to the d functions $g_{i,a,1},\ldots,g_{i,a,d}$, is $d/(4\log k)$. We use the Chernoff bound to estimate the probability that user 1 is a suspect for s_i.

Set $X_j = 1$ if user 1 is marked with respect to $g_{i,a,d}$, and $X_j = 0$ otherwise. Then $Pr(X_j = 1) \le 1/(4\log k)$. By the version of the Chernoff bound mentioned above, with $p = 1/(4\log k)$ and $\beta = 4$,

$$Pr\left(\frac{1}{d}\sum_{j=1}^{d} X_j \ge \frac{1}{\log k}\right) < \left(\frac{e^3}{4^4}\right)^{d/4\log k} \le 2^{-3d/4\log k} .$$

Setting $d = 2\log^2 k$, the conditional probability that user 1 is a suspect for s_i is at most $2^{-3\log k/2} < 1/16k$. The probability of the condition not happening is at most $1/16k$. So overall, the total (unconditional) probability that user 1 is the suspect for s_i is at most $1/8k$.

For $i = 1, \ldots, \ell$, let $Y_i = 1$ if 1 is the suspect for s_i, and $Y_i = 0$ otherwise. Then

$$Pr\left(\frac{1}{\ell}\sum_{i=1}^{\ell} X_i \geq \frac{1}{2k}\right) < \left(\frac{e^7}{8^8}\right)^{\ell/8k} < 2^{-\ell/k}.$$

So with probability at least $1 - 2^{-\ell/k}$, user 1 is a suspect for fewer than ℓ/k of the s_i.

For every s_i $(i = 1, \ldots, \ell)$, at least one member of T is a suspect for s_i. T contains k traitors, and so there must be one or more traitor who is a suspect for at least ℓ/k s_i's. Therefore the probability that user 1 is mistakenly identified as a traitor is smaller than $2^{-\ell/k}$. The probability that for one of the $\binom{n}{k}$ possible coalitions T of size k, some good user is mistakenly identified, is smaller than $n \cdot \binom{n}{k} \cdot 2^{-\ell/k}$. Setting $\ell = k^2 \log n$, this probability is smaller than 1. This means that there exists a choice of hash functions h_i and $g_{i,a,j}$ such that a good user is never mistakenly identified as a traitor. The resulting open k-traceability scheme has parameters $m = \ell d = 2k^2 \log^2 k \log n$ and $r = 2k\ell d \log^2 k = 4k^3 \log^4 k \log n$.

3.3 A Secret One Level Scheme

We simplify the construction and improve its costs by using a secret scheme. The proposed scheme is one level, and the hash values of users are kept secret. The major source of saving is that it suffices to map the n users into a set of $4k$ keys (rather than k^2 keys as in the simple one level scheme). A coalition of size k will contain the key of any specific user with constant probability. However, as the traitors do not know which key this is, any key they choose to insert into the pirate decoder will miss the key of the authorized user (with high probability).

Initialization: Each user u $(u \in \{1, 2, \ldots, n\})$ is assigned a random name n_u from a universe $\mathcal{U}$ of size exponential in n. These names are kept secret. A set of ℓ hash functions $h_1, h_2, \ldots, h_\ell$ are chosen independently at random. Each hash function h_i maps $\mathcal{U}$ into a set of $4k$ random keys $S_i = \{s_{i,1}, s_{i,2}, \ldots, s_{i,4k}\}$. The hash functions are kept secret as well. User u receives, upon initialization, ℓ keys $\{h_1(n_u), h_2(n_u), \ldots, h_\ell(n_u)\}$.

Distributing a Key: For each i $(i = 1, 2, \ldots, \ell)$ the data supplier encrypts a key s_i under each of the $4k$ keys in S_i. The final key is the bitwise XOR of the s_i's (the j-th bit is $s^{(j)} = XOR_{i=1}^{\ell} s_i^{(j)}$). Each authorized user has one key from S_i, so he can decrypt every s_i, and thus compute s.

Parameters: The memory required per user is $m = \ell$ keys. The total number of broadcasts, used in distributing the key s, is $r = 4k\ell$.

Fraud: The k traitors can get together and expose their own keys. Given these keys, they chose one key per set S_i $(i = 1, 2, \ldots, \ell)$. These ℓ keys are put together in a pirate decoder. This set of keys F, enables the purchaser of such decoder to decrypt every s_i, and thus compute s.

Detection of Traitors: Upon confiscation of a pirate decoder, the set of keys in it, F, is exposed. F contains ℓ keys, one per set S_i. Denote these keys by $f_i \in S_i$. For each i, the users in $h_i^{-1}(f_i)$ are identified and marked. The user with largest number of marks is exposed.

Goal: We want to show that for all (almost all) coalitions, the probability of exposing a user who is not a traitor is negligible.

Clearly, at least one of the traitors contributes at least ℓ/k of the keys to the pirate decoder. We want to show that the probability that a good user is marked ℓ/k times is negligible. Consider a specific user, say 1, and a specific coalition T of k traitors (which does not include 1). As the name assigned to user 1 is random and the hash functions are random, the value $a_i = h_i(n_1)$ is uniformly distributed in S_i, even given the k values hashed by h_i from the names of the coalition members. The probability that the value f_i, chosen by the coalition to the pirate decoder, equals a_i is therefore $q = 1/4k$. Let X_i be a zero-one random variable, where $X_i = 1$ if $a_i = f_i$. The mean value of $\sum_{i=1}^{\ell} X_i$ is $\ell/4k$. By the Chernoff bound

$$Pr\left(\frac{1}{\ell}\sum_{i=1}^{\ell} X_i \geq 4q\right) < \left(\frac{e^3}{4^4}\right)^{q\ell} < 2^{-3\ell/4k}.$$

In order to overcome all but p of the $\binom{n}{k}$ coalitions and all n choices of users, we choose ℓ satisfying $n \cdot 2^{-3\ell/4k} < p$. That is, $4k\log(n/p)/3 < \ell$, which gives

Theorem 5. *There is a (p,k)-resilient secret traceability scheme, where a user personal key consists of $m = 4k\log(n/p)/3$ decryption keys, and an enabling block consists of $16k^2\log(n/p)/3$ key encryptions.*

4 Lower Bounds

In this section we derive lower bounds for the case where incrimination has to be absolute, i.e. with no error probability. We assume that the keys the data supplier distributes to the users are unforgeable. This is not accurate, since there is always the small chance that the adversary guesses the keys of the user it wants to incriminate. However, we distinguish between the probability of guessing the keys (which is exponentially small in the length of the key) and the probability of incrimination for other reasons which we would like to be zero. Our view of the system is therefore as follows: let the set of keys used be $S = \{s_1, s_2, \ldots s_r\}$ and let each user i obtain a subset $U_i \subset S$ of size m.

Claim 1 *If no coalition of k users $i_1, i_2, \ldots i_k$ should be able to incriminate a user $i_0 \notin \{i_1, i_2, \ldots i_k\}$, then for all such $i_0, i_1, i_2, \ldots i_k$ we should have that*

$$U_{i_0} \not\subset \cup_{j=1}^{k} U_{i_j}$$

Proof. Suppose not, i.e. there exist $i_0, i_1, \ldots i_k$ such that $U_{i_0} \subset \cup_{j=1}^{k} U_{i_j}$, then the coalition of $i_1, i_2, \ldots i_k$ can reconstruct the keys of U_{i_0} and put them in the pirate decoder for sale. Anyone examining the contents of the box will have to deduce that i_0 is the traitor that generated it.

Luckily, the issue of set systems obeying the conditions of Claim 1 has been investigated by Erdös, Frankl and Füredi [4]. From Theorem 3.3 and Proposition 3.4 there we can deduce that r is $\Omega(\min\{n, k^2 \frac{\log n}{\log\log n}\})$ and from Proposition 2.1 there we get that $m \geq k\frac{\log n}{\log r}$. Hence we have:

Theorem 6. *In any open k-resilient traceability scheme distributing every one of the n user m keys out of r we have that r is* $\Omega(\min\{n, k^2 \frac{\log n}{\log\log n}\})$ *and* $m \geq k\frac{\log n}{\log r}$.

Note that the lower bounds on both r and m are roughly a factor of k smaller than the best construction we have for an open traceability system.

Acknowledgments

We are very grateful to Alain Catrevaux, Christophe Declerck, and Henri Joubaud for educating us on the issues of pay television and for motivating all of this work. We also thank to Amos Beimel for his comments on earlier drafts of this manuscript.

References

1. N. Alon, J. Bruck, J. Naor, M. Naor and R. Roth, *Construction of Asymptotically Good Low-Rate Error-Correcting Codes through Pseudo-Random Graphs*, IEEE Transactions on Information Theory, vol. 38 (1992), pp. 509-516.
2. N. Alon and J. Spencer, **The Probabilistic Method**, Wiley, 1992.
3. J. L. Carter and M. N. Wegman, *Universal Classes of Hash Functions*, Journal of Computer and System Sciences 18 (1979), pp. 143-154.
4. P. Erdös, P. Frankl, Z. Füredi, *Families of finite sets in which no set is covered by the union of r others*, Israel J. of math. **51**, 1985, pp. 79–89.
5. M.L. Fredman, J. Komlós and E. Szemerédi, *Storing a Sparse Table with $O(1)$ Worst Case Access Time*, Journal of the ACM, Vol 31, 1984, pp. 538–544.
6. A. Fiat and N. Naor, *Broadcast Encryption*, Proc. Advances in Cryptology - Crypto '93, 1994, pp. 480–491.
7. K. Mehlhorn, **Data Structures and Algorithms: Sorting and Searching**, Springer-Verlag, Berlin Heidelberg, 1984.
8. F. J. MacWilliams and N. J. A. Sloane, **The theory of error correcting codes**, North Holland, Amsterdam, 1977.
9. M. N. Wegman and J. L. Carter, *New Hash Functions and Their Use in Authentication and Set Equality*, Journal of Computer and System Sciences 22, pp. 265-279 (1981).

Towards the Equivalence of Breaking the Diffie-Hellman Protocol and Computing Discrete Logarithms

Ueli M. Maurer

Institute for Theoretical Computer Science
ETH Zürich
CH-8092 Zürich, Switzerland
Email address: maurer@inf.ethz.ch

Abstract. Let G be an arbitrary cyclic group with generator g and order $|G|$ with known factorization. G could be the subgroup generated by g within a larger group H. Based on an assumption about the existence of smooth numbers in short intervals, we prove that breaking the Diffie-Hellman protocol for G and base g is equivalent to computing discrete logarithms in G to the base g when a certain side information string S of length $2\log|G|$ is given, where S depends only on $|G|$ but not on the definition of G and appears to be of no help for computing discrete logarithms in G. If every prime factor p of $|G|$ is such that one of a list of expressions in p, including $p-1$ and $p+1$, is smooth for an appropriate smoothness bound, then S can efficiently be constructed and therefore breaking the Diffie-Hellman protocol is equivalent to computing discrete logarithms.

1. Introduction

Two challenging open problems in cryptography are to prove or disprove that breaking the Diffie-Hellman protocol [5] is computationally equivalent to computing discrete logarithms in the underlying group and that breaking the RSA system [17] is computationally equivalent to factoring the modulus. In this paper we take a significant step towards the solution of the first of these problems.

Let H be a finite group (written multiplicatively), and for $g \in H$, let $G = \langle g \rangle$ be the cyclic subgroup generated by g. The discrete logarithm problem for the group H (or G) can be stated as follows: Given g and $a \in G$, find the unique integer x in the interval $[0, |G|-1]$ such that $g^x = a$, where x is called the discrete logarithm of a to the base g. The discrete logarithm problem is sometimes also defined as the generally easier problem of finding any x satisfying $g^x = a$, but if $|G|$ is known then the two problems are equivalent.

The Diffie-Hellman protocol allows two parties Alice and Bob connected by an authenticated but otherwise insecure channel (for instance an insecure telephone line where Alice and Bob authenticate each other by speaker recognition) to generate a mutual secret key which is computationally infeasible to determine for a passive eavesdropper overhearing the entire conversation between Alice and Bob.

The protocol works as follows. Let $G = \langle g \rangle$ be a cyclic group generated by g for which the discrete logarithm problem is computationally infeasible. (It should be pointed out that it is unknown whether such a group exists.) Specific groups that have been proposed for application in this protocol are the multiplicative groups of large finite fields (prime fields [5] or extension fields), the multiplicative group of residues modulo a composite number [10, 11], elliptic curves over finite fields, the Jacobian of a hyperelliptic curve over a finite field and the class group of imaginary quadratic fields [2].

In order to generate a mutual secret key, Alice and Bob secretly choose integers x_A and x_B, respectively, at random from the interval $[0, |G|-1]$. Then they compute secretly $y_A = g^{x_A}$ and $y_B = g^{x_B}$, respectively, and exchange these group elements over the insecure public channel. Finally, they compute $z_{AB} = y_B^{x_A} = g^{x_A x_B}$ and $z_{BA} = y_A^{x_B} = g^{x_B x_A}$, respectively. Note that $z_{AB} = z_{BA}$, and hence this quantity can be used as a secret key shared by Alice and Bob. More precisely, they need to apply a function mapping elements of G to the key space of a cryptosystem.

In contrast to digital signature schemes based on the discrete logarithm problem (e.g., [6],[19]), it is not required for the Diffie-Hellman protocol that the order of the group be known. In this case, x_A and x_B are chosen from a sufficiently large interval. In fact, it has been pointed out (e.g., see [11]) that using groups with unknown order may be advantageous, and the non-interactive public-key scheme of [10] relies crucially on the fact that the group order is unknown.

2. Computing discrete logarithms and breaking the Diffie-Hellman protocol

An eavesdropper knowing y_A and y_B can in principle compute z_{AB} by computing the discrete logarithm x_A of y_A to the base g and then computing $z_{AB} = y_B^{x_A}$. It is unknown in general whether there exists a faster method for computing z_{AB} from y_A and y_B. This paper investigates the relation between the two problems. More precisely, we investigate in which cases an efficient subroutine breaking the Diffie-Hellman protocol for the group G and base g could be used for computing discrete logarithms to the base g in G.

Definition 1. A *Diffie-Hellman oracle* (DH-oracle for short) for a group G with generator g takes as inputs two elements $a, b \in G$ (where $a = g^x$ and $b = g^y$) and returns the element g^{xy}.

A DH-oracle hence allows to multiply two logarithms without knowing them explicitly but also without receiving the result explicitly. For instance, when given g^x but not x one can use the oracle to compute $g^{(x^2)}$, and more generally $g^{P(x)}$ for any polynomial P with integer coefficients. Multiplications and additions in the exponent are performed by using the oracle and the normal group multiplication, respectively. Multiplication of the exponent with -1 can be achieved by an inversion operation in the group. When $|G|$ is known and $\gcd(x, |G|) = 1$ one can compute g^z from g^x such that $z = x^{-1} \pmod{|G|}$ by computing $g^z = g^{(x^{|G|-1})}$ using $O(\log |G|)$ calls to the DH-oracle. Hence the DH-oracle allows one to compute $g^{f(x)}$ for any rational function $f(x)$ with integer coefficients. More generally, such a DH-oracle can be used to perform any algorithm on implicitly given (but hidden) logarithms, provided that the algorithm uses only addition, subtraction, multiplication and makes decisions only based on testing equality of intermediate results.

For instance, one can compute g^z from g^x where $z^2 = x \pmod{|G|}$ by using the algorithm of [14]. (A more efficient but unpublished algorithm is due to Massey [9].) For the case $|G| = p$ with p prime and $p \equiv 3 \pmod 4$ one can compute $g^z = g^{(x^{(p+1)/4})}$. Our proof techniques will exploit these facts.

When proving a reduction of a problem A to another problem B it is important to state precisely what type of instances of B are generated by the reduction process. If, in the process of solving problem A, the instances of problem B that need to be solved are very special, then the reduction from A to B is not satisfactory because it is conceivable that these special instances are easy to solve even if the general problem B is nevertheless infeasible. This problem is one of the reasons for the limited applicability to cryptography of the theory of NP-completeness. One can show that this problem does not arise in our case because a Diffie-Hellman oracle that answers correctly for a fraction ϵ of the inputs can be transformed into a uniform Diffie-Hellman oracle.

We are only interested in groups G for which the discrete logarithm problem is believed to be intractable. The fastest algorithm for general groups, which is attributed to Shanks and referred to as the baby-step giant-step algorithm, runs in time $O(\sqrt{n} \log n)$ and requires space $O(\sqrt{n})$, where n is a known upper bound on $|G|$. If $|G|$ is known, an algorithm of Pollard with essentially the same running time but almost no space requirement can be used. However, its running time has not been proven rigorously. Furthermore, it is well-known [15] that for an arbitrary group G, discrete logarithms can be computed in time $O(\sqrt{q})$ where q is the largest prime factor of the order $|G|$ of G. For certain specific groups, such as the multiplicative group modulo p, there exist algorithms which run much faster than the generic algorithms. We refer to [12] for a detailed discussion of the discrete logarithm problem and algorithms for solving it.

The first published result on the equivalence between computing discrete logarithms and breaking the Diffie-Hellman protocol is due to den Boer [4]. He proved the equivalence for the group Z_q^* when q is a prime such that $\varphi(q-1)$ has only small prime factors. In order to avoid any confusion it should be pointed out that this is not equivalent to the condition that $q-1$ be smooth, and that no efficient discrete logarithm algorithm is known for Z_q^* for primes q of the described special form. In the following we give a generalized description of den Boer's idea which will serve as an introductory example for our proof technique.

For simplicity, assume that G is a group with prime order $|G| = p$ where $p - 1 = \prod_{j=1}^{r} q_j$ with $q_j \leq B$ for all j for some smoothness bound B, where the q_j's are pairwise relatively prime. The arguments can easily be generalized to arbitrary groups with known factorization of the group order, which also need not be square-free if a DH-oracle for subgroups of G is also available. The group G can be arbitrary, for instance a subgroup of a multiplicative group modulo a larger prime (as suggested in [19] and in the recent NIST proposal for a digital signature standard), or an elliptic curve. Let $a = g^x$ be given. The case $x = 0$ is easily detected because in this case a is simply the neutral element of G. Let c be a primitive element of $\mathbf{F}_p$. Note that such a c can easily be found when the factorization of $p-1$ is known. If $x \neq 0$, then we have

$$x \equiv c^w \pmod{p}$$

for some w satisfying $0 \leq w < p-1$. Instead of computing x directly we will compute w by computing w modulo all the q_j and using the Chinese remainder theorem, i.e., by computing $w_1, \ldots, w_r$ where

$$w \equiv w_j \pmod{q_j}$$

and $0 \leq w_j < q_j$ for $j = 1, \ldots, r$. We have

$$w \cdot \frac{p-1}{q_j} \equiv w_j \cdot \frac{p-1}{q_j} \pmod{(p-1)}$$

and hence

$$x^{\frac{p-1}{q_j}} \equiv c^{w \cdot \frac{p-1}{q_j}} \equiv c^{w_j \cdot \frac{p-1}{q_j}} \pmod{p}.$$

Thus w_j can be determined by computing

$$g^{(c^{t\frac{p-1}{q_j}})}$$

for $t = 0, 1, 2, \ldots$ until this group element is equal to $g^{(x^{\frac{p-1}{q_j}})}$. The latter group element can be computed by $O(\log p)$ applications of the DH-oracle.

The total computational effort for computing x corresponds to $O(\log p)$ applications of the DH-oracle and $O(B(\log p)^2/\log B)$ group operations. For any $u < B$ (e.g., $u = \sqrt{B}$), a baby-step giant-step-type time-memory tradeoff allows to reduce the number of group operations by a factor u at the expense of increasing the number of calls to the DH-oracle by a factor u and further requiring a table of size u, which must also be sorted.

3. Towards an equivalence proof for all groups

The arguments of the previous section only apply to groups G for which every large prime factor p of $|G|$ occurs only once and is of the special form that $p-1$ is smooth with respect to a small bound B. Although no DL-algorithm is known for groups of this type that is faster than for general groups of the same size, it appears to be questionable whether it is secure to use such groups only for the benefit of being able to prove the equivalence.

In this section we present a proof technique that applies to any group G. It can be viewed as a generalization of the technique discussed in the previous section in a similar sense as the elliptic curve factoring algorithm [8] is a generalization of the $(p-1)$-factoring algorithm [16]. In other words, we exploit the fact that there exist collections of groups defined algebraically over $\mathbf{F}_p$ whose orders vary over a certain interval.

Let the order of the group G be given by

$$|G| = \prod_{i=1}^{s} p_i^{e_i}. \quad (1)$$

In the (generally unlikely) case that the square of a large prime p_i divides $|G|$ (i.e., $e_i > 1$), one needs a DH-oracle not only for the group G with generator g but also for subgroups of G of the form $\langle g^{p_i^z} \rangle$, for $z = 1, \ldots, e_i - 1$. This case will not be discussed further in this paper.

In the sequel, consider a cyclic group $G = \langle g \rangle$ and any single prime divisor p of $|G|$, i.e., let

$$|G| = p \cdot h$$

where $\gcd(p, h) = 1$. For instance, $|G|$ may be given by (1) where $p = p_i$ and $e_i = 1$ for some i. Let $a \in G$ with $a = g^x$ be given and consider the problem of computing x modulo p, i.e., computing the unique x' satisfying

$$x \equiv x' \pmod{p} \quad (2)$$

and $0 \leq x' < p$.

For any two group elements g^z and $g^{z'}$ we can test whether $z \equiv z' \pmod{|G|}$ simply by testing equality of g^z and $g^{z'}$ in G. In order to test the more general condition

$$z \equiv z' \pmod{p}$$

we note that this condition is equivalent to

$$hz \equiv hz' \pmod{|G|}$$

which is satisfied if and only if

$$(g^z)^h = g^{hz} = (g^{z'})^h = g^{hz'}. \quad (3)$$

Equality of logarithms modulo p can thus be tested by two exponentiations with exponent h and a comparison in the group.

For the purpose of illustration we describe our proof techniques by applying an elliptic curve over the field $\mathbf{F}_p$, but it can be generalized to other groups defined algebraically over $\mathbf{F}_p$ such as the multiplicative group of an extension field or certain subgroups thereof, elliptic curves over extension fields of $\mathbf{F}_p$ or the Jacobian of hyperelliptic curves of $\mathbf{F}_p$.

Assume now that we know the parameters A and B of a cyclic elliptic curve $E_{A,B}(\mathbf{F}_p)$ over $\mathbf{F}_p$ which is defined as the set of points $\{(x, y) \in \mathbf{F}_p \times \mathbf{F}_p : y^2 =$

$x^3 + Ax + B\}$ together with the point $\mathcal{O}$ at infinity, where the order of the curve is given by

$$T := \#E_{A,B}(\mathbf{F}_p) = \prod_{j=1}^{r} q_j^{f_j} \tag{4}$$

with $q_j \leq B$ for $1 \leq j \leq r$. It is well-known that

$$p - 2\sqrt{p} + 1 \leq \#E_{A,B}(\mathbf{F}_p) \leq p + 2\sqrt{p} + 1 \tag{5}$$

and that all orders in this range are taken on for some parameters A and B. Furthermore, a theorem of Rück [18] implies that for each order in this interval there exists a *cyclic* elliptic curve. We also refer to [13] for an introduction to elliptic curves.

Very little is known about the existence of smooth numbers in the interval (5) of interest for a given prime p. However, it is known [3] that for every fixed u,

$$\psi(n, n^{1/u})/n = u^{-(1+o(u))u}$$

where $\psi(n, y)$ denotes the number of integers $\leq n$ with no prime divisor $\geq y$. This fact suggests that every integer n has the property P_c defined below for some $c > 1$.

Definition 2. An integer n has *property* P_c if there exists an integer b satisfying $n - 2\sqrt{n} + 1 \leq b \leq n + 2\sqrt{n} + 1$ with no prime divisor greater than $n^{c/u}$, where u is defined by $u^{2u} = n$.

Conjecture. There exists a constant $c > 1$ such that all sufficiently large n have property P_c.

An even stronger conjecture is that the above conjecture holds for any $c > 1$. For example, let n be a 100-digit number and note that $10^{100} \approx 33^{66}$. One can therefore expect to find an integer in the interval $[n - 10^{50}, n + 10^{50}]$ with no prime divisor greater than $10^{c \cdot 100/33}$ for some c. For $c = 1.1$ and $c = 2$ this bound is approximately 2000 and 10^6, respectively.

If p has a special form for which an elliptic curve with smooth order can be constructed efficiently, then our proof technique described below allows to prove that computing x' can be reduced efficiently to breaking the Diffie-Hellman protocol. If p has no special form but has property P_c for some small c, our proof technique allows to prove that for a general group G for which p divides $|G|$ there exists a small fixed piece of information (the elliptic curve parameters) which, when given, allows to reduce the problem of computing x' to breaking the Diffie-Hellman protocol.

One can explicitly construct certain super-singular elliptic curves with known order. For example, the curves defined over $\mathbf{F}_p$ for $p \equiv 3 \pmod 4$ by the equation $y^2 = x^3 + ax$ have order $p+1$ as do the curves defined over $\mathbf{F}_p$ for $p \equiv 2 \pmod 3$ by the equation $y^2 = x^3 + b$. An alternative group for exploiting the smoothness of $p+1$ is the subgroup of order $p+1$ of $\mathbf{F}_{p^2}$. This group can also be constructed for that quarter of the primes (namely those with $p \not\equiv 1 \pmod{12}$) for which no appropriate super-singular elliptic curve can be obtained.

Let us assume for now that a cyclic elliptic curve $E = E_{A,B}(\mathbf{F}_p)$ with smooth order $T = \prod_{j=1}^{r} q_j^{f_j}$ is given and that we wish to compute x' according to (2) where $a = g^x$ in G. Let $P = (u, v)$ be a generator of E. We can consider x' to be the x-coordinate of a point on $E_{A,B}(\mathbf{F}_p)$. If there exists no such point, i.e., if $x^3 + Ax + B$ is a quadratic non-residue modulo p, then an expected number of only two random choices $d \in \mathbf{F}_p$ is required until for $x'' = x + d$, $(x'')^3 + Ax'' + B$ is a quadratic residue modulo p. Let y be one of the corresponding y-coordinates, i.e., let (x'', y) be a point on E. This point can be written as some multiple of the generator P of E:

$$(x'', y) = w \cdot P. \tag{6}$$

If we can determine w, then we can compute x'' and hence x' because we know d.

We now describe an efficient algorithm using the DH-oracle for computing w from $a = g^x$. Note first that we can compute $g^z = g^{x^3+Ax+B}$ by two applications of the DH-oracle and $O(\log A + \log B) = O(\log p)$ group operations. Here z is a quadratic residue modulo p if and only if $z^{(p-1)/2} \equiv 1 \pmod{p}$. This condition is equivalent to

$$hz^{(p-1)/2} \equiv h \pmod{|G|}$$

and thus also to

$$g^{hz^{(p-1)/2}} = \left(g^{(z^{(p-1)/2})}\right)^h = g^h.$$

Testing quadratic residuosity of z modulo p is thus equivalent to testing equality in G of two elements of G, which can be computed from g^z by $O(\log p)$ applications of the DH-oracle and $O(\log h) = O(\log |G|)$ group operations. If z is not a quadratic residue modulo p we can perform the same check for g^{x+d} for randomly selected d's, until it is successful. For the first d for which $z = (x+d)^3 + A(x+d) + B$ is a quadratic residue modulo p we can compute g^y from g^z where $y^2 \equiv z \pmod{p}$ by using the DH-oracle in a modular square root algorithm [9, 14]. Hence $(x + d, y)$ is a point on E. This step requires $O(\log p)$ calls to the DH-oracle.

We further note that for given pairs (g^{u_1}, g^{v_1}) and (g^{u_2}, g^{v_2}), where (u_1, v_1) and (u_2, v_2) are points on E (not known explicitly), we can compute the pair (g^{u_3}, g^{v_3}) such that $(u_3, v_3) = (u_1, v_1) + (u_2, v_2)$ on E. This is achieved by using a standard algorithm for addition on elliptic curves [13], where multiplications modulo p are replaced by calls to the DH-oracle. Note that u_1, v_1, u_2 and v_2 need not be, and that u_3 and v_3 generally will not be, in reduced form modulo p. If the points on E are represented in affine coordinates as shown here, one such hidden elliptic curve addition requires $O(\log p)$ calls to the DH-oracle. However, if the points on E are represented in projective coordinates, only a constant number of oracle calls are needed. The conversion from projective to affine coordinates requires $O(\log p)$ oracle calls.

In order to compute x'' (which satisfies $x + d \equiv x'' \pmod{p}$) we compute w defined by (6) where y is such that $(x + d, y)$ is a point on E (see above). Let w_{jk} for $1 \le j \le r$ and $0 \le k < f_j$ be defined uniquely by $0 \le w_{jk} < q_j$ and

$$w \equiv \sum_{k=0}^{f_j-1} w_{jk} q_j^k \pmod{q_j^{f_j}}.$$

The number w can easily be computed from the w_{jk}'s using the Chinese remainder theorem.

Consider a specific j for which we wish to determine $w_{j0}, \ldots, w_{j,f_j-1}$. We have

$$\frac{T}{q_j}\, w \equiv \frac{T}{q_j}\, w_{j0} \pmod{T}$$

and therefore

$$(u', v') := \frac{T}{q_j} \cdot (x'', y) = \left(\frac{T}{q_j}\, w\right) \cdot P = \left(\frac{T}{q_j}\, w_{j0}\right) \cdot P$$

on E. We can thus compute $(g^{u'}, g^{v'})$ from $(g^{x''}, g^{y})$ by using $O(\log p)$ oracle calls (for projective coordinates, $O(\log^2 p)$ for affine coordinates). Using "normal" group operations we can now compute $(g^{hu'}, g^{hv'})$. We further compute $(g^{hu''_t}, g^{hv''_t})$ for $t = 0, 1, 2, \ldots$, where

$$(u''_t, v''_t) = \left(\frac{T}{q_j}\, t\right) \cdot P$$

on E, using normal operations in G, and compare the pairs $(g^{hu''_t}, g^{hv''_t})$ and $(g^{hu'}, g^{hv'})$ as suggested by (3), until a match is found for some t, which is set equal to w_{j0}. Note that $t < q_j$, that is at most q_j trials are needed.

The numbers w_{jm} for $m \geq 1$ can be computed by a generalization of the described method. This allows to prove the following theorem.

Theorem 1. *Let $G = \langle g \rangle$ be an arbitrary cyclic group with order $|G| = \prod_{i=1}^{r} p_i^{e_i}$. If for each prime p_i the parameters A_i and B_i of a cyclic elliptic curve $E_{A_i,B_i}(p_i)$ with smooth order for a smoothness bound B are given, then discrete logarithms in G can be computed using $O(\log^2 |G|)$ calls to the DH-oracle and $O((B/\log B)$ $\log^2 |G|)$ group operations. If $e_i > 1$ for some i, then a DH-oracle for subgroups of G is also required.*

Corollary 2. *For groups whose order is such that for every p_i one can construct an elliptic curve according to Theorem 1 (or another cyclic group with smooth order defined algebraically over $\mathbf{F}_{p_i}$), breaking the Diffie-Hellman protocol is equivalent to computing discrete logarithms. Among these groups are those for which either $p_i - 1$ or $p_i + 1$ is smooth for all i.*

Corollary 3. *If the stated number-theoretic conjecture is true, then for every group $G = \langle g \rangle$ with known order $|G|$ there exists a side information string S of length at most $2 \log |G|$ such that when given S, breaking the Diffie-Hellman protocol for G and base g is polynomial-time equivalent to computing discrete logarithms in G to the base g. If $|G|$ contains multiple prime factors greater than $(\log |G|)^k$ for some fixed k, then the equivalence only holds with respect to breaking the Diffie-Hellman protocol for certain subgroups of G.*

Remarks.
(1) The string S consists of appropriate elliptic curve parameters for all prime divisors of $|G|$. It need not be assumed in Corollary 3 that $|G|$ be known because

it can also be computed from S. Note that the order of G and its factorization are known in many proposed cryptographic applications. In fact, it is often suggested to use a group (or subgroup) of prime order.
(2) When the order of G is not known, it is conceivable that giving $|G|$ could be of some help in computing discrete logarithms in G. However, in those cases where $|G|$ is known, there seems to be no reason to believe that knowledge of S could reduce the difficulty of the discrete logarithm problem, but this has not been proved.

4. Concluding remarks

Let p be a prime factor of $|G|$. The proof technique presented in this paper applies to any cyclic group F (with generator f) with smooth order which is defined algebraically over $\mathbf{F}_p$ and whose elements are represented by vectors over $\mathbf{F}_p$, provided it is possible to determine (by an algebraic computation) explicitly an element of F when the value of one of the coordinates is fixed. The idea is to assign to this coordinate the (hidden) logarithm x implicitly given by g^x and to perform a computation using the DH-oracle to obtain the hidden values of the other coordinates. This results in the disguised version of an element P of F. The next step is to compute certain powers of f (in disguised form) using normal group operations in G, and to compare this disguised element of F with another disguised element of F obtained from P by calls to the DH-oracle. These tests allow to compute explicitly the logarithm w of P (in F) to the base f. Given w and f we can explicitly compute the group element containing x as one of the coordinates. Examples considered in this paper were the multiplicative group of $\mathbf{F}_p$ proposed initially in [4] and elliptic curves over $\mathbf{F}_p$. The equivalence holds in a strict sense (without side information S) only if the group F with smooth order can be constructed explicitly.

It is conceivable that the application of hyperelliptic curves or some higher-degree Abelian varieties could allow to remove the plausible but unproven number-theoretic assumption from Theorem 3 because the relative sizes of the corresponding intervals for the orders of the groups are much larger than for elliptic curves. This would be similar to the generalization of the Goldwasser-Kilian elliptic curve primality test [7] to hyperelliptic curves [1], which allowed to settle the last unproven details in [7] and resulted in the first rigorously proven polynomial-time primality test.

Corollary 2 implies that if one could explicitly construct elliptic curves with smooth order for a given prime, then breaking the Diffie-Hellman protocol would be equivalent to computing discrete logarithms for all groups, without side information S. However, because the solution of the same problem for *composite* moduli would immediately yield an efficient factoring algorithm based on [8], it appears quite unlikely that this problem can be solved efficiently for prime moduli.

The results of this paper suggest to construct groups of prime order p for use in the Diffie-Hellman protocol in a manner that an explicit group (defined modulo p) with smooth order is known. Note that the description of the group need not be published but only built into the system design in a secret manner. However, it appears questionable whether using a group G with prime order $|G| = p$ such that $p - 1$ or $p + 1$ is smooth is a good idea, although no efficient discrete logarithm algorithms are known in this case. To find such an algorithm is suggested as an open problem.

Acknowledgment

I would like to thank Jim Hafner, Jürg Kramer, Kevin McCurley, Alfred Menezes, and Carl Pomerance for answering my questions on the state of knowledge in number theory about the existence of smooth numbers in certain intervals, and for other helpful comments.

References

1. L.M. Adleman and M.A. Huang, Primality testing and abelian varieties over finite fields, *Lecture Notes in Mathematics*, vol. 1512, Springer Verlag, 1992.
2. J. Buchmann and H.C. Williams, A key-exchange system based on imaginary quadratic fields, *Journal of Cryptology*, vol. 1, no. 2, pp. 107-118, 1988.
3. E.R. Canfield, P. Erdös and C. Pomerance, On a problem of Oppenheim concerning "Factorisatio Numerorum", *J. Number Theory*, vol. 17, pp. 1-28, 1983.
4. B. den Boer, Diffie-Hellman is as strong as discrete log for certain primes, *Advances in Cryptology – CRYPTO '88*, Lecture Notes in Computer Science, vol. 403, pp. 530-539, Berlin: Springer-Verlag, 1989.
5. W. Diffie and M.E. Hellman, New directions in cryptography, *IEEE Transactions on Information Theory*, vol. 22, no. 6, pp. 644-654, 1976.
6. T. El-Gamal, A public key cryptosystem and a signature scheme based on the discrete logarithm, *IEEE Transactions on Information Theory*, vol. 31, no. 4, pp. 469-472, 1985.
7. S. Goldwasser and J. Kilian, Almost all primes can be quickly certified, *Proc. of the 18th Annual ACM Symposium on the Theory of Computing*, pp. 316-329, 1986.
8. H.W. Lenstra, Jr., Factoring integers with elliptic curves, *Annals of Mathematics*, vol. 126, pp. 649-673, 1987.
9. J.L. Massey, Advanced Technology Seminars Short Course Notes, Zurich, 1993, pp 6.66-6.68.
10. U.M. Maurer and Y. Yacobi, Non-interactive public-key cryptography, *Advances in Cryptology - EUROCRYPT '91*, Lecture Notes in Computer Science, Berlin: Springer-Verlag, vol. 547, pp. 498-507, 1991.
11. K.S. McCurley, A key distribution system equivalent to factoring, *Journal of Cryptology*, vol. 1, no. 2, pp. 95-105.
12. K.S. McCurley, The discrete logarithm problem, in *Cryptology and computational number theory*, C. Pomerance (ed.), Proc. of Symp. in Applied Math., vol. 42, pp. 49-74, American Mathematical Society, 1990.
13. A. Menezes, Elliptic curve public key cryptosystems, Kluwer Academic Publishers, 1993.

14. R. Peralta, A simple and fast probabilistic algorithm for computing square roots modulo a prime number, *IEEE Trans. on Information Theory*, vol. 32, no. 6, pp. 846-847, 1986.
15. S.C. Pohlig and M.E. Hellman, An improved algorithm for computing logarithms over $GF(p)$ and its cryptographic significance, *IEEE Transactions on Information Theory*, vol. 24, no. 1, pp. 106-110, 1978.
16. J.M. Pollard, Theorems on factorization and primality testing, *Proceedings of the Cambridge Philosophical Society*, vol. 76, pp. 521-528, 1974.
17. R.L. Rivest, A. Shamir, and L. Adleman, A method for obtaining digital signatures and public-key cryptosystems, *Communications of the ACM*, vol. 21, no. 2, pp. 120-126, 1978.
18. H. Rück, A note on elliptic curves over finite fields, *Math. Comp.*, vol. 49, pp. 301-304, 1987.
19. C.P. Schnorr, Efficient identification and signatures for smart cards, Advances in Cryptology – CRYPTO '89, Lecture Notes in Computer Science, vol. 435, pp. 239-252, Berlin: Springer-Verlag, 1990.

Fast Generation of Provable Primes Using Search in Arithmetic Progressions

Preda Mihailescu

Union Bank of Switzerland, CH 8021 Zürich

Abstract. Many cryptographic algorithms use number theory. They share the problem of generating large primes with a given (fixed) number n of bits. In a series of articles, Brandt, Damgard, Landrock and Pomerance address the problem of optimal use of probabilistic primality proofs for generation of cryptographic primes. Maurer proposed using the Pocklington lemma for generating provable primes. His approach loses efficiency due to involved mechanisms for generating close to uniform distribution of primes. We propose an algorithm which generates provable primes and can be shown to be the most efficient prime generation algorithm up to date. This is possible at the cost of a slight reduction of the set of primes which may be produced by the algorithm. However, the entropy of the primes produced by this algorithm is assymptotically equal to the entropy of primes with random uniform distribution. Primes are sought in arithmetic progressions and proved by recursion. Search in arithmetic progressions allows the use of Eratosthenes sieves, which leads finaly to saving 1/3 of the psuedo prime tests compared to random search.

1 Introduction

Primality testing has a long history and has undergone a radical development during the last decade. We shall not go into details of this development but begin with mentioning that prime tests split into compositeness proofs (which are polynomial) and primality proofs, which are more complex. The fastest known primality proof [8] which has been implemented by Atkins and Morain [16], is polynomial, but in practice slower than superpolynomial algorithms such as the cyclotomy test [4].

Algorithms proposed for the specific purpose of generating primes for cryptographic use are different due to the technological motivation behind them. Efficiency and simplicity are the main concerns and - as a consequence of the domain of application - "security" of primes is an issue of concern too. Security of a prime p basically means that the discrete logarithm base p and factorization of p out of a product of usually two primes should be, if possible, harder than average. The last condition raised the wish to have large prime factors in $p \pm 1$; this was formalized in conditions for so called "Gordon secure primes" [10]. It was obvious that with the information about factors of $p \pm 1$ available for a Gordon prime, a proof by Pocklington's lemma was almost for free. The first to make use of this observation was Maurer in [14]. By the date of Maurer's paper,

Gordon primes were de facto outdated by the ECM ([12]) factoring method: Gordon primes were designed to reduce the odds of $p \pm 1$ factoring methods [19], but cannot provide any specific protection against elliptic curve factoring. This facts are reflected in [21]. For technological applications, Maurer's algorithm lacks both the ease of implementation and the efficiency requirements - being, by the authors statement ([15]), p19 60% slower than Rabin - Miller tests which are also simpler to implement. Maurer does state in [15],p29, that his algorithm is more efficient than probable primes, but this statement is not supported by facts and is in contradiction with the rest of the paper. The paper [14] is important for having first indicated that for the specific purpose of cryptographic use, provable primes were worth considering. We propose an algorithm that seeks primes in arithmetic sequences for which a factorization of the ratio is known - this implies a recursive use of the algorithm. For the trial division phase we use Eratosthenes sieve methods, which lead to an increase of the optimal length of trial division; optimal trial division was also evaluated in [14], [2]. In the last paper, the authors also consider means of speeding up Maurer's method and notice that the 'almost' uniform distribution required by Maurer has a negative impact on the performance. An elegant analysis of the information entropy of primes produced by "incremental search" - and proved with Rabin - Miller tests - is provided in [1]. Incremental search is a special case of arithmetic progressions - when the ratio is 2. We shall adapt some of the methods of that article for the analysis of our alogithm. Probabilistic methods like Rabin - Miller may falsely declare composites for primes and the probability for this event decreases with the iteration of the same test for different bases. In [3] the authors show that this probability also depends on the size of the primes produced; tables for the error probability $q_{t,k}$ that a composite with k - binary digits is declared prime after t independent Rabin - Miller tests are quoted also in [1]; the tables show that the probability of failure of the Rabin - Miller test after t rounds is substantially lower than the 4^{-t} initially proved by Rabin [20]. These results suggest to the authors the conclusion that "using the Rabin test remains in many cases the most practical approach" [1],p.1. We show that recursive primality proofs combined with search in arithmetic progressions is preferable to Rabin - Miller tests, also from the point of view of efficiency, in the range of sizes of primes covered by the tables in [1] and probably for primes up to at least 1000 bits. For larger primes, the difference becomes negligeable.

2 Outline of the Algorithm

Let $B > 0$ and $n > 0$ be positive integers and $s, c > 1$ real constants. We propose following algorithm for generating provable primes of n digits, together with their certifiactes:

Algorithm AP:
Input: n.
Constants: B, c, s.

- **Step 1**. If $n < B$ return a random prime with n bits (generated by trial division).
- **Step 2**. Produce an integer F with $2^{\epsilon n} < F < 2^{c\epsilon n}$ which is completely factored, by recursive use of this algorithm. The constant ϵ may be 1/2, or 1/3 , depending on the prime ceritification used (see bellow).
- **Step 3**. Draw a random number $t \in (2^{n-2}/F, 2^{n-1}/F - sn)$.
- **Step 4**. Find a prime in the arithmetic progression $P = \{N \mid N = N_0 + ia; N_0 = ta + 1; a = 2F; 0 \leq i \leq s\}$.

The primality test for Step 4 splits naturally in three parts which are applied to prime candidates in the progression P until a prime is found - together with a corresponding certificate, or the bound s is passed and the algorithm returns failure.

Part I : **Trial division by primes** $< A$, where A is a given upper bound. calculate $a'_p = a \bmod p, \forall p < A$. calculate $N'_p = N_0 \bmod p, \forall p < A$. initiate an array tab of length s. For all $p < A$, set $tab[i] = 1$ for all solutions of the modular equation $N'_p + ia'_p = 0 \bmod p$. This is obviously an Eratosthenes sieve.

Part II : **Compositeness test** using Rabin - Miller. traverse the table tab starting with $i = 0$ and perform a Rabin - Miller test with base 2, for the i-th elements in the progression P, such that $tab[i] = 0$. Continue if the test declares "composite", go to Part III otherwise.

Part III : **Primality proof** using the Pocklington lemma.

We recall the Pocklington lemma:

Proposition 1. : *Let N be an integer, such that $N - 1 = F \cdot R$, where F is completely factored. Suppose that,*

$$\forall q \mid F, \quad \exists \alpha_q \text{ with } \quad (\alpha_q^{(N-1)/q}, N) = 1 \text{ and } \quad \alpha_q^{N-1} = 1 \bmod N \tag{1}$$

Then all primes $p \mid N$ are of the shape $p = kF + 1$. In particular, if $F > \sqrt{N}$, then N is prime.

Proof. Let $\alpha = \prod_{q|F} \alpha_q$. Then

$$(\alpha^{(N-1)/q}, N) = 1 \quad \text{and} \quad \alpha^{N-1} = 1 \bmod N \quad \forall q \mid F \tag{2}$$

If $r||N$ is a prime, the above equations hold a fortiori mod r. The multilpicative group mod r has thus a subgroup of index F generated by α thus proving the first assertion: $F \mid (r - 1)$ and, a fortiori, $r \geq F + 1$. If $F > \sqrt{N}$, it follows that $r > \sqrt{N}, \forall r \mid N, r$ prime. This is a contradiction, since N must have at least one prime factor $r \leq \sqrt{N}$. Consequently, N must be prime, which ends the proof.

In part III of the algorithm, for each $q||F$, a base α_q verifying (3.4) is sought among the first primes, starting with 2. If for some base $\alpha_q^{N-1} = 1 \bmod N$ does not hold, N is composite (little Fermat does not hold). If for some q, no base verifying (1) is found before a fixed upper bound R of trials, no primality or compositeness proof can be provided and N is incremented (Part II is proceeded). Finally, if bases α_q are found $\forall q \mid F$, N is prime. The prime certificate consists of the pairs $\{(\alpha_q, q) \mid q \mid F\}$ together with certificates for each prime q, recursively [18]. Providing all these pairs together with the certificates for q being prime gives a proof of N's primality by the above lemma. In [5] a certification method using an $O(n^{1/3})$ factored part of $n-1$ is provided. It requires proving that a certain discriminant is not a perfect square. Suppose $n = Ft + 1$ and $F > \sqrt[3]{n}$, and bases α_q verifying (3.4) have been found for all $q \mid F$. If n is composite, it is built up of two primes $p_i = k_i \cdot F + 1$. Let $n = aF^2 + bF + 1$; by comparing $n = p_1 p_2$ we get $k_1 k_2 = a$ and $k_1 + k_2 = b$. It is then easy to see that n is composite iff $\Delta = b^2 - 4a$ is a perfect square. This again can be checked by verifying if Δ is a perfect square modulo a set of primes $\{p \mid \prod p > F\}$; if this is the case, $(k_i \bmod F) = k_i < F$. Otherwise n is prime. One can take in general $\{p \mid p < logF\}$ as set of primes.

3 Optimal Trial Division and Run Time Analysis

We assume a machine with wordlength B, the time for a short integer multiplication is t and let $m = n/B$. The time involved in the trial division is: $T_t = t((1+\epsilon)m + \log A)\frac{A}{\log A}$ and is build up of $(1+\epsilon)m$ short divisions for calculating a' and N' and $\log A$ divisions for calculating a modular inverse, for each prime; we shall write $T_t = tm(1+\epsilon+\delta)\frac{A}{\log A}$, with $\delta = \frac{\log A}{m} \longrightarrow 0$ for $m \longrightarrow \infty$. Note that the trial divisions are performed only once and do not have to be repeated for all n candidates. Assuming that the prime candidates for AP are equally distributed among the residue classes $\bmod p$ for all small test primes p, the fraction of prime candidates after this step is, by Mertens' theorem $f = \frac{e^{-\gamma}}{\log A}$. The second part then takes in average:

$$T_R = t \cdot f \cdot d \cdot w^2 \cdot m^4 = t \cdot \frac{f \cdot d \cdot n^4}{w^2}, \tag{3}$$

where the length of the expected seek interval is $d \cdot n$, and naive multiplication is used for exponentiation. We shall see that $d = 1$, so we subsequently neglect this factor. Optimal table length A may be found by writing $g_P(A) = T_t(A) + T_R(A)$ and taking the derivative. The optimal value found is $A_P = \frac{kn^2m}{\log(n^2m)}$, with a constant $k = \frac{e^{-\gamma}}{1+\epsilon+\delta}$ close to 0.4. As a term of comparison, optimal table length in [14] is $O(nm)$. Actually, Maurer's argument goes like this: adding a prime A to the table saves a pseudoprime test with probability $1/A$, so optimal length is reached when the expected saving T_e/A balances the expected extra division which he sets to $1 \cdot T_d$; the argument is incorrect, since both values should be averaged over the whole seek interval. It does however yield a good estimate of the optimal value $A_R = \frac{2e^{-\gamma}}{\alpha(A)} nm$, with $\alpha(A)$ defined in [14], as found by taking the derivative of the adequate function g_R (indices P stay for progression search, R for random search). In [2] the optimal table length is even smaller: $O\left(\frac{nm}{\log(nm)}\right)$. An alternative to the sieve method consists in computing the current remainders $N_c = N_i + ia \bmod p$. This yields the same assymptotics as A_P, with δ replaced by a constant $\alpha = 2wt_a$ where t_a is the ratio between the time for a short addition and a short multiplication. Beside the smaller constant, this approach has the disadvantage that the remainders $\bmod p$ must be stored (three values for each test prime p). Comparing A_P to A_R shows that the use of the Eratosthenes sieve sensibly increases optimal table length, leading to the elimination of more candidates in the trial division phase. One can evaluate this performance improvement: since the number of pseudoprime tests is in average fn both for random search and for search in progressions, in the last case the exponentiations drop by a factor of $f_p = \frac{\log A_R}{\log A_P} = O\left(\frac{\log(nm)}{\log\left(n^2 m / \log(n^2 m)\right)}\right) \to \frac{2}{3}$ for $n \to \infty$. This improvement appears to be more substantial than what can be saved in a probabilistic proof approach by better error estimates, like in [3]. For $n = 1024$ the number of primes for the trial division phase is about 2^{15} (with $A \approx 2^{19}$) and this may be argued to be too large for some machines. Using the incremental difference storing for the primes, this requires 32 KB of memory which should be afordable on most current PC's. The saving factor is 0.72 and it results in about 10 pseudo prime tests less! It is theoretically intersting to observe that $A_P = O\left(\frac{n^4}{(\log n)^2}\right)$ can be achieved by using a divide and conquer method. The idea is following: group the test primes in $A/(\log A\sqrt{m})$ products of length $\sqrt{m}$ and take first the remainders modulo these products and then the remainders of these remainders modulo the single primes; this yields: $T_t = 2t\sqrt{m}(1+\epsilon+\delta)\frac{A}{\log A}$. Iterating this method leads to the stated result, when the factor $2\sqrt{m}$ is replaced by $(\log m) \cdot m^{1/\log m} = O(\log m)$. This means taking products of length $m/2, m/4, \ldots, 2$. Of course, the building of the factors belongs to a precomputation phase and the storage space explodes to $O(n^2 m^2)$ - the method is not meant for implementation! Note that the base 2 is a suitable base for the Pocklington lemma, being a q-th power nonresidue with probability $(1-1/q)$. This is remarcable, since exponentiation base 2 is particularly efficient ([2]).We shall consider here the version of the algorithm where F is built up of a single prime. With this, the time for part III is $1-\epsilon$ exponentiations, since for proving that 2 is an adequate F-th power nonresidue, one only raises to the power $(n-1)/F$, the exponent thus having length $n(1-\epsilon)$. With probability $1/F$ a further base should be tested: we do not take this event into account, since its probability is exponentially low. If $\epsilon = 1/3$, checking that Δ is not a perfect square takes $\sum \log p = O(\log F) = O(n)$ short opertations, which are negligeable. We now want to motivate our claim that this algorithm is more efficient than pseudo - prime algorithms, for the range of values of n of current interest. Of course, the sieve method can also be used with pseudo - primes in incremental search, although this method was not proposed before. Some implementation might also use suboptimal tablelength, due to limited memory.

We thus do not consider the advantages of trial division as specific for AP. The advantage of our algorithm compared to pseudo - prime algorithms will thus relay in the shorter proof, the disadvantage, in recursion. It is these two factors we shall have to compare. If R is the number of Rabin - Miller tests performed until a pseudoprime is found and T_e is the time for an n-bit exponentiation, the time for the proof is, as seen above, $(1-\epsilon)T_e$ and the time involved in recursion is upper bounded by $T_r = (\frac{\epsilon^4}{1-\epsilon^4}) \cdot RT_e$. The total overhead for proving a prime is thus: $T_r = (\frac{\epsilon^4}{1-\epsilon^4} \cdot R + (1-\epsilon)) \cdot T_e$. With $n = 384$ - which is currently standard for RSA - for a weak implementation, with $\epsilon = 1/2$ and $A = 1000$ - thus $R \approx 20$, the overhead is $T_p \approx \frac{11}{6}T_e$. If $\epsilon = 1/3$, $T_p < T_e$, even without increasing the tables. With optimal tables, $T_p < T_e$, up to more than 1000 bits. A standard implementation of our algorithm needs thus less than the time for one additional exponentiation as overhead for a primality proof, for $n \leq 1000$ bits. Compared to that, a Rabin - Miller test with probability of error $< 2^{-48}$ requires three additional exponentiations for 384 bits and two for 600 bits; no values for $n = 1000$ are provided in [1]. Extrapolating the values from [1], two Rabin - Miller tests for $n = 1000$ bits will provide an error probability p_e with $2^{-60} < p_e < 2^{-48}$. For primes with more than 1000 bits our algorithm may become more expensive than a pseudo - prime algorithm, while keeping the advantage of providing a primality proof. It must be noted that for large primes, the number R of Rabin tests which eliminate composite candidates increases (linearly with n) and the overhead for the primality proof becomes negligeable: a good approximation of R is $R(n, A) = \frac{n\ e^{-\gamma} \log 2}{\log A} \approx 0.392 \frac{n}{\log A}$, the logarithm being the natural one. From this and the estimate for the overhead T_p above, it follows,

$$f(n) = \frac{T_p}{R\ T_e} = \frac{\epsilon^4}{1-\epsilon^4} + \frac{0.392(1-\epsilon)\log A}{n} \approx \frac{1}{80} + \frac{2/3}{n/\log(n)}. \tag{4}$$

and $f(n) < 2\%$ for $n > 1000$. Thus AP is never more than 2% slower than probabilistic primes methods, if at all!

4 Analysis of the Algorithm

This algorithm has two different and independent reasons for producing non uniformly distributed primes. The first is that while searching primes in arithmetical progressions, primes at the end of long gaps are chosen with higher probability. This very phenomenon was investigated by Brandt and Darmgard in [1], for progressions with ratio 2. Their results relay upon the prime r-tuple conjecture of Hardy - Littlewood. This is an assympotoic formula for the number $\pi_{\mathbf{d}}(N)$ of positive integers $n \leq N$ such that $n + d_1, n + d_2, \ldots, n + d_r$ are simultaneously primes. The vector $\mathbf{d}$ is the r-tuple $(d_1, d_2, \ldots, d_r)$. The conjecture says:

$$\pi_{\mathbf{d}}(N) \sim S_{\mathbf{d}} \frac{N}{(\log N)^r} \qquad \text{for } N \to \infty. \tag{5}$$

with $S_{\mathbf{d}} = \prod_p \frac{p^{r-1}(p-\nu_{\mathbf{d}}(p))}{(p-1)^r}$, provided $S_{\mathbf{d}} \neq 0$ and where $\nu_{\mathbf{d}}(p)$ is the number of distinct residue classes modulo p occupied by the numbers in the r-tuple $\mathbf{d}$. Gallagher [9] uses this conjecture to prove that:

$$\sum_{\mathbf{d} \in H'} S_{\mathbf{d}} \sim h^r \quad \text{for } h \to \infty \quad \text{where } H' = \{(d_1, d_2, \ldots, d_r) \mid d \in N, d_i \neq d_j\} \tag{6}$$

In order to use Gallagher's result for primes in arithmetic progressions with ratio F, we must assume the extended Riemann conjecture and $F = o(\sqrt{n})$, for proving the necessary distribution of primes in arithmetic progressions [6]. With this, Gallagher's result can be generalized to following statement. The number $Q_k(h, N)$ of integers $n < N$ for which the interval $(n, n + hF)$ contains exactly k primes in the progression $1 + tF$ verifies:

$$Q_k(h, N) \sim N \frac{e^{-\lambda} \lambda^k}{k!} \quad \text{for } N \to \infty,\ h \sim \lambda \frac{\varphi(F)}{F} \log N \tag{7}$$

and lemma 5 from [1] can be restated:

Proposition 2. *Let $G_h(x)$ denote the number of primes p in the progression $1 + tF$, such that $p < x$ and $p - q \leq hF$, where q is the prime preceding p in the progression. Assuming the r-tuple conjecture, the ERH and with $F = o(\sqrt{n})$, for any constant λ,*

$$G_{\lambda \varphi(F) \log(x)/F}(x) = \frac{xF}{\varphi(F) \log(x)} (1 - e^{-\lambda}) \cdot (1 + o(1)). \tag{8}$$

The factor $\varphi(F)/F$ stems from the fact that $\varphi(F)$ out of F residue classes modulo F contain primes. The proof of this lemma uses (6) and Dirichlet's theorem on distribution of primes in arithmetic progressions [6]. Let M_k be the set of primes with k bits and for $k \leq B$, let $M'_k = M_k$. For $k > B$, we put: $M'_k = \{p \in M \mid p = tq + 1, q \in M'_{[k/3]+1}\}$. Then M'_k is a subset of the primes that can be produced using AP. Let H''_k be the uncertainty about primes sought in arithmetic porgressions by AP and H'_k be the uncertainty of randomly chosen primes in M'_k. Using (7) and methods like in [1], one can prove that:

$$\frac{H''_k}{H'_k} \sim 1 \quad \text{for } k \to \infty. \tag{9}$$

It can also be shown with the same means that the expected length of the seek interval for primes in M'_k is k. This result is not trivial. The average distance between two primes in an arithmetic progression is k, but a prime laying at the end of an "average length interval" is found after only $k/2$ trials, since the starting point of the seek-sequence may lay anywhere between this prime and its predecessor. However, the expected value of the length of the seek interval is not $k/2$ but k and this is a consequence of the fact that primes at the end of longer intervals are chosen with higher probability. So, incidentaly, the average number of candidates to be tested for primality is the same in random choice as in progressions, asuming the r-tuple conjecture! In [2], this result is stated for incremental search, as an empiric evidence. The paper [1] provides the tools for such a proof, but does not address the topic. We state a theorem on arithmetic progressions, which contains the empirical result in [2] as a subcase, when the ratio of the progression is $F = 2$.

Theorem 3. *Let $P_k = \{p = \lambda F + u\} \cap M_k$ be the set of k-bit numbers in an arithmetic progression, with $(u, F) = 1$. Assuming the r-tuple conjecture, the extended Riemann conjecture and if $F = o(\sqrt{2}^k)$, then the expected search length for finding a prime by incremental search in P_k, with random starting point is $E(l) = \log(2^k)\frac{\varphi(F)}{F}$.*

Proof. Let $l(p) = \frac{p-q}{F}$ be the distance of the prime p to its prime predecessor q in P_k. The probability that a prime p is chosen is $P(p) = \frac{l(p)}{N/(2F)}$: in fact, $l(p)$ out of $N/(2F)$ possible startpoints lead to the choice of p, so $E(l) = \sum_{p \in P_k} P(p)\frac{l(p)}{2} = \sum_{p \in P_k} \frac{l^2(p)}{N/F}$. Let $\mathbf{L} = \{0 < \lambda_1 < \lambda_2 < \ldots < \lambda_m < \infty\}$ be a partition and k such that the term $o(1)$ in Proposition 2 is $< 1/m^2$. Then

$$E(l) = \sum_{i=0}^{m} \sum_{l\lambda_i < l(p) < l\lambda_{i+1}} \frac{l^2(p)}{N/F} + \sum_{l\lambda_{i+1} < l(p)} \frac{l^2(p)}{N/F}, \quad \text{with } l = \frac{\varphi(F)\log N}{F}. \tag{10}$$

By Proposition 2, the number of primes with $l\lambda_i < l(p) < l\lambda_{i+1}$ is

$$\frac{NF}{2\varphi(F)\log N}\left(e^{-\lambda_i} - e^{-\lambda_{i+1}}\right)(1+\delta), \tag{11}$$

with $\delta < 2/m^2$. It follows that $E(l) = (\log N)\frac{\varphi(F)}{2F} f(\mathbf{L}) + O(1/m)$, where $f(\mathbf{L})$ is a Riemann sum of the function $h(x) = x^2 e^{-x}$; letting m grow together with k, so that the condition on the error term stays valid, $f(\mathbf{L})$ will converge to the integral $\int_0^\infty f(x) = 2$, which finishes our proof.

This result needs some comments. It shows that introducing small factors in F will decrease the expected seek length; this seems to suggest a means to reduce the number of pseudoprime tests in our algorithm. However, if $p \mid F$ then $(1 + tF, p) = 1$, so p will be ineffective in the trial division stage. Overall, what one gains in the seek interval length, one loses in the trial division. One may thus aswell choose F to be a prime and this explains our statement about the expected seek interval. The second reason for nonuniform distribution is more dramatic: certain primes with n bits are not produced at all (their probability is 0). We finally want to prove that this does not affect the entropy in the sense that the uncertainty about a prime randomly chosen in M_k' is assymptotically equal to the uncertainty of a prime randomly chosen in M_k. For random choice, it is easy to see that the entropy is $H_k = \log(\|M_k\|)$ resp. $H_k' = \log(\|M_k'\|)$. We need to give an estimate of $||M'_k||$. By definition, $M'_k = \{p = 2tq + 1 | q \in M'_{[k/3]+1}\} \cap M_k$ and it follows that:

$$\frac{2^{\frac{2k}{3}-3}}{k \log 2} \leq \frac{||M'_k||}{||M'_{k_1}||} \leq \frac{2^{\frac{2k}{3}-2}}{k \log 2}, \tag{12}$$

where $k_i = \left[\frac{k_{i-1}}{3}\right] + 1, k_0 = k, i > 0$. By induction we have: $\frac{k}{3^i} + \frac{1}{2} \leq k_i \leq \frac{k}{3^i} + \frac{3}{2}$ and we let m be the least index for which $k_m < B$, so that $||M'_m|| = \frac{2^{k_m - 1}}{k_m \log 2}$. This anchors the recursion and we have, after some manipulations of the above inequalities:

$$\frac{2^k}{(8 \log 2\sqrt{3}kB)^m} \leq ||M'_k|| \leq \frac{2^k}{(2 \log 2\sqrt{k}B)^m}, \tag{13}$$

By definition of m we have $m = \left[\frac{\log(k/B)}{\log 3}\right] + 1$. Putting $x = 2^k$, (13) implies:

$$H'_k = \log x - (\log \log x)^2 \left(\frac{1}{2 \log 3} + o(1)\right). \tag{14}$$

Since $H_k = \log\left(\frac{x}{2 \log x}\right)$, we have a forteriori $\frac{H'_k}{H_k} \sim 1$ for $k \to \infty$; together with (12), this yields:

$$\frac{H''_k}{H_k} \sim 1 \quad \text{for } k \to \infty, \tag{15}$$

which is a measure for the distribution of the primes produced by AP.

5 Cryptographic Security and Related Topics

It has already been remarked [21] that facing recent factorization and discrete logarithm mehtods, concepts of secure primes like the ones of Gordon [10] become irrelevant. Actual questions for security of prime generating algorithms remain following. The distribution of the produced primes, which was measured in the preceding chapter by their entropy. This was shown to be assymptotically equal to the entropy of randomly chosen k-bit primes. The fact that the produced primes are collission free is also a consequence of the above. Finally, the iterated encryption attack for the RSA algorithm seems to suggest that p-1 should have some larger factor ([14]) - although this fact was also relativized by [21]. However, cryptography remains a somewhat subjective field. The author of [23] has been cited for saying that, while knowing that 'secure primes' are not harder to brake, he would still prefer Gordon primes for the systems he uses. An argument for this atitude may be the fact that algorithms against which Gordon primes offer no additional security may be regarded as more difficult to implement, the crew of possible crackers being thus reduced... The particularities of AP are particularly favourable for efficient generation of primes with special structure. This can be required in algorithms like the [7] or the new signature scheme of Nyberg and

Rüppel [17], which require factors of $p-1$ of fixed length. It is also the case for 'secure primes'. We end this chapter with an application of AP for finding Gordon strong primes. Let M be a magnitude considered infeasible for algorithms on computers - say, $M = 2^{64}$. We want to produce primes p, such that, there are primes l and l' with: $l \mid (p-1), l' \mid (p+1)$ and $l, l' > M$. Let u_0 and v_0 be the minimal solutions of the equation $vl' - ul = 1$. Following modification in the definition of the progression P, in step 4 of AP gives a solution to the problem: $P = \{N \big| N = N_0 + \lambda a; N_0 = ta + 2u_0 l + 1; a = 2l \cdot l'; \lambda \geq 0\}$. The choice of t in step 3 must also be adjusted accordingly , so that $t \in (\frac{2^{n-1} - 2u_0 l}{a}, \frac{2^n - 2u_0 l}{a} - sn)$ It is easy to verify that a prime in the progression P has the desired properties.

6 Performance

The simple version of AP has been implemented on a SUN/IPX machine. It uses suboptimal table length ($A = 2^{15}$) and naive multiplication. Following table shows the performance of this implementation, with averaged values (over 100 generated primes) for different prime lengths.

Table 1. Performance **AP**

n (bits)	Runtime (sec)	$r(n) = t(2n)/t(n)$	Seek Interval	Psp tests	SeekInt/ log(x)	SeekInt/ #Psp.Tests
256	2.2		173	9.3	0.98	18.6
332	4.0		238	13.0	1.03	18.5
384	7.0		260	14.0	0.98	18.5
512	17.5	8.4	331	17.8	0.93	18.5
664	49.1	11.8	495	26.9	1.07	18.4
768	70.1	11.0	472	25.7	0.89	18.4
1024	213.8	12.2	662	35.7	0.93	18.5

This table reflects the behaviour of predicitions in practice. The most stable predicted value, is the ratio (length of the seek - interval)/ (number of pseudo prime tests): it balances around 18.5, whereas the expected value is $\frac{\log A}{e^{-\gamma}} \approx 18.52$. The length of the seek interval also balances around $\log(x)$, but the variance is larger. The $O(n^4)$ behaviour of the algorithm is less well reflected, certainly because of the overhead which is independent of the length of the primes found. The ratio $r = t(2n)/t(n)$ - where $t(x)$ is the average time for finding a x-bits prime - has used corrected times, proportional to the relative seek interval. The table reflects a monotonous increase of the apparent specific exponent of the run time. Obviously the data is insufficient and this exponent will approach the expected value 4 for larger primes. Actually this behaviour only confirms the known rule that subquadratic multiplication algorithms are not recommendable for small lengths of the multiplicants.

7 Conclusions

We proposed an algorithm for generating provable primes using incremental search in arithmetic progressions. We showed that trial division can be performed using the Eratosthenes sieve method, which increases the number of prime candidates eliminated by the trial division step by an assymptotic factor of 3/2. Independently of the search approach used, we showed that our algorithm is more efficient than probable prime algorithms for primes of at least up to 1000 bits length, whereas for larger primes the loss in efficiency is not more than 2%, while the primes produced are always provided with a certificate. The advantages of our algorithm are more substantial when the primes produced are required to have special prime divisors of $p \pm 1$, since this feature can be incorporated for free in the algorithm. The advantages of this algorithm are at the cost of a reduction of the set of primes which can be produced and their nonuniform distribution. We proved though that the information entropy of the primes produced by the algorithm is assymptotically equal to the entropy of randomly chosen primes. We finally presented the performance of a - non optimal - implementation of the algortihm on a SUN/IPX machine.

Acknowledgement:
I wish to thank P. Landrock and I.Damgard for valuable discussions, informations and literature.

References

1. Brandt, J.; Damgard, I.: *On Generation of Probable Primes by Incremental Search*, Proceedings CRYPTO'92, Lecture Notes in Computer Science Vol. **740** pp. 358-71
2. Brandt, J.; Damgard I.; Landrock, P.: *Speeding up Prime Number Generation*, Proc. of Asiacrypt 91, Lecture Notes in Computer Science, Vol **739**, pp. 440-50.
3. Damgard I.; Landrock, P.; Pomerance, C.: *Average Case Bounds for the Strong Probable Prime Test*, Math. Comp., Vol **61**, Oct 1993, pp. 177-195,
4. Bosma, W.; VanderHulst, L.: *Primality Test Using Cyclotomy*, Phd, University of Amsterdam, 1990
5. Couvreur, C.; Quisquater, J.J.: *An introduction to fast generation of large primes*, Philips Journal of Research, vol. **37**, pp 231-264, 1982
6. Davenport, Harold: *Multiplicative Number Theory*, Springer 1980, Second Edtion.
7. *Digital Signature Algorithm*, Federal Information Processing Standards Publication XX, 1993 February 1.
8. Goldwasser,S. and Kilian, J.: *Almost all primes can be quickly certified*, Proc. of the 18th Annual ACM Symposium on the Theory of Computing, 1986, pp. 316-329.
9. Gallagher, P.X.: *On the distribution of primes in short interevals*, Mathematika, vol. **23**, 1976, pp. 4-9
10. Gordon, J.: *Strong primes are easy to find*, Advances in Cryptology - EUROCRYPT '84, Lecture Notes in Computer Science, vol. **209**, 1984, pp. 216-223.
11. Knuth, Donald E.: *The Art of Computer Programming*, Addison-Wesley, 1981

12. Lenstra, H.W.: *Factoring Integers With Elliptic Curves*, Annals of Mathematics, Vol. **126**, 1987, pp649-673.
13. Lenstra, A.K. and Lenstra, H.W.,Jr.: *Algorithms in number theory*, Technical Report 87-008, May 1987, University of Chicago, Dept. of Computer Science.
14. Maurer, Ueli M.: *Fast Generation of Prime Numbers and Secure Public-Key Cryptographic Parameters*, Internal Report, Dept. of Comp. Science, Princeton University, 1991.
15. Maurer, Ueli M.: *Fast Generation of Prime Numbers and Secure Public-Key Cryptographic Parameters*, Internal Report, Dept. Informatik, ETH Zrich, 1993.
16. Morain, F.: *Distributed primality proving and the primality of (23539+1)/3*, Advances in Cryptology, EUROCRYPT '90, Lecture Notes in Computer Science, vol. **473**, 1990, pp. 110-123.
17. Nyberg,K.; Rüppel,R.:*A New Signature Scheme Based on the DSA Giving Message Recovery*, Preprint, 1993.
18. Plaisted, D. A.: *Fast verification, testing and generation of large primes*, Theoretical Computer Science, vol **9**, 1979, pp. 1-17.
19. Pollard, J.M.:*Theorems on factorization and primality testing*, Proceedings of the Cambridge Philosophical society, Vol. **76**, 1974, pp. 521-528.
20. Rabin, M.O.: *Probabilistic algorithm for testing primality*, Journal of number theory, vol **12**, 1980, pp. 128-138
21. Rivest, R.: *Are 'Strong' Primes needed for RSA ?*, preprint, 1991
22. Rivest, R.L., Shamir, A., Adleman, L.: *A method for obtaining digital signatures and public-key cryptosystems*, Communications of the ACM, Vol. **21**, No. 2, 1978, pp.120-126.
23. Schneier, B.: *Applied Cryptography*, Wiley & Sons, 1993

Attack on the Cryptographic Scheme NIKS-TAS

Don Coppersmith

IBM Research, T.J. Watson Research Center
Yorktown Heights, NY 10598, USA

Abstract. The NIKS-TAS scheme, proposed by Tsujii, Araki, and Sekine in 1993, is an ID-based cryptographic key sharing scheme. We present an algebraic method for attacking this scheme, requiring the cooperation of a small number of collaborators to discover the key shared by any two parties.

1 Introduction

The NIKS-TAS scheme, proposed by Tsujii, Araki, and Sekine [2] in 1993, is an ID-based cryptographic key sharing scheme. That is, to each user (User A) there corresponds a publicly known quantity ID_A, closely related to his personal identification; for example, this could consist of name, address, and telephone number. User A also owns some secret information, which we temporarily abbreviate $\mathbf{X}_A$. Suppose Users A and B wish to communicate secretly. User A executes the NIKS-TAS protocol, using as inputs his own secret information $\mathbf{X}_A$ and User B's public information ID_B, to prepare a cryptographic key $K_{AB}=f(\mathbf{X}_A, ID_B)$. Simultaneously, User B uses $\mathbf{X}_B$ and ID_A to compute the key $K_{BA}=f(\mathbf{X}_B, ID_A)$. The NIKS-TAS scheme is designed such that $K_{AB}=K_{BA}$. Thus both A and B can compute the same cryptographic key, and use it in private communication with each other.

The scheme shares some aspects with the Diffie and Hellman key exchange scheme [1], but differs in one crucial place. In the Diffie and Hellman scheme, public information includes a large prime p and a generator g of the multiplicative group of integers modulo p, called $\mathbf{Z}_p{}^*$. The user's public information y_A is a member of the multiplicative group $\mathbf{Z}_p{}^*$, while the secret information x_A lies in the additive group of exponents or logarithms. The security lies in the fact that computing a group element from its logarithm is easy, but computing the logarithm from the group element seems to be difficult.

$$\begin{array}{rl} \text{public:} & y_A \in \mathbf{Z}_p{}^* \\ \text{secret:} & x_A \in \mathbf{Z}_{p-1} \\ & y_A = g^{(x_A)} \in \mathbf{Z}_p{}^* \\ \text{easy:} & x_A \rightarrow y_A \\ \text{hard:} & y_A \rightarrow x_A \end{array} \qquad \text{(DH)}$$

By contrast, in the NIKS-TAS scheme the public information is used to operate in the additive group of exponents, and some of the secret information lies in the multiplicative group $\mathbf{Z}_p{}^*$.

$$\begin{array}{rl} \text{public:} & ID_A \rightarrow i_{A,j,1} \in \{0,1\} \text{ operating on } \mathbf{Z}_{p-1} \\ \text{secret:} & \mathbf{X}_A \rightarrow G_{Ak} \in \mathbf{Z}_p{}^* \end{array} \qquad \text{(NIKS-TAS)}$$

This reversal comes about because of the requirements of ID-based cryptography. For example, if one tried to make the Diffie-Hellman scheme ID-based, the public information y_A would be predetermined (by the user's name, etc.) and someone (a trusted key distribution center, perhaps) would have to derive the secret information x_A from this public information y_A. But the premise of the Diffie-Hellman scheme is that this task is computationally infeasible, even for the key distribution center. So, in the NIKS-TAS, in order to accommodate the ID-based requirements, the roles are reversed.

When the NIKS-TAS scheme was announced, its inventors offered a reward of $3,000 (US) for showing "theoretically substantial defects proving that our scheme cannot overcome the collusion problem." The present paper represents a cryptanalysis of this scheme and an attempt to claim the reward.

We show here that a small number of colluding users can pool their secret information and obtain enough knowledge that they can easily find the key shared by any two other users. For a typical value of the parameters of the system, $N = 20$, 43 users can usually conspire and break the system.

The attack on the system is made possible by the reversal of roles from the Diffie-Hellman system: the fact that the public information is involved in the ring $\mathbf{Z}_{p-1}$ instead of the multiplicative group $\mathbf{Z}_p{}^*$. Where the legitimate user computes a key in $\mathbf{Z}_p{}^*$, the attacker mirrors this computation with computations in the group of exponents $\mathbf{Z}_{p-1}$, where he has greater flexibility in the computations available to him, and thus is able to carry out an attack not envisioned by the system's designers. The attack uses differential techniques: User A considers keys K_{AB} and $K_{AB'}$ which he would share with two users B and B' whose public information ID_B and $ID_{B'}$ differ in only one bit, and uses the difference between K_{AB} and $K_{AB'}$ to probe the secret information that should be inaccessible to him.

In the balance of the paper, we present the NIKS-TAS scheme, condensed from [2], followed by our attack, and a worked example.

2 The NIKS-TAS Scheme

This description is copied from [2], except that indices have been reordered for ease of understanding (particularly on the quantities $i_{A,j,m}$), and background material has been omitted.

Public information common for all entities

The trusted key distribution center publishes a large prime number p such that the discrete logarithm problem $(\bmod p)$ is inaccessible. It determines the parameters of the system, $\pounds$, N, and L; although not spelled out in the proposal, we can imagine $N = 20$, $L = 50$ and $\pounds = 200$. The center chooses and publishes a one-way hash function which sends $\pounds$ bits to $K = NL$ bits.

Public information for a single entity

Each entity, User A, publicizes his $\pounds$-bit IDCODE (which may be name and address) to all other entities, and registers the IDCODE at a trusted key distribution center. This code is processed by the publicly known hash function to obtain the K-bit ID vector of a typical entity A, called ID_A:

$$ID_A = \mathrm{Hash}(IDCODE_A)$$
$$ID_A = [i_{A,1,1}, i_{A,1,2}, ..., i_{A,N,L}] = [\mathbf{i}_{A1}, \mathbf{i}_{A2}, ..., \mathbf{i}_{AN}]$$
$$i_{A,j,m} \in \{0,1\}, \quad j = 1, ..., N, \quad m = 1, ..., L$$
$$\mathbf{i}_{Aj} \in \{0,1\}^L, \quad j = 1, ..., N$$
$$\mathbf{i}_{Aj} = [\, i_{A,j,1}, \; i_{A,j,2}, \; ..., \; i_{A,j,L} \,]$$

This data is publicly known, derived from the $\pounds$-bit IDCODE by the public hash function.

Center Secrets common for all entities

Meanwhile, the key distribution center chooses the following integers as secret parameters of its own:

$$g\text{: a generator of } \mathbf{Z}_p{}^*$$
$$\gamma \in \mathbf{Z}_{p-1}$$
$$\mathbf{x}_j, \mathbf{y}_j, \mathbf{u}_j, \mathbf{v}_j \in \mathbf{Z}^L_{p-1}, \quad j = 1,2,\ldots,N$$
$$X = [x_{1,1}, x_{1,2}, \ldots, x_{N,L}] = [\mathbf{x}_1, \mathbf{x}_2, \ldots, \mathbf{x}_N]$$
$$Y = [y_{1,1}, y_{1,2}, \ldots, y_{N,L}] = [\mathbf{y}_1, \mathbf{y}_2, \ldots, \mathbf{y}_N]$$
$$U = [u_{1,1}, u_{1,2}, \ldots, u_{N,L}] = [\mathbf{u}_1, \mathbf{u}_2, \ldots, \mathbf{u}_N]$$
$$V = [v_{1,1}, v_{1,2}, \ldots, v_{N,L}] = [\mathbf{v}_1, \mathbf{v}_2, \ldots, \mathbf{v}_N]$$

where $\mathbf{Z}^L_{p-1}$ designates the set of L-dimensional vectors whose components belong to $\mathbf{Z}_{p-1}$. $\mathbf{x}_j$, $\mathbf{y}_j$, $\mathbf{u}_j$, and $\mathbf{v}_j$ are global parameters with which the individual secret keys are generated.

Individual secrets kept in center

The center produces random numbers for each entity; namely the center produces $\alpha_{Aj\ell}$, $\beta_{Aj\ell}$, c_{Ak}, and $r_{Aj\ell}$ for entity A.

$$\alpha_{Aj\ell}, \beta_{Aj\ell} \in \mathbf{Z}^L_{p-1}, \quad j = 1,2,\ldots,N, \quad \ell = 1,2,\ldots,2(N+1-j)$$
$$c_{Ak} \in \mathbf{Z}_{p-1}, \quad k = 1,2,\ldots,N+1$$
$$r_{Aj\ell} \in \mathbf{Z}_{p-1}, \quad j = 1,2,\ldots,N, \quad \ell = 1,2,\ldots,2(N+1-j)$$

subject to the following relations:

$$c^{(0)}_{A,i} = c_{A,i}, \quad i = 1,2,\ldots,N+1$$
$$c^{(k)}_{A,i} = c^{(k-1)}_{A,i} r_{A,k,2i-1} + c^{(k-1)}_{A,i+1} r_{A,k,2i}, \quad k = 1,2,\ldots,N, \quad i = 1,2,\ldots,N-k+1$$
$$c^{(k-1)}_{A,i} \alpha_{A,k,2i-1} + c^{(k-1)}_{A,i+1} \alpha_{A,k,2i} = 0, \quad k = 1,2,\ldots,N, \quad i = 1,2,\ldots,N-k+1$$
$$c^{(k-1)}_{A,i} \beta_{A,k,2i-1} + c^{(k-1)}_{A,i+1} \beta_{A,k,2i} = 0, \quad k = 1,2,\ldots,N, \quad i = 1,2,\ldots,N-k+1$$
$$\gamma = c^{(N)}_{A1}$$

The center calculates the following parameters for each entity in secret and keeps these parameters out of reach of any entity:

$$S_{Aj} = \mathbf{x}_j \bullet \mathbf{i}_{Aj} + \mathbf{y}_j \bullet \mathbf{i}^c_{Aj}$$
$$T_{Aj} = \mathbf{u}_j \bullet \mathbf{i}_{Aj} + \mathbf{v}_j \bullet \mathbf{i}^c_{Aj}$$

where $\mathbf{i}^c_{Aj}$ is the complement of $\mathbf{i}_{Aj}$, and $\mathbf{x}_j \bullet \mathbf{i}_{Aj}$ is the inner product.

Individual secrets distributed to each entity

The center calculates the following parameters for each entity, then distributes to each entity his own secret parameters through a secure channel:

$$\pi_A \in S_N \quad \text{(a permutation)}$$
$$G_{Ak} = g^{c_{Ak}}, \quad k = 1,\ldots,N+1$$
$$\mathbf{X}_{Aj\ell} = r_{Aj\ell}(S_{Aj'}\mathbf{x}_{j'} + T_{Aj'}\mathbf{u}_{j'}) + \alpha_{Aj\ell}$$
$$\mathbf{Y}_{Aj\ell} = r_{Aj\ell}(S_{Aj'}\mathbf{y}_{j'} + T_{Aj'}\mathbf{v}_{j'}) + \beta_{Aj\ell}$$
$$j = 1,2,\ldots,N, \quad \ell = 1,2,\ldots,2(N+1-j)$$

Here π_A is a permutation on N elements, and we define $j' = \pi_A(j)$.

Key Sharing Algorithm

Two users A and B each compute a common key. User A sets

$$G^{(0)}_{A,B,i} = G_{A,i}, \; i = 1,\ldots,N+1.$$

For each $k = 1,2,\ldots,N$ in turn, User A calculates:

$$Z^{(B)}_{A,k,j} = \mathbf{X}_{A,k,j} \bullet \mathbf{i}_{Bk'} + \mathbf{Y}_{A,k,j} \bullet \mathbf{i}^c_{Bk'}, \quad j = 1,2,\ldots,2(N+1-k)$$
$$= r_{A,k,j}(S_{Ak'}S_{Bk'} + T_{Ak'}T_{Bk'}) + \alpha_{Akj} \bullet \mathbf{i}_{Bk'} + \beta_{Akj} \bullet \mathbf{i}^c_{Bk'}$$
$$G^{(k)}_{A,B,i} = (G^{(k-1)}_{A,B,i})^{(Z^{(B)}_{A,k,2i-1})} \times (G^{(k-1)}_{A,B,i+1})^{(Z^{(B)}_{A,k,2i})}$$
$$= g^{\{c^{(k)}_{A,i} \prod^{k}_{j=1}(S_{Aj'}S_{Bj'} + T_{Aj'}T_{Bj'})\}}, \quad i = 1,2,\ldots,(N-k+1).$$

Verifying these equalities is an exercise in algebra, and depends on the relations among the quantities α, etc.

After the Nth step we have

$$K_{AB} = G^{(N)}_{A,B,1} = g^{\{\gamma \prod_{j=1}^{N} (S_{Aj} S_{Bj} + T_{Aj} T_{Bj})\}},$$

recalling $\gamma = c^{(N)}_{A,1} = c^{(N)}_{B,1}$.

Meanwhile User B performs analogous calculations using his secret quantities and A's public quantities, to produce K_{BA}. One can check [2] that the two keys computed by these two users are the same:

$$K_{AB} = K_{BA}.$$

User A and User B can use this common quantity $K_{AB} = K_{BA}$ as a cryptographic key for secret communication with each other.

3 The Attack

We show how $2N + 3$ collaborators can usually break the scheme. If $N = 20$, then $M = 43$ collaborators should suffice.

Notation: Users A,B will be collaborators, and users D,E,F will be victims. We will use public and secret information from the collaborators A,B, and we will use only public information from the victims D,E,F; in fact, we will fabricate public information for D,E,F (invent new users with new public information, but no secret information) when it suits our purpose.

For any collaborator A and user E (who may be victim or collaborator), let A follow the computation of key K_{AE} from $G_{Ak} = g^{c_{Ak}}$, $1 \le k \le N + 1$. Instead of computing K_{AE}, A will compute exponents $e^{(k)}_{AE} \in \mathbf{Z}_{p-1}$, $1 \le k \le N + 1$, such that

$$K_{AE} = \prod_{k=1}^{N+1} (G_{Ak})^{e^{(k)}_{AE}} = g^{\{\Sigma_k c_{Ak} e^{(k)}_{AE}\}}.$$

He does this by keeping track of powers of G_{Ak} in the computation of K_{AE}. Each $e^{(k)}_{AE}$ is a sum of products of the quantities $Z^{(E)}_{A,j,\ell}$.

Of special interest is the case where both A and $B = E$ are collaborators. In this case, from the requirement $K_{AB} = K_{BA}$ we get:

$$\sum_{k=1}^{N+1} c_{Ak} e_{AB}^{(k)} = \sum_{k=1}^{N+1} c_{Bk} e_{BA}^{(k)} \qquad (*)$$

in the ring $\mathbf{Z}_{p-1}$. The $e_{AB}^{(k)}$ are known, while c_{Ak} are unknown.

As we let $\{A,B\}$ range over all unordered pairs of collaborators, we obtain $M(M-1)/2$ such linear equations (*), one for each unordered pair $\{A,B\}$. There are $M(N+1)$ unknowns c_{Ak}, $1 \le k \le N+1$, $A \in \{\text{collaborators}\}$. Because we chose $M \ge 2N+3$, we get $M(M-1)/2 \ge M(N+1)$, so that there are at least as many equations as unknowns. These are homogeneous linear equations in c_{Ak}. We solve them to get the ratios $c_{Ak}:c_{Bj}$. The example in the next section gives reason to believe that the rank of this system of homogeneous linear equations is $M(N+1)-1$, allowing for unique solution; and possible remedies for the opposite case.

Fix one collaborator $\underline{B}$ as a reference point, and set

$$\hat{g} = G_{\underline{B}1} = g^{c_{\underline{B}1}}.$$

We can calculate $\hat{g}$, and from now on we use it instead of the center's secret generator g, which we cannot calculate. We have gotten exponents $\hat{c}_{Ak}$ such that

$$G_{Ak} = \hat{g}^{\hat{c}_{Ak}}, \quad \forall A \in \{\text{collaborators}\}, \quad 1 \le k \le N+1.$$

Now we can work with these exponents $\hat{c}_{Ak}$ and solve algebraic equations in the exponents.

Set $f_{AB} = \Sigma_k \hat{c}_{Ak} e_{AB}^{(k)}$ so that $K_{AB} = \hat{g}^{f_{AB}}$, and $f_{AB} = f_{BA}$. We can calculate f_{AB} if we know A's private information and B's public information. (B need not be a collaborator; in fact we can fabricate B's public information.)

Explanation. The data $\hat{c}_{Ak}$ is gotten from the collaboration and from A's private information. The data $e_{AB}^{(k)}$ is gotten from A's private information and B's public information. So if A is a collaborator and E is a victim, A can compute f_{AE} such that $f_{AE} = \sum_k \hat{c}_{Ak} e_{AE}^{(k)}$ and $K_{AE} = \hat{g}^{f_{AE}}$.

We have

$$f_{AE} = \hat{\gamma} \prod_{j=1}^{N} (S_{Aj} S_{Ej} + T_{Aj} T_{Ej})$$

for $\hat{\gamma}$ unknown, fixed, and independent of A,E. In fact, $\hat{\gamma} = \gamma / c_{\underline{B}1}$.

Now fix two collaborators, A_1 and A_2, and a victim, D. Let A_1 compute

$$f_{A_1 D}, f_{A_1 D'_1}, f_{A_1 D'_2}, \ldots, f_{A_1 D'_L},$$

where D'_j is a fabricated user with the same public information as D except for bit i_{D1j}. That is,

$$i_{D'_j, a,b} = i_{D,a,b} \text{ if } (a,b) \neq (1,j)$$
$$i_{D'_j, 1,j} = 1 - i_{D,1,j}$$

Set $d_{A_1 D1} = \hat{\gamma} \pi_{i \neq 1} (S_{A_1 i} S_{Di} + T_{A_1 i} T_{Di})$. As it happens, we do not know $d_{A_1 D1}$, but it is the same as $d_{A_1 D'_j 1}$. We can compute

$$\pm (f_{A_1 D'_j} - f_{A_1 D}) = d_{A_1 D1} \{ S_{A_1 1} (x_{1j} - y_{1j}) + T_{A_1 1} (u_{1j} - v_{1j}) \},$$

where $x_{1j} \in \mathbf{Z}_{p-1}$ is the jth entry of $\mathbf{x}_1$. The sign $\pm$ is determined by the bit $i_{D,1,j}$. By combining these values for the different j, we get some (unknown) linear combination of the two vectors $(\mathbf{x}_1 - \mathbf{y}_1)$ and $(\mathbf{u}_1 - \mathbf{v}_1)$; call it $\mathbf{q}_1$:

$$\mathbf{q}_1 = (d_{A_1 D1} S_{A_1 1})(\mathbf{x}_1 - \mathbf{y}_1) + (d_{A_1 D1} T_{A_1 1})(\mathbf{u}_1 - \mathbf{v}_1).$$

Repeat for another collaborator A_2 instead of A_1 to get a second (unknown) linear combination of $(\mathbf{x}_1 - \mathbf{y}_1)$ and $(\mathbf{u}_1 - \mathbf{v}_1)$; call this one $\mathbf{r}_1$:

$$\mathbf{r}_1 = (d_{A_2 D1} S_{A_2 1})(\mathbf{x}_1 - \mathbf{y}_1) + (d_{A_2 D1} T_{A_2 1})(\mathbf{u}_1 - \mathbf{v}_1).$$

Note: We can compute the vectors $\mathbf{q}_1$ and $\mathbf{r}_1$, but not the quantities $d_{A_1 E1}$, $S_{A_1 1}$, $\mathbf{x}_1$, $\mathbf{y}_1$, $T_{A_1 1}$, $\mathbf{u}_1$, or $\mathbf{v}_1$.

The vectors $\mathbf{q}_1$, $\mathbf{r}_1$ span the same space as $(\mathbf{x}_1 - \mathbf{y}_1)$ and $(\mathbf{u}_1 - \mathbf{v}_1)$, a 2-dimensional subspace of $\mathbf{Z}_{p-1}^{L}$. We have computed $\mathbf{q}_1$ and $\mathbf{r}_1$. Put another way, we have computed the subspace spanned by $(\mathbf{x}_1 - \mathbf{y}_1)$ and $(\mathbf{u}_1 - \mathbf{v}_1)$.

Letting A be an arbitrary collaborator and E an arbitrary victim, we have that $(S_{A1} S_{E1} + T_{A1} T_{E1})$ is an inhomogeneous bilinear form in

$\{(\mathbf{x}_1 - \mathbf{y}_1) \bullet \mathbf{i}_{A1},\ (\mathbf{u}_1 - \mathbf{v}_1) \bullet \mathbf{i}_{A1}\}$ (on the one hand) and $\{(\mathbf{x}_1 - \mathbf{y}_1) \bullet \mathbf{i}_{E1},\ (\mathbf{u}_1 - \mathbf{v}_1) \bullet \mathbf{i}_{E1}\}$ (on the other hand), namely

$$\begin{aligned}(S_{A1}S_{E1} + T_{A1}T_{E1}) = \\ &= [(\Sigma_j y_{1j}) + (\mathbf{x}_1 - \mathbf{y}_1) \bullet \mathbf{i}_{A1}] \times [(\Sigma_j y_{1j}) + (\mathbf{x}_1 - \mathbf{y}_1) \bullet \mathbf{i}_{E1}] + \\ &+ [(\Sigma_j v_{1j}) + (\mathbf{u}_1 - \mathbf{v}_1) \bullet \mathbf{i}_{A1}] \times [(\Sigma_j v_{1j}) + (\mathbf{u}_1 - \mathbf{v}_1) \bullet \mathbf{i}_{E1}]\end{aligned}$$

It is symmetric in the interchange of (A,E).

Imagine a change of basis from $\{(\mathbf{x}_1 - \mathbf{y}_1) \bullet \mathbf{i}_{A1},\ (\mathbf{u}_1 - \mathbf{v}_1) \bullet \mathbf{i}_{A1}\}$ to $\{\mathbf{q}_1 \bullet \mathbf{i}_{A1},\ \mathbf{r}_1 \bullet \mathbf{i}_{A1}\}$. (We do not know the coordinates of this basis change.) Then we have that $(S_{A1}S_{E1} + T_{A1}T_{E1})$ is an inhomogeneous bilinear form in $\{\mathbf{q}_1 \bullet \mathbf{i}_{A1},\ \mathbf{r}_1 \bullet \mathbf{i}_{A1}\}$ (on one hand) and $\{\mathbf{q}_1 \bullet \mathbf{i}_{E1},\ \mathbf{r}_1 \bullet \mathbf{i}_{E1}\}$ (on the other hand), again symmetric with respect to (A,E). So we may write it as

$$\begin{aligned}(S_{A1}S_{E1} + T_{A1}T_{E1}) = \alpha_1 &+ \beta_1(\mathbf{q}_1 \bullet \mathbf{i}_{A1} + \mathbf{q}_1 \bullet \mathbf{i}_{E1}) + \gamma_1(\mathbf{r}_1 \bullet \mathbf{i}_{A1} + \mathbf{r}_1 \bullet \mathbf{i}_{E1}) + \\ &+ \delta_1(\mathbf{q}_1 \bullet \mathbf{i}_{A1})(\mathbf{q}_1 \bullet \mathbf{i}_{E1}) + \varepsilon_1(\mathbf{r}_1 \bullet \mathbf{i}_{A1})(\mathbf{r}_1 \bullet \mathbf{i}_{E1}) + \\ &+ \tau_1[(\mathbf{q}_1 \bullet \mathbf{i}_{A1})(\mathbf{r}_1 \bullet \mathbf{i}_{E1}) + (\mathbf{r}_1 \bullet \mathbf{i}_{A1})(\mathbf{q}_1 \bullet \mathbf{i}_{E1})]\end{aligned}$$

where we do not know $\alpha_1, \beta_1, \gamma_1, \delta_1, \varepsilon_1, \tau_1 \in \mathbf{Z}_{p-1}$. (We should not confuse α_1 with the previously defined $\alpha_{Aj\ell}$.)

By repeating with E'_j instead of E, and seeing the dependence of $f_{AE'_j}$ on $\mathbf{q}_1 \bullet \mathbf{i}_{E'_j1}$ and $\mathbf{r}_1 \bullet \mathbf{i}_{E'_j1}$, user A can establish the *ratios* of the coefficients of 1, $\mathbf{q}_1 \bullet \mathbf{i}_{E1}$, $\mathbf{r}_1 \bullet \mathbf{i}_{E1}$, respectively:

$$\begin{aligned}&[\alpha_1 + \beta_1\mathbf{q}_1 \bullet \mathbf{i}_{A1} + \gamma_1\mathbf{r}_1 \bullet \mathbf{i}_{A1}] : [\beta_1 + \delta_1\mathbf{q}_1 \bullet \mathbf{i}_{A1} + \tau_1\mathbf{r}_1 \bullet \mathbf{i}_{A1}], \\ &[\alpha_1 + \beta_1\mathbf{q}_1 \bullet \mathbf{i}_{A1} + \gamma_1\mathbf{r}_1 \bullet \mathbf{i}_{A1}] : [\gamma_1 + \tau_1\mathbf{q}_1 \bullet \mathbf{i}_{A1} + \varepsilon_1\mathbf{r}_1 \bullet \mathbf{i}_{A1}].\end{aligned}$$

Call this data $(**)_A$. When other collaborators A', A'' repeat the experiment with their different values of $\mathbf{q}_1 \bullet \mathbf{i}_{A'1}, \mathbf{r}_1 \bullet \mathbf{i}_{A'1}, \mathbf{q}_1 \bullet \mathbf{i}_{A''1}, \mathbf{r}_1 \bullet \mathbf{i}_{A''1}$, but using the same basis vectors $\mathbf{q}_1$ and $\mathbf{r}_1$, the data $(**)_A$ and $(**)_{A'}$ and $(**)_{A''}$ suffices to calculate the *ratios*

$$\alpha_1 : \beta_1 : \gamma_1 : \delta_1 : \varepsilon_1 : \tau_1.$$

Without loss of generality we can divide through by α_1 (and change $\hat{\gamma}$ accordingly), and so we calculate

$$(\alpha_1 = 1),\ \beta_1,\ \gamma_1,\ \delta_1,\ \varepsilon_1,\ \tau_1.$$

Repeat with $k = 2,3,\ldots,N$ replacing 1, to calculate

$$(\alpha_k = 1),\ \beta_k,\ \gamma_k,\ \delta_k,\ \varepsilon_k,\ \tau_k,\quad k = 1,2,\ldots,N.$$

Then we have expressed

$$f_{AE} = \tilde{\gamma} \prod_{j=1}^{N} [\alpha_j + \beta_j(\mathbf{q}_j \bullet \mathbf{i}_{Aj} + \mathbf{q}_j \bullet \mathbf{i}_{Ej}) + \gamma_j(\mathbf{r}_j \bullet \mathbf{i}_{Aj} + \mathbf{r}_j \bullet \mathbf{i}_{Ej}) + \delta_j(\mathbf{q}_j \bullet \mathbf{i}_{Aj})(\mathbf{q}_j \bullet \mathbf{i}_{Ej}) + $$
$$+ \varepsilon_j(\mathbf{r}_j \bullet \mathbf{i}_{Aj})(\mathbf{r}_j \bullet \mathbf{i}_{Ej}) + \tau_j\{(\mathbf{q}_j \bullet \mathbf{i}_{Aj})(\mathbf{r}_j \bullet \mathbf{i}_{Ej}) + (\mathbf{r}_j \bullet \mathbf{i}_{Aj})(\mathbf{q}_j \bullet \mathbf{i}_{Ej})\}]$$

with $\tilde{\gamma}$ unknown, and $\alpha_j = 1$, and the rest of the data $\mathbf{q}_j$, β_j, etc. known. We can compute $\tilde{\gamma}$ from one known key $f_{A_0B_0}$, so we know the complete expression.

This expression depends only on the public information $\mathbf{i}_{Aj}$, $\mathbf{i}_{Ej}$ of A and E, so it holds as well for victims F,E:

$$f_{FE} = \tilde{\gamma} \prod_{j=1}^{N} [\alpha_j + \beta_j(\mathbf{q}_j \bullet \mathbf{i}_{Fj} + \mathbf{q}_j \bullet \mathbf{i}_{Ej}) + \gamma_j(\mathbf{r}_j \bullet \mathbf{i}_{Fj} + \mathbf{r}_j \bullet \mathbf{i}_{Ej}) + \delta_j(\mathbf{q}_j \bullet \mathbf{i}_{Fj})(\mathbf{q}_j \bullet \mathbf{i}_{Ej}) + $$
$$+ \varepsilon_j(\mathbf{r}_j \bullet \mathbf{i}_{Fj})(\mathbf{r}_j \bullet \mathbf{i}_{Ej}) + \tau_j\{(\mathbf{q}_j \bullet \mathbf{i}_{Fj})(\mathbf{r}_j \bullet \mathbf{i}_{Ej}) + (\mathbf{r}_j \bullet \mathbf{i}_{Fj})(\mathbf{q}_j \bullet \mathbf{i}_{Ej})\}]$$

and

$$K_{FE} = \hat{g}^{(f_{FE})}.$$

We have used the private information of $2N + 3$ collaborators and the public information $\mathbf{i}_{Fj}$, $\mathbf{i}_{Ej}$ of the victims F,E, and obtained the key shared by victims F,E.

Remark: It is interesting to note that the act of collusion may be even easier than the legitimate use of the system, because the colluders use addition and multiplication, respectively, in $\mathbf{Z}_{p-1}$ while the legitimate users are using multiplication and exponentiation, respectively, in $\mathbf{Z}_p^*$.

4 Example

Take the simple case $N = 1$. Here we can ignore the permutations π_A, which give just minor complications. We have $M = 2N + 1 = 5$ users, A,B,C,D,E, colluding to break the scheme. They determine the following $M(M-1)/2 = 10$ equations:

$$e_{AB}^{(1)}c_{A1} + e_{AB}^{(2)}c_{A2} \equiv e_{BA}^{(1)}c_{B1} + e_{BA}^{(2)}c_{B2} \bmod p-1$$
$$e_{AC}^{(1)}c_{A1} + e_{AC}^{(2)}c_{A2} \equiv e_{CA}^{(1)}c_{C1} + e_{CA}^{(2)}c_{C2} \bmod p-1$$
$$\cdots$$
$$e_{DE}^{(1)}c_{D1} + e_{DE}^{(2)}c_{D2} \equiv e_{ED}^{(1)}c_{E1} + e_{ED}^{(2)}c_{E2} \bmod p-1$$

Notice that $e_{AB}^{(1)} = Z_{A,1,1}^{(B)}$.

Starting with the $[M(M-1)/2] \times [M(N+1)] = 10 \times 10$ matrix of coefficients $\pm e_{AB}^{(1)}$, delete one column to make it an inhomogeneous system, and delete one row to make it square. We obtain the 9×9 matrix F:

$$\begin{bmatrix}
e_{AB}^{(1)} & e_{AB}^{(2)} & -e_{BA}^{(1)} & -e_{BA}^{(2)} & 0 & 0 & 0 & 0 & 0 \\
e_{AC}^{(1)} & e_{AC}^{(2)} & 0 & 0 & -e_{CA}^{(1)} & -e_{CA}^{(2)} & 0 & 0 & 0 \\
0 & 0 & e_{BC}^{(1)} & e_{BC}^{(2)} & -e_{CB}^{(1)} & -e_{CB}^{(2)} & 0 & 0 & 0 \\
0 & 0 & e_{BD}^{(1)} & e_{BD}^{(2)} & 0 & 0 & -e_{DB}^{(1)} & -e_{DB}^{(2)} & 0 \\
0 & 0 & 0 & 0 & e_{CD}^{(1)} & e_{CD}^{(2)} & -e_{DC}^{(1)} & -e_{DC}^{(2)} & 0 \\
0 & 0 & 0 & 0 & e_{CE}^{(1)} & e_{CE}^{(2)} & 0 & 0 & -e_{EC}^{(1)} \\
0 & 0 & 0 & 0 & 0 & 0 & e_{DE}^{(1)} & e_{DE}^{(2)} & -e_{ED}^{(1)} \\
-e_{AD}^{(1)} & -e_{AD}^{(2)} & 0 & 0 & 0 & 0 & e_{DA}^{(1)} & e_{DA}^{(2)} & 0 \\
-e_{AE}^{(1)} & -e_{AE}^{(2)} & 0 & 0 & 0 & 0 & 0 & 0 & e_{EA}^{(1)}
\end{bmatrix}$$

Define vectors **C**, **e** as

$$\mathbf{C} = (c_{A1}, c_{A2}, c_{B1}, c_{B2}, c_{C1}, c_{C2}, c_{D1}, c_{D2}, c_{E1})^{\mathrm{T}}$$

$$\mathbf{e} = (0,0,0,0,0,e_{EC}^{(2)}, e_{ED}^{(2)}, 0, -e_{EA}^{(2)})^{\mathrm{T}}.$$

We are then faced with solving the 9×9 system of inhomogeneous linear equations $FC = \mathbf{e}c_{E2}$. To facilitate the solution of this system, we depend on F having full rank, that is, $\det(F)$ is a unit in $\mathbf{Z}_{p-1}$. To verify that F generally has full rank, we perform some elementary column operations. Letting F_j denote the jth column, we set $F_2 \leftarrow (c_{A1}F_1 + c_{A2}F_2)\hat{\gamma}/\gamma$, $F_4 \leftarrow (c_{B1}F_3 + c_{B2}F_4)\hat{\gamma}/\gamma$, $F_6 \leftarrow (c_{C1}F_5 + c_{C2}F_6)\hat{\gamma}/\gamma$, and $F_8 \leftarrow (c_{D1}F_7 + c_{D2}F_8)\hat{\gamma}/\gamma$. The new entry in the first row, second column, is now

$$\begin{aligned}(c_{A1}e_{AB}^{(1)} + c_{A2}e_{AB}^{(2)})\hat{\gamma}/\gamma &= (c_{A1}Z_{A11}^{(B)} + c_{A2}Z_{A12}^{(B)})\hat{\gamma}/\gamma \\ &= \gamma(S_{A1}S_{B1} + T_{A1}T_{B1})\hat{\gamma}/\gamma = \hat{\gamma}(S_{A1}S_{B1} + T_{A1}T_{B1}) = f_{AB},\end{aligned}$$

recalling $f_{AB} = \hat{\gamma}(S_{A1}S_{B1} + T_{A1}T_{B1}) = f_{BA}$. The altered matrix is then seen to be

$$\begin{bmatrix}
\underline{e_{AB}^{(1)}} & f_{AB} & -e_{BA}^{(1)} & -f_{BA} & 0 & 0 & 0 & 0 & 0 \\
e_{AC}^{(1)} & f_{AC} & 0 & 0 & \underline{-e_{CA}^{(1)}} & -f_{CA} & 0 & 0 & 0 \\
0 & 0 & \underline{e_{BC}^{(1)}} & f_{BC} & -e_{CB}^{(1)} & -f_{CB} & 0 & 0 & 0 \\
0 & 0 & e_{BD}^{(1)} & f_{BD} & 0 & 0 & -e_{DB}^{(1)} & -f_{DB} & 0 \\
0 & 0 & 0 & 0 & e_{CD}^{(1)} & f_{CD} & \underline{-e_{DC}^{(1)}} & -f_{DC} & 0 \\
0 & 0 & 0 & 0 & e_{CE}^{(1)} & f_{CE} & 0 & 0 & -e_{EC}^{(1)} \\
0 & 0 & 0 & 0 & 0 & 0 & e_{DE}^{(1)} & f_{DE} & \underline{-e_{ED}^{(1)}} \\
-e_{AD}^{(1)} & -f_{AD} & 0 & 0 & 0 & 0 & e_{DA}^{(1)} & f_{DA} & 0 \\
-e_{AE}^{(1)} & -f_{AE} & 0 & 0 & 0 & 0 & 0 & 0 & e_{EA}^{(1)}
\end{bmatrix}$$

Consider the five underlined entries. A typical one is

$$e_{BC}^{(1)} = r_{B11}(S_{B1}S_{C1} + T_{B1}T_{C1}) + \alpha_{B11} \cdot \mathbf{i}_{C1} + \beta_{B11} \cdot \mathbf{i}_{C1}^{c}$$

It depends on the random vector α_{B11}, as do the other entries $e_{BX}^{(1)}$ in that column. But the dot product $\alpha_{B11} \cdot \mathbf{i}_{C1}$ is linearly independent of the other such products $\alpha_{B11} \cdot \mathbf{i}_{X1}$. So, by varying α_{B11}, the entry $e_{BC}^{(1)}$ can vary without changing any other matrix entries. (*Note:* $e_{BC}^{(2)}$ also depended on $\alpha_{B11} \cdot \mathbf{i}_{c1}$, through the dependence between α_{B11} and α_{B12}. But k_{BC} is independent of the choice of α_{B11}. This is why we did the elementary column operations, to remove this interdependence between entries.)

The cofactor of the five underlined entries is the 4×4 matrix

$$\begin{bmatrix} 0 & f_{BD} & 0 & -f_{DB} \\ 0 & 0 & f_{CE} & 0 \\ -f_{AD} & 0 & 0 & f_{DA} \\ -f_{AE} & 0 & 0 & 0 \end{bmatrix}$$

whose determinant, $f_{BD}f_{CE}f_{AD}f_{AE}$, is in general nonzero. Thus $\det(F)$ is not identically 0. This gives evidence that the linear equations will usually be solvable.

It may become necessary to solve the equations mod q for each prime q dividing $p-1$, and piece together with the Chinese remainder theorem. In particular, for small primes q the determinant might vanish accidentally and we might have to compute logarithms mod q by brute force.

Consider now the general case $N \geq 1$. Now the general matrix entry $e_{AB}^{(k)}$ is not just $Z_{A,k,j}^{(B)}$ but rather a sum of products of such Z values. Nonetheless there are enough randomly chosen values $\alpha_{Aj\ell}$ that an independence argument can be made similar to the case $N = 1$.

Select $M \geq 2N + 1$ conspirators randomly, and build an $[M(M-1)/2] \times [M(N+1)]$ matrix F corresponding to the example F; we have again omitted one column to make an inhomogeneous system, but this time we do not drop any rows. Also M may be somewhat larger than $2N + 3$.

If the rank of F is fully $M(N + 1) - 1$, we solve as before.

Suppose that, as M grows, the rank of F is always at least $M(N+1)-I$, where I is independent of M. Then, instead of representing all c_{Ai} as multiples of a single $c_{\underline{B}1}$ and using the corresponding group element $\hat{g}$, we select I different reference points $c_{\underline{B}i}$ and use corresponding group elements $\hat{g}_i$. This is more tedious but still possible. One or two more conspirators might be necessary in this case.

Suppose that the rank of F is always bounded by $M(N+1-J)$, $J \geq 1$. This suggests that J linear relations exist among the $N+1$ columns of $e_{AX}^{(i)}$ corresponding to each user A, and that the quantities c_{Ai} could vary with J degrees of freedom without affecting the keys. (This would mean that the quantities described as independent variables were actually being chosen in a dependent fashion.) Detect these linear combinations and describe the system in terms of a reduced set of $N+1-J$ variables c_{Ai}. Then continue as before.

Finally, any systematically low rank of $F \bmod q$ would indicate systematic relations among the keys K_{AB}, which we should be able to exploit directly.

References

1. W. Diffie and M. E. Hellman, "New directions in cryptography," *IEEE Trans. Informat. Theory,* vol. **IT-22**, pp. 644-654, Nov. 1976.

2. S. Tsujii, K. Araki, and T. Sekine, "A new scheme of non interactive ID-based key sharing with explosively high degree of separability (second version)," Technical report of Department of Computer Science, Tokyo Institute of Technology, 93TR-0020, July 1993.

Note: the NIKS-TAS scheme has since appeared in the 1993 Korea-Japan Joint Workshop on Information Security and Cryptology. A partial attack, in the same workshop, was "An Attack on an ID-Based Key Sharing System," by V. Luchangco and K. Koyama.

On the Risk of Opening Distributed Keys

Mike Burmester

Department of Mathematics, RH – University of London,
Egham, Surrey TW20 OEX, U.K.
e-mail uhah205@vax.rhbnc.ac.uk

Abstract. We describe an insider known-key attack on key distribution systems which are based on public keys. This is of a general type and applies to the key distribution system presented by Yacobi at Crypto '90, the Goss system, the Günther system presented at Eurocrypt '89 and the key exchange version of COMSET, based on a system presented by Brandt et al. at Crypto '89. The attack is primarily theoretical, in the sense that it assumes that some session keys are leaked or lost. Well designed systems will prevent this. However it could have practical consequences with certain applications (e.g. negotiation of contracts or poor implementations). We discuss the implications and ways to prevent the attack.

1 Introduction

Matsumoto-Takashima-Imai [21], Yacobi-Shmuely [29] and Yacobi [28] presented a family of variants of the Diffie-Hellman [13] key distribution system which are based on public keys. These are two-party systems which are non-interactive, have a low communication and computation overhead, and offer a good level of security. In particular, the 'non-paradoxical' key distribution system is provably secure against passive attacks and certain types of known-key attacks [28] (provided the Composite Diffie-Hellman problem is hard). In this paper we show that this system is not secure against general known-key attacks. We use an insider attack which is two-pronged, consisting of a passive attack followed by an active known-key attack. In the passive attack the adversary eavesdrops on a conversation. Then the adversary uses a known-key attack to obtain two chosen keys from subsequent sessions. From these, and from the calls exchanged, the adversary computes the key of the eavesdropped session. The attack is general and applies to many other systems including the Goss system [19], the Günther system [20], and the key exchange version of COMSET [9], when the session key is the bitwise exclusive-or of the partial keys.

Known (chosen) key attacks [28] are analogous to chosen-ciphertext attacks for encryptions. With these the cryptanalyst has access to some session keys. As with encryptions, they are the hardest to foil. Various scenarios for known-key attacks are discussed in [12]. These are,

- *Negotiation of contracts*: After a contract is signed, there is no need to keep secrecy and the keys used for privacy may be revealed.

- *Verification of treaties*: To reduce the possibility of hiding a covert message, the key is revealed immediately after the message is authenticated [2, p. 33].
- *Jealous spouse*: Alice wants to get the secret session key K_{bc} which her spouse Bob used to encrypt a file sent to Carol. She expects to receive a large file from Bob, but will not adhere to the rules of the key exchange protocol. As a result she will not know the session key K_{ba} which she will exchange with Bob.[1] So when she gets the encrypted file under the session key K_{ba} from Bob she cannot decrypt it. A few days later she claims to have lost the key K_{ba}. Bob has no reason to refuse to give her *this* key.

The robustness of a protocol against the loss of a session key has been the subject of many recent investigations (e.g. [28, 3, 24]). Yacobi linked it to the "paradox" for signature schemes [28]. Bellare and Rogaway [3] observed that it is necessary for secure authenticated key exchange. They pointed out that even if an adversary gets hold of a session key, this should effect only the session which that key protects. In particular, it should not be any easier for the adversary to compute another session key.

Earlier attacks on key distribution systems such as the mafia attack, or intruder-in-the-middle attack (*e.g.* [25, 4, 14]), and interleaving attacks [5] (see also [3]), fail to take advantage of the scope of known-key attacks. Typically, they lead to disclosure of secret information which subsequently can be used to obtain session keys, or to impersonate. Our attack is different. The adversary is an insider who succeeds in computing the session key of an earlier conversation, from subsequent sessions with the parties involved in the conversation.

Van Oorschot [27] compared the formal goals of the Goss and Günther systems[2] by using an extended version of the authentication logic of Burrows, Abadi and Needham [7, 1] (see also [18, 8, 17, 16]) and found that they are essentially the same. One of the main features highlighted was the fact that these systems provide only *implicit* key authentication, i.e. the session key is not authenticated until both parties prove knowledge of it by using it in a subsequent communication. It may seem that that this is not a serious threat [27, p. 241], since even though an adversary can replay old messages and complete fraudulent protocol runs, no real advantage is gained because the resulting key cannot be computed.[3] But if some keys are leaked or lost then it is a threat. Indeed with our attack the adversary uses knowledge gained from leaked keys of fraudulent runs to compute the session key of an earlier honest run.

The organization of this paper is as follows. In Section 2 we describe our attack and show how it can be applied to the Yacobi system. In Section 3 we extend it to the Goss [19] system, the Günther [20] system, and to the key exchange version of COMSET [9]. In Section 4 we analyze the attack and discuss ways to foil it. We conclude with remarks.

---

[1] By doing so she hopes to gain some knowledge about K_{bc}. We shall see that she may succeed in some cases.

[2] Yacobi's scheme was not analyzed, but it is similar.

[3] Diffie–Oorschot–Wiener warn about a threat to the Station-to-Station protocol when the signatures are not encrypted [14, p. 116]. However in this case an intruder can easily hijack the key and we do not have key authentication.

Remark. It is important to differentiate between theoretical and practical attacks. The attack described in this paper is primarily theoretical since well designed implementations of key distribution systems will prevent session keys being disclosed or lost. Indeed in virtually all real systems the users do not know their encryption keys, and only interact through well defined interfaces. However one should be wary, particularly with poor implementations, or with applications in which the session keys are eventually disclosed (e.g. with negotiations of contracts).

2 Cryptanalysis of the Yacobi System

We first consider the 'non-paradoxical' key distribution system presented by Yacobi [28] (see also [21]) and show how our attack applies to it. In the following sections we will show that the attack can be extended to other systems.

2.1 The Yacobi System

This system uses a discrete logarithm setting with composite modulus $m = pq$, p, q large secret primes, and base $\alpha \in Z_m^*$ of large order. Each user U_ℓ has a secret key $s_\ell \in Z_m$ and a public key $P_\ell = \alpha^{s_\ell} \bmod m$. When U_i, U_j want to establish a common session key they select random exponents $e_i, e_j \in Z_m$, respectively, and exchange calls $R_i = \alpha^{e_i} \bmod m$, $R_j = \alpha^{e_j} \bmod m$. The session key is $K_{ij} \equiv R_i^{s_j} \cdot P_i^{e_j} \equiv R_i^{s_j} \cdot R_j^{s_i} \equiv P_j^{e_i} \cdot R_j^{s_i} \pmod m$, which both U_i and U_j can compute. Observe that the session key is the modular product of the 'partial keys' $R_i^{s_j} \bmod m$ and $R_j^{s_i} \bmod m$, which both parties can compute.

The Yacobi system is secure against ciphertext-only attacks by a passive eavesdropper if the Composite Diffie-Hellman problem is hard [28]. Under the same cryptographic assumption it is also secure against certain types of known-key attacks [28]. However these are rather limited in scope [12].

2.2 Cryptanalysis of the Yacobi System

Let A, B and C be users whose secret and public keys are (s_a, P_a), (s_b, P_b), and (s_c, P_c) respectively. $\widetilde{A}$ is a dishonest user who does not follow the prescribed protocol, but knows the secret key s_a of A. B and C execute the prescribed protocol.

In the following attack $\widetilde{A}$ eavesdrops on a conversation of B and C. Subsequently, from separate conversations with B and C, and by using a known-key attack, she will compute the session key of B and C.

2.3 The Triangle Attack

Phase 1. *$\widetilde{A}$ eavesdrops on a conversation of B and C.* Let R_b, R_c be the calls exchanged and let $K_{bc} \equiv R_b^{s_c} \cdot R_c^{s_b} \pmod{m}$ be the session key. $\widetilde{A}$ wants to compute *this* key. She gets R_b, R_c by eavesdropping.

Phase 2. *B exchanges a key with $\widetilde{A}$.* $\widetilde{A}$ uses the call R_c. Let R_b' be the call of B and let $K_{ba} \equiv R_b'^{s_a} \cdot R_c^{s_b} \pmod{m}$ be the session key of B.

Phase 3. *C exchanges a key with $\widetilde{A}$.* $\widetilde{A}$ uses the call R_b. Let R_c' be the call of C and let $K_{ca} \equiv R_c'^{s_a} \cdot R_b^{s_c} \pmod{m}$ be the session key of C.

Event 1. *Opening the key K_{ba}.* Suppose that the session key of B, K_{ba}, is revealed to $\widetilde{A}$.

Event 2. *Opening the key K_{ca}.* Suppose that the session key of C, K_{ca}, is revealed to $\widetilde{A}$.

Computation. *$\widetilde{A}$ computes the key of B and C.* $\widetilde{A}$ now computes:

- the 'partial key' $R_c^{s_b} \equiv R_b'^{-s_a} \cdot K_{ba} \pmod{m}$,
- the 'partial key' $R_b^{s_c} \equiv R_c'^{-s_a} \cdot K_{ca} \pmod{m}$, and finally
- the session key $K_{bc} \equiv R_b^{s_c} \cdot R_c^{s_b} \pmod{m}$.

Observe that in Phase 2 and Phase 3, $\widetilde{A}$ cannot compute the keys K_{ba} and K_{ca} by herself, unless the Composite Diffie-Hellman problem is feasible (she does not know the discrete logarithms of R_c and R_b).

Remark. The conventional definition of a known key attack relates to a situation in which an old session key has leaked or is lost. For example, a session key that was used to encrypt a file is stored in a place that was supposed to be safe, but has leaked. Prior to using the session key, the parties use a handshake protocol to authenticate it. In the triangle attack, B and C initiate Phase 2 and Phase 3 respectively because they want to send files to $\widetilde{A}$. The session keys they 'exchange' are then used to encrypt the files. These keys are leaked to $\widetilde{A}$, who then computes the session key of B and C. Observe that the Yacobi 'handshake' is non-interactive and provides only implicit authentication ($\widetilde{A}$ has sent no file and is not authenticated).

2.4 Hiding the Attack

The attack can be prevented by having the parties keep a cache of received old calls R_i, and requesting fresh calls for each new session. We discuss two methods to hide the triangle attack, based on a variant of Moore's attack for RSA signatures [10, 11, 22]).

Modification 1

Instead of using the old calls R_c and R_b in Phase 2 and Phase 3, $\widetilde{A}$ uses

$$\hat{R}_c \equiv R_c \cdot \alpha^{t_c} \pmod{m} \text{ and } \hat{R}_b \equiv R_b \cdot \alpha^{t_b} \pmod{m}, \tag{1}$$

where t_c, t_b are randomly selected in Z_m by $\widetilde{A}$. Then $K_{ba} \equiv R_b'^{s_a} \cdot \hat{R}_c^{s_b} \pmod{m}$ and $K_{ca} \equiv R_c'^{s_a} \cdot \hat{R}_b^{s_c} \pmod{m}$, where R_b' and R_c' are the calls of B and C. In this case $\widetilde{A}$ computes $\hat{R}_c^{s_b} \equiv R_b'^{-s_a} \cdot K_{ba} \pmod{m}$ and $\hat{R}_b^{s_c} \equiv R_c'^{-s_a} \cdot K_{ca} \pmod{m}$. From these, and from (1), she gets the partial keys

$$R_b^{s_c} \equiv \hat{R}_b^{s_c} . \alpha^{-t_b s_c} \equiv \hat{R}_b^{s_c} . P_c^{-t_b} \pmod{m}, \; R_c^{s_b} \equiv \hat{R}_c^{s_b} . \alpha^{-t_c s_b} \equiv \hat{R}_c^{s_b} . P_b^{-t_c} \pmod{m},$$

from which she computes the session key $K_{bc} \equiv R_b^{s_c} . R_c^{s_b} \pmod{m}$ of B and C.

Modification 2

If the modulus is a prime p, then instead of using the calls R_c, R_b in Phase 2 and Phase 3, $\widetilde{A}$ may use $\bar{R}_c = R_c^{t_c} \bmod p$ and $\bar{R}_b = R_b^{t_b} \bmod p$, where t_c, t_b are randomly selected in Z_{p-1}^*. Then $K_{ba} \equiv R_b'^{s_a} \cdot \bar{R}_c^{s_b} \pmod{p}$ and $K_{ca} \equiv R_c'^{s_a} \cdot \bar{R}_b^{s_c} \pmod{p}$, where R_b' and R_c' are the calls of B and C. So $\widetilde{A}$ can compute $\bar{R}_b^{s_c} \equiv R'^{-s_a}_c . K_{ca} \pmod{p}$ and $\bar{R}_c^{s_b} \equiv R'^{-s_a}_b . K_{ba} \pmod{p}$, and hence the partial keys $R_b^{s_c} \equiv R'^{-s_a t_b^{-1}}_c \cdot K_{ca}^{t_b^{-1}} \pmod{p}$ and $R_c^{s_b} \equiv R'^{-s_a t_c^{-1}}_b \cdot K_{ba}^{t_c^{-1}} \pmod{p}$, from which she gets K_{bc}.

3 Cryptanalysis of Other Key Distribution Systems

3.1 The Goss System

This system [19] is similar to the Yacobi system except that the modulus is a prime p, and the session key is the bitwise exclusive-or of the partial keys, instead of their product. That is $K_{ij} = (R_i^{s_j} \bmod p) \oplus (R_j^{s_i} \bmod p)$.

The Attack

This is essentially the same as for the Yacobi system. In this case $\widetilde{A}$ computes the session key K_{bc} of B and C from the partial keys $R_b^{s_c} \bmod p = K_{ca} \oplus (R'^{s_a}_c \bmod p)$ and $R_c^{s_b} \bmod p = K_{ba} \oplus (R_b'^{s_a} \bmod p)$. Both modifications in Section 2.4 apply to this scheme as well.

3.2 The Günther System

This system [20] uses ElGamal signatures [15] for authentication. It is similar to the Yacobi system but with a prime modulus p. A trusted center T with secret key x, $1 \le x \le p-1$, and public key $Y = \alpha^x \bmod p$, issues each user U_ℓ with a distinguishing identifier D_ℓ and an ElGamal signature (P_ℓ, s_ℓ) of D_ℓ. Thus $P_\ell = \alpha^{k_\ell} \bmod p$, where $k_\ell \in Z_{p-1}$ is a random exponent with $\gcd(k_\ell, p-1) = 1$, and s_ℓ is obtained by solving $h(D_\ell) \equiv x P_\ell + k_\ell s_\ell \mod (p-1)$, where $h(\cdot)$ is a suitable hash function. P_ℓ is the public key of U_ℓ and s_ℓ the secret key. Observe that anyone can compute

$$P_\ell^{s_\ell} \equiv \alpha^{k_\ell s_\ell} \equiv \alpha^{h(D_\ell) - x \cdot P_\ell} \equiv \alpha^{h(D_\ell)} Y^{-P_\ell} \pmod{p}. \tag{2}$$

When U_i wants to establish a session key with U_j it selects a random exponent e_i, computes $R_i = P_j^{e_i} \bmod p$, and sends D_i, P_i and R_i to U_j. Then U_j selects a random e_j, computes $R_j = P_i^{e_j} \bmod p$, and sends D_j, P_j and R_j to U_i. The session key is $K_{ij} \equiv R_i^{s_j}.(P_i^{s_i})^{e_j} \equiv R_i^{s_j}.R_j^{s_i} \equiv (P_j^{s_j})^{e_i}.R_j^{s_i} \pmod p$, which both parties can compute by (2).

The Attack

As with the Yacobi system, $\widetilde{A}$ eavesdrops on a conversation of B and C with calls R_b, R_c and secret key $K_{bc} \equiv R_b^{s_c} \cdot R_c^{s_b} (\bmod p)$. Subsequently she gets R'_b, R'_c, and computes the partial keys $R_b^{s_c} \bmod p$ and $R_c^{s_b} \bmod p$, and hence the session key K_{bc}. Again both modifications apply. Furthermore, it is easy to see that the attack extends to the 'perfect forward secrecy' variant [20, pp. 34–35].

3.3 COMSET (the Key Exchange Version)

Each user U_ℓ of COMSET [9, 6] selects secretely two appropriate primes $p_\ell \equiv 3 (\bmod 8)$ and $q_\ell \equiv 7 (\bmod 8)$ and publishes $n_\ell = p_\ell q_\ell$ as its public key. The encryption function of U_ℓ is $z \rightarrow z^2 \bmod n_\ell$, and the decryption function is $z \rightarrow z^{d_\ell} \bmod n_\ell$, where $d_\ell = ((p_\ell - 1)(q_\ell - 1) + 4)/8$. To establish a secret key with U_i, U_j selects a random $x_j \in Z_{n_i}$ and computes $m_j \equiv x_j^2 \bmod n_i$. Then U_j sends U_i the encryption $r_j \equiv m_j^2 \bmod n_i$ together with a validator v_j of m_j which consists of the $|n_i|/4$ least significant bits of m_j. U_i decrypts r_j to get $m_j' = r_j^{d_i} \bmod n_i$. It checks the validators of m_j' and $n_i - m_j' (\bmod n_i)$, and takes m_j'' to be the one whose validator is v_j. Then it sends U_j the next $|n_i|/4$ significant bits of m_j''. U_j checks these. If they are the same as the corresponding bits of m_j then the authentication of U_i is successful. The secret (partial) key is $[m_j] = [m_j'']$, where $[z_j]$ is an appropriate substring of z_j. For mutual authentication U_j must also be authenticated. The same protocol is used (the processes can be executed in parallel). The common session key is the bitwise exclusive-or $K_{ij} = [m_i] \oplus [m_j]$ of the partial keys.

The Attack

As with the Yacobi system, $\widetilde{A}$ eavesdrops on a conversation of B and C with calls $r_b \equiv m_b^2 \bmod n_c$ and $r_c \equiv m_c^2 \bmod n_b$, and secret key $K_{bc} = [m_b] \oplus [m_c]$ in Phase 1. In Phase 2 she gets the call $r_b{}^* = m_b^{*2} \bmod n_a$. Since n_a is her own key, she can compute $m_b{}^*$ and hence send B the appropriate substring for authentication. In Phase 3 she gets the call $r_c{}^* = m_c^{*2} \bmod n_a$, computes $m_c{}^*$, and sends C the appropriate substring for authentication. Then she gets the keys $K_{ca} = [m_c^*] \oplus [m_b]$ and $K_{ba} = [m_b{}^*] \oplus [m_c]$. Finally she computes the partial key $[m_b]$ and the partial key $[m_c]$ from which she gets the session key K_{bc} of B and C. For this scheme the modifications in Section 2.4 do not apply.

4 Repairing the Systems

When a cryptographic algorithm is used to solve a security problem, it is implemented within the framework of a protocol which ensures that certain levels of security required by the system are attained. If this goal is not met, this is either because the cryptographic algorithm is defective or because the protocol procedures are inappropriate. Many of the proposed Diffie-Hellman non-interactive variants have succumbed to attacks by passive and/or active adversaries. However their failure can be attributed to a large extent to the protocol procedures, and can be controlled by specifying a different implementation.

The simplest way to foil the triangle attack is to refuse to willingly reveal session keys to anyone. A sensible precaution is to ensure that session keys are destroyed immediately after the session is ended. This is one of the fundamental design principles for secure key management (and has been widely recognized). Alternatively, the users can be prevented from knowing their own session keys. To implement this, the key distribution algorithm can be encapsulated in a tamper-resistant device [4]. To enforce security the encryption algorithm must also be encapsulated.

Another way to foil the triangle attack is to modify the key distribution systems so as to get key confirmation. This can be done by having each user send an additional message which employs the established session key, e.g. via encryption or a MAC [14]. Then $\widetilde{A}$ cannot complete Phase 2 and Phase 3 of the attack in Section 2.3, and therefore B and C will destroy the session keys. However this is at the cost of making the key distribution systems interactive: the messages the users send to each other are now dependent (COMSET is already interactive).

Alternatively one can take the session key to be the hash $h(K_{ij})$, where $h(\cdot)$ is a suitable hash function, or, the hash of the concatenation of the partial keys. This will foil the triangle attack. However, it must be pointed out that the fact that this attack is foiled does not guarantee that the system is secure (e.g. from other known key attacks). The problem with these modifications is that the adversary gains some knowledge from fraudulent runs when their session keys are leaked, later on (this is not the case with honest runs when a uniform distribution is used, and the 'view' of the adversary can be simulated [12]). It is sensible to prevent such runs e.g. by requiring key confirmation.

Finally one can strengthen the algorithm. This problem is addressed in [12]. The solution proposed there is to require, additionally, that the users U_i, U_j prove to each other that they *know* the discrete logarithm of their calls R_i, R_j, by using an interactive zero-knowledge proof. This will prevent fraudulent runs (B and C will not compute their session key unless they are convinced that $\widetilde{A}$ knows the discrete logarithms of her calls), and furthermore no knowledge about the session keys is leaked to the adversary ($\widetilde{A}$ can simulate her view).

5 Conclusion

We have described a general known-key attack for key distribution systems for which the parties compute the session key from partial keys which are established non-interactively, and for which, given the session key and one partial key, the other partial key can be recovered. From our discussion it follows that:

- Revealing keys to legitimate holders may be harmful.
- Ensuring that session keys are deleted at the end of a session is a sensible precaution.
- Implicit key authentication can be dangerous.
- Taking the session key to be the bitwise exclusive-or, or product, of partial keys may be risky.
- It should not be possible to complete a fraudulent session.
- If files are to be stored, then the plaintext should be encrypted with a different key.

The following remains an open problem:

- Do there exist practical non-interactive proven secure[4] key distribution systems?

Acknowledgements

The author would like to thank Kaisa Nyberg for many helpful (and enjoyable) discussions, and in particular for pointing out that the triangle attack extends to COMSET. Many others have contributed in different ways to this investigation, in particular, Ivan Damgård, Whit Diffie, Fred Piper, Peter Landrock, Yacov Yacobi, and Paul van Oorschot. The author is particularly grateful to the referees for several constructive comments on the submitted version.

References

1. M. Abadi, M. Tuttle: A semantics for a logic of authentication. In: Proc. ACM Symp. on Principles of Distributed Computing. ACM Press 1991, pp. 201–216
2. J.A. Adam: Ways to verify the U.S.-Soviet arms pact. IEEE Spectrum, pp. 30–34 (1988)
3. M. Bellare and P. Rogaway: Entity authentication. To appear in: Crypto '93, Proceedings, Lecture Notes in Computer Science. Berlin: Springer
4. S. Bengio, G. Brassard, Y. G. Desmedt, C. Goutier, J-J. Quisquater: Secure implementations of identification systems. Journal of Cryptology 4(3), 175–183 (1991)
5. R. Bird, I. Gopal, A. Herzberg, P. Jansen, S. Kutten, R. Molva, and M. Yung: Systematic design of two-party authentication protocols. In: J. Feigenbaum (ed.): Advances in Cryptology – Crypto '91, Proceedings. Lecture Notes in Computer Science 576. Berlin: Springer 1992, pp. 44–61
6. J. Brandt, I. Damgård, P. Landrock, and T. Pedersen: Zero-knowledge authentication scheme with secret key exchange. In: S. Goldwasser (ed): Advances in Cryptology – Crypto '88, Proceedings. Lecture Notes in Computer Science 403. Berlin: Springer 1990, pp. 583–588

---

[4] A system is proven secure if breaking it (e.g. by using a known key attack) is as hard as solving a (believed to be) hard cryptographic 'reference' problem such as, factoring, the discrete logarithm, the Diffie-Hellman problem, etc. For practical systems this seems to be the best achievable level of security.

7. M. Burrows, M. Abadi, R. M. Needham: A logic of authentication. ACM Trans. Computer Systems 8, 18–36 (1990)
8. P-C. Cheng, V. Gligor: On the formal specification and verification of multiparty session protocols. In: Proc. IEEE Sym. on Research in Security and Privacy. IEEE Press 1990, pp. 216–233
9. COMSET. Ripe Integrity Primitives. Final Report of RACE Integrity Primitives Evaluation (R1040). June 1992, pp. 191–201
10. G.I. Davida: Chosen signature cryptanalysis of the RSA (MIT) public key cryptosystem. Tech. Report TR-CS-82-2, University of Wisconsin-Milwaukee, October 1982.
11. D.E.R. Denning: Digital signatures with RSA and other public-key cryptosystems. Comm. ACM 27(4), 388–392 (1984)
12. Y. Desmedt, M. Burmester: Towards practical 'proven secure' authenticated key distribution. In: Proceedings 1st ACM Conference on Computer and Communication Security, Fairfax, Virginia. ACM Press, November 1993, pp. 228–231
13. W. Diffie, M.E. Hellman: New directions in cryptography. IEEE Trans. Inform. Theory, IT–22(6), 644–654 (1976)
14. W. Diffie, P.C. van Oorschot, M.J. Wiener: Authentication and authenticated key exchanges. Designs, Codes and Cryptography 2, 107–125 (1992)
15. T. ElGamal: A public key cryptosystem and a signature scheme based on discrete logarithms. IEEE Trans. Inform. Theory 31, 469–472 (1985)
16. K. Gaarder, E. Snekkens: Applying a formal analysis technique to the CCITT X.509 strong two-way authentication protocol. J. Cryptology 3(1), 81–98 (1991)
17. V. Gligor, R. Kailar, S. Stubblebine, L. Gong: Logics for cryptographic protocols – virtues and limitations. In: Proc. IEEE 1991 Computer Security Foundations Workshop. Franconia, New Hampshire 1991
18. L. Gong, R. Needham, R. Yalom: Reasoning about belief in cryptographic protocols. In: Proc. 1990 IEEE Symp. on Security and Privacy, Oakland, California. IEEE Press 1990, pp. 234–248
19. K.C. Goss: Cryptographic method and apparatus for public key exchange with authentication. U.S. Patent 4,956,863, Granted Sept 11, 1990.
20. C.G. Günther: An identity-based key exchange protocol. In: J-J. Quisquater, J. Vandewalle (eds.): Advances in Cryptology, Eurocrypt '89. Lecture Notes in Computer Science 434. Berlin: Springer 1990, pp. 29–37
21. T. Matsumoto, Y. Takashima, H. Imai: On seeking smart public key distribution systems. The Transactions of the IECE of Japan E69(2), 99–106 (1986)
22. J.H. Moore: Protocol failures in cryptosystems. Proc. IEEE 76(5), 594–602 (1988)
23. K. Nyberg, R. Rueppel: A new signature scheme based on the DSA giving message recovery. In: Proceedings 1st ACM Conference on Computer and Communication Security, Fairfax, Virginia. ACM Press 1993, pp. 58–61
24. K. Nyberg, R. Rueppel. Weaknesses in some recent key agreement protocols. Electronics Letters, 30(1), 26–27 (1994)
25. R. L. Rivest, A. Shamir: How to expose an eavesdropper. Comm. ACM 27(4), pp. 393–395 (1984)
26. G.J. Simmons: How to ensure that data aquired to verify treaty compliance are trustworthy. In: G.J. Simmons (ed.): Contemporary Cryptology. New York: IEEE Press 1992, pp. 615–630.
27. P.C. van Oorschot: Extending cryptographic logics of belief to key agreement protocols. In: Proc. 1st ACM Conference on Computer and Communications Security, Fairfax, Virginia. ACM Press 1993, pp. 232-243

28. Y. Yacobi: A key distribution paradox. In: A.J. Menezes, S.A. Vanstone (eds.): Advances in Cryptology – Crypto '90, Proceedings. Lecture Notes in Computer Science 537. Berlin: Springer 1991, pp. 268–273
29. Y. Yacobi, Z. Shmuely: On key distribution systems. In: G. Brassard (ed.): Advances in Cryptology – Crypto '89, Proceedings. Lecture Notes in Computer Science 435. Berlin: Springer 1990, pp. 344–355

Cryptanalysis of Cryptosystems based on Remote Chaos Replication

Th. Beth, D. E. Lazic, A. Mathias

Universität Karlsruhe, Fakultät für Informatik,
Institut für Algorithmen und Kognitive Systeme,
Am Fasanengarten 5, D–76 128 Karlsruhe, Germany

Abstract. In the last five years, many cryptosystems based on the chaos phenomenon have been proposed. Most of them use chaotic maps, i. e., the discrete–time chaos. The recent announcement of a cryptosystem based on continuous–time chaos that is generated by a very simple electronic circuit known as Chua's circuit passed unrecognized by a large part of the cryptographic community. It is an analog to the VERNAM–cipher system, but uses auto–synchronization through remote replication of the chaotic masking signal. After the introductory description of continuous–time chaotic systems and their synchronization a general definition and discussion of cryptosystems based on remote chaos replication is given. A cryptanalytic attack for these systems is developed that can break the cryptosystem using Chua's circuit for all types of information–bearing signals.

1 Introduction

Analog scrambling devices have been a part of classical cryptography ever since secure transmission by wire and radio have been used. Amongst the many procedures for such tasks the method of adding synchronous noise is thoroughly understood from the information theoretic point of view. The well known VERNAM–cipher system [Ver26] as a digital counterpart of this method which has been defined for secure transmission of binary strings was proved unbreakably secure by SHANNON [Sha49], provided the noise sequence has maximal entropy and a secure key/synchronization channel is used. While digital cryptosystems of this kind have successfully been used during the last seven decades, the recent announcement of a by far cheaper and more efficient analog scrambling cryptosystem realized as a simple electronic circuit by L. CHUA and his coworkers [KHE92] passed unrecognized by a large part of the cryptographic community.

The idea, widely acclaimed in the community of control engineers, is that with a cryptosystem based on the continuous–time chaos phenomenon not only secure communication can be guaranteed, but also, what is more, it is implicitly claimed that this is achievable without the need for key management and external synchronization. The principle of this system resembles a modification of the VERNAM–cipher system where the additive scrambling noise is coming from a chaotic analog signal generator. The information–bearing signal is covered in chaotic pseudo noise of high amplitude giving a waveform with a very small signal–to–noise (S/N) ratio thus masked for the interceptor. At the receiver's side, this waveform is used to drive a replicating circuit that is equally tuned as the chaotic signal generator, the parameters being unknown to the interceptor. The auto–synchronizing replicating circuit produces a quite accurate copy of the chaotic noise which is then subtracted from the incoming waveform, thus revealing the buried signal.

In this paper we do not only show how such a system can be broken by well–adapted methods of signal processing. Also, a short discussion about the behavior of this type of cryptosystem in a real communication environment with channel noise and distortions points out the doubtfulness of a simple realization, especially in heavily disturbed communication channels. Besides, the implicated key–free–ness of this auto–synchronizing device is also questionable, although the common parameters of the chaos generating and replicating circuits may eventually be used as a key space. However, even the introduction of such a key doesn't significantly improve the security of the proposed system. Actually, the dominating power of the chaotic noise designed for auto–synchronizing purposes is from an information theoretic point of view the reason why this type of system is generally breakable [Sim79], [Sha93].

2 Continuous Time Chaos

When the behavior of some physical system is well understood, it is often possible to model it in terms of a set of state variables $SV = \{x_1(t), \ldots, x_N(t)\}$ varying in time. The most familiar systems using this approach are dynamical systems. A very common representation of a continuous time dynamical system is that of a system of N simultaneous first-order ordinary differential equations (ODE):

$$\dot{\mathbf{x}} = \mathbf{f}(\mathbf{x}, \mathbf{d}, \boldsymbol{\lambda}, t), \tag{1}$$

where

$$\begin{aligned}
&\mathbf{x} = \mathbf{x}(t) = (x_1(t), \ldots, x_N(t))\,; \qquad x_n(t) \in SV, n = 1, \ldots, N,\\
&\dot{\mathbf{x}} = \dot{\mathbf{x}}(t) \;=\; (\mathrm{d}x_1(t)/\mathrm{d}t, \ldots, \mathrm{d}x_N(t)/\mathrm{d}t)\,,\\
&\mathbf{f} = \mathbf{f}\,(\mathbf{x}, \ldots): \qquad U \longrightarrow \mathbb{R}^N, \qquad U \subseteq \mathbb{R}^N,\\
&\mathbf{d} = \mathbf{d}(t) = (d_1(t), \ldots, d_M(t))\,, \qquad d_m(t) \in DF = \{d_1(t), \ldots, d_M(t)\}\,,\\
&\boldsymbol{\lambda} = (\lambda_1, \ldots, \lambda_L) \in \mathbb{R}^L, \qquad \lambda_l \in SP = \{\lambda_1, \ldots, \lambda_L\}\,,\\
&t \in I = (a, b) \subseteq \mathbb{R}.
\end{aligned}$$

Here, $x_n(t)$, $n = 1, \ldots, N$ are unknown real-valued functions of a real variable t (time) and $\mathbf{f}$ is a known *vector field* which depends on $\mathbf{x}$ and, but not necessarily, on a set DF of real-valued drive functions, on a set SP of real-valued system parameters and on the time t. If $\dot{\mathbf{x}} = \mathbf{f}\,(\mathbf{x}, \boldsymbol{\lambda}, t)$, the dynamical system is non-driven, and if $\dot{\mathbf{x}} = \mathbf{f}\,(\mathbf{x}, \boldsymbol{\lambda})$, it is autonomous.

Solving an initial value problem (IVP) of (1) for some given $\boldsymbol{\lambda}_0 \in \mathbb{R}^L$ of system parameters consists of finding N real-valued functions

$$\mathbf{X}(t) = (X_1(t), \ldots, X_N(t)) \tag{2}$$

satisfying (1) and the initial conditions

$$\mathbf{X}(t_0) = \mathbf{X}_0 = (X_1^0, \ldots, X_N^0) \in U \subset \mathbb{R}^N, \tag{3}$$

where $X_1^0, \ldots, X_N^0$ represent a chosen set of initial states of the continuous-time dynamical system at some chosen initial time $t_0 \in I$. The set of points $\{\mathbf{X}(t) \mid \mathbf{X}(t_0) = \mathbf{X}_0, t \in I\}$ is called the *trajectory* through $\mathbf{X}_0$ in the *state space* $\mathbb{R}^N$. If the vector field satisfies certain reasonable conditions, then an IVP has a unique solution in I. In general, different initial conditions at t_0 lead to different IVP solutions. An important property of autonomous systems is that if the IVP: $\dot{\mathbf{x}} = \mathbf{f}(\mathbf{x}, \boldsymbol{\lambda}_0), \mathbf{X}(t_0 = 0) = \mathbf{X}_0$ has the solution $\mathbf{X}(t)$, the IVP: $\dot{\mathbf{x}} = \mathbf{f}(\mathbf{x}, \boldsymbol{\lambda}_0), \mathbf{X}(t_0 \neq 0) = \mathbf{X}_0$ has the solution $\mathbf{X}(t - t_0)$ [GH90], [Per91].

The term *steady state* refers to the asymptotic behavior of IVP solutions as $t \to \infty$ and defines the *limit set* in the state space. For continuous-time dynamical systems there exist four kinds of steady state behaviors: equilibrial, periodical, quasiperiodical and chaotic. The corresponding limit sets are: equilibrium point, closed curve, two-torus and the strange attractor. Chaotic behavior arises only in nonlinear systems for $N \geq 3$, for particular vector fields. A chaotic vector field has chaotic behavior only for particular parameter values, i.e., in *chaotic regime*, and usually has many different strange attractors characterized by the corresponding parameter values.

State variables in the chaotic regime are nonperiodic with a continuous broadband spectrum and exhibit sensitive dependence on initial conditions. Two identical autonomous chaotic systems started at initial conditions arbitrarily close to one another have trajectories which quickly become uncorrelated. Practically, it is impossible to construct two identical, independent chaotic systems with synchronized trajectories. There is always some error in measuring or specifying the initial conditions. Due to sensitive dependence, these errors, however small, will almost always alter the macroscopic behavior of a chaotic system [PC89]. Thus, in a very real sense, chaotic systems are unpredictable, i.e., their IVP solutions in chaotical regime behave similar to random processes. For this reason they are attractive for cryptographic applications. Until now, many cryptosystems based on chaos are proposed, most of them relying on discrete dynamical systems (time discrete variants of dynamical systems where ODE's are replaced by difference equations).

3 Synchronization in Chaotic Dynamical Systems

The vector field of an N–dimensional autonomous dynamical system

$$\dot{\mathbf{x}} = \mathbf{f}(\mathbf{x}, \boldsymbol{\lambda}) \tag{4}$$

with state variables in SV and with system parameters in SP is *drive decomposable* if it can be split up into the *drive* and *response* subsystems. This means that there is a dimension $N_1 < N$ and corresponding functions $\mathbf{f}^{(1)} : U_1 \longrightarrow$

$\mathbb{R}^{N_1}, U_1 \subseteq \mathbb{R}^{N_1}$ and $\mathbf{f}^{(2)}: U_2 \longrightarrow \mathbb{R}^{N_2}, U_2 \subseteq \mathbb{R}^{N_2}$, $N_2 = N - N_1$ that give two separate subsystems

$$\dot{\mathbf{x}}^{(1)} = \mathbf{f}^{(1)}\left(\mathbf{x}^{(1)}, \boldsymbol{\lambda}^{(1)}\right) \tag{5a}$$

$$\dot{\mathbf{x}}^{(2)} = \mathbf{f}^{(2)}\left(\mathbf{x}^{(2)}, \mathbf{d}^{(2)} = \mathbf{x}^{(1)}, \boldsymbol{\lambda}^{(2)}\right), \tag{5b}$$

where

$$\begin{aligned}
&\mathbf{x}^{(1)} = \mathbf{x}^{(1)}(t) = \left(x_1^{(1)}(t), \ldots, x_{N_1}^{(1)}(t)\right); \quad \left\{x_i^{(1)}(t)\right\}_{i=1}^{N_1} = SV_1 \subset SV, \\
&\mathbf{x}^{(2)} = \mathbf{x}^{(2)}(t) = \left(x_1^{(2)}(t), \ldots, x_{N_2}^{(2)}(t)\right); \quad \left\{x_j^{(2)}(t)\right\}_{j=1}^{N_2} = SV_2 \subset SV, \\
&SV_1 \cap SV_2 = \emptyset; \qquad SV_1 \cup SV_2 = SV; \qquad N_1 + N_2 = N, \\
&\boldsymbol{\lambda}^{(1)} = \left(\lambda_1^{(1)}, \ldots, \lambda_{L_1}^{(1)}\right); \quad \left\{\lambda_i^{(1)}\right\}_{i=1}^{L_1} = SP_1 \subseteq SP, \\
&\boldsymbol{\lambda}^{(2)} = \left(\lambda_1^{(2)}, \ldots, \lambda_{L_2}^{(2)}\right); \quad \left\{\lambda_j^{(2)}\right\}_{j=1}^{L_2} = SP_2 \subseteq SP, \\
&SP_1 \cup SP_2 = SP, \qquad L_1 + L_2 \geq L.
\end{aligned}$$

The two subsystems are coupled such that the behavior of the second (5b) is dependent on the behavior of the first (5a), but the first is not influenced by the behavior of the second. The first subsystem is called the *drive* and the second the *response*. Theoretically, for every $\boldsymbol{\lambda} \in \mathbb{R}^L$ and every $\mathbf{X}(0) = \mathbf{X}_0 \in U$ the solution of the IVP of (4) should always be equal to the solution of the IVP of the corresponding decoupled system (5) if the initial states $\mathbf{X}^{(1)}(0) = \mathbf{X}_0^{(1)}$, $\mathbf{X}^{(2)}(0) = \mathbf{X}_0^{(2)}$ and system parameters $\boldsymbol{\lambda}^{(1)} \cup \boldsymbol{\lambda}^{(2)}$ of (5) are identical to the corresponding ones in (4).

The concept of drive decomposition of an autonomous dynamical system makes a spatial separation of drive and response possible. If the dynamical system models an electronic circuit, the corresponding drive and response can be connected with a communication link that transmits the set of drive functions SV_1 as signals from the transmitting point (drive) to the receiving point (response). If the regime of the whole system is chaotic, such a communication system would represent a source which generates chaotic signals SV_2 driven by remotely generated chaotic signals SV_1 (SV_1 are all different from SV_2). Furthermore, if a second *inverse dynamical subsystem* (circuit)

$$\dot{\mathbf{x}}^{(I)} = \mathbf{f}^{(I)}\left(\mathbf{x}^{(I)}, \mathbf{d}^{(I)} = \mathbf{x}^{(II)}, \boldsymbol{\lambda}^{(I)}\right), \tag{6}$$

where

$$\begin{aligned}
&\mathbf{x}^{(I)} = \mathbf{x}^{(I)}(t) = \left(x_1^{(I)}(t), \ldots, x_{N_I}^{(I)}(t)\right); \quad \left\{x_i^{(I)}(t)\right\}_{i=1}^{N_I} = SV_I \subseteq SV_1, \\
&\mathbf{x}^{(II)} = \mathbf{x}^{(II)}(t) = \left(x_1^{(II)}(t), \ldots, x_{N_{II}}^{(II)}(t)\right); \quad \left\{x_j^{(II)}(t)\right\}_{j=1}^{N_{II}} = SV_{II} \subseteq SV_2, \\
&\boldsymbol{\lambda}^{(I)} = \left(\lambda_1^{(I)}, \ldots, \lambda_{L_I}^{(I)}\right); \quad \left\{\lambda_i^{(I)}\right\}_{i=1}^{L_I} = SP_I \subseteq SP,
\end{aligned}$$

can be constructed at the receiving point which is driven with $SV_{II} \subseteq SV_2$ and generates one or more signals $SV_I \subseteq SV_1$, a remote replica of chaotic signals would be theoretically possible.

For practical realization of such chaotic, synchronized subsystems, the stability of the response and inverse circuit is required. That means,

$$\lim_{t \to 0} \left| x_i^{(I)}(t) - x_i^{(1)}(t) \right| = 0 \qquad i = 1, \ldots, N_I, \tag{7}$$

1. independent of initial conditions $\mathbf{X}_0^{(2)}$ and $\mathbf{X}_0^{(I)}$,
2. when response and inverse circuit have common parameters, and parameters corresponding to the drive are slightly different, and
3. when there is an additive disturbance $\delta(t) = (\delta_1(t), \ldots, \delta_{N_1}(t))$ in the communication channel, i. e., $d_i^{(2)}(t) = x_i^{(1)}(t) + \delta_i(t)$; $i = 1, \ldots, N_1$, with disturbance–to–chaos power ratios D/C_i, $i = 1, \ldots, N_1$, smaller than some upper limit value $\overline{D/C}$.

We will call these three stability conditions *initial value*, *parameter* and *channel disturbance* stability. In practical applications, (7) should converge relatively rapid.

Recently, Pecora and Carroll [PC90], [PC91] practically showed that if the response is driven only by some subset of the driven system's state variables, it is possible to construct decoupled systems whose common state variables are synchronized even in chaotic regime. This gives a subdivision of the system (4) into three subsystems:

$$\dot{\mathbf{x}}^{(1)} = \mathbf{f}^{(1)} \left(\mathbf{x}^{(1)},\ \mathbf{d}^{(1)} = \mathbf{x}^{(2)},\ \boldsymbol{\lambda}^{(1)} \right), \tag{8a}$$

$$\dot{\mathbf{x}}^{(2)} = \mathbf{f}^{(2)} \left(\mathbf{x}^{(2)},\ \mathbf{d}^{(2)} = \mathbf{x}^{(1)},\ \boldsymbol{\lambda}^{(2)} \right), \tag{8b}$$

$$\dot{\mathbf{x}}^{(3)} = \mathbf{f}^{(3)} \left(\mathbf{x}^{(3)},\ \mathbf{d}^{(3)} = \mathbf{x}^{(1)},\ \boldsymbol{\lambda}^{(3)} \right), \tag{8c}$$

where $\mathbf{f}^{(1)}$ is N_1–dimensional, $\mathbf{f}^{(2)}$ is N_2–dimensional and both together represent the drive, while $\mathbf{f}^{(3)}$ is the N_3–dimensional response $(N_1 + N_2 + N_3 = N)$. The special case of (8), in which $\mathbf{f}^{(2)} = \mathbf{f}^{(3)}$ (and thus $N_2 = N_3$), leads to the concept of synchronization of chaotic subsystems. This special case is called ***homogeneous driving***. The construction of this system consists of dividing the initial system (4) into two subsystems $\mathbf{f}^{(1)}$ and $\mathbf{f}^{(2)}$. Then the subsystem $\mathbf{f}^{(2)}$ (which will not be used for driving $\mathbf{f}^{(3)}$) is duplicated. This duplicate is applied as response. How to divide the drive is determined by calculating the conditional *Lyapunov exponents* [PC91], [Sch89]. If the conditional Lyapunov exponents of $\mathbf{f}^{(2)}$ driven by $\mathbf{x}^{(1)}$ are all negative, the state variables of $\mathbf{f}^{(2)}$ and $\mathbf{f}^{(3)}$ synchronize.

Pecora and Carroll apply the idea of homogeneous driving to the Lorenz and Rössler chaotic systems as well as to the hysteretic electronic circuit and its numerical model [PC91]. In all these threedimensional systems, chaotic synchronization is achieved with respect to the initial value and parameter stability conditions. The channel noise stability was not considered.

Based on the homogeneous driving principle and using an inverse subsystem, a model of remote chaos replication can be constructed, as shown in figure 1.

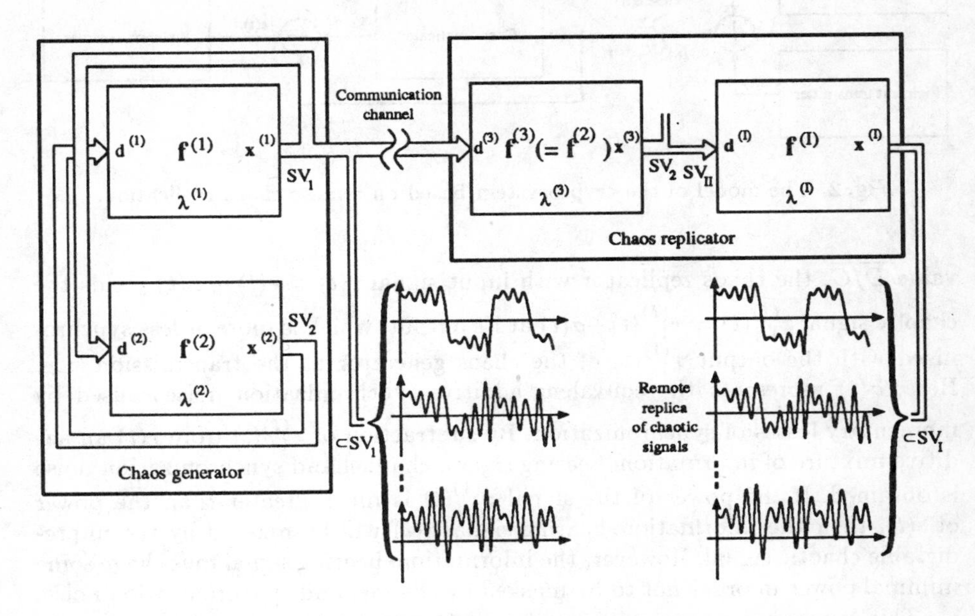

Fig. 1. The model of remote chaos replication.

4 Cryptosystems based on Remote Chaos Replication

Using the model of remote chaos replication, a secret key cryptosystem for analog communication channels can be formulated. For this purpose, only one replica of the chaotic signals in SV_1 is required, i.e., $N_1 = 1$, $\mathbf{x}^{(1)}(t) = x_1^{(1)}(t)$, $\mathbf{x}^{(I)}(t) = x_1^{(I)}(t)$ and $\lim_{t\to\infty} \left| x_1^{(I)}(t) - x_1^{(1)}(t) \right| = 0$. At the transmitter side, this system is an analog variant of the VERNAM–cipher system, because the chaotic signal $x_1^{(1)}(t)$ is added (real addition) to the information–bearing signal $s(t)$ so that their sum $c(t)$ represents the channel input signal (see figure 2). The random key space could be the subset $\boldsymbol{\lambda}_k \subseteq \boldsymbol{\lambda}^{(2)} \cup \boldsymbol{\lambda}^{(I)}$ of all values of parameters of $(\mathbf{f}^{(2)} \cup \mathbf{f}^{(I)}) \cap (\mathbf{f}^{(2)} \cup \mathbf{f}^{(1)})$ that cause chaotic regime in the chaos generator with mutually different strange attractors.

At the receiver side, this system principally differs from VERNAM–cipher systems. Here, the channel disturbance stability of the subsystem $\mathbf{f}^{(3)}$ in the chaos replicator enables the auto–synchronization and the recovery of the information–bearing signal $s(t)$. The additive disturbance $\delta(t) = \delta_1(t)$ in the communication channel is $\delta_1(t) = s(t) + n(t)$, where $n(t)$ represents the equivalent additive channel noise. If the actual disturbance–to–chaos ratio D/C is smaller than the limit

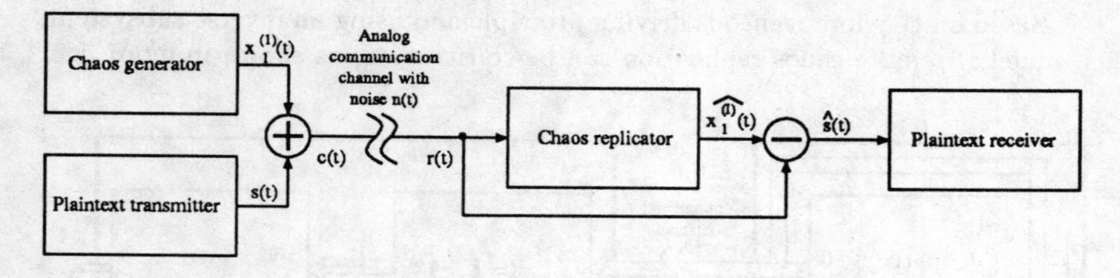

Fig. 2. The model of the cryptosystem based on remote chaos replication.

value $\overline{D/C}$, the chaos replicator with input signal $r(t) = c(t) + n(t)$ yields the chaotic signal $\widehat{x_1^{(I)}}(t) = x_1^{(I)}(t) + \sigma(t)$ at its output, which is more or less synchronized with the output $x_1^{(1)}(t)$ of the chaos generator at the transmission side. Here, $\sigma(t)$ represents the equvalent additive synchronization noise caused by momentary losses of synchronization. By subtraction of $\widehat{x_1^{(I)}}(t)$ from $r(t)$ an additive mixture of information–bearing signal, channel and synchronization noise is obtained. If the power of the signal $x_1^{(1)}(t)$ is much greater than the power of $s(t)$, then the information–bearing signal $s(t)$ will be masked by the unpredictable chaotic signal. However, the information–bearing signal must have some minimal power in order not to be masked by channel and synchronization noise.

This cryptosystem was proposed by A. OPPENHEIM and his coworkers without specifying what is the key, if any, in the system [OWIC92]. An example of masking and recovering a segment of a speech signal using the chaotic LORENTZ–system was demonstrated by computer simulation. The influence of the channel noise and the parameter stability of the system was not considered.

An experimental demonstration of the cryptosystem based on remote chaos replication was recently realized using Chua's circuit as chaos generator [KHE92]. This circuit is a very simple and robust electronic circuit built up with four linear elements and one nonlinear element called Chua's diode (CD). The circuit is shown in figure 3(a), and the state equations are given by

$$\dot{\mathbf{x}} = \mathbf{f}(\mathbf{x}, \boldsymbol{\lambda}) = \begin{cases} \dot{x}_1 = f_1(x_1, x_2, \lambda_1, \lambda_4, \lambda_5, \lambda_6, \lambda_7) = \frac{1}{\lambda_1}\left(\frac{x_2 - x_1}{\lambda_4} - h\right), \\ \dot{x}_2 = f_2(x_1, x_2, x_3, \lambda_2, \lambda_4) \qquad = \frac{1}{\lambda_2}\left(\frac{x_2 - x_1}{\lambda_4} + x_3\right), \\ \dot{x}_3 = f_3(x_2, \lambda_3) \qquad\qquad = \frac{1}{\lambda_3}(-x_2), \end{cases} \tag{9}$$

where $h = h(x_1, \lambda_5, \lambda_6, \lambda_7)$ is the piecewise linear characteristic of Chua's diode, as shown in figure 3(b). Here, x_1 is the voltage across the capacitor $C_1 = \lambda_1$, x_2 is the voltage across the capacitor $C_2 = \lambda_2$, and x_3 is the current through the inductor $L = \lambda_3$. The parameter λ_4 is the resistance R, while λ_5 and λ_6 are slopes of the inner and outer regions of h, and λ_7 indicates the breakpoints of h (for details see [Ken92]).

Using the principle of homogeneous driving, the Chua's circuit was decomposed in the following way: $\mathbf{f}^{(1)} = f_1$, $\mathbf{f}^{(2)} = (f_2, f_3)$, and $\mathbf{f}^{(I)} = f_1$, so that

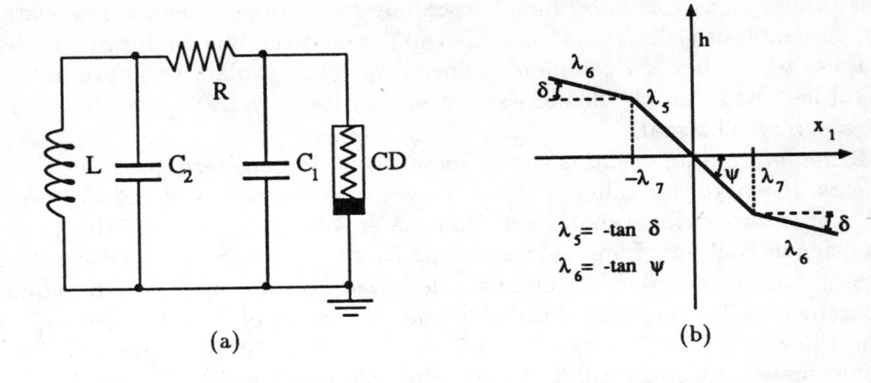

Fig. 3. (a) Chua's circuit, (b) the characteristic of Chua's diode.

$x_1(t) = x_1^{(1)}(t)$ (voltage across C_1) was used as a masking signal. The information–bearing signal $s(t)$ was a sine wave of frequency taken from the interval between 10% and 90% of the natural frequency of the RÖSSLER–type attractor. With an S/N–ratio (in this case D/C–ratio) of approximately −6 dB, the signal loss at the receiver was limited to less than −2 dB (−4 dBV). This signal loss is affected by momentary losses of synchronization which occur during some transitions between two Rössler–type attractors. The influence of the channel noise and the parameter stability was not considered.

Using these arguments and the fact that the sine wave $s(t)$ was not recognizable in the spectrum of $x_1(t)$, the authors of [KHE92] believe that this is sufficient to demonstrate a secure communication. It was not considered whether there should be a key in the proposed system, what it should consist of, and how it would be managed.

5 Cryptanalysis of Cryptosystems based on Remote Chaos Replication

There are basically three possibilities for the cryptanalysis of cryptosystems based on remote chaos replication:

- The extraction of the information–bearing signal $s(t)$ from the channel output signal $r(t)$,
- the extraction of the chaotic masking signal $x_1(t)$ from $r(t)$, and
- the estimation of parameters of the chaos replicator, which are chosen from the key space $\boldsymbol{\lambda}_k$ of the cryptosystem.

The extraction of the information–bearing signal is generally possible if $s(t)$ is a periodic signal or consists of periodic frames with sufficient duration, e. g., different types of low-rate digital modulations. Detection by autocorrelation and crosscorrelation used in radar signal processing and various other communication systems enables an effective extraction of $s(t)$ even at very low S/N–ratio [Lee64]. As these techniques are commonly known, we won't explain their use here. It should be noted that the demonstration of security in [KHE92] was done using just a periodical signal.

If the information–bearing signal doesn't contain long–term periodical components, the next possibility is the extraction of the masking signal $x_1(t)$ from $r(t)$. Lower–dimensional continuous time systems in chaotic regime always generate signals containing unpredictable repetitions of similar signal patterns that can generally be described as one parameterized function with a small number of parameters. Finding such a function and estimation of its actual parameters using the channel output signal $r(t)$ is the main task at this approach. From an information theoretic point of view, this parameter estimation can be done with sufficient accuracy since the S/N–ratio for the masking signal $x_1(t)$ is very high so that this communication system operates far below the channel capacity which makes it unsecure as a cryptosystem. The approach is demonstrated for the cryptosystem based on Chua's circuit.

Since the Chua's circuit contains the piecewise linear Chua's diode, the chaotic masking signal $x_1(t)$ behaves in a linear fashion as long as $x_1(t)$ remains within one linear segment of the characteristic shown in figure 3(b). For each of these time frames, the Laplace transform of $x_1(t)$ can be calculated and is of the following form:

$$X_1(s) = \frac{P(s)}{Q(s)} = \frac{s^3 a_3 + s^2 a_2 + s a_1 + a_0}{s(s-s_1)(s-s_2)(s-s_3)} \qquad (10)$$
$$= \frac{s^3 a_3 + s^2 a_2 + s a_1 + a_0}{s(s^3 + s^2 b_2 + s b_1 + b_0)},$$

where

$$s_1 = U + V - \frac{b_2}{3},$$
$$s_2 = -\frac{3U + 3V + 2b_2}{6} + i\sqrt{3}\frac{U-V}{2},$$
$$s_3 = -\frac{3U + 3V + 2b_2}{6} - i\sqrt{3}\frac{U-V}{2},$$

$$U = \sqrt[3]{-\frac{q}{2} + \sqrt{D}}, \quad V = \sqrt[3]{-\frac{q}{2} - \sqrt{D}}, \quad D = \frac{p^3}{27} + \frac{q^2}{4},$$

$$p = b_1 - \frac{b_2^2}{3}, \quad q = \frac{2b_2^3}{27} - \frac{b_2}{3} + b_0,$$

$$b_0 = \frac{\lambda^* + \lambda_4}{\lambda_1\lambda_2\lambda_3\lambda_4\lambda^*}, \quad b_1 = \frac{\lambda_1\lambda_4\lambda^* + \lambda_3}{\lambda_1\lambda_2\lambda_3\lambda_4\lambda^*}, \quad b_2 = \frac{\lambda_2\lambda^* + \lambda_1\lambda^* + \lambda_2\lambda_4}{\lambda_1\lambda_2\lambda_4\lambda^*}.$$

Here, λ^* represents a time–variable parameter (Chua's diode inner resistance) which depends on the momentary value of the voltage $x_1(t)$, i. e.,

$$\lambda^* = \begin{cases} \frac{1}{\lambda_6} & x_1(t) < -\lambda_7, \\ \frac{1}{\lambda_5} & -\lambda_7 \leq x_1(t) < \lambda_7, \\ \frac{1}{\lambda_6} & x_1(t) \geq \lambda_7. \end{cases}$$

Accordingly, the masking signal $x_1(t)$ can be viewed as a sequence of three separate types of time frames: lower, middle and upper signal time frame with corresponding variable durations τ_l, τ_m, and τ_u in varying order, as shown in fig. 4.

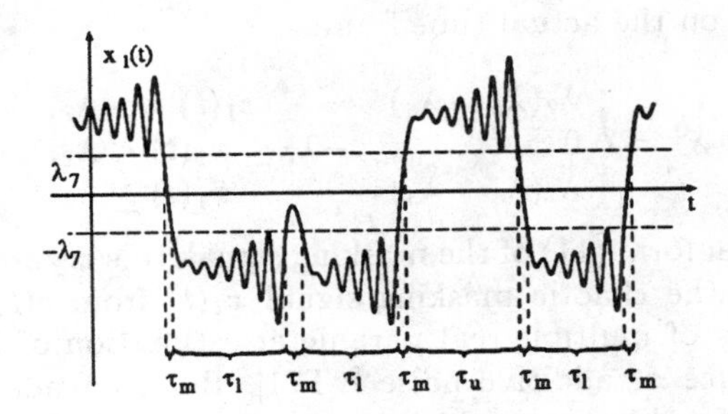

Fig. 4. A typical sample of the chaotic masking signal $x_1(t)$.

In all three types of time frames, the masking signal is, according to the inverse Laplace transform of (10), of the following form

$$x_1(t) = Ae^{at}\cos(\omega t - \phi) + Be^{bt} + g, \tag{11}$$

if $D > 0$ and s_1, s_2 and s_3 are mutually different and different from zero, which is the necessary condition for a chaotic regime. The exponents a and b and the frequency ω depend only on the poles of (10) and thus on λ_1, λ_2, λ_3, λ_4, λ_5 and λ_6

$$a = Re(s_2), \quad \omega = Im(s_2), \quad b = s_1, \tag{12}$$

while the amplitudes

$$A = 2\frac{P(s_2)}{Q'(s_2)}, \quad B = 2\frac{P(s_1)}{Q'(s_1)}, \quad g = \frac{a_0}{b_0} \tag{13}$$

and the initial phase

$$\phi = \arctan \frac{Im\frac{P(s_2)}{Q'(s_2)}}{Re\frac{P(s_2)}{Q'(s_2)}} \tag{14}$$

additionally depend on coefficients in the numerator of (10)

$$a_0 = -\frac{\lambda^0}{\lambda_1\lambda_2\lambda_3}, \tag{15}$$

$$a_1 = \frac{\lambda_1\lambda_4 x_1(0) + \lambda_3 x_3(0) - \lambda_3\lambda^0}{\lambda_1\lambda_2\lambda_3\lambda_4},$$
$$a_2 = \frac{\lambda_1 x_1(0) - \lambda_2\lambda_4\lambda^0 + \lambda_2 x_2(0)}{\lambda_1\lambda_2\lambda_4},$$
$$a_3 = x_1(0),$$

and thus on $x_1(0)$, $x_2(0)$ and $x_3(0)$ representing initial values of state variables at the beginning of each new time frame. λ^0 is, like λ^*, a time–variable parameter (Chua's diode constant current) which depends on the momentary value of the voltage $x_1(t)$, i.e., on the actual time frame

$$\lambda^0 = \begin{cases} \lambda_7(\lambda_5 - \lambda_6) & x_1(t) < -\lambda_7, \\ 0 & -\lambda_7 \le x_1(t) < \lambda_7, \\ \lambda_7(\lambda_6 - \lambda_5) & x_1(t) \ge \lambda_7. \end{cases}$$

Having a general form (11) of the masking signal, this cryptanalysis approach (the extraction of the chaotic masking signal $x_1(t)$ from $r(t)$) reduces to the classical techniques of multiple real parameter estimation of the known signal form in the presence of additive noise [vT71]. Here, an adaptive suboptimal estimation method will be used which is relatively simple but accurate enough to demonstrate the extraction of the masking signal.

By iterated differentiation of (11) one finds the following series of equations:

$$-(a^2+\omega^2)b(x_1(t)-g) + (a^2+\omega^2+2ab)x_1^{(1)}(t) + \\ -(2a+b)x_1^{(2)}(t) + x_1^{(3)}(t) = 0 \quad (16a)$$

$$-(a^2+\omega^2)bx_1^{(k)}(t) + (a^2+\omega^2+2ab)x_1^{(k+1)}(t) + \\ -(2a+b)x_1^{(k+2)}(t) + x_1^{(k+3)}(t) = 0, \quad (16b)$$

where $x_1^{(k)}(t)$ is the k–th order derivative of $x_1(t)$. Since the power of $x_1(t)$ dominates that of $s(t)$ in order to allow auto–synchronization of the chaos replicator, the parameters of $x_1(t)$ can be estimated quite accurately with appropriately smoothed derivatives of the sampled channel output signal $r(t) = x_1(t) + s(t)$, so that

$$\widehat{x_1^{(k)}}(t) = r(t) * g^{(k)}(t) = r^{(k)}(t) * g(t), \quad (17)$$

where

$$g(t) = \frac{s}{\sqrt{2\pi}} e^{-\frac{(st)^2}{2}}$$

is the Gaussian bell curve controlled by the scaling factor s, $s > 0$, and $*$ denotes convolution. Using (16b) for three consecutive values of k yields the following system of linear equations

$$\begin{pmatrix} \widehat{x_1^{(k)}} & \widehat{x_1^{(k+1)}} & \widehat{x_1^{(k+2)}} \\ \widehat{x_1^{(k+1)}} & \widehat{x_1^{(k+2)}} & \widehat{x_1^{(k+3)}} \\ \widehat{x_1^{(k+2)}} & \widehat{x_1^{(k+3)}} & \widehat{x_1^{(k+4)}} \end{pmatrix} \begin{pmatrix} \hat{b}(\hat{a}^2+\hat{\omega}^2) \\ -\hat{a}^2-\hat{\omega}^2-2\hat{a}\hat{b} \\ 2\hat{a}+\hat{b} \end{pmatrix} = \begin{pmatrix} \widehat{x_1^{(k+3)}} \\ \widehat{x_1^{(k+4)}} \\ \widehat{x_1^{(k+5)}} \end{pmatrix}, \quad (18)$$

whose solution is

$$\begin{pmatrix} \hat{b}(\hat{a}^2+\hat{\omega}^2) \\ -\hat{a}^2-\hat{\omega}^2-2\hat{a}\hat{b} \\ 2\hat{a}+\hat{b} \end{pmatrix} = \begin{pmatrix} \alpha \\ \beta \\ \gamma \end{pmatrix} = \tag{19}$$

$$= \frac{1}{-\widehat{x_1^{(k+2)}} + 2\widehat{x_1^{(k+1)}}\,\widehat{x_1^{(k+2)}}\,\widehat{x_1^{(k+3)}} - \widehat{x_1^{(k)}}\,\widehat{x_1^{(k+3)}}^2 - \widehat{x_1^{(k+1)}}^2\,\widehat{x_1^{(k+4)}} + \widehat{x_1^{(k)}}\,\widehat{x_1^{(k+2)}}\,\widehat{x_1^{(k+4)}}}.$$

$$\begin{pmatrix} -\widehat{x_1^{(k+3)}}^2 + \widehat{x_1^{(k+2)}} * \widehat{x_1^{(k+4)}} & \widehat{x_1^{(k+2)}}\,\widehat{x_1^{(k+3)}} - \widehat{x_1^{(k+1)}}\,\widehat{x_1^{(k+4)}} & -\widehat{x_1^{(k+2)}}^2 + \widehat{x_1^{(k+1)}}\,\widehat{x_1^{(k+3)}} \\ \widehat{x_1^{(k+2)}}\,\widehat{x_1^{(k+3)}} - \widehat{x_1^{(k+1)}}\,\widehat{x_1^{(k+4)}} & -\widehat{x_1^{(k+2)}}^2 + \widehat{x_1^{(k)}}\,\widehat{x_1^{(k+4)}} & \widehat{x_1^{(k+1)}}\,\widehat{x_1^{(k+2)}} - \widehat{x_1^{(k)}}\,\widehat{x_1^{(k+3)}} \\ -\widehat{x_1^{(k+2)}}^2 + \widehat{x_1^{(k+1)}}\,\widehat{x_1^{(k+3)}} & \widehat{x_1^{(k+1)}}\,\widehat{x_1^{(k+2)}} - \widehat{x_1^{(k)}}\,\widehat{x_1^{(k+3)}} & -\widehat{x_1^{(k+1)}}^2 + \widehat{x_1^{(k)}}\,\widehat{x_1^{(k+2)}} \end{pmatrix} \cdot \begin{pmatrix} \widehat{x_1^{(k+3)}} \\ \widehat{x_1^{(k+4)}} \\ \widehat{x_1^{(k+5)}} \end{pmatrix}$$

Solutions of the system of nonlinear equations (18) in $\hat{b}$, $\hat{a}$, and $\hat{\omega}$ represent the estimations of b, a and ω

$$\hat{b} = u + v + \frac{\gamma}{3}, \tag{20}$$

$$\hat{a} = \frac{-\hat{b}-\gamma}{2}, \tag{21}$$

$$\hat{\omega} = \sqrt{-\hat{a}^2 - 2\hat{a}\hat{b} - \beta}, \tag{22}$$

where

$$f = -\beta - \frac{\gamma^2}{3}, \quad h = -\frac{2\gamma^3}{27} + \frac{\gamma}{3} - \alpha,$$

$$d = \frac{f^3}{27} + \frac{h^2}{4}, \quad u = \sqrt[3]{-\frac{h}{2} + \sqrt{d}}, \quad v = \sqrt[3]{-\frac{h}{2} - \sqrt{d}}.$$

The DC term g is estimated by substitution of $\hat{a}$, $\hat{b}$ and $\hat{\omega}$ in equation (16a). The equations

$$\chi = Ae^{at}\,\widehat{\cos(\omega t - \phi)} = \frac{(\hat{b}^2 - 2\hat{a}\hat{b})(\widehat{x_1} - \hat{g}) + 2\hat{a}\widehat{x_1^{(1)}} - \widehat{x_1^{(2)}}}{(\hat{a}-\hat{b})^2 + \hat{\omega}^2}$$

$$\sigma = Ae^{at}\,\widehat{\sin(\omega t - \phi)} =$$

$$\frac{\hat{b}(\hat{a}\hat{b} - \hat{a}^2 + \hat{\omega}^2)(\widehat{x_1} - \hat{g}) + \widehat{x_1^{(1)}}(\hat{a}^2 - \hat{b}^2 - \hat{\omega}^2) + \widehat{x_1^{(2)}}(\hat{b} - \hat{a})}{\hat{\omega}((\hat{a}-\hat{b})^2 + \hat{\omega}^2)}$$

obtained from (11) and its derivatives, allow the estimations of ϕ, Ae^{at} and Be^{bt}

$$\hat{\phi} = \arctan\frac{\sigma}{\chi}, \tag{23}$$

$$\widehat{Ae^{at}} = \sqrt{\sigma^2 + \chi^2} \tag{24}$$

$$\widehat{Be^{bt}} = \frac{(\widehat{x_1} - \hat{g})(\hat{a}^2 + \hat{\omega}^2) - 2\hat{a}\widehat{x_1^{(1)}} + \widehat{x_1^{(2)}}}{(\hat{a}-\hat{b})^2 + \hat{\omega}^2}. \tag{25}$$

With these results, a program with graphic interaction was written that reads in the sampled channel output signal, performs the calculations and displays the estimated parameters of (11). It then synthesizes a signal which is subtracted from the input signal, leaving the signal for which masking was attempted with considerably reduced chaotic masking component. Experiments with this program using speech signals 15 dB below the masking signal that were initially not audible reproduced the speech signals in clearly understandable quality. The influence of the channel noise and parameter stability of the system was not considered. It should be noted that in the upper and lower signal time frames the exponent b is always negative and the component Be^{bt} is rather weak. Its neglection leads to a much simpler estimation process and gives relatively good results, too.

The third mentioned cryptanalysis approach consisting of the estimation of the chaos replicator's parameters directly follows from the presented results. It was not addressed here since the described method gives sufficient results for the cryptanalysis of the cryptosystem based on Chua's circuit. We only point out that the back–substitution of the estimated parameters in (10), (12), (13) and (14) or the coefficient comparison of (9) in its third order differential equation form with (16) makes the estimation of the unknown parameters from the key space $\boldsymbol{\lambda}_k$ possible.

6 Conclusions

The presented cryptanalysis method of extracting the chaotic masking signal from the channel output signal breaks the cryptosystem using Chua's circuit for all types of information–bearing signals. This method is applicable for all other cryptosystems based on the remote chaos replication principle where the functional form of the masking signal or its dominating components can be revealed from the structure of the chaos generator. For three–dimensional and some other lower–dimensional chaotic continuous–time systems this is always possible by using adequate linearization techniques. Having the functional form, the dominating power of the chaotic masking signal necessary for the auto–synchronization of the cryptosystem always enables a sufficiently accurate estimation of remaining unknown parameters. Cryptosystems of this type based on higher–dimensional continuous–time chaotic systems would require much higher cryptanalytic effort. For such systems however, the realization of auto–synchronization based on only one state variable (the masking signal) is still an open problem [PC91].

Acknowledgements

We would like to thank Michael Pal, student at our institute, for his valuable aid in the verification of the presented methods.

References

[BVKC93] Belykh, V. N., Verichev, N. N., Kocarev, Lj., and Chua, L. O., "On chaotic synchronization in a linear array of Chua's circuits", *Journal of Circuits, Systems and Computers*, Vol. 3, no. 2, pp. 579–589, 1993.

[CKEI92] Chua, L. O., Kocarev, Lj., Eckert, K., and Itoh, M., "Experimental chaos synchronization in Chua's circuit", *IBID*, pp. 705–708, 1992.

[GH90] Guckenheimer, L. and Holmes, P., *Nonlinear Oscillations, Dynamical Systems and Bifurcations of Vector Fields*, Applied Mathematical Sciences, Springer-Verlag, 1990.

[Ken92] Kennedy, M. P., "Robust op amp realization of Chua's circuit", *Frequenz*, Vol. 46, no. 3–4, pp. 66–80, 1992.

[KHE92] Kocarev, Lj., Halle, K. S., Eckert, K., Chua, L. O., and Parlitz, U., "Experimental demonstration of secure communications via chaotic synchronization", *International Journal of Bifurcation and Chaos*, Vol. 2, no. 3, pp. 709–713, 1992.

[Lee64] Lee, Y. W., *Statistical Theory of Communication*, John Wiley & Sons, 1964.

[OWIC92] Oppenheim, A. V., Wornell, G. W., Isabelle, S. H., and Cuomo, K. M., "Signal processing in the context of chaotic signals", *Proc. 1992 IEEE ICASSP*, Vol. IV, pp. 117–120, 1992.

[PC89] Parker, T. S. and Chua, L. O., *Practical Numerical Algorithms for Chaotic Systems*, Springer-Verlag, 1989.

[PC90] Pecora, L. M. and Carroll, T. L., "Synchronization in chaotic systems", *Physical Review*, Vol. 64, no. 8, pp. 821–824, 1990.

[PC91] Pecora, L. M. and Carroll, T. L., "Driving system with chaotic signals", *Physical Review*, Vol. 44, no. 4, pp. 2374–2383, 1991.

[PCK92] Parlitz, U., Chua, L. O., Kocarev, Lj., Halle, K. S., and Shang, A., "Transmission of digital signals by chaotic synchronization", *International Journal of Bifurcation and Chaos*, Vol. 2, no. 4, pp. 937–977, 1992.

[Per91] Perko, L., *Differential Equations and Dynamical Systems*, Texts in Applied Mathematics, Springer–Verlag, 1991.

[Sch89] Schuster, H. G., *Deterministic Chaos*, VCH, Germany, 1989.

[Sha49] Shannon, C. E., "Communication theory of secrecy systems", *Bell System Technical Journal*, Vol. 28, pp. 656–715, 1949.

[Sha93] Shannon, C. E., "Analogue of the Vernam system for continuous time series", In Sloane, N. J. A. and Wyner, A. D., editors, *Claude Elwood Shannon: Collected Papers*, Memorandum MM 43–110–44, Bell Laboratories, 1943, pp. 144–147, IEEE Press, 1993.

[Sim79] Simmons, G. J., "The mathematics of secure communication", *Math. Intell. 1*, pp. 233–246, 1979.

[Ver26] Vernam, G. S., "Cipher printing telegraph systems for secret wire and radio telegraphic communications", *J. Am. Inst. Elec. Eng.*, Vol. 55, pp. 109–115, 1926.

[vT71] van Trees, H. L., *Detection, Estimation and Modulation Theory*, Number Part I, II and III, John Wiley and Sons, Inc., New York, 1968, 1971, 1971.

A Fourier Transform Approach to the Linear Complexity of Nonlinearly Filtered Sequences

James L. Massey and Shirlei Serconek*
Signal and Information Processing Laboratory
Swiss Federal Institute of Technology
ETH-Zentrum
CH-8092 Zürich, Switzerland

Abstract

A method for analyzing the linear complexity of nonlinear filterings of PN-sequences that is based on the Discrete Fourier Transform is presented. The method makes use of "Blahut's theorem", which relates the linear complexity of an N-periodic sequence in $GF(q)^N$ and the Hamming weight of its frequency-domain associate. To illustrate the power of this approach, simple proofs are given of Key's bound on linear complexity and of a generalization of a condition of Groth and Key for which equality holds in this bound.

1 Introduction

Fourier transforms in a Galois field play an important role in the study and processing of $GF(q)$-valued signals, particularly in coding theory. By revisiting many topics by way of the *frequency-domain*, deeper understanding and alternative encoding and decoding techniques can be found (Blahut [1]).

By exploiting "Blahut's theorem", which states that the linear complexity of an N-periodic sequence in $GF(q)^N$ and the Hamming weight of its *frequency-domain* associate are equal, we use Discrete Fourier Transform (DFT) techniques here to study the linear complexity of nonlinear filterings of PN(pseudo-noise)-sequences. To illustrate the power of this approach, we give a simple and transparent proof of a generalization of a result of Groth [2] and Key [3].

Groth applied second-order boolean functions to the stages of an LFSR with a primitive connection polynomial. No stage was allowed to be used more than once. But Groth, as well as Key [3], as pointed out by Rueppel [7], "limited

*This work was done while the author was on leave from CEPESC, Cx Postal 02976, Brasília, DF, BRASIL, CEP 70610-200

himself to consider only those sequences which are available at the stages of an LFSR, in particular he allowed only phase differences of at most the length of LFSR. But when the speed of the LFSR is taken as an additional parameter, this theory must be extended to allow arbitrary phase differences. But then even the result of Key's which might be considered as his most solid one, namely, that a 2nd order product of 2 distinct phases of the same sequence never degenerates, is no longer true." Using the DFT method that we develop in section 4, we show that Key's non-degeneration result always holds when the length of the LFSR is a prime; the restrictions imposed by Groth and Key on the phase differences are unnecessary in this case.

We will discuss only the binary case $q = 2$ in this paper, but the Discrete Fourier Transform properties hold in the general case and all our results readily generalize to any finite field.

2 Period and Linear Complexity of Sequences

The linear complexity, $\mathcal{L}(\tilde{s})$, of the sequence $\tilde{s} = s_0, s_1, \cdots$, $s_i \in F$, F an arbitrary field, is the length L of the shortest linear feedback shift-register (LFSR) that can generate $\tilde{s}$ when the first L digits of $\tilde{s}$ are initially loaded in the register. Equivalently, the linear complexity is defined to be the smallest nonnegative integer L such that there exist coefficients $c_1, c_2, \cdots, c_L$ in F such that

$$s_j + c_1 s_{j-1} + \cdots + c_L s_{j-L} = 0 , \qquad j \geq L .$$

Linear complexity is very useful in the study of stream ciphers; a necessary condition for security of a running-key generator is that it produce a sequence with large linear complexity.

The sequence $\tilde{s}$ will be called N-periodic if N is a positive integer such that $s_i = s_{i+N}$ for all $i \geq 0$. If we characterize a periodic sequence $\tilde{s} = s_0, s_1, \cdots$ by its D-transform

$$S(D) = s_0 + s_1 D + s_2 D^2 + \cdots ,$$

then it is well known [5] that

$$S(D) = \frac{P(D)}{C(D)} ,$$

where

$$C(D) = 1 + c_1 D + c_2 D^2 + \cdots + c_L D^L , \qquad c_L \neq 0$$

and

$$\deg P(D) < \deg C(D) .$$

The polynomial $C(D)$ is the connection polynomial of an LFSR of length L that generates $\tilde{s}$ when its initial state is $[s_0, s_1, \cdots, s_{L-1}]$. If $\gcd(P(D), C(D)) = 1$, then this is the unique shortest LFSR that can generate the sequence $\tilde{s}$.

3 The Discrete Fourier Transform (DFT)

3.1 Definition of the DFT

We write $\mathbf{a}^N$ to denote the N-tuple $[a_0, a_1, \cdots, a_{N-1}] \in F^N$ where F is an arbitrary field. Under the assumption that there exists a primitive N-th root of unity ω in the field F, the Discrete Fourier Transform (DFT) of a *time-domain* N-tuple $\mathbf{a}^N = [a_0, a_1, \cdots, a_{N-1}]$ is defined to be the *frequency-domain* N-tuple $\mathbf{A}^N = [A_0, A_1, \cdots, A_{N-1}]$, where the components of $\mathbf{A}^N$ are given by

$$A_j = \sum_{i=0}^{N-1} a_i \omega^{ij} \quad for\ j = 0, 1, \cdots N-1.$$

The sequence $\mathbf{a}^N$ can be recovered from $\mathbf{A}^N$ by the inverse DFT in the manner

$$a_j = \frac{1}{N^*} \sum_{i=0}^{N-1} A_i \omega^{-ij} \quad for\ j = 0, 1, \cdots N-1,$$

where

$$N^* = \begin{cases} N\ modulo\ p & \text{, if the characteristic of } F \text{ is a prime } p \text{ .} \\ N & \text{, if the characteristic of } F \text{ is } 0 \text{ .} \end{cases}$$

The components of the N-tuples $\mathbf{a}^N$ and $\mathbf{A}^N$ can be used as coefficients to form two polynomials $a(X)$ and $A(X)$ in the indeterminant X whose degrees are at most $N-1$, namely

$$a(X) = a_0 + a_1 X + a_2 X^2 + \cdots + a_{N-1} X^{N-1}$$

$$A(X) = A_0 + A_1 X + A_2 X^2 + \cdots + A_{N-1} X^{N-1} \quad .$$

In this polynomial representation, the DFT relations can be written simply as

$$A_j = a(\omega^j) \ \ and \ \ a_j = \frac{1}{N^*} A(\omega^{-j}) \quad for\ j = 0, 1, \cdots (N-1).$$

3.2 Properties of the DFT

The following properties of the DFT are valid in any field F containing a primitive N-th root of unity ω. Proofs may be found in [1]. A double parenthesis (()) about an integer denotes that this integer should be taken modulo N.

3.2.1 The Shifting Property

Assume that the time-domain N-tuple $\mathbf{b}^N = [b_0, b_1, \cdots, b_{N-1}]$ is formed by shifting the N-tuple $\mathbf{a}^N = [a_0, a_1, \cdots, a_{N-1}]$ cyclically to the *left* by k positions, i.e.,

$$b_i = a_{((i+k))} \quad for\ i = 0, 1, \cdots N-1,$$

then the components of the frequency-domain N-tuple $\mathbf{B}^N$ are given by

$$B_i = A_i.\omega^{-ki} \quad for\ i = 0, 1, \cdots N-1.$$

Equivalently,

$$B(X) = A(\omega^{-k}X) .$$

3.2.2 The Convolution Property

If the time-domain N-tuple $\mathbf{c}^N = [c_0, c_1, \cdots, c_{N-1}]$ is given by the component-wise product $c_i = a_i.b_i$ $(i = 0, 1, \cdots N-1)$ of N-tuples $\mathbf{a}^N$ and $\mathbf{b}^N$, then the frequency-domain N-tuple $\mathbf{C}^N = [C_0, C_1, \cdots, C_{N-1}]$ can be found by cyclically convolving the N-tuples $\mathbf{A}^N$ and $\mathbf{B}^N$ in the manner

$$C_j = \frac{1}{N^*} \sum_{k=0}^{N-1} A_{((j-k))} B_k \quad for\ j = 0, 1, \cdots N-1 .$$

Equivalently,

$$C(X) = \frac{1}{N^*} A(X)B(X) \ mod\ X^N - 1.$$

3.2.3 The Conjugacy Constraints Property

Let the elements of the frequency-domain N-tuple $\mathbf{A}^N$ belong to the finite field $GF(q^m)$ where N divides $q^m - 1$, which condition is necessary and sufficient to ensure that $GF(q^m)$ contains primitive N-th roots of unity. Then the elements of the time-domain N-tuple $\mathbf{a}^N$ belong to the subfield $GF(q)$ if and only if the elements of the frequency-domain N-tuple satisfy the conjugacy constraints:

$$A_j^q = A_{((qj))} \quad 0 \leq j < N .$$

Note that the set of indices $\{((qj)) \mid 0 \leq j < N\}$ is a cyclotomic coset modulo N and that its cardinality must be a divisor of N.

3.2.4 Blahut's Theorem

Consider the frequency-domain vector $\mathbf{S}^N = [S_0, S_1, \cdots, S_{N-1}]$ associated with the time-domain vector $\mathbf{s}^N = [s_0, s_1, \cdots, s_{N-1}]$ by the DFT defined by a primitive N-th root of unity ω in F. Then, for $\omega = \alpha^{-1}$,

$$s_j = \frac{1}{N^*} \sum_{i=0}^{N-1} S_i \omega^{-ij} = \frac{1}{N^*} \sum_{i=0}^{N-1} S_i \alpha^{ij}$$

holds for all $j \geq 0$ since replacing j by $j + N$ leaves the sum unchanged. Thus the D-transform of $\tilde{s}$ is

$$\sum_{j=0}^{\infty} s_j D^j = \frac{1}{N^*} \sum_{j=0}^{\infty} \sum_{i=0}^{N-1} S_i \alpha^{ij} D^j$$

$$= \frac{1}{N^*} \sum_{i=0}^{N-1} S_i \sum_{j=0}^{\infty} (\alpha^i D)^j$$

$$= \frac{1}{N^*} \sum_{\substack{i=0 \\ S_i \neq 0}}^{N-1} S_i \frac{1}{1 - \alpha^i D} .$$

Because the roots of the denominator polynomials are $\alpha^{-i} = \omega^i$ for those i such that $0 \leq i < N$ and $S_i \neq 0$, these roots are all distinct. Thus, letting $w(.)$ denote the Hamming weight (i.e., the number of nonzero components) of the enclosed N-tuple), we have

$$\sum_{j=0}^{\infty} s_j D^j = \frac{P(D)}{C(D)}$$

where

$$C(D) = \prod_{\substack{i=0 \\ S_i \neq 0}}^{N-1} (1 - \alpha^i D) \tag{1}$$

has degree $w(\mathbf{S}^N)$, where $\deg P(D) < \deg C(D)$ and $\gcd(P(D), C(D)) = 1$. By (1) and the last comment in section 2 we now have

$$\mathcal{L}(\tilde{s}) = w(\mathbf{S}^N) , \tag{2}$$

which relationship we will refer as *Blahut's Theorem*, since, as has been pointed out in [6], this relation was implicitly used by Blahut.

Example 3.1:
Let $F = GF(2)$ and let $\tilde{s} = s_0, s_1, s_2, \cdots$ be the characteristic phase of the PN-sequence defined by the primitive element α of $GF(2^3)$ whose minimum polynomial is $h(X) = X^3 + X + 1$, i.e., $\tilde{s}$ is defined by

$$s_i = Tr(\alpha^i) , \;\; i \geq 0$$

where Tr is the trace operator from $GF(2^3)$ to $GF(2)$, i.e., $Tr(\alpha) = \alpha + \alpha^2 + \alpha^4$. Then $\tilde{s} = 1, 0, 0, 1, 0, 1, 1, 1, 0, 0, \cdots$. The sequence $\tilde{s}$ is periodic of period $N = 7$ and is generated by the LFSR with connection polynomial $C(D) = D^3 + D^2 + 1$ and initial state $[1, 0, 0]$. Consider now the time-domain vector $\mathbf{s}^N = [1, 0, 0, 1, 0, 1, 1]$ and its associated frequency-domain vector $\mathbf{S}^N = [S_0, S_1, \cdots, S_{N-1}]$ for the DFT defined by the primitive element $\omega = \alpha^{-1}$. Then $S_j = s(\omega^j)$, for $0 \leq j < N$, where $s(X) = 1 + X^3 + X^5 + X^6$ is the time-domain polynomial. Hence $\mathbf{S}^N = [0, 1, 1, 0, 1, 0, 0]$ and the frequency-domain polynomial is $S(X) = X + X^2 + X^4$, which is a so-called linearized polynomial [4].

4 PN-Sequences and their linearized polynomials

Suppose that α is a primitive element of $GF(2^L)$ whose minimum polynomial is $h(X) = X^L + c_1X^{L-1} + \cdots + c_L$. Then the characteristic phase of the PN-sequence defined by $h(X)$ is the sequence $\tilde{s}$ such that

$$s_i = Tr(\alpha^i) = \alpha + \alpha^2 + \cdots + \alpha^{2^{L-1}} , \quad i \geq 0 . \tag{3}$$

The sequence $\tilde{s}$ is generated by the LFSR with connection polynomial $C(D) = 1 + c_1D + \cdots + c_LD^L$ and initial state $[s_0, s_1, \cdots, s_{L-1}]$. Observe that $C(D)$ is the reciprocal polynomial of $h(X)$, so $\omega = \alpha^{-1}$ is (primitive and) a root of $C(D)$.
Thus

$$C(D) = \prod_{i \in \mathcal{C}} (1 - \alpha^i D)$$

where $\mathcal{C} = \{1, 2, 4, \cdots, 2^{L-1}\}$ is the main cyclotomic coset modulo $2^L - 1$. It follows from (3) and the inverse DFT relation

$$s_i = (\frac{1}{N*})S(\omega^{-i}) = S(\alpha^i)$$

that

$$S(X) = X + X^2 + X^4 + \cdots + X^{2^{L-1}} .$$

5 Second Order Nonlinear Filterings of PN-sequences

Let $\tilde{s} = s_0, s_1, s_2, \cdots$ be the characteristic phase of a PN-sequence generated by a maximal-length LFSR of length L and let $N = 2^L - 1$. Let $\tilde{s}_i$ denote the shifted version $s_{i-1}, s_i, s_{i+1}, \cdots$ for $i = 1, \cdots, N$ (in particular $\tilde{s} = \tilde{s}_1$). Note that, for $1 \leq i \leq L$, $\tilde{s}_i$ is the sequence that one would observe at the i-th stage of the LFSR when the LFSR was clocked repeatedly. Let $\mathbf{s}_i \in GF(2)^N$ be the binary vector corresponding to the first period of $\tilde{s}_i$. Let $w_2(i)$ denote the Hamming weight of the radix-2 form of the integer i.

We are interested in the linear complexity of sequences $\tilde{z}$ obtained by the product of the sequences produced from two different initial states of the maximal LFSR. Let $\tilde{t} = \tilde{s}_i\tilde{s}_j$ be the Hadamard product of $\tilde{s}_i$ and $\tilde{s}_j, i \neq j$, i.e., the bit-by-bit product of these two sequences. As was shown in section 4, defining the DFT by $\omega = \alpha^{-1}$ associates the frequency-domain polynomial $S(X) = X + X^2 + X^4 + \cdots + X^{2^{L-1}}$ with $\tilde{s}_1$. The shifting property 3.2.1 applied to $\tilde{s}_k$, $k \geq 1$, shows that the frequency-domain polynomial associated with $\tilde{s}_k$, $k \geq 1$, is $S(\alpha^{k-1}X)$, where α is the primitive element of $GF(2^L)$ used to define the characteristic phase $\tilde{s}$ of the PN-sequence and $\omega = \alpha^{-1}$ is used to define DFT.

We use the notation $\tilde{s}_k \leftrightarrow S(\omega^{k-1}X)$ for the association given by the DFT. Then, for two distinct phases (say, $j > i$) $\tilde{s}_i$ and $\tilde{s}_j$ of $\tilde{s}$ we have $\tilde{s}_i \leftrightarrow S(\alpha^{i-1}X)$ and $\tilde{s}_j \leftrightarrow S(\alpha^{j-1}X)$. The convolution property (3.2.2), because $N^* = 1$, gives

$$\tilde{t} = \tilde{s}_i\tilde{s}_j \leftrightarrow T(X) = S(\alpha^{i-1}X)S(\alpha^{j-1}X) \ mod \ X^N - 1 .$$

The linearized polynomial $S(\alpha^{i-1}X)$ has non-zero coefficients precisely for those powers e of X such that $w_2(e) = 1$ and $1 \leq e < 2^L$. The product $S(\alpha^{i-1}X)S(\alpha^{j-1}X)$ thus has its potentially non-zero coefficients for those powers e of X such that $w_2(e) = 1$ or $w_2(e) = 2$ and $1 \leq e < 2^L$. By Blahut's Theorem, the linear complexity of the sequence $\tilde{z}$ is just the number of non-zero coefficients of $T(X)$. Because $N = 2^L - 1$, all elements of a cyclotomic coset modulo N have the same Hamming weight since their L-bit radix-2 forms differ by just a cyclic shift. Thus

$$T(X) = \sum_{e \in I_2} T_e X^e$$

where I_2 is the union of the cyclotomic cosets $\mathcal{C}_s$ with $1 \leq w_2(s) \leq 2$. [For example, for $N = 7$, $\mathcal{C}_1 = \{1,2,4\}$ and $\mathcal{C}_3 = \{3,6,5\}$ are the only cyclotomic cosets mod N with weights 1 and 2.] Thus, because there are $\binom{L}{j}$ integers i, $1 \leq i < 2^L$, such that $w_2(i) = j$, we have immediately the bound

$$\mathcal{L}(\tilde{s}_i\tilde{s}_j) \leq L + \binom{L}{2},$$

which is Key's bound for second order filtering [3].

To find $\mathcal{L}(\tilde{t})$ exactly, it follows from the conjugacy constraints property (3.2.3) that it is enough to compute the coefficient T_e for those e that are representatives of the cyclotomic cosets modulo N. We now examine the coefficients T_e where $e = 1$ and $e = 2^{e_1} + 1$, $1 \leq e_1 < L$, which are the representatives of the cyclotomic coset of weight 1 and those of weight 2. If we make the change of variables $Y = \alpha^{i-1}X$, then

$$T(\alpha^{-i+1}Y) = S(Y)S(\alpha^{j-i}Y) \ mod \ Y^N - 1 .$$

Define the coefficients A_i, $i \in \mathcal{C}$, by

$$S(\alpha^{j-i}Y) = \sum_{i \in \mathcal{C}} A_i Y^i$$

where $\mathcal{C}$ is the main coset modulo N, i.e., $\mathcal{C} = \{1, 2, 4, \cdots, 2^{L-1}\}$.

The coefficient of Y in $S(Y)S(\alpha^{j-i}Y) \ mod \ Y^N - 1$ and in $T(\alpha^{-i+1}Y)$ is

$$A_{2^{L-1}} = (A_1)^{2^{L-1}} = (\alpha^{j-i})^{2^{L-1}} \neq 0 \text{ and } T_1\alpha^{-i+1},$$

respectively. This implies that T_1 is non zero. For the cosets of weight 2, we consider the coefficient of $Y^{2^{e_1}+1}$ in $S(Y)S(\alpha^{j-i}Y)$ *mod* $Y^N - 1$ and in $T(\alpha^{-i+1}Y)$, which is

$$A_1 + A_1^{2^{e_1}} \text{ and } \alpha^{-i+1}T_{2^{e_1}+1} .$$

We conclude that $T_{2^{e_1}+1}$ will be zero if and only if $A_1 = \alpha^{j-i} \neq 1$ is a zero of the polynomial $X^{2^{e_1}} + X$, which happens if and only if e_1 divides L and $e_1 > 1$

We have proved the following result:

Proposition 1 *Let $\tilde{s}_i$ and $\tilde{s}_j$, $1 \leq i < j < 2^L - 1$ be distinct phases of the same sequence $\tilde{s}$ whose minimal polynomial $h(X) \in GF(2)[X]$ is primitive and has degree L, L a prime. Let $\alpha \in GF(2^L)$ denote a root of $h(X)$, and let $\tilde{t} = \tilde{s}_i\tilde{s}_j$ be the Hadamard product of the two distinct phases. Then*

$$\mathcal{L}(\tilde{t}) = L + \binom{L}{2} .$$

The above arguments show in general that any sum of Hadamard products of order n of the sequences $\tilde{s}_i$, $i = 1, 2, \cdots, N$, is a sequence $\tilde{t}$ for which $T(X)$ has $\binom{L}{n}$ potentially non-zero coefficients. Thus $\tilde{t}$ has linear complexity at most $\binom{L}{n}$. The linear complexity of a sequence $\tilde{t}$ obtained by k-th order nonlinear filtering of $\tilde{s}$ then satisfies

$$\mathcal{L}(\tilde{t}) = \sum_{n=1}^{k} \binom{L}{n} ,$$

which is Key's general bound.

6 Conclusion

We have given a method for analyzing the linear complexity of nonlinear filterings of PN-sequences that is based on the Discrete Fourier Transform.
As an application to show the usefulness of our approach, we gave a simple proof of Key's upper bound on linear complexity of nonlinearly filtered PN-sequences and we further showed that Key's result, viz. that products of two distinct phases of the same PN-sequence of period $2^L - 1$ always have the maximum possible linear complexity $\binom{L}{1} + \binom{L}{2}$ for such products when L is prime; the restrictions on phase differences imposed by this author for this nondegeneracy of linear complexity are unnecessary.

References

[1] R. E. Blahut, *Theory and Practice of Error Control Codes*, Addison-Wesley Publishing Company, 1983.

[2] E. J. Groth "Generation of binary sequences with controllable complexity," *IEEE Trans. Inform. Theory*, vol. IT-17, no. 3, pp.288-296, May 1971.

[3] E. L. Key "An analysis of the structure and complexity of nonlinear binary sequence generators," *IEEE Trans. Inform. Theory*, vol. IT-22, no. 6, pp.732-736, Nov. 1976.

[4] R. Lidl and H. Niederreiter, *Introduction to Finite Fields and their Applications*, Cambridge: Cambridge Univ. Press, 1986.

[5] J. L. Massey, "Shift-register synthesis and BCH decoding," *IEEE Trans. Inform. Theory*, vol. IT-15, pp. 122-127, Jan. 1969.

[6] J. L. Massey, "Review of R. E. Blahut, *Theory and Practice of Error Control Codes*", *IEEE Trans. Inform. Theory*, vol. IT-31, no. 4, p. 553, Jul. 1985.

[7] R. A. Rueppel, *Analysis and Design of Stream Ciphers*, Berlin: Springer-Verlag, 1986.

The Security of Cipher Block Chaining

Mihir Bellare[1] and Joe Kilian[2] and Phillip Rogaway[3]

[1] Advanced Networking Laboratory, IBM T.J. Watson Research Center, PO Box 704, Yorktown Heights, NY 10598, USA. e-mail: mihir@watson.ibm.com

[2] NEC Research Institute, 4 Independence Way, Princeton, NJ 08540, USA. e-mail: joe@research.nj.nec.com

[3] Department of Computer Science, University of California at Davis, Davis, CA 95616, USA. e-mail: rogaway@cs.ucdavis.edu

Abstract. The Cipher Block Chaining – Message Authentication Code (CBC MAC) specifies that a message $x = x_1 \cdots x_m$ be authenticated among parties who share a secret key a by tagging x with a prefix of

$$f_a^{(m)}(x) \stackrel{\text{def}}{=} f_a(f_a(\cdots f_a(f_a(x_1)\oplus x_2)\oplus \cdots \oplus x_{m-1})\oplus x_m),$$

where f is some underlying block cipher (eg. f = DES). This method is a pervasively used international and U.S. standard. We provide its first formal justification, showing the following general lemma: that cipher block chaining a pseudorandom function gives a pseudorandom function. Underlying our results is a technical lemma of independent interest, bounding the success probability of a computationally unbounded adversary in distinguishing between a random ml-bit to l-bit function and the CBC MAC of a random l-bit to l-bit function.

1 Introduction

1.1 The problem: Is the CBC MAC secure?

Message authentication lets communicating partners who share a secret key verify that a received message originates with the party who claims to have sent it. This is one of the most important and widely used cryptographic tools. It is most often achieved using a "message authentication code," or MAC. This is a short string $\text{MAC}_a(x)$ computed on the message x to be authenticated and the shared secret key a. The sender transmits $\langle x, \text{MAC}_a(x)\rangle$ and the receiver, who gets $\langle x', \sigma'\rangle$, verifies that $\sigma' = \text{MAC}_a(x')$.

The most common MAC is built using the idea of "cipher block chaining" some underlying block cipher. To discuss this we first need some notation. Given a function $f\colon \{0,1\}^l \to \{0,1\}^l$ and a number $m \geq 1$ we denote by $f^{(m)}\colon \{0,1\}^{ml} \to \{0,1\}^l$ the function which maps an ml-bit input $x = x_1 \cdots x_m$ ($|x_i| = l$) to the l-bit string

$$f^{(m)}(x_1 \cdots x_m) = f(f(\cdots f(f(x_1)\oplus x_2)\oplus \cdots \oplus x_{m-1})\oplus x_m).$$

We call $f^{(m)}$ the *(m-fold) cipher block chaining* of f.[4] Now, a block cipher F (with key length k and block size l) specifies a family of permutations f_a: $\{0,1\}^l \to \{0,1\}^l$, one for each k-bit key a. The CBC MAC constructed from F has an associated parameter $s < l$ which is the number of bits it outputs. The CBC MAC is then defined for any ml-bit string $x = x_1 \ldots x_m$ by

$$\text{CBC-MAC}_a^F(x_1 \ldots x_m) \stackrel{\text{def}}{=} \text{The first } s \text{ bits of } f_a^{(m)}(x_1 \cdots x_m)\,.$$

The CBC MAC is an International Standard [13]. The most popular and widely used special case uses $F =$ DES (so $k = 56$ and $l = 64$) and $s = 32$, in which case we recover the definition of the corresponding U.S. Standard [1]. These standards are extensively employed in the banking sector and in other commercial sectors. Given this degree of usage and standardization, you might expect that there would be a large body of work aimed at learning if the CBC MAC is secure. Yet this has not really been the case. To the best of our knowledge, it was seen as entirely possible that F could be a perfectly secure block cipher even though CBC-MAC^F might be a completely insecure MAC. There was no reason to be sure that the internal structure of F couldn't "interact badly" with the specifics of cipher block chaining in exactly such a way as to defeat the CBC MAC.

1.2 Our approach

In this paper we will show that CBC MAC construction is secure if the underlying block cipher is secure. To make this statement meaningful we need first to discuss what we mean by security in each case.

What does it mean to assume DES is secure?

To describe the security of a block cipher we adopt the viewpoint introduced by Luby and Rackoff [15, 16] with regard to DES. They suggest that a block cipher should be assumed to be a pseudorandom function (PRF) with respect to "practical" computation. The notion of a PRF is in turn due to Goldreich, Goldwasser and Micali [9]. Roughly said, a function family F is pseudorandom if any reasonable adversary is unable to distinguish the following two types of objects, based on their input/output behavior: a black-box for $f_a(\cdot)$, on a random key a; a black-box for a "truly random" function $f(\cdot)$.

What does it mean for a MAC to be secure?

Our notion of security for a message authentication code adopts the viewpoint of Goldwasser, Micali and Rivest [11] with regard to signature schemes; namely, a secure MAC must resist existential forgery under adaptive message attack. However, what we will show is actually stronger: if F is a pseudorandom function family then $F^{(m)}$, the family of functions $f^{(m)}$ for $f \in F$, is *itself* shown to be

[4] Notice that here and in what follows we require the input to consist of exactly m blocks, not at most m. See Section 1.4 for a discussion of length variability.

a pseudorandom function family. That a PRF automatically makes a secure message authentication code is a well-known observation due to [9, 10]—see Section 6 for details.

Exact security

We wish to obtain results which are meaningful for practice. In particular, in our setting we need to say something about the correct or incorrect use of DES, where there are no asymptotics present. This demands not only that we avoid asymptotics and address security "exactly," but also that we strive for security reductions which are as efficient as possible.

We will only talk about finite families of functions and the resources needed to "learn" things about these finite function families. We will describe the resources necessary to "break" the finite family F given an adversary of specified resources who succeeds in breaking $F^{(m)}$. The parameters of interest are the running time t of the adversary; the number of queries q which she makes to an oracle which is her only point of access to $f(\cdot)$-values for the given $f \in F$; and the adversary's advantage, ϵ, over simple guessing. We emphasize the importance of keeping t and q separate: in practice, oracle queries correspond to observations or interaction with a system whose overall structure often severely limits q (e.g., the system might limit the amount of plaintext encrypted before the key is changed); but t corresponds to off-line computation by the adversary, and so is much less under the good guys control.

Assume that adversary A can (t, q, ϵ)-break $F^{(m)}$. This means she runs in time t, makes q queries of her oracle, and succeeds with advantage ϵ in distinguishing a random member of $F^{(m)}$ from a random function of ml-bits to l-bits. Our results specify (t', q', ϵ') (as functions of t, q, ϵ, m, l) such that there exists an adversary A' (a simple modification of A) that (t', q', ϵ')-breaks F.

Exact security is not new. It is true that most theoretical works only provide asymptotic security guarantees of the form "the success probability of a polynomially bounded adversary is negligible" (everything measured as a function of the security parameter), but the exact security can usually be derived from examination of the proof. However, a lack of concern with the exactness means that in many cases the reductions are very inefficient, and the results are not useful for practice.

Previous works which address exact security explicitly and strive for efficient reductions are [8, 12, 19, 6, 14, 4], the last four on the more practical side.

1.3 Main result

Of course the power of results such as those indicated above depends on what values of t', q', ϵ' one can prove. Our analysis is directed at achieving the best values possible. Our main lemma is stated formally as Lemma 5. Informally, it says the following. Suppose there is an adversary A who (t, q, ϵ)-breaks $F^{(m)}$. Then A can be turned into an adversary A' of comparable size and time complexity to A which, making qm oracle queries, achieves advantage $\epsilon' = \epsilon - 3q^2m^2 \cdot 2^{-l-1}$.

Current knowledge gives us values t', q', ϵ' for which it seems safe to rule out (t', q', ϵ')-breaks on DES. From this we can derive values of t, q, ϵ for which (t, q, ϵ)-breaks of $\mathrm{DES}^{(m)}$ are effectively ruled out. Thus, we reduce the security of $\mathrm{DES}^{(m)}$ to that of DES in a constructive and useful way.

The brunt of the proof addresses the information-theoretic case of the above lemma. Here we consider the problem of distinguishing a random ml-bit to l-bit function from the m-fold CBC of a random l-bit to l-bit function. We prove an absolute bound of $3q^2m^2 \cdot 2^{-l-1}$ on the advantage an adversary can derive. (The bound holds irrespective of the adversary's running time and depends only on m and the number q of queries she makes.) The lemma appears in Section 3.

The proof of this information-theoretic case of the CBC Lemma is not easy. For whatever reasons, it seems quite susceptible to specious arguments and to a general difficulty in moving from intuition to proof.

Section 6 completes the picture by showing that the standard construction of a MAC from a PRF has tight security. In this light, we view our main results as being those discussed above.

1.4 Extensions and corollaries

The CBC Lemma provides an efficient method to produce a PRF to $\leq l$-bits when the input is of fixed length ml. But often the input lengths may vary. We exhibit in Section 5 some simple extensions to the CBC MAC which allow one to correctly authenticate words of arbitrary length. We also demonstrate that a mechanism which is commonly employed —setting $\mathrm{MAC}_f(x_1 \cdots x_m) = f^{(m+1)}(x_1 \cdots x_m\, m)$— does *not* work to generate a secure message authentication code.

Pseudorandom functions are basic tools in cryptography. In addition to shedding light on the security of the CBC MAC our work provides a method of building secure PRFs which can be used in a wide range of applications, in the following way.

Practice readily provides PRFs on fixed input lengths, in the form of block ciphers like DES. On the other hand PRFs are very useful in applications, but one typically needs PRFs on long strings. Our CBC Lemma provides a provably-good way of extending the basic PRFs (which work on short inputs) to PRFs which work on longer inputs. It was based on these facts that PRFs were suggested by [3] as the tools of choice for practical applications, particularly entity authentication and key distribution.

1.5 History and related work

The lack of any theorem linking the security of f to that of $f^{(m)}$ lead previous users of the CBC-MAC to view $f^{(m)}$, and not f, as the basic primitive. Thus for example in Bird et. al. [5], when the authors require a practical message authentication code in order to achieve their higher-level goal of entity authentication they made appropriate assumptions about the CBC MAC.

The cryptanalytic approach to the problem of the security of the CBC MAC is to attack the CBC MAC construction for a particular block cipher F. Refer to [17] for an attempt to directly attack the DES CBC MAC using differential cryptanalysis.

Another approach to studying MACs is rooted in the examination of protocols which use them. Stubblebine and Gligor [20] find flaws in the use of the CBC MAC in some well-known protocols. But as the authors make clear, it is not the CBC MAC itself which is at fault for the indicated protocol failures—it is the manner in which the containing protocols incorrectly embed the CBC MAC. The authors go on to correct some protocols by having them properly use the CBC MAC.

Cipher block chaining is not the only method of constructing a MAC out of a block cipher. Amongst the other methods that have been proposed we note that of [2]. There the authors concern was to provide a construction which, unlike cipher-block chaining, is parallelizable. Their constructions are simple and efficient. The security is analyzed exactly and the bounds achieved are actually somewhat better than what we prove here for cipher block chaining.

1.6 Discussion and open questions

Block ciphers like DES are in fact *permutations*. One open question is whether the permutativity of the block cipher could be exploited to prove a stronger reduction than that in our main lemma. The fact that one typically outputs a number of bits $s < l$ seems relevant and useful in strengthening the bounds that would otherwise be achieved.

Feige and Naor [7] observe that the dependence on q in the bound in Lemma 1 is optimal up to a constant: they can show that there is an adversary who achieves an advantage of $O(mq^2 \cdot 2^{-l})$. An open question is whether our analysis can be tightened to meet this lower bound, or whether the latter can be improved to meet our upper bound of $O(m^2q^2 \cdot 2^{-l})$.

2 Preliminaries

A *finite function family* is a finite set of strings, called keys, each of which names a function according to a fixed and specified manner. To pick a function f at random from a finite function family F means to pick a random key and let f be the corresponding function. Note that two keys can name the same function.

We let $\mathcal{R}_{a\to l}$ denote the set of all functions from $\{0,1\}^a$ to $\{0,1\}^l$. The name of each function $f \in \mathcal{R}_{a\to l}$ is the string which describes its truth table. We let $\mathcal{R}_{*l\to l}$ denote the set of all functions from $\cup_{1\le i<2^l}\{0,1\}^{il}$ to $\{0,1\}^l$. The name of a function $f \in \mathcal{R}_{*l\to l}$ is the string which describes its truth table.

Let A be a Turing machine with access to an oracle $\mathcal{O}$. We say that A is a q-adversary if A makes at most q queries to $\mathcal{O}$. We say that A is a (t,q)-adversary if A runs in at most t steps and makes at most q queries to $\mathcal{O}$.

If F is a finite function family we write $E[A^F] = \Pr[A^f = 1]$ for the probability that A answers 1 when A's oracle is selected to be a random function from F. (If A is probabilistic, the probability is over A's coins as well.) We let

$$advantage_A(F, F') = \frac{E[A^F] - E[A^{F'}]}{2}$$

denote the advantage of A in distinguishing F from F'. Here, following [9], we are considering the following game, or statistical test. Algorithm A is provided as oracle a function g chosen at random from either F or from F', the choice being made at random according to a bit b. The algorithm is trying to predict b. The advantage is $\Pr[A^g = b] - 1/2$, the amount that the probability of A is correct is bounded away from the guessing probability 1/2.

For $0 \le \epsilon \le 1$, we say that A ϵ-distinguishes F and F' if $advantage_A(F, F') \ge \epsilon$. We say that A (q, ϵ)-distinguishes F and F' if A is a q-adversary and A ϵ-distinguishes F and F'. We say that A (t, q, ϵ)-distinguishes F and F' if A is a (t, q)-adversary which ϵ-distinguishes F and F'. Let $d^q(F, F')$ be the supremum of all numbers ϵ, $0 \le \epsilon \le 1$, such that there exists a q-adversary A that ϵ-distinguishes F and F'.

Suppose $F \subseteq \mathcal{R}_{a \to b}$. We say that A (q, ϵ)-breaks F if A is a q-adversary who (q, ϵ)-distinguishes F from $\mathcal{R}_{a \to b}$. We say that A (t, q, ϵ)-breaks F if A is a (t, q)-adversary who (t, q, ϵ)-distinguishes F from $\mathcal{R}_{a \to b}$.

Let $m > 0$ be a number and let F be a finite function family whose keys name functions in $\mathcal{R}_{l \to l}$. We define the finite function family $F^{(m)}$ the keys of which are precisely F but the interpretation of $f^{(m)} \in F^{(m)}$ is as a function $f^{(m)} \in \mathcal{R}_{ml \to l}$ defined by

$$f^{(m)}(x_1 \cdots x_m) = f(f(\cdots f(f(x_1) \oplus x_2) \oplus \cdots \oplus x_{m-1}) \oplus x_m) .$$

Let $m > 0$ be a number and let F be a finite function family whose keys name functions in $\mathcal{R}_{l \to l}$. The m-fold CBC-MAC based on F, denoted CBC-MAC^{F,m}, is defined by CBC-MAC$_f^{F,m}(x_1 \ldots x_m) = f^{(m)}(x_1 \ldots x_m)$ for all $f \in F$ and all $x_1, \ldots x_m \in \{0,1\}^l$.

3 The CBC Lemma: Information-theoretic case

3.1 Statement

The information-theoretic case of the CBC lemma considers an adversary A of unrestricted computational power. She is faced with the following problem. She is given an oracle to a function g chosen in one of the following ways: either g is a random function of ml bits to l bits; or $g = f^{(m)}$ for a random function f of l bits to l bits. The choice between these two possibilities is made according to a hidden coin flip. What is A's advantage in figuring out which type of oracle she has? The answer is specified by a tradeoff which says how A's advantage grows with the number q of queries she makes. The formal statement is made in terms of the distance function defined previously.

Lemma 1. (CBC Lemma: Information-theoretic case.) Let $l \geq 1$ and $m \geq 1$ be integers. Suppose $qm \leq 2^{(l+1)/2}$. Then:

$$d^q\left(\mathcal{R}_{ml \to l},\ \mathcal{R}_{l \to l}^{(m)}\right) \leq \frac{3q^2m^2}{2^{l+1}}.$$

In other words an adversary making $q \leq 2^{(l+1)/2}/m$ queries cannot hope to have an advantage exceeding $3q^2m^2 \cdot 2^{-l-1}$. Thinking of m as being small compared to $q, 2^l$, this means that as long as the total number of queries is roughly $\sqrt{2^l}$, the incremental advantage from the adversary's i-th query is bounded by roughly $i \cdot 2^{-l}$.

3.2 Proof

Fix an adversary A. Since we are not restricting computation time a standard argument shows that we may assume without loss of generality that A is deterministic. The bulk of the proof will be devoted to seeing what happens when A is supplied with a g be chosen at random from $\mathcal{R}_{l \to l}^{(m)}$. We begin with some definitions. The connection of these definitions to the game we are considering will be made later.

Query sequences and labelings. Call the 2^l-ary rooted tree of depth m the *full* tree. A sequence $x_1 \ldots x_i$ of l-bit strings ($1 \leq i \leq m$) names a node at depth i in the natural way. The root is denoted Λ. A sequence of distinct non-root nodes $X_1, \ldots, X_n$ is a *query sequence* if for every i there is a $j < i$ such that the parent of X_i is either X_j or Λ. The *query tree* associated to a query sequence $X_1, \ldots, X_n$ is the (rooted) subtree of the full tree induced by the nodes $\{\Lambda, X_1, \ldots, X_n\}$; it consists of a collection of root emanating paths. Nodes at depth m are called *border* nodes.

A labeling of a query sequence is a map assigning an l-bit string to each node (equivalently, a map assigning an l-bit string to each non-root node of the query tree). A function f: $\{0,1\}^l \to \{0,1\}^l$ induces a labeling Z_f of a query sequence $X_1, \ldots, X_n$ as follows. Let Λ, x_1, x_1x_2, $\ldots$, $x_1x_2\ldots x_i$ be any root emanating path in the query tree. Set

$$Z_f(x_1) = f(x_1)\ ;\ Z_f(x_1 \ldots x_j) = f(Z_f(x_1 \ldots x_{j-1}) \oplus x_j)) \quad \text{for } j = 2, \ldots, i\,.$$

A labeling Z of $X_1, \ldots, X_n$ induces another labeling Y of the same query sequence, defined as follows. On any root emanating path x_1, x_1x_2, $\ldots$, $x_1x_2\ldots x_i$ set:

$$Y(x_1) = x_1\ ;\ Y(x_1 \ldots x_j) = Z(x_1 \ldots x_{j-1}) \oplus x_j \quad \text{for } j = 2, \ldots, i\,.$$

The following is the fundamental relation between Z_f and its induced labeling Y_f:

$$Z_f(X_i) = f(Y_f(X_i)) \quad \text{for all } i = 1, \ldots, n\,.$$

For this reason, the induced labeling is called the *input* labeling.

A labeling of the query sequence $X_1, \ldots, X_n$ is said to be *collision free* if the n values $Z(X_1), \ldots, Z(X_n)$ are distinct and also the n values $Y(X_1), \ldots, Y(X_n)$ are distinct, where Y is the input labeling induced by Z. A border labeling of a query sequence is a map assigning an l-bit string to each border node in the query tree. A labeling Z is consistent with a border labeling $\hat{Z}$ if the two agree on the border nodes.

A NEW VIEW OF THE GAME. A query $x_1 \ldots x_m$ of the adversary to the g-oracle can be thought of as specifying a root to border path in the full tree. Now imagine a slightly different game in which the adversary has more power. She can sequentially make qm queries, each a node in the full tree, with the restriction that her queries form a query sequence $X_1, \ldots, X_{qm}$ according to the above definition. She receives no answer to queries which are internal nodes of the full tree, but when she queries a border node she receives its Z_f value. It is easy to see that it suffices to prove the lemma for this game.

THE BASIC RANDOM VARIABLES. The query sequence, its Z_f-labeling, and the values returned to the adversary are all random variables over the random choice of $f \in \mathcal{R}_{l \to l}$. We will denote by $\mathsf{X}_1, \ldots, \mathsf{X}_{qm}$ the random variables which are the queries of A. We will denote by Z_n (resp. $\widehat{\mathsf{Z}}_n$) the labeling of $\mathsf{X}_1, \ldots, \mathsf{X}_n$ (resp. of the border nodes of the query tree associated to $\mathsf{X}_1, \ldots, \mathsf{X}_n$) specified by Z_f. The input labeling induced by Z_n is denoted Y_n. The *view* of A after her n-th query is the random variable $\mathsf{View}_n = (\mathsf{X}_1, \ldots, \mathsf{X}_n; \widehat{\mathsf{Z}}_n)$. The term labeling usually refers to a value of Z_n; when we want to discuss the induced labeling we talk of the induced or input labeling.

EQUI-PROBABILITY OF COLLISION-FREE LABELINGS. The following lemma fixes the number n of queries that A has made. It then fixes a particular view $(X_1, \ldots, X_n ; \hat{Z})$ of A. It now examines the distribution on labelings from the point of view of A. It says that as far as A can tell, all collision free labelings of $X_1, \ldots, X_n$ consistent with her current view are equally likely.

Lemma 2. Let $1 \leq n \leq qm$ and let $X_1, \ldots, X_n$ be a query sequence. Let Z_n^1 and Z_n^2 be collision free (output) labelings of $X_1, \ldots, X_n$ which are consistent with a border labeling $\hat{Z}_n$ of $X_1, \ldots, X_n$. Then

$$\Pr\left[\, \mathsf{Z}_n = Z_n^1 \mid \mathsf{View}_n = (X_1, \ldots, X_n ; \hat{Z}_n) \,\right]$$
$$= \Pr\left[\, \mathsf{Z}_n = Z_n^2 \mid \mathsf{View}_n = (X_1, \ldots, X_n ; \hat{Z}_n) \,\right] .$$

The proof is given in Appendix A.1.

MORE DEFINITIONS. Let $X_1, \ldots, X_n$ be a query sequence. We will discuss labelings $\mathbf{z}$ which assign values only to some specified subset S of this sequence. The input labeling induced by $\mathbf{z}$ assigns values to all nodes of $X_1, \ldots, X_n$ which are at level one and all nodes whose parents are in S. We can discuss collision freeness of such labelings, or their consistency with a border labeling, in the usual way. We denote by Z_n^S the labeling of S given by restricting Z_n to S. Let $\mathsf{ColFree}(Z)$ be true if labeling Z is collision free.

Unpredictability of internal labels. The following lemma fixes the number n of queries that A has made, as well as a particular view $X_1, \ldots, X_n ; \hat{Z}$ of A. It now makes the assumption that the current labeling Z_n is collision free; think of this fact as being known to A. Given all this, it examines the distribution on labels from the point of view of A. Some labels are known: for example, the Z_n values of border nodes and the Y_n values of nodes at depth one. The lemma says that all other labels are essentially unpredictable. First, it considers a node $x_1 \ldots x_i x_{i+1}$ which is at depth at least two, and says that even given the output labels (i.e. Z_n values) of all nodes except its parent $x_1 \ldots x_i$, the Y_n value of $x_1 \ldots x_i x_{i+1}$ is almost uniformly distributed. Second, it considers a node $x_1 \ldots x_i$ which is not a border node, and says that even given the output labels of all other nodes, the Z_n value of $x_1 \ldots x_i$ is almost uniformly distributed. For technical reasons the lemma requires a bound on the number n of queries that have been made.

Lemma 3. Let $1 \leq n \leq qm-1$ and suppose $n^2/4+n-1 \leq 2^l/2$. Let $X_1, \ldots, X_n$ be a query sequence and let $\hat{Z}$ be a labeling of the border nodes of $X_1, \ldots, X_n$. Let

$$\Pr_n[\cdot] = \Pr\left[\cdot \mid \mathsf{View}_n = (X_1, \ldots, X_n ; \hat{Z}) \wedge \mathsf{ColFree}(\mathsf{Z}_n)\right] .$$

Suppose $x_1 \ldots x_i \in \{X_1, \ldots, X_n\}$ is a non-border node and let $S = \{X_1, \ldots, X_n\} - \{x_1 \ldots x_i\}$. Suppose $\mathbf{z}$: $S \rightarrow \{0,1\}^l$ is a collision free labeling of S which is consistent with $\hat{Z}$.

(1) Let $x_1 \ldots x_i x_{i+1} \in S$ be a child of $x_1 \ldots x_i$. Then for any $y^* \in \{0,1\}^l$:

$$\Pr_n\left[\mathsf{Y}_n(x_1 \ldots x_i x_{i+1}) = y^* \mid \mathsf{Z}_n^S = \mathbf{z}\right] \leq 2 \cdot 2^{-l} .$$

(2) For any $z^* \in \{0,1\}^l$:

$$\Pr_n\left[\mathsf{Z}_n(x_1 \ldots x_i) = z^* \mid \mathsf{Z}_n^S = \mathbf{z}\right] \leq 2 \cdot 2^{-l} .$$

The proof is given in Appendix A.2.

Bounding the probability of collisions. The following lemma fixes the number n of queries that A has made, as well as a particular view $X_1, \ldots, X_n ; \hat{Z}$ of A. It now makes the assumption that the current labeling Z_n is collision free; think of this fact as being known to A. Given all this, it considers A's adding a new node X_{n+1} to the tree. It says that the labeling is likely to retain its collision freeness; that is, Z_{n+1} is collision free with high probability. The same technical condition on n as in the previous lemma is required.

Note that X_{n+1} is determined by $X_1, \ldots, X_n ; \hat{Z}$. The value $\mathsf{Z}_{n+1}(X_{n+1})$ has not yet been returned to A, and it makes sense to discuss the distribution of this value given $X_1, \ldots, X_n ; \hat{Z}$.

Lemma 4. Let $1 \leq n \leq qm-1$ and suppose $n^2/4+n-1 \leq 2^l/2$. Let $X_1, \ldots, X_n$ be a query sequence and let $\hat{Z}$ be a labeling of the border nodes of $X_1, \ldots, X_n$. Then

$$\Pr\left[\neg\mathsf{ColFree}(\mathsf{Z}_{n+1}) \mid \mathsf{View}_n = (X_1, \ldots, X_n ; \hat{Z}) \wedge \mathsf{ColFree}(\mathsf{Z}_n)\right] \leq 3n \cdot 2^{-l} .$$

The proof is given in Appendix A.3.

CONCLUDING THE PROOF. We now complete the proof of Lemma 1 given the lemmas above. We need to show that $advantage_A(\mathcal{R}_{ml\to l}, \mathcal{R}_{l\to l}^{(m)}) \le 3q^2m^2 2^{-l-1}$. We first consider running A in the above game, with $g = f(m)$ for f chosen at random from $\mathcal{R}_{l\to l}$]. As long as the current labeling Z_n which A has is collision-free, the value of a border node returned to A is a random l bit string distributed independently of anything else. Thus the distribution on A's view is the same as if she were replied to by a function from $\mathcal{R}_{l\to l}$. On the other hand if the labeling Z_n has a collision we pessimistically declare that A has won and stop the game. Thus, if $\Pr_{\mathcal{R}}[\cdot]$ denotes the probability function when g is drawn randomly from $\mathcal{R}_{ml\to l}$ and $\Pr_{\mathcal{C}}[\cdot]$ denotes the probability function when g is drawn randomly from $\mathcal{R}_{l\to l}^{(m)}$ then for each $b \in \{0,1\}$ we have

$$
\begin{aligned}
&\Pr_{\mathcal{R}}[A^g = b] \\
&\le \Pr_{\mathcal{C}}[A^g = b \mid \mathsf{ColFree}(\mathsf{Z}_{qm})] + \Pr_{\mathcal{R}}[\neg\mathsf{ColFree}(\mathsf{Z}_{qm})] \\
&\le \Pr_{\mathcal{C}}[A^g = b \mid \mathsf{ColFree}(\mathsf{Z}_{qm})] \\
&\qquad + \textstyle\sum_{n=1}^{qm-1} \Pr_{\mathcal{R}}[\neg\mathsf{ColFree}(\mathsf{Z}_{n+1}) \mid \mathsf{ColFree}(\mathsf{Z}_n)] \,.
\end{aligned}
$$

The first term in the above equals $\Pr_{\mathcal{C}}[A^g = b]$. One can check that our assumption $qm \le 2^{(l+1)/2}$ implies $n^2/4+n-1 \le 2^l/2$ for all $n = 1, \ldots, qm-1$. Thus we can apply the previous lemmas to argue that the second term above is bounded by

$$\textstyle\sum_{n=1}^{qm-1} 3n \cdot 2^{-l} \;\le\; 3q^2m^2 \cdot 2^{-l-1} \,.$$

Thus $|\Pr_{\mathcal{R}}[A^g = 1] - \Pr_{\mathcal{C}}[A^g = 1]| \le 3q^2m^2 \cdot 2^{-l}$. Under our definition, adversary A's advantage is given by half of the last bound.

4 The CBC Lemma: The computational case

Let F be a family of functions of l bits to l bits. Think of it as a family of pseudorandom functions. (For concreteness, we could consider $F = \{\mathrm{DES}_a\}_{|a|=56}$ to be the family of functions specified by the DES algorithm; each individual function is specified by a 56 bit key.) Thus it is "hard" to distinguish a random member of F from a random function of l bits to l bits. With f drawn randomly from F, we want to see how $f^{(m)}$ compares to a random function of ml bits to l bits. We'd like to say that $f^{(m)}$ is also pseudorandom. The lemma that follows implies this. But the actual statement is much stronger. It says exactly how the security of $F^{(m)}$ relates to that of F.

Lemma 5. (CBC Lemma: Computational Case). There is an algorithm U and a constant c as follows. Let $l \ge 1$ and $m \ge 1$ be integers. Let $F \subseteq R^{l\to l}$ be a given function family. Suppose $qm \le 2^{(l+1)/2}$. Let A be an algorithm which (t,q,ϵ)-breaks $F^{(m)}$. Then $U^A(l,m)$ (t',q',ϵ')-breaks F, where

$$t' = t + cqml\,, \quad q' = qm\,, \quad \text{and} \quad \epsilon' = \epsilon - 3q^2m^2 \cdot 2^{-l-1}\,.$$

The constant c is a small number which depends only on the underlying machine model. One should think of t as being much larger than $cqml$ (this is apparent from the definition of the U below) and so the additive $cqml$ term is effectively irrelevant.

Proof. We may assume $\epsilon > \delta \stackrel{\text{def}}{=} 3q^2m^2 \cdot 2^{-l-1}$ since otherwise there is nothing to prove. Let A be an adversary which (t, q, ϵ)-distinguishes $\mathcal{R}_{ml \to l}$ from $F^{(m)}$. From the triangle inequality at least one of the following must be true:

(1) A (t, q, δ)-distinguishes $\mathcal{R}_{ml \to l}$ from $\mathcal{R}_{l \to l}^{(m)}$; or

(2) A $(t, q, \epsilon - \delta)$-distinguishes $\mathcal{R}_{l \to l}^{(m)}$ from $F^{(m)}$.

However given the assumed bound on qm and the Information Theoretic CBC Lemma (Lemma 1), the first option is ruled out. Thus the second must be true. To complete the proof we now construct an adversary A' which (t', q', ϵ')-distinguishes $\mathcal{R}_{l \to l}$ from F. A' is given access to an oracle for a function f: $\{0,1\}^l \to \{0,1\}^l$. Observe that A' can compute the function $f^{(m)}$ and doing this at a point $x_1 \cdots x_m$ costs A' a total of m queries to its f-oracle and time proportional to ml. Algorithm A''s procedure is to run A and answer its oracle queries according to $f^{(m)}$. Finally A' takes A's prediction as its own. We leave to the reader to check that

$$advantage_{A'}(\mathcal{R}_{l \to l}, F) = advantage_{A}(\mathcal{R}_{l \to l}^{(m)}, F^{(m)}) .$$

This completes the proof. □

5 Length Variability

For simplicity, let us assume throughout this section that strings to be authenticated have length which is a multiple of l bits. This restriction is easy to dispense with by using simple and well-known padding methods: for example, always append a "1" and then append the minimal number of 0's to make the string a multiple of l bits.

THE CBC MAC DOESN'T HANDLE VARIABLE-LENGTH INPUTS. The CBC MAC does not directly give a method to authenticate messages of variable input lengths. In fact, if the length of strings is allowed to vary, it is easy to "break" the basic CBC MAC construction. (This fact is well-known.) As an example, if you request $f_a^{(1)}$ of b, getting back t_b, and then you request $f_a^{(1)}(t_b)$, getting back t_{t_b}, then you have just learned the authentication tag $f_a^{(2)}(b\ 0) = t_{t_b}$ for $b\ 0$—a string for which you have not asked the authentication tag.

APPENDING THE LENGTH DOESN'T WORK. One possible attempt to authenticate messages of varying lengths is to append to each string $x = x_1 \cdots x_m$ the number m, properly encoded as the final l-bit block and then CBC MAC the resulting string $m+1$ blocks. (Of course this imposes a restriction that $m < 2^l$, not likely to be a serious concern.) Let us define $\bar{f}^*(x_1 \cdots x_m) = f^{(m+1)}(x_1 \cdots x_m\ m)$.

We show that $\bar{f}^*$ is not a secure MAC. Take arbitrary l-bit words b, b' and c, $b \neq b'$. It is easy to check that given

(1) $t_b = \bar{f}^*(b)$,
(2) $t_{b'} = \bar{f}^*(b')$, and
(3) $t_{b1c} = \bar{f}^*(b\ 1\ c)$

the adversary has in hand $\bar{f}^*(b'\ 1\ t_b \oplus t_{b'} \oplus c)$ —the authentication tag of a string she has (almost certainly) not asked about before— since this is precisely t_{b1c}.

METHODS WHICH DO WORK. Despite the failure of the method which appends the message length, there are many methods which are almost as simple and which work correctly. We describe three. In each, let F be a block cipher on l bits.

(1) *Input-length key separation.* Set $\dot{f}^*(x)$, where $x = x_1 \cdots x_m$, to be $f_{a_m}^{(m)}(x)$, where $a_m = F_a(m)$. The corresponding finite function family $\dot{F}^*$ is not only a MAC, it can be shown to be computationally close to $\mathcal{R}_{*l \to l}$
(2) *Two-step MAC.* Set $\mathrm{MAC}_a(x)$, where $x = x_1 \cdots x_m$, to be $\langle f_{a'}^{(m)}(x), f_{a''}^{(2)}(m\ f_{a'}^{(m)}(x))\rangle$, where $a' = f_a(0)$ and $a'' = f_a(1)$.
(3) *Length-prepend MAC.* Define $f^*(x_1 \cdots x_m) = f^{(m+1)}(m\ x_1 \cdots x_m)$. The corresponding finite function family F^* is not only a MAC, it can be shown to be computationally close to $\mathcal{R}_{*l \to l}$.

The third of these claims has the most involved proof. We know of no argument which does not involve modifying and verifying that the proof of the CBC Lemma goes through after making this extension.

6 From PRFs to MACs

Recall that justifying the CBC-MAC was the primary motivation of this paper. To formally complete this project we need one more step—to show that pseudorandom functions make good message authentication codes. As we remarked in the introduction the reduction is standard [9, 10]. But we need to see what is the exact security. The following shows that the reduction is almost tight— security hardly degrades at all.

Let G be a finite function family whose keys name functions in $\mathcal{R}_{k \to l}$. Let MAC^G be defined by $\mathrm{MAC}_g^G(y) = g(y)$ for all $g \in G$ and all $y \in \{0,1\}^k$.

Security of MAC^G is discussed via the notion of chosen message attack [11]. An adversary B attacks MAC^G via the following experiment. Pick g at random from G and provide $\mathrm{MAC}_g^G(\cdot)$ to B as an oracle. Suppose B makes q oracle queries and runs for time t, halting with an output (y, σ), where $y \in \{0,1\}^k$ and y is different from any string which B has queried of its oracle. We say that B is successful if $\mathrm{MAC}_g^G(y) = \sigma$. We say that B (t, q, ϵ)-breaks MAC^G if it is successful with probability at least ϵ. The Proposition that follows is the exact security version of the standard reduction of [9, 10].

Proposition 6. There is an algorithm U and a constant c as follows. Let $l \geq 1$ and $k \geq 1$ be integers. Let $G \subseteq R^{k \to l}$ be a given function family. Suppose $q < 2^k$.

Let B be an algorithm which (t, q, ϵ)-breaks MAC^G. Then $U^B(k,l)$ (t', q', ϵ')-breaks G, where

$$t' = t + cq(k+l)\ , \quad q' = q+1\ , \quad \text{and} \quad \epsilon' = (\epsilon - 2^{-l})/2\ .$$

Proof. We provide A such that $advantage_A(G, \mathcal{R}_{k\to l}) \geq \epsilon'$. A has oracle access to g: $\{0,1\}^k \to \{0,1\}^l$. It runs B and answers an oracle query x of B by invoking its own oracle to return $g(x)$. B eventually outputs (y, σ). A makes a last oracle query of y to get $\sigma^* = g(y)$. It outputs 1 if $\sigma^* = \sigma$ and 0 otherwise. It is easy to see that, on the one hand, $E[A^{\mathcal{R}_{k\to l}}] \leq 2^{-l}$ and, on the other hand, $E[A^G]$ is just the success probability of B and hence is at least ϵ. So the advantage is as claimed. The program of A can be easily implemented in the stated complexity by a machine with oracle access to B. □

Let $m > 0$ be a number and let F be a finite function family whose keys name functions in $\mathcal{R}_{l\to l}$. Combining Lemma 5 with the above proposition tells us exactly how secure is the CBC-MAC $\text{CBC-MAC}^{F,m}$ based on F.

Acknowledgments

We thank Uri Feige and Moni Naor for their assistance in the proof of Lemma 1, and also for comments on the paper.

References

1. ANSI X9.9, "American National Standard for Financial Institution Message Authentication (Wholesale)," American Bankers Association, 1981. Revised 1986.
2. M. Bellare, R. Guérin and P. Rogaway, "Fully parallelizable message authentication," Manuscript, April 1994.
3. M. Bellare and P. Rogaway, "Entity authentication and key distribution," *Advances in Cryptology – Crypto 93 Proceedings*, Lecture Notes in Computer Science Vol. 773, Springer-Verlag, D. Stinson, ed., 1994.
4. M. Bellare and P. Rogaway, "Optimal asymmetric encryption," *Advances in Cryptology – Eurocrypt 94 Proceedings*, 1994.
5. R. Bird, I. Gopal, A. Herzberg, P. Janson, S. Kutten, R. Molva and M. Yung, "Systematic design of two-party authentication protocols," *Advances in Cryptology – Crypto 91 Proceedings*, Lecture Notes in Computer Science Vol. 576, Springer-Verlag, J. Feigenbaum, ed., 1991.
6. S. Even, O. Goldreich and S. Micali, "On-line/Off line digital signatures," Manuscript, March 1994. Preliminary version in Crypto 89.
7. U. Feige and M. Naor, Private communication, April 1994.
8. O. Goldreich and L. Levin, "A hard predicate for all one-way functions," *Proceedings of the Twenty First Annual Symposium on the Theory of Computing*, ACM, 1989.
9. O. Goldreich, S. Goldwasser and S. Micali, "How to construct random functions," *Journal of the ACM*, Vol. 33, No. 4, 210–217, (1986).

10. O. Goldreich, S. Goldwasser and S. Micali, "On the cryptographic applications of random functions," *Advances in Cryptology – Crypto 84 Proceedings*, Lecture Notes in Computer Science Vol. 196, Springer-Verlag, B. Blakley, ed., 1985.
11. S. Goldwasser, S. Micali and R. Rivest, "A digital signature scheme secure against adaptive chosen-message attacks," *SIAM Journal of Computing*, 17(2):281–308, April 1988.
12. R. Impagliazzo, L. Levin and M. Luby, "Pseudo-random generation from one-way functions," *Proceedings of the Twenty First Annual Symposium on the Theory of Computing*, ACM, 1989.
13. ISO/IEC 9797, "Data cryptographic techniques – Data integrity mechanism using a cryptographic check function employing a block cipher algorithm," 1989.
14. T. Leighton and S. Micali, "Provably fast and secure digital signature algorithms based on secure hash functions," Manuscript, March 1993.
15. M. Luby and C. Rackoff, "How to construct pseudorandom permutations from pseudorandom functions," *SIAM J. Computation*, Vol. 17, No. 2, April 1988.
16. M. Luby and C. Rackoff, "A study of password security," *Advances in Cryptology – Crypto 87 Proceedings*, Lecture Notes in Computer Science Vol. 293, Springer-Verlag, C. Pomerance, ed., 1987.
17. K. Ohta and M. Matsui, "Differential attack on message authentication codes," *Advances in Cryptology – Crypto 93 Proceedings*, Lecture Notes in Computer Science Vol. 773, Springer-Verlag, D. Stinson, ed., 1994.
18. R. Rivest, "The MD5 message-digest algorithm," IETF Network Working Group, RFC 1321, April 1992.
19. C. Schnorr, "Efficient identification and signatures for smart cards," *Advances in Cryptology – Crypto 89 Proceedings*, Lecture Notes in Computer Science Vol. 435, Springer-Verlag, G. Brassard, ed., 1989.
20. S. Stubblebine and V. Gligor, "On message integrity in cryptographic protocols," Proceedings of the 1992 IEEE Computer Society Symposium on Research in Security and Privacy. May 1992.

A Proof of Lemmas

We present here the proofs of the lemmas in Section 3.2.

A.1 Proof of Lemma 2

The proof is by induction on n. The Lemma holds vacuously for $n = 1$. Assuming the lemma for $1, \ldots, n-1$ we now prove it for n. Let Z^i_{n-1} be the restriction of Z^i_n to $X_1, \ldots, X_{n-1}$ $(i = 1, 2)$. Let $\hat{Z}_{n-1}$ be the restriction of $\hat{Z}_n$ to the border nodes of $X_1, \ldots, X_{n-1}$. Let V_i be the event $\mathrm{View}_i = (X_1, \ldots, X_i\ ;\ \hat{Z}_i)$ and let $\Pr_i[\cdot] = \Pr[\cdot \mid V_i]$, for $i = n-1, n$. Let Y^i_j denote the input labeling induced by Z^i_j for $j = n-1, n$ and $i = 1, 2$. We consider two cases.

Case 1. X_n is not a border node.

For $i = 1, 2$ we have:

$$\begin{aligned} &\Pr_n\left[\mathsf{Z}_n = Z_n^i\right] \\ &= \Pr_{n-1}\left[\mathsf{Z}_n = Z_n^i\right] \\ &= \Pr_{n-1}\left[\mathsf{Z}_{n-1} = Z_{n-1}^i\right] \cdot \Pr_{n-1}\left[\mathsf{Z}_n(X_n) = Z_n^i(X_n) \mid \mathsf{Z}_{n-1} = Z_{n-1}^i\right] \\ &= \Pr_{n-1}\left[\mathsf{Z}_{n-1} = Z_{n-1}^i\right] \cdot 2^{-l}\,. \end{aligned}$$

The proof is concluded by using the inductive hypothesis. We now justify the above equations. Since X_n is not a border node, it is determined by $X_1, \ldots, X_{n-1}$; $\hat{Z}_{n-1}$. This means that $\Pr_n[\cdot]$ equals $\Pr_{n-1}[\cdot]$ which justifies the first line. The second is just conditioning. Since Z_n^i is collision free, Y_n^i differs from all the points $Y_{n-1}^i(X_1), \ldots, Y_{n-1}^i(X_{n-1})$ on which the underlying randomly chosen f has been evaluated so far. But $\mathsf{Z}_n(X_n) = f(Y_n^i(X_n))$. So the second term in the product in the second line above is indeed 2^{-l}.

Case 2. X_n is a border node.

Both Z_n^1 and Z_n^2 are by assumption consistent with $\hat{Z}_n$. But since X_n is a border node, the value $\hat{\zeta} \stackrel{\text{def}}{=} \mathsf{Z}_n(X_n)$ is contained in $\hat{Z}_n$, and $\hat{\zeta} = Z_n^1(X_n) = Z_n^2(X_n)$. Now for $i = 1, 2$ we have:

$$\begin{aligned} &\Pr_n\left[\mathsf{Z}_{n-1} = Z_{n-1}^i\right] \\ &= \Pr_{n-1}\left[\mathsf{Z}_{n-1} = Z_{n-1}^i \mid \mathsf{Z}_n(X_n) = \hat{\zeta}\right] \\ &= \Pr_{n-1}\left[\mathsf{Z}_n(X_n) = \hat{\zeta} \mid \mathsf{Z}_{n-1} = Z_{n-1}^i\right] \cdot \frac{\Pr_{n-1}\left[\mathsf{Z}_{n-1} = Z_{n-1}^i\right]}{\Pr_{n-1}\left[\mathsf{Z}_n(X_n) = \hat{\zeta}\right]} \\ &= 2^{-l} \cdot \frac{\Pr_{n-1}\left[\mathsf{Z}_{n-1} = Z_{n-1}^i\right]}{\Pr_{n-1}\left[\mathsf{Z}_n(X_n) = \hat{\zeta}\right]}\,. \end{aligned}$$

The first equality is because the events V_n and $V_{n-1} \wedge (\mathsf{Z}_n(X_n) = \hat{\zeta})$ are the same. The second line is Bayes rule. That the first term of the product in the second line above is indeed 2^{-l} is argued as in Case 1 based on the fact that Z_n^i is collision free. Now note the denominator in the fraction above is independent of $i \in \{1, 2\}$. Thus, applying the inductive hypothesis, we conclude

$$\Pr_n\left[\mathsf{Z}_{n-1} = Z_{n-1}^1\right] = \Pr_n\left[\mathsf{Z}_{n-1} = Z_{n-1}^2\right]. \tag{1}$$

Now for $i = 1, 2$:

$$\begin{aligned} \Pr_n\left[\mathsf{Z}_n = Z_n^i\right] &= \Pr_n\left[\mathsf{Z}_{n-1} = Z_{n-1}^i\right] \cdot \Pr_n\left[\mathsf{Z}_n(X_n) = \hat{\zeta} \mid \mathsf{Z}_{n-1} = Z_{n-1}^i\right] \\ &= \Pr_n\left[\mathsf{Z}_{n-1} = Z_{n-1}^i\right] \cdot 1\,. \end{aligned}$$

That the second term in the above product is 1 is because V_n contains $\hat{\zeta}$ as the value of $\mathsf{Z}_n(X_n)$. The proof for this case is concluded by applying Equation 1.

A.2 Proof of Lemma 3

Let $x_1 \ldots x_i x_{i+1}^u$ $(u = 1, \ldots, s)$ be the children of $x_1 \ldots x_i$. Let

$$\text{CHILDREN}(x_1 \ldots x_i) = \{x_1 \ldots x_i x_{i+1}^1, \ldots, x_1 \ldots x_i x_{i+1}^s\} .$$

Let $\mathbf{y}$: $\{X_1, \ldots, X_n\} - \text{CHILDREN}(x_1 \ldots x_i) \to \{0,1\}^l$ be the input labeling induced by $\mathbf{z}$. We prove the two claims in turn.

Proof of (1). Let's begin by giving some intuition for the proof. We observe that with $\mathbf{z}$ given, if we assign an input label $y \in \{0,1\}^l$ to $x_1 \ldots x_i x_{i+1}$ then the value of Z_n at the parent node $x_1 \ldots x_i$ is determined; given this, the values of Y_n at the other children of $x_1 \ldots x_i$ are also determined. Thus, both Z_n and Y_n are now fully determined for all nodes $X_1, \ldots, X_n$. We will show that there is a large set $S(\mathbf{z})$ of these y values for which the determined labeling is collision free. Moreover, all collision free labelings have this form and are equally likely by Lemma 2; thus as far as A can tell, the value at $x_1 \ldots x_i x_{i+1}$ is equally likely to be anything from the set $S(\mathbf{z})$. The formal proof follows.

Assume wlog that $x_1 \ldots x_i x_{i+1}^1 = x_1 \ldots x_i x_{i+1}$. Let $y \in \{0,1\}^l$ be some fixed string. Now define the labeling $Z_{\mathbf{z},y}$: $\{X_1, \ldots, X_n\} \to \{0,1\}^l$ by:

$$Z_{\mathbf{z},y}(X_j) = \begin{cases} \mathbf{z}(X_j) & \text{if } X_j \neq x_1 \ldots x_i \\ y \oplus x_{i+1}^1 & \text{otherwise.} \end{cases}$$

Let $Y_{\mathbf{z},y}$ denote the input labeling induced by $Z_{\mathbf{z},y}$, and observe that it is given by

$$Y_{\mathbf{z},y}(X_j) = \begin{cases} \mathbf{y}(X_j) & \text{if } X_j \notin \text{CHILDREN}(x_1 \ldots x_i) \\ y \oplus x_{i+1}^1 \oplus x_{i+1}^u & \text{if } X_j = x_1 \ldots x_i x_{i+1}^u \text{ for some } 1 \le u \le s. \end{cases}$$

Let $S(\mathbf{z})$ be the set of all strings y such that $Z_{\mathbf{z},y}$ is a collision free labeling. We leave to the reader to check that $y \notin S(\mathbf{z})$ if and only if one of the following two conditions is satisfied:

(1) Either $y \oplus x_{i+1}^1 \in \{\, \mathbf{z}(X_j) : 1 \le j \le n \text{ and } X_j \neq x_1 \ldots x_i \,\}$; or
(2) For some $u \in \{1, \ldots, s\}$ it is the case that $y \oplus x_{i+1}^1 \oplus x_{i+1}^u \in \{\, \mathbf{y}(X_j) : 1 \le j \le n \text{ and } X_j \notin \text{CHILDREN}(x_1 \ldots x_i) \,\}$.

This implies that $|\{0,1\}^l - S(\mathbf{z})| \le (n-1) + (n-s)s \le n - 1 + n^2/4 \le 2^l/2$. So $|S(\mathbf{z})| \ge 2^l - 2^l/2 \ge 2^l/2$. Now observe that any collision free labeling equals $Z_{\mathbf{z},y}$ for some $\mathbf{z}, y$ as above. Furthermore by Lemma 2 all collision free labelings are equally likely. From this one can prove the desired statement.

Proof of (2). The idea is very similar to the above. This time, observe that with $\mathbf{z}$ given, if we assign an output label $z \in \{0,1\}^l$ to $x_1 \ldots x_i$ then the values of both Z_n and Y_n are fully determined for all nodes $X_1, \ldots, X_n$. We show as before that there is a set $S(\mathbf{z})$ of these z values for which the determined labeling is collision free, and conclude as before using the equi-probability of collision free labelings. The formal proof follows.

Let $z \in \{0,1\}^l$ be some fixed string. Now define the labeling $Z_{\mathbf{z},z}$: $\{X_1, \ldots, X_n\} \to \{0,1\}^l$ by:

$$Z_{\mathbf{z},z}(X_j) = \begin{cases} \mathbf{z}(X_j) \text{ if } X_j \neq x_1 \ldots x_i \\ z \quad \text{otherwise.} \end{cases}$$

Let $Y_{\mathbf{z},z}$ denote the input labeling induced by $Z_{\mathbf{z},z}$, and observe that it is given by

$$Y_{\mathbf{z},z}(X_j) = \begin{cases} \mathbf{y}(X_j) & \text{if } X_j \notin \text{CHILDREN}(x_1 \ldots x_i) \\ z \oplus x_{i+1}^u & \text{if } X_j = x_1 \ldots x_i x_{i+1}^u \text{ for some } 1 \leq u \leq s. \end{cases}$$

Let $S(\mathbf{z})$ be the set of all strings z such that $Z_{\mathbf{z},z}$ is a collision free labeling. We leave to the reader to check that $z \notin S(\mathbf{z})$ if and only if one of the following two conditions is satisfied:

(1) Either $z \in \{ \mathbf{z}(X_j) \,:\, 1 \leq j \leq n \text{ and } X_j \neq x_1 \ldots x_i \}$; or

(2) For some $u \in \{1, \ldots, s\}$ it is the case that $z \oplus x_{i+1}^u \in \{ \mathbf{y}(X_j) \,:\, 1 \leq j \leq n \text{ and } X_j \notin \text{CHILDREN}(x_1 \ldots x_i) \}$.

This implies that $|\{0,1\}^l - S(\mathbf{z})| \leq (n-1) + (n-s)s \leq n - 1 + n^2/4 \leq 2^l/2$. So $|S(\mathbf{z})| \geq 2^l/2$. Now observe that any collision free labeling equals $Z_{\mathbf{z},z}$ for some $\mathbf{z}, z$ as above. Furthermore by Lemma 2 all collision free labelings are equally likely. From this one can prove the desired statement.

A.3 Proof of Lemma 4

We'll use the following notation:

$$\Pr_n[\cdot] = \Pr\left[\cdot \mid \mathsf{View}_n = (X_1, \ldots, X_n \,;\, \hat{Z}) \wedge \mathsf{ColFree}(\mathsf{Z}_n)\right].$$

Case 1. X_{n+1} is at level one.

Let $X_{n+1} = \bar{x}_1$. Note its input label is by definition $\bar{x}_1$. For each $t = 1, \ldots, n$ we claim that

$$\Pr_n[\mathsf{Y}_n(X_t) = \bar{x}_1] \leq 2 \cdot 2^{-l}. \tag{2}$$

To see why this is true, consider two cases. First, if X_t is at level one then $\Pr_n[\mathsf{Y}_n(X_t) = \bar{x}_1] = 0$ by definition. On the other hand suppose X_t is at depth at least two. Then $X_t = x_1 \ldots x_i x_{i+1}$ is the child of some $x_1 \ldots x_i \in \{X_1, \ldots, X_n\}$. Equation 2 now follows by Part 1 of Lemma 3.

Given Equation 2 we can bound the probability of a collision as follows:

$$\begin{aligned} \Pr_n[\neg\mathsf{ColFree}(\mathsf{Z}_{n+1})] \leq\ & \Pr_n[\bar{x}_1 \in \{\mathsf{Y}_n(X_1), \ldots, \mathsf{Y}_n(X_n)\}] + \\ & \Pr_n[\mathsf{Z}_{n+1}(X_{n+1}) \in \{\mathsf{Z}_n(X_1), \ldots, \mathsf{Z}_n(X_n)\} \\ & \quad \mid \bar{x}_1 \notin \{\mathsf{Y}_n(X_1), \ldots, \mathsf{Y}_n(X_n)\}] \\ \leq\ & \frac{2n}{2^l} + \frac{n}{2^l} \leq \frac{3n}{2^l}. \end{aligned}$$

Case 2. X_{n+1} is not at level one.

Then $X_{n+1} = x_1 \ldots x_i x_{i+1}$ is the child of some $x_1 \ldots x_i \in \{X_1, \ldots, X_n\}$. Let $S = \{X_1, \ldots, X_n\} - \{x_1 \ldots x_i\}$. We first claim that for any $X_t \in \{X_1, \ldots, X_n\}$:

$$\Pr_n[\mathsf{Y}_{n+1}(X_{n+1}) = \mathsf{Y}_n(X_t)] \leq 2 \cdot 2^{-l}. \qquad (3)$$

To see why this is true, consider two cases. First, if X_t is a sibling of X_{n+1} then

$$\Pr_n[\mathsf{Y}_{n+1}(X_{n+1}) = \mathsf{Y}_n(X_t)] = 0$$

by definition. On the other hand suppose X_t is not a sibling of X_{n+1}. Then a collision free labeling $\mathbf{z}$ of S determines $\mathsf{Y}_n(X_t)$. Using this and Part 2 of Lemma 3 we have the following: (The sum here is over all collision free labelings $\mathbf{z}$ of S which are consistent with $\hat{Z}$.)

$$\begin{aligned}
&\Pr_n[\mathsf{Y}_{n+1}(X_{n+1}) = \mathsf{Y}_n(X_t)] \\
&= \textstyle\sum_{\mathbf{z}} \Pr_n\left[\mathsf{Y}_{n+1}(X_{n+1}) = \mathsf{Y}_n(X_t) \mid \mathsf{Z}_n^S = \mathbf{z}\right] \cdot \Pr_n\left[\mathsf{Z}_n^S = \mathbf{z}\right] \\
&= \textstyle\sum_{\mathbf{z}} \Pr_n\left[\mathsf{Z}_n(x_1 \ldots x_i) = \mathsf{Y}_n(X_t) \oplus x_{i+1} \mid \mathsf{Z}_n^S = \mathbf{z}\right] \cdot \Pr_n\left[\mathsf{Z}_n^S = \mathbf{z}\right] \\
&\leq \frac{2}{2^l} \cdot \textstyle\sum_{\mathbf{z}} \Pr_n\left[\mathsf{Z}_n^S = \mathbf{z}\right] \leq \frac{2}{2^l}.
\end{aligned}$$

Thus Equation 3 is again established.

Given Equation 3 we can bound the probability of a collision:

$$\begin{aligned}
\Pr_n[\neg\mathsf{ColFree}(\mathsf{Z}_{n+1})] &\leq \Pr_n[\mathsf{Y}_{n+1}(X_{n+1}) \in \{\mathsf{Y}_n(X_1), \ldots, \mathsf{Y}_n(X_n)\}] + \\
&\quad \Pr_n[\mathsf{Z}_{n+1}(X_{n+1}) \in \{\mathsf{Z}_n(X_1), \ldots, \mathsf{Z}_n(X_n)\} \\
&\qquad \mid \mathsf{Y}_{n+1}(X_{n+1}) \notin \{\mathsf{Y}_n(X_1), \ldots, \mathsf{Y}_n(X_n)\}] \\
&\leq \frac{2n}{2^l} + \frac{n}{2^l} \leq \frac{3n}{2^l}.
\end{aligned}$$

This completes the proof of Lemma 4.

A Chosen Plaintext Attack of the 16-round Khufu Cryptosystem

Henri Gilbert and Pascal Chauvaud

France Télécom-CNET
PAA-TSA-SRC
38-40 Rue du Général Leclerc, 92131 Issy-les-Moulineaux
France

Abstract. In 1990, Merkle proposed two fast software encryption functions, Khafre and Khufu, as possible replacements for DES [1]. In 1991, Biham and Shamir applied their differential cryptanalysis technique to Khafre [2], and obtained an efficient attack of the 16-round version and some bounds on the 24-round version. However, these attacks take advantage of the fact that the S-boxes used for Khafre are public; they cannot be applied to Khufu, which uses secret S-boxes, and no attack of Khufu has been proposed so far. In this paper, we present a chosen plaintext attack of the 16-round version of Khufu, which is based on differential properties of this algorithm. The derivation of first information concerning the secret key requires about 2^{31} chosen plaintexts and 2^{31} operations. Our estimate of the resources required for breaking the entire scheme is about 2^{43} chosen plaintexts and about 2^{43} operations.

1 Description of Khufu

Khufu is an iterated blockcipher with a 64-bit blocksize. The keys used in the 16-round version (the single one considered in this paper) are the following :

- four 32-bit words K1, K2, K3, K4, used before the first round and after the last round (initial and final xors);
- four secret permutations p_0, p_1, p_2 and p_3 of the [0..255] set (the columns of the first S-box introduced in [1]), which provide the functions used in rounds 1 to 8;
- four secret permutations q_0, q_1, q_2 and q_3 of the [0..255] set (the columns of the second S-box introduced in [1]), which provide the functions used in rounds 9 to 16.

The 16-round version of Khufu is depicted in Figure 1, which represents the encryption of one 64-bit block consisting of two 4-byte halves $L = (l_0,l_1,l_2,l_3)$ and $R = (r_0,r_1,r_2,r_3)$. We are using the following notations :

- π_i (i = 0 to 3) denotes the projection :

$$\pi_i : [0..255]^4 \rightarrow [0..255]$$
$$(x_0, x_1, x_2, x_3) \mapsto x_i$$

– $[p_{i_0}, p_{i_1}, p_{i_2}, p_{i_3}]$ (where (i_0, i_1, i_2, i_3) is a circular permutation of (0,1,2,3)) denotes an S-box the columns of which are provided by the $p_{i_0}, p_{i_1}, p_{i_2}$ and p_{i_3} permutations :

$$\begin{aligned} [p_{i_0}, p_{i_1}, p_{i_2}, p_{i_3}] : [0..255] &\rightarrow [0..255]^4 \\ x &\mapsto (p_{i_0}[x], p_{i_1}[x], p_{i_2}[x], p_{i_3}[x]) \end{aligned}$$

The representation of Figure 1 is slightly different from the one provided in [1]: in order to avoid swapping the right and left halves R and L and rotating the R half at each round, we are using a different round function at each round. It is however easy to check that both representations are strictly equivalent.
For a more detailed presentation of the Khufu algorithm, see [1].

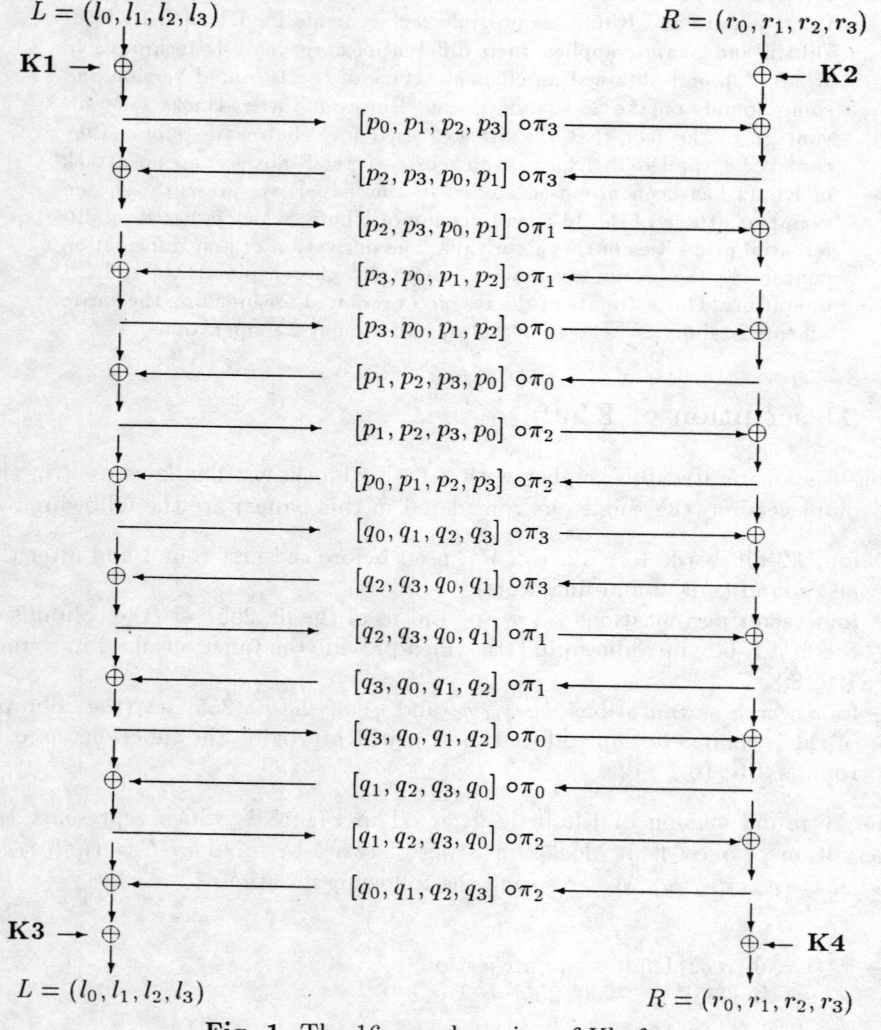

Fig. 1. The 16-round version of Khufu

2 Differential properties used in our attack

At the various steps of our attack, sets of n (P,P') pairs of 64-bit plaintext blocks are considered. In practice, these sets are defined by a combination of conditions on :

- the plaintext difference P $\oplus$ P';
- some bytes of P.

Let S be such a set of n (P,P') pairs. After r rounds (where $r \in [0,..,16]$), a (P,P') pair provides a $(C[r], C'[r])$ pair of intermediate blocks, of xor value : $\Delta[r] = C[r] \oplus C'[r]$. We denote by $\Delta S[r]$ the set of $\Delta[r]$ values derived from the n (P,P') pairs in S after r rounds.

All the differential properties we are using in our attack can be expressed in terms of the cardinal $\mid \Delta S[r] \mid$ of the $\Delta S[r]$ set of the distinct difference values after r rounds for an appropriately selected set S of plaintext pairs.

The following Proposition will be useful for the initial step of our attack:

Proposition 1. *Let λ_1, λ_3 and ρ_3 be arbitrary constants in* $[0..255]$.
For every $\alpha \in]0..255]$, there exist four constants δ_0, δ_1, δ_2 and $\beta \in [0..255]$ such that, if $S_{\lambda_1,\lambda_3,\rho_3,\alpha}$ denotes the following set of plaintext pairs :
$S_{\lambda_1,\lambda_3,\rho_3,\alpha} = \{((L,R);(L',R')) \mid L \oplus L' = (0,\alpha,0,0);\ R \oplus R' = (\delta_0,\delta_1,\delta_2,0);$
$l_1 = \lambda_1;\ l_3 = \lambda_3;\ r_3 = \rho_3\}$
then $\Delta S_{\lambda_1,\lambda_3,\rho_3,\alpha}[9] = ((0,\alpha,0,0),(0,0,0,\beta))$,
i. e. the difference for the various pairs of the $S_{\lambda_1,\lambda_3,\rho_3,\alpha}$ set is constant after 9 rounds.

Proof : After two rounds, the (l_1, l_1') pair is fixed and equal to the two constant bytes :
$c_1 = \lambda_1 \oplus K1_1 \oplus p_3\,[\rho_3 \oplus K2_3 \oplus p_3[\lambda_3 \oplus K1_3]]$; $c_1' = c_1 \oplus \alpha$;

furthermore, these l_1 and l_1' values act as inputs to the $[p_2, p_3, p_0, p_1]$ S-box at round 3. It suffices to select $\delta_0, \delta_1, \delta_2$ as to "compensate" the difference between the two $[p_2, p_3, p_0, p_1]$ outputs to obtain a constant xor value for the six subsequent rounds. The encryption of a $S_{\lambda_1,\lambda_3,\rho_3,\alpha}$ pair is depicted in Figure 2 (where the difference value at each round is provided).
The $\delta_0, \delta_1, \delta_2$ and β constants are given by the relation :
(1) : $(\delta_0, \delta_1, \delta_2, \beta) = [p_2, p_3, p_0, p_1][c_1] \oplus [p_2, p_3, p_0, p_1][c_1']$ where c_1 and c_1' are defined above.

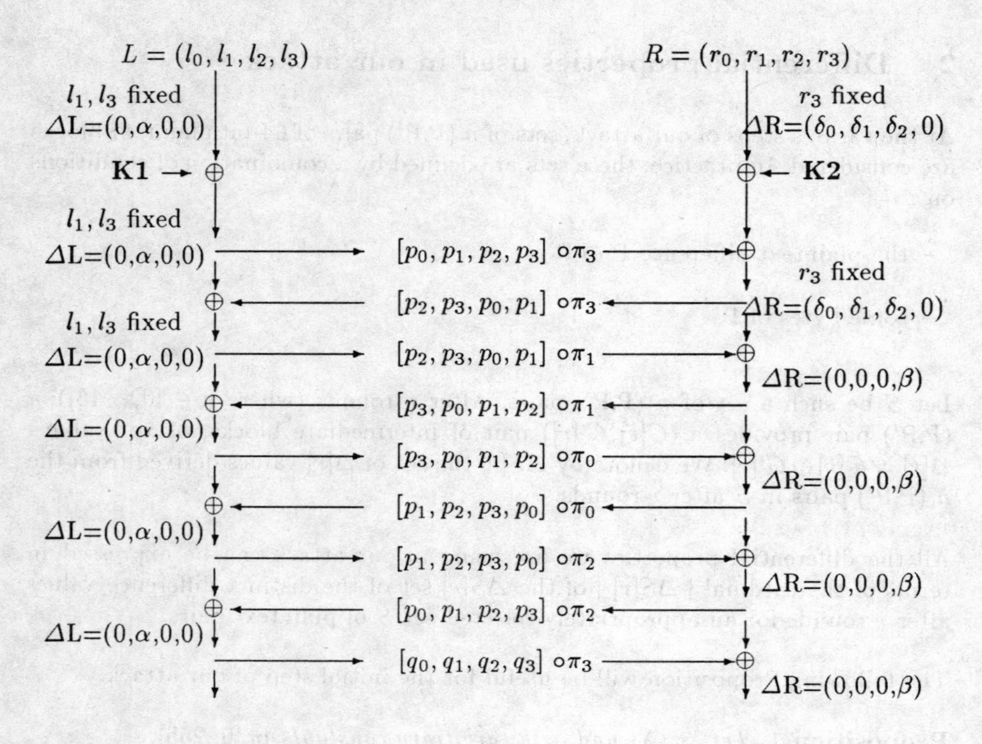

Fig. 2. Encryption of an $S_{\lambda_1,\lambda_3,\rho_3,\alpha}$ pair

The next proposition, which is very similar to Proposition 1, will be useful for the subsequent steps of our attack.

Proposition 2. *Let λ_3 be arbitrary constant $\in [0..255]$. For every $\alpha \in]0..255]$, there exist four constants $\delta_0, \delta_1, \delta_2$ and $\delta_3 \in [0..255]$ such that, if $S_{\lambda_3,\alpha}$ denotes the following set of plaintext pairs :*
$S_{\lambda_3,\alpha} = \{((L,R);(L',R')) \mid L \oplus L' = (0,0,0,\alpha); R \oplus R' = (\delta_0, \delta_1, \delta_2, \delta_3); l_3 = \lambda_3\}$
then $\Delta S_{\lambda_3,\alpha}[8] = \{((0,0,0,\alpha),(0,0,0,0))\}$, *i. e. the difference for the various pairs of the $S_{\lambda_3,\alpha}$ set is constant after 8 rounds.*

Proof : The proof is similar to the proof of Proposition 1. The l_3 and l'_3 inputs to the $[p_0,p_1,p_2,p_3]$ S-box in the first round are the constant bytes: $c_3 = \lambda_3 \oplus K1_3$ and $c'_3 = c_3 \oplus \alpha$.
It suffices to take for $\delta_0, \delta_1, \delta_2$ and δ_3 the constants given by the relation :
(2) : $(\delta_0, \delta_1, \delta_2, \delta_3) = [p_0, p_1, p_2, p_3]\ [c_3] \oplus [p_2, p_3, p_0, p_1]\ [c'_3]$ to obtain the announced result.

The next proposition essentially states that for a given set of plaintext pairs, there are some bounds limiting the increase of the number of distinct difference values after r rounds. When used in conjonction with Propositions 1 and 2, Proposition 3 provides non-trivial differential properties for the whole 16-round algorithm.

Proposition 3. *If S is a set of n (P,P') plaintext pairs and if $\Delta S[r]$ is defined as above, then :*
$\forall\, r \in [0..15],\ |\Delta S[r+1]| \leq 128 \times |\Delta S[r]|$

Proof : Let $\Delta[r+1]$ be a $\Delta S[r+1]$ value. There exists a (P,P') pair in S leading to the $\Delta[r+1]$ xor value after $r+1$ rounds. Let us denote by $\Delta[r]$ the xor value for this pair after r rounds, and by $\delta[r]$ the byte of $\Delta[r]$ which position is picked up as an input to the S-box used in round $(r+1)$; by a the input to the S-box of round $(r+1)$ in the encryption of P.
Depending on the parity of r, we have either :

(*) : $\Delta[r+1] = \Delta[r] \oplus ((Sbox[a] \oplus Sbox[a \oplus \delta[r]]), (0,0,0,0))$
or
(**) : $\Delta[r+1] = \Delta[r] \oplus ((0,0,0,0), (Sbox[a] \oplus Sbox[a \oplus \delta[r]]))$.

Moreover, $\Delta[r] \in \Delta S[r]$ by definition.

The results now follow from the obvious property that, for a fixed value of $\Delta[r]$ (which determines a fixed value for $\delta[r]$), we have :
$|\ \{\ Sbox[a] \oplus Sbox[a \oplus \delta[r]]\ |\ a \in [0..255]\ \}\ | \leq 128$.

3 An attack of the 16-round Khufu

In this Section, we describe how to use the differential properties presented above for deriving the secret key (i.e. four 32-bit words and 8 permutations of the [0..255] set) from about 2^{43} chosen plaintexts.

The proposed attack is divided in four main steps. Only the first Step (the derivation of first information about the secret key) is developed in some detail. The purpose of the description of the subsequent steps is only to give some evidence that the information obtained at Step 1 can be used for breaking the entire scheme.

Step 1 is based on Propositions 1 and 3. We are using the notation of Proposition 1. We fix four arbitrary constants λ_1, λ_3 and ρ_3 and $\alpha \neq 0$ in [0..255].
The purpose of Step 1 is to find the three corresponding bytes $\delta_0, \delta_1, \delta_2$.
We select a fixed arbitrary byte λ_2. For 64 different values of l_0 (for instance the 0 to 63 values) we perform the following computations :

• We encrypt the X_{l_0} set of 2^{24} plaintext values $((l_0, \lambda_1, \lambda_2, \lambda_3), (r_0, r_1, r_2, \rho_3))$, where $(r_0, r_1, r_2) \in [0..255]^3$, thus obtaining 2^{24} encrypted values $((ll_0, ll_1, ll_2, ll_3), (rr_0, rr_1, rr_2, rr_3))$;

• We encrypt the X'_{l_0} set of 2^{24} plaintext values $((l_0, \lambda_1 \oplus \alpha, \lambda_2, \lambda_3), (r'_0, r'_1, r'_2, \rho_3))$ where $(r'_0, r'_1, r'_2) \in [0..255]^3$; thus obtaining 2^{24} encrypted values $((ll'_0, ll'_1, ll'_2, ll'_3), (rr'_0, rr'_1, rr'_2, rr'_3))$;

• We now want to find all the (P,P') pairs in $(X_{l_0} \times X'_{l_0})$ such that $ll_2 = ll'_2$; $rr_0 = rr'_0$; $rr_2 = rr'_2$ (i.e. we want to filter those pairs of $(X_{l_0} \times X'_{l_0})$ for which the inputs and outputs of the S-boxes in the three last rounds of the encryption are equal), and group these pairs according to the $(r_0 \oplus r'_0, r_1 \oplus r'_1, r_2 \oplus r'_2)$ difference value. This can be done efficiently, in about 2^{25} operations in average, as follows :
- we group the X_{l_0} plaintexts according to the (ll_2, rr_0, rr_2) value, thus obtaining a list of X_{l_0} plaintexts of average size 1 for each (ll_2, rr_0, rr_2) triplet;
- we group the X'_{l_0} plaintexts according to the (ll'_2, rr'_0, rr'_2) value, thus obtaining a list of X'_{l_0} plaintexts of average size 1 for each (ll'_2, rr'_0, rr'_2) triplet;
- for each (ll_2, rr_0, rr_2), we consider all the (P,P') pairs in the crossproduct of the corresponding lists of X_{l_0} and X'_{l_0} (average number of pairs : 1);
- we group the (P,P') pairs obtained as above according to the $(r_0 \oplus r'_0, r_1 \oplus r'_1, r_2 \oplus r'_2)$ difference, thus obtaining a list of average size 1 for each $(r_0 \oplus r'_0, r_1 \oplus r'_1, r_2 \oplus r'_2)$ value.

For each $(r_0 \oplus r'_0, r_1 \oplus r'_1, r_2 \oplus r'_2)$ triplet, we thus obtain, by merging the lists obtained for each l_0 value, a list S of 64 (P,P') pairs in average. Let us consider the ΔS set of output xors for the (P,P') pairs contained in such a list :

Claim :

(i) : If $(r_0 \oplus r'_0, r_1 \oplus r'_1, r_2 \oplus r'_2) = (\delta_0, \delta_1, \delta_2)$ *then* $|\Delta S| \leq 16$.

(ii) : If $(r_0 \oplus r'_0, r_1 \oplus r'_1, r_2 \oplus r'_2) \neq (\delta_0, \delta_1, \delta_2)$ *then* $|\Delta S| \approx 64$.

Proof (heuristic arguments) : In the first case, S is a subset of the $S_{\lambda_1, \lambda_3, \rho_3, \alpha}$ considered in Proposition 1. Moreover, $\Delta S = \Delta S[13]$, because the three last rounds have no effect on the output xor of an S pair and $|\Delta S[13]| \leq |\Delta S_{\lambda_1, \lambda_3, \rho_3, \alpha}[13]|/2^{24}$ because $\Delta S[13]$ contains only $\Delta S_{\lambda_1, \lambda_3, \rho_3, \alpha}[13]$ elements $((\delta_{l_0}, \delta_{l_1}, \delta_{l_2}, \delta_{l_3}), (\delta_{r_0}, \delta_{r_1}, \delta_{r_2}, \delta_{r_3}))$ such that $\delta_{l_2} = 0$ and $\delta_{r_0} = 0$ and $\delta_{r_2} = 0$.
The first part of the claim now results from :
$|\Delta S_{\lambda_1, \lambda_3, \rho_3, \alpha}[9]| = 1$ (by Proposition 1)
and
$|\Delta S_{\lambda_1, \lambda_3, \rho_3, \alpha}[13]| \leq 128^4 |\Delta S_{\lambda_1, \lambda_3, \rho_3, \alpha}[9]|$ (by Proposition 3).
The second part of the claim follows from the assumption that in other cases,

the final xor values for pairs in the S set simply behave as random elements of the set of $((\delta_{l_0}, \delta_{l_1}, \delta_{l_2}, \delta_{l_3}), (\delta_{r_0}, \delta_{r_1}, \delta_{r_2}, \delta_{r_3}))$ differences such that $\delta_{l_2} = 0$ and $\delta_{r_0} = 0$ and $\delta_{r_2} = 0$.
The above claim was confirmed by computer experiments, where (i) was checked. Because of the difference between the behaviours *(i)* and *(ii)*, the $(\delta_0, \delta_1, \delta_2)$ triplet can be detected, as the one leading to a ΔS set of output *xors* of size less than 16.

Step 1 requires the encryption of about 2^{31} chosen plaintext blocks, and about 2^{31} operations.

Step 2 : we repeat Step 1 for all possible $\alpha \neq 0$ values (without modifying the λ_1, λ_3 and ρ_3 constants). We are using the notation of Proposition 1 : c_1 denotes the common input to the S-box of round 3 for all the plaintexts of the various X_{l_0} sets.
For every α, we obtain the 24-bit word :
$(\delta_0, \delta_1, \delta_2)(\alpha) = [p_2, p_3, p_0][c_1] \oplus [p_2, p_3, p_0][c_1 \oplus \alpha]$.

In summary, after Step 2, the p_2, p_3 and p_0 permutations are entirely determined, up to the four unknown bytes c_1, $p_2[c_1]$, $p_3[c_1]$ and $p_0[c_1]$. Steps 1 and 2 require to encrypt about 2^{39} plaintext blocks, and the computational cost of steps 1 and 2 is about 2^{39} operations.

Step 3 is based on Propositions 2 and 3. We fix an arbitrary constant $\lambda_3 \in [0..255]$. We are using the notations of Proposition 2. The purpose of Step 3 is to find for every $\alpha \in [1..255]$ the four corresponding bytes δ_0, δ_1, δ_2 and δ_3, such that :

$$(\delta_0, \delta_1, \delta_2, \delta_3)(\alpha) = [p_0, p_1, p_2, p_3][c_3] \oplus [p_0, p_1, p_2, p_3][c_3 \oplus \alpha].$$

We are using the fact that since $[p_2,p_3,p_0]$ is known up to the unknown bytes $c_1, p_2[c_1], p_3[c_1]$ and $p_0[c_1], \delta_0, \delta_2$, and δ_3 are known up to the single unknown byte $c_1 \oplus c_3$: there are only 256 possible values for the $(\delta_0, \delta_2, \delta_3)$ triplet.

Step 3 is divided in two substeps, numbered 3.1 and 3.2.

Substep 3.1 : We want to find $(\delta_0, \delta_1, \delta_2, \delta_3)$ for a first fixed $\alpha \neq 0$ value. We are doing an exhautive search on the $c_1 \oplus c_3$ byte. For each $c_1 \oplus c_3$ assumption, the candidate values δ_0, δ_2 and δ_3 are determined, and we efficiently test the 256 candidate δ_1 values by a method close to the one of Step 1. For that purpose, we select fixed arbitrary bytes ρ_0, ρ_2 and ρ_3; for 2^{24} different values of the (l_0, l_1, l_2) triplet, we perform the following computations :

- We encrypt the $X_{l_0 l_1 l_2}$ set of 2^8 plaintext values $((l_0, l_1, l_2, \lambda_3), (\rho_0, r_1, \rho_2, \rho_3))$, where $(r_1 \in [0..255])$, thus obtaining 2^8 encrypted values

$((ll_0, ll_1, ll_2, ll_3), (rr_0, rr_1, rr_2, rr_3))$.(Note: this can be done once for all for all $c_1 \oplus c_3$ assumptions).

- We encrypt the X'$_{l_0 l_1 l_2}$ set of 2^8 plaintext values $((l_0, l_1, l_2, \lambda_3 \oplus \alpha), (\rho_0 \oplus \delta_0, r'_1, \rho_2 \oplus \delta_2, \rho_3 \oplus \delta_3))$ where $r'_1 \in [0..255]$; thus obtaining 2^8 encrypted values $((ll'_0, ll'_1, ll'_2, ll'_3), (rr'_0, rr'_1, rr'_2, rr'_3))$;

- We now want to find all the (P,P') pairs in (X$_{l_0 l_1 l_2}$ × X'$_{l_0 l_1 l_2}$) such that $ll_2 = ll'_2$; $rr_2 = rr'_2$ (i.e. we want to "filter" those pairs of (X$_{l_0 l_1 l_2}$ × X'$_{l_0 l_1 l_2}$) for which the inputs and outputs of the S-boxes in the two last rounds of the encryption are equal), and group these pairs according to the $r_1 \oplus r'_1$ difference value. This can be done efficiently, in about 2^8 operations in average, and provides in average one (P,P') pair after filtering.

For each $\delta_1 = r_1 \oplus r'_1$ candidate difference, we thus obtain, by merging the contributions obtained for each $l_0 l_1 l_2$ value, a list S of 2^{16} (P,P') pairs in average. Let us consider the ΔS set of output xors for the (P,P') pairs contained in such a list :

Claim :

(i) : If the $(\delta_0, \delta_1, \delta_2, \delta_3)$ candidate is correct, then the ΔS differences are picked from a set of size less than $128^6 \;/\; 256^2$.

(ii) : If the $(\delta_0, \delta_1, \delta_2, \delta_3)$ candidate is wrong, then the ΔS differences are picked from a set of size about 256^6.

(iii) : The size of the S sets (about 2^{16} pairs) is sufficient to test $(\delta_0, \delta_1, \delta_2, \delta_3)$, by comparing the sizes of S and ΔS, i.e. by counting how many collisions occur between the output xors for S pairs.

Proof (heuristic arguments) : In the first case, S is a subset of the $S_{\lambda_3,\alpha}$ considered in Proposition 2. Moreover, $\Delta S = \Delta S[14]$, because the two last rounds have no effect on the output xor of an S pair. The $\Delta S[14]$ elements are picked from a set of size about $|\Delta S_{\lambda_3,\alpha}[14]|/256^2$ because $\Delta S[14]$ contains only $\Delta S_{\lambda_3,\alpha}[14]$ elements $((\delta_{l_0}, \delta_{l_1}, \delta_{l_2}, \delta_{l_3}), (\delta_{r_0}, \delta_{r_1}, \delta_{r_2}, \delta_{r_3}))$ such that $\delta_{l_2} = 0$ and $\delta_{r_2} = 0$.
The first part of the claim now results from :
$|\Delta S_{\lambda_3,\alpha}[8]| = 1$ (by Proposition 2)
and
$|\Delta S_{\lambda_3,\alpha}[14]| \leq 128^6 |\Delta S_{\lambda_3,\alpha}[8]|$ (by Proposition 3).

The second part of the claim follows from the assumption that in other cases, the final xor values for pairs in the S set simply behave as random elements of the set of $((\delta_{l_0}, \delta_{l_1}, \delta_{l_2}, \delta_{l_3}), (\delta_{r_0}, \delta_{r_1}, \delta_{r_2}, \delta_{r_3}))$ differences such that $\delta_{l_2} = 0$ and $\delta_{r_2} = 0$.

The third part of the claim follows from the facts that since $2^{16} \gg (128^6/256^2)^{1/2}$ many collisions will occur in the first case, whereas few collisions will occur in the second case, since $2^{16} \ll (256^6)^{1/2}$.

Because of the difference between behaviours (i) and (ii), Substep 3.1 provides $c_1 \oplus c_3$, (as the value for which one δ_1 value behaves according to (i)), and also the $\delta_0\ \delta_1\ \delta_2\ \delta_3$ difference for the considered $\alpha \neq 0$ value. Substep 3.1 requires to encrypt about 2^{40} chosen plaintexts, and about 2^{40} operations.

Substep 3.2 : We are doing the same search as at Substep 3.1 for each of the 254 remaining $\alpha \neq 0$ values, but the $c_1 \oplus c_3$ byte has no longer to be exhaustively searched, since it has been determined at Substep 3.1.

Step 3 requires to encrypt about 2^{41} plaintext blocks, and the computational cost of Step 3 is about 2^{41} operations. After Step 3, the p_0, p_1, p_2 and p_3 permutations are entirely determined, up to the five unknown bytes c_3, $p_0[c_3]$, $p_1[c_3]$, $p_2[c_3]$, and $p_3[c_3]$, and the output of round 1 is known up to 8 constant bytes.

Step 4 is based on the results of Step 3.
We define four permutations $\overline{p}_0, \overline{p}_1, \overline{p}_2, \overline{p}_3$, by the relations :
$(\overline{p}_0, \overline{p}_1, \overline{p}_2, \overline{p}_3)[\lambda_3] = 0$;
$(\overline{p}_0, \overline{p}_1, \overline{p}_2, \overline{p}_3)[\lambda_3 \oplus \alpha] = (\delta_0; \delta_1, \delta_2, \delta_3)(\alpha)$ for $\alpha \neq 0$.
We thus define a $[\overline{p}_0, \overline{p}_1, \overline{p}_2, \overline{p}_3]$ S-box which is intended to be the basis for the construction of an equivalent representation of the first half of Khufu.
If we set $\overline{K1}_3 = 0$, the assumptions $[p_0, p_1, p_2, p_3] = [\overline{p}_0, \overline{p}_1, \overline{p}_2, \overline{p}_3]$ and $K1_3 = \overline{K1}_3$ provide the output of round 1 up to 8 unknown constant bytes for every input block.

The purpose of Step 4 is to gradually derive, one after the other, seven additional bytes $\overline{K2}_3, \overline{K1}_1, \overline{K2}_1, \overline{K1}_0, \overline{K2}_0, \overline{K1}_2, \overline{K2}_2$ such that :

- the assumptions $[p_0, p_1, p_2, p_3] = [\overline{p}_0, \overline{p}_1, \overline{p}_2, \overline{p}_3]$, $K1_3 = \overline{K1}_3$; $K2_3 = \overline{K2}_3$ provide the output of round 2 up to 8 unknown constant bytes for every input block;

⋮

- the assumptions $[p_0, p_1, p_2, p_3] = [\overline{p}_0, \overline{p}_1, \overline{p}_2, \overline{p}_3]$; $K1_3 = \overline{K1}_3$; $K2_3 = \overline{K2}_3$; $K1_1 = \overline{K1}_1$; $K2_1 = \overline{K2}_1$; $K1_0 = \overline{K1}_0$; $K2_0 = \overline{K2}_0$; $K1_2 = \overline{K1}_2$; $K2_2 = \overline{K2}_2$ provide the output of round 8 up to 8 unknown constant bytes for every input block.

Our estimate of the cost of the derivation of the above equivalent key bytes is at most 2^{40} plaintext blocks per key byte (2^{32} per test of an assumption on such a byte), using differential techniques similar to the one of Proposition 2 : introduction of a fixed difference in the S-box inputs of round r and compensation of the resulting difference for the S-box inputs of the subsequent rounds until round 8, with $r = 1$ in Proposition 2 and $r = 2$ to 8 here. Once the seven above equivalent key bytes have been derived, a differential attack on the second half of Khufu (i.e. rounds 8 to 16 and the final xor with auxiliary key bits) can be mounted, at no substantial additional expense.

4 Aknowledgements

The attack presented in this paper is based on a preliminary investigation of Khufu in cooperation with Sean Murphy.
We also want to thank Thierry Baritaud for his help in the elaboration of the Latex version of this paper.

5 Conclusion

We have shown in some detail that first information concerning the secret S-box used in the first half of the scheme can be derived with about 2^{31} chosen plaintexts, and about 2^{31} operations (Step 1). Our estimate of the resources required for breaking the whole scheme is about 2^{43} chosen plaintexts and 2^{43} operations (Steps 2 to 4). However, further verifications (in particular computer experiments) are required to make sure that the figures announced in the description of Steps 3 and 4 are valid. Although the proposed attack is far from being realistic, because of the required amount of chosen plaintext, it suggests that the security of the 16-round Khufu might be too low for providing a suitable replacement to DES. In fairness to Merkle, it should be noticed however that some warnings concerning the choice of the 16-round version as compared with the 24-round and 32-round versions are already contained in [1], in particular the remark that the "safety factor" in Khufu with 16 rounds is less than that of DES.
We have found no similar attack for the 24-round version of Khufu.

References

1. Ralph Merkle, "Fast Software Encryption Functions", Advances in Cryptology - Crypto'90, Springer Verlag.
2. E. Biham and A. Shamir, "Differential Cryptanalysis of Snefru, Khafre, REDOC-II, LOKI, and Lucifer". Advances in Cryptology - Crypto'91, Springer Verlag.
3. E. Biham and A. Shamir, "Differential Cryptanalysis of the Data Encryption Standart". Springer Verlag, 1993, Chapter 7.

Ciphertext Only Attack for One-way function of the MAP using One Ciphertext

Yukiyasu TSUNOO Eiji OKAMOTO Tomohiko UYEMATSU

School of Information Science
Japan Advanced Institute of Science and Technology
15 Asahidai Tatsunokuchi Nomi Ishikawa, 923-12, Japan
Tel.:+81-761-51-1287, Fax.:+81-761-51-1338, Email:okamoto@jaist.ac.jp

Abstract. FEAL-N proposed by NTT is an N-round block cipher which is well suited for a fast software execution. We research the method which improves FEAL algorithm against cryptanalysis based on the careful analysis of data randomizer. Since the round function f used as data randomizer has an inverse function f^{-1}, we have already shown that there exists a key-independent relation between plaintext and ciphertext.
In this paper, we apply our attack to the ciphertext only attack for one-way function of the MAP using only one ciphertext. Computer implementation of the proposed attack shows that the improvement of the computational complexity to obtain the plaintext is decreased and given by $O(2^{32})$.

1 Introduction

FEAL-N proposed by NTT is a N-round block cipher which is well suited for a fast software implementation on 8 or 16 bit microprocessors. FEAL is a DES-like secret-key cryptosystem[1][2]. There are many kinds of attacks of FEAL, such as a differential attack[3], a statistical attack[4], an attack using a linear approximation of S-boxes[5], an attack using bit-equations of data randomizer[6] and so on[7]. We research the method which improves FEAL algorithm against cryptanalysis by based on the careful analysis of data randomizer. We have studied an attack that employs an inverse function f^{-1} of the intermediate message blocks, where f is a round function of FEAL. Our proposed attack is divided into two steps. First step is to search out intermediate message blocks using a key-independent relation of a FEAL data randomizer between ciphertext blocks and corresponding plaintext blocks. Second step is to calculate the key using the above inverse function f^{-1} of the intermediate message blocks.

In this paper, we apply our method to the ciphertext only attack for a one-way function that employs the mapping function called MAP[1]. The algorithm of MAP function is very similar to that of FEAL data randomizer in the shape of the round function f. By virtue of the simplification of the MAP algorithm, fairly simple related equations between ciphertext blocks and corresponding plaintext blocks are given. Only one 64-bit ciphertext is needed to calculate the corresponding plaintext and its computational complexity is $O(2^{32})$. This implies that the computational complexity to obtain the plane text is decreased and given by $O(2^{32})$.

The contents of this paper is given in three parts; first part relates to the specification of one-way function of the MAP, second part concerns our proposed ciphertext only attack and final part describes concluding remarks.

2 One-way function of the MAP

The one-way function of the MAP is a 4-round Feistel block cipher operating on 64-bit plaintext and ciphertext blocks. A round function depends on the output of the previous round only and has nothing to do with enciphering/deciphering key. The one-way function of the MAP is shown in Fig.1.

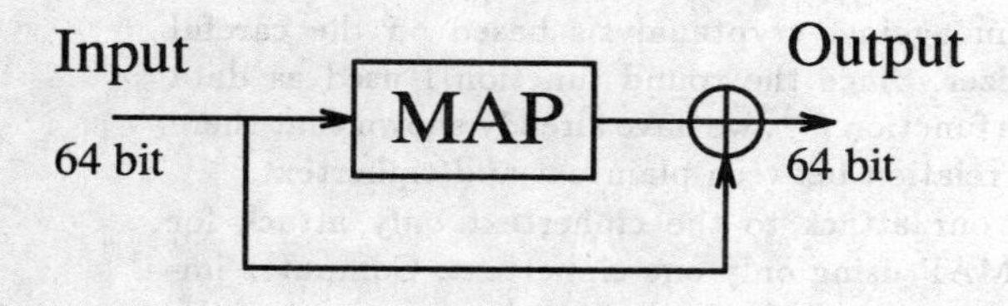

Fig. 1. One-way function of the MAP

The MAP algorithm is based on iterating a round function f four times. The round function f is based on nonlinear S-box functions. The MAP algorithm, S-box and f functions are defined in the following sections.

2.1 MAP Algorithm

The MAP algorithm is shown in Fig.2. In the mapping process, a 64-bit input block is partitioned into eight 8-bit subblocks, and transformed into eight 8-bit output subblocks. The MAP algorithm performs as follows:
Suppose we compute a 64-bit input block. At first, we split the input block into left and right halves of 32-bit long, and compute the exclusive-or (hereafter exclusive-or is referred as XOR) of the 32-bit left half and the 32-bit right half as shown in Fig.2. Second, we compute the XOR of the 32-bit left half and a 32-bit random number calculated by function f shown in Fig.3. We then interchange both the 32-bit left and the 32-bit

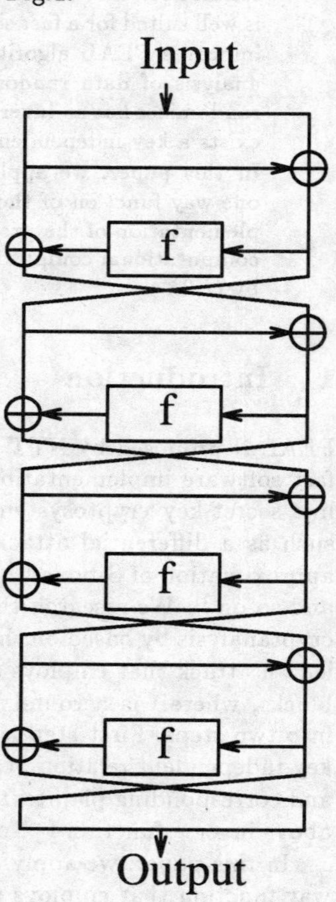

Fig. 2. MAP algorithm

right halves. In order to get a 64-bit output block, we perform these processes four times. Obviously, this MAP algorithm has the inverse algorithm MAP^{-1}, because we can perform the same procedure above in reverse. Since the computational complexity of calculating output blocks (depicted as Output) from input blocks (depicted as Input) shown in Fig.2 is equal to that of calculating input blocks from output blocks, this MAP algorithm cannot be regarded as a one-way function without any changes. Thus, the one-way function using MAP algorithm shown in Fig.1 may be considered as a 64-bit block cipher based on a 64-bit key where the key is a 64-bit input block. Thus, in Fig.1, the computational complexity of calculating input blocks from output blocks is given by $O(2^{64})$.

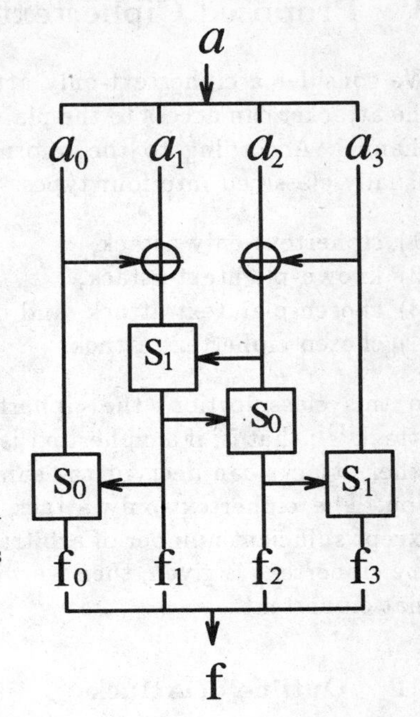

Fig. 3. Function f

2.2 Function S-BOX

The S-box function has two 1-byte inputs, x_1 and x_2, and outputs one 1-byte block, such that

$$S_d(x_1, x_2) = rot_2(x_3) \quad \text{and} \tag{2.1}$$

$$x_3 = x_1 + x_2 + d \mod 256, \tag{2.2}$$

where $d \in \{0, 1\}$, and $rot_2(x_3)$ denotes a 2-bit left rotation on 8-bit data x_3.

2.3 Function f

The round function f has one 4-byte input and outputs one 4-byte block. Let $a = (a_0, a_1, a_2, a_3)$ be a 4-byte input block and $f = (f_0, f_1, f_2, f_3)$ be a 4-byte output block. Then, the function f is defined as the following equations:

$$f_1 = S_1(a_1 \oplus a_0, a_2 \oplus a_3) \tag{2.3}$$

$$f_2 = S_0(f_1, a_2 \oplus a_3) \tag{2.4}$$

$$f_0 = S_0(a_0, f_1) \tag{2.5}$$

$$f_3 = S_1(a_3, f_2) \tag{2.6}$$

where symbol $\oplus$ denotes the XOR operation. Function f can be illustrated in Fig.3. (In FEAL algorithm, equation $f_1 = S_1(a_1 \oplus a_0 \oplus \beta_0, a_2 \oplus a_3 \oplus \beta_1)$ and $f_2 = S_0(f_1, a_2 \oplus a_3 \oplus \beta_1)$, where β_0 and β_1 denote a 8-bit key respectively, are applied to definitions of f_1 and f_2 instead of equation (2.3) and (2.4)).

3 Proposed Ciphertext Only Attack

We consider a ciphertext-only attack. In general, assumption of cryptography, the attacker can access to the plaintext/ciphertext transmitted over the insecure channel. According to the information obtained by the attacker, attacks are usually classified into four types:

(1) ciphertext-only attack,
(2) known-plaintext attack,
(3) chosen-plaintext attack and
(4) chosen-ciphertext attack.

In this classification, the ciphertext-only attack includes the other types of attack[11]; that is, if a ciphertext is decrypted by the ciphertext-only attack, then other attacks can decrypt the same ciphertext without any additional information. The ciphertext-only attack means that the attacker has no information except sufficient number of arbitrary ciphertexts. In our proposed attack, if only one ciphertext is given, then we will manage to compute plaintext depending on that ciphertext.

3.1 Outline of Attack

As shown in 2.1, as for the one-way function of the MAP shown in Fig.1, we can compute input blocks from output blocks with the computational complexity of $O(2^{64})$. But $O(2^{64})$ of computational complexity means the intractable numerical problem and input blocks may not be computed from output blocks.

Our proposed attack needs the computational complexity of the size $O(2^{32})$ and is divided into three steps.

(1) Simplify the MAP algorithm,
(2) Find out the relation between plaintext and ciphertext
(3) Compute plaintext by means of the relation (2).

The computation of step (3) can be performed by a brute force attack with computational complexity of $O(2^{32})$.

3.2 Simplification of the MAP Algorithm

In this subsection, we simplify the MAP algorithm shown in Fig.2

Fig.4 shows the equivalent algorithm to Fig.2. In Fig.4, both the second round and the forth round of function f are applied in the opposite direction, instead of the interchanging between the 32-bit left half and the 32-bit right half calculated by previous rounds of function f. In what follows, the 32-bit left half and the 32-bit right half of Input blocks are described as A_0 and B_0, the 32-bit left half and the 32-bit right half of blocks calculated by the second round (hereafter call these blocks intermediate message) are described as A_2 and B_2 and the 32-bit left half and the 32-bit right half of Output blocks are described as A_4 and B_4.

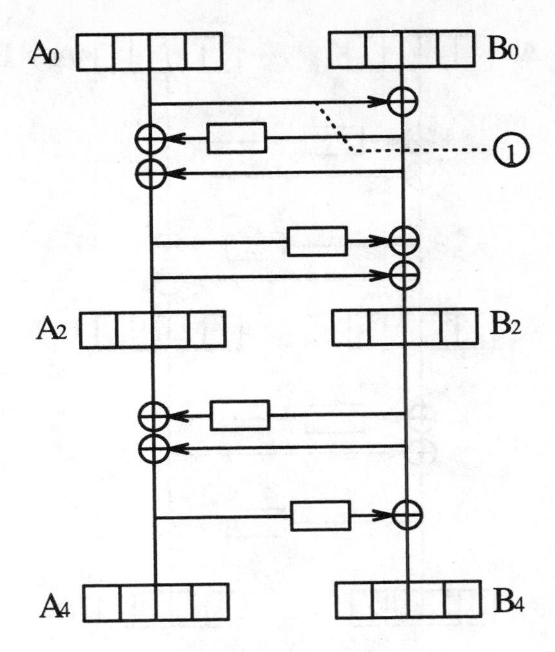

Fig. 4. 1st Simplification

We also define in following Figures, A_{ij} and B_{ij}, $(i = 0, 2, 4)$, $(j = 0, 1, 2, 3)$, denote the 1-byte block defined by following equations:
$A_i = (A_{i0}, A_{i1}, A_{i2}, A_{i3})$, $B_i = (B_{i0}, B_{i1}, B_{i2}, B_{i3})$.

The simplification procedure of the MAP algorithm is shown below.

(1) In Fig.4, the first XOR at the circle number one (at① with a dotted line) can be omitted, if input blocks are given by (A_0 , $A_0 \oplus B_0$) instead of (A_0 , B_0).
Thus, Fig.4 and Fig.5 are equivalent.

(2) In Fig.5, without affecting of all the value of A_i and B_i, $(i = 0, 2, 4)$, the XOR just after the first round function f can be moved just before the first round one, and the XOR just after the third round function f can also be moved just before the third round one, (at ② and ③ with a dotted curve in Fig.6, respectively).
Thus, Fig.5 and Fig.6 are equivalent.

(3) In Fig.6, the XOR after the movement depicted by the circle number two (at ② with a dotted curve) can be omitted, if input blocks are given by (B_0 , $A_0 \oplus B_0$) instead of (A_0 , $A_0 \oplus B_0$).
Thus, Fig.6 and Fig.7 are equivalent.

(4) In Fig.7, let us estimate the value of the intermediate messages indicated by the circle number four, five, six and seven (see ④ ,⑤ ,⑥ and ⑦ with a dotted message blocks, respectively). The value of intermediate message at

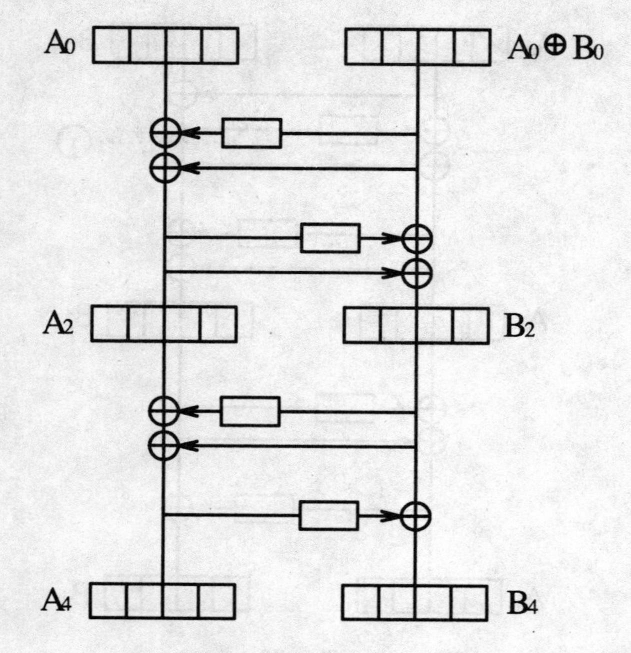

Fig. 5. 2nd Simplification

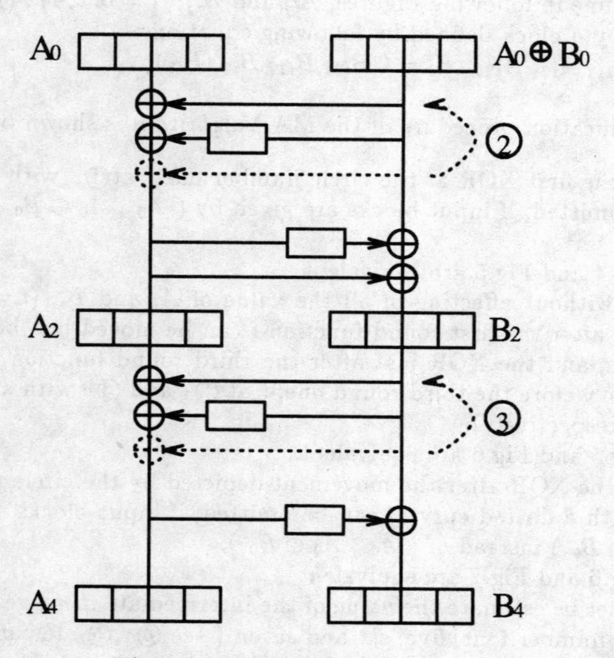

Fig. 6. 3rd Simplification

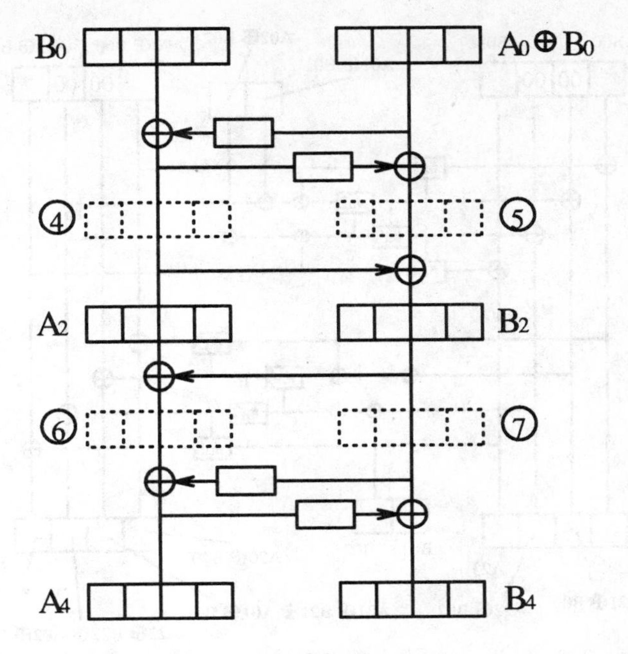

Fig. 7. 4th Simplification

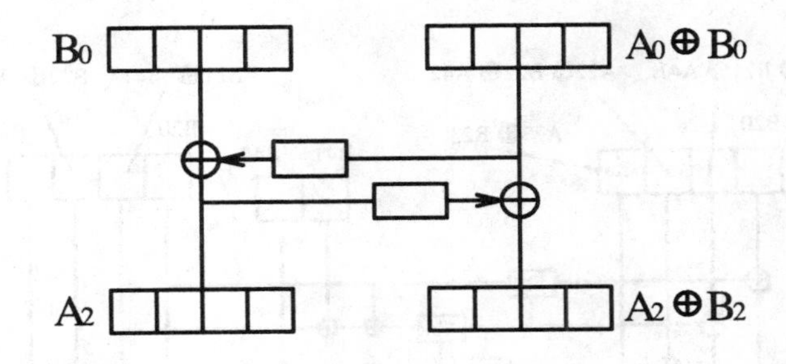

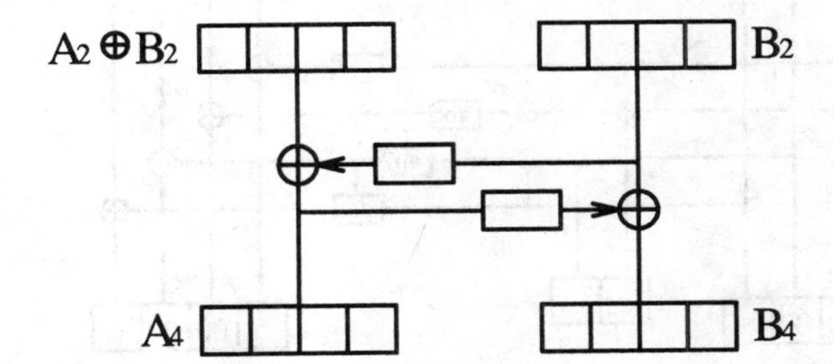

Fig. 8. 5th Simplification

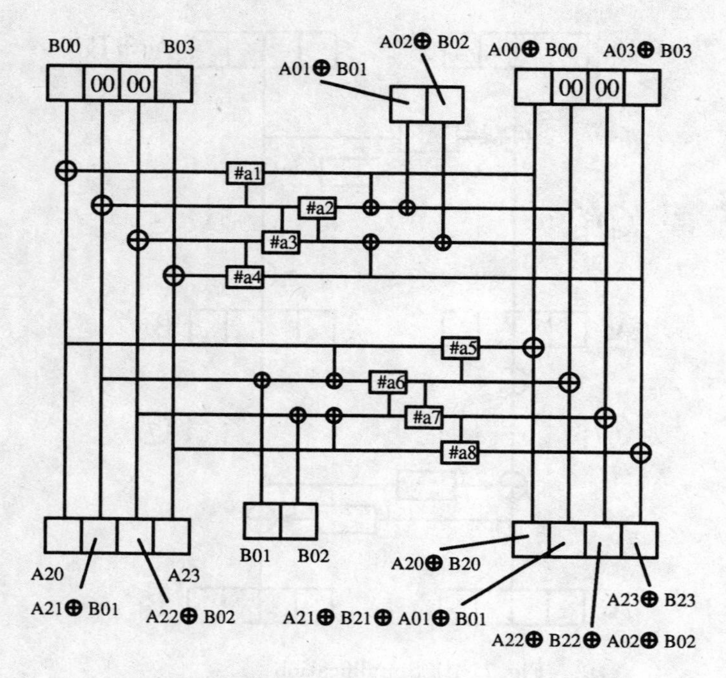

Fig. 9. 6th Simplification by S-Box (a)

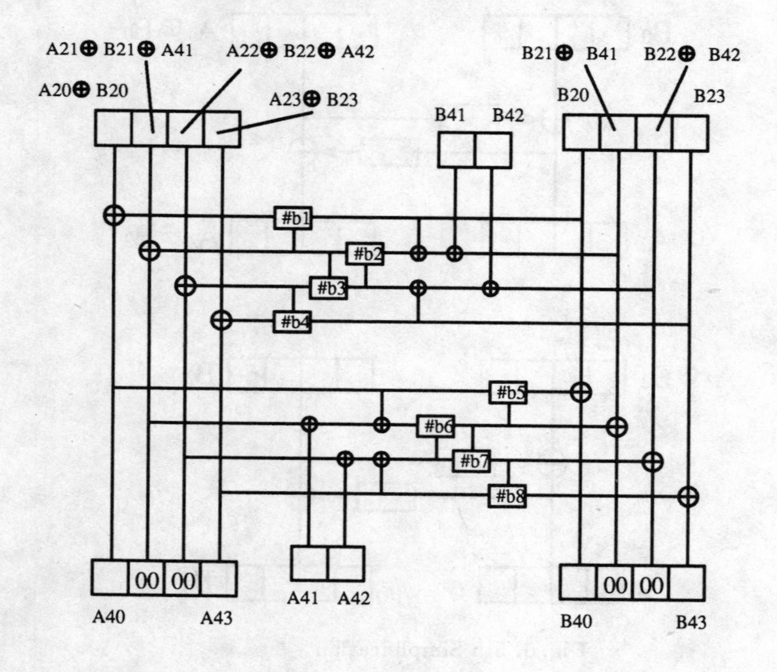

Fig. 10. 6th Simplification by S-Box (b)

④ is equivalent to the value of A_2, the value of intermediate message at⑦ is equivalent to the value of B_2, the value of intermediate message at⑤ and at⑥ are equivalent to the value of ($A_2 \oplus B_2$).
Thus, Fig.7 and Fig.8 are equivalent.

(5) Now, let a new mechanism introduce into Fig.8. The new mechanism exchanges the value of the 1-byte blocks at B_{01}, B_{02}, $A_{01} \oplus B_{01}$, $A_{02} \oplus B_{02}$, A_{41}, A_{42}, B_{41} and B_{42} with the value of "00". In order to maintain the value of these 1-byte blocks, we introduce two XORs into each function f shown in the pair of Fig.9 and Fig.10. Fig.8 is equivalent to the pair of Fig.9 and Fig.10. Let notice as follows:
In Fig.9, the new XOR function of $(A_{01} \oplus B_{01}, A_{02} \oplus B_{02})$ and (B_{01}, B_{02}) are introduced into the input blocks of function S-Box denoted by $(\#a2, \#a3)$ and $(\#a6, \#a7)$ and the value of the 1-byte blocks at $(A_{21}\ ,\ A_{22})$ and $(A_{21} \oplus B_{21}\ ,\ A_{22} \oplus B_{22})$ are replaced by $(A_{21} \oplus B_{01}\ ,\ A_{22} \oplus B_{02})$ and $(A_{21} \oplus B_{21} \oplus A_{01} \oplus B_{01},\ A_{22} \oplus B_{22} \oplus A_{02} \oplus B_{02})$, respectively. Also, in Fig.10, new XOR function of (B_{41}, B_{42}) and (A_{41}, A_{42}) are introduced into the input blocks of function S-Box denoted by $(\#b2, \#b3)$ and $(\#b6, \#b7)$ and the value of the 1-byte blocks at $(A_{21} \oplus B_{21}\ ,\ A_{22} \oplus B_{22})$ and $(B_{21}\ ,\ B_{22})$ are replaced by $(A_{21} \oplus B_{21} \oplus A_{41},\ A_{22} \oplus B_{22} \oplus A_{42})$ and $(B_{21} \oplus B_{41},\ B_{22} \oplus B_{42})$, respectively.

According to the above simplification procedure, Fig.4 is equivalent to the pair of Fig.9 and Fig.10. Therefore, Fig.2 can be simplified into Fig.9 and Fig.10.

3.3 Calculation of plaintext

In this subsection, we give some equations that can compute input blocks from output blocks shown in Fig.1. By using these equations, the plaintext ((A_{00}, A_{01}, A_{02}, A_{03}), (B_{00}, B_{01}, B_{02}, B_{03})) can be computed from a given ciphertext (C_0, C_1, C_2, C_3, C_4, C_5, C_6, C_7).

From now on, let us call the one-way function of the MAP in Fig.1 the target function and call the MAP algorithm in Fig.2 simply the MAP. Let us consider the computational complexity of ciphertext only attack for the target function, which depends on the computational complexity of the brute force attack. The computational complexity of this attack can be described as the number of tried decryptions. Seeing Fig.4, obviously, the MAP does not have the one-way property, because the value of both A_0 blocks and B_0 blocks can be easily calculate from the value of both A_4 blocks and B_4 blocks. Seeing Fig.1, however, the target function seems to be practically secure against brute force attack, because the value of output blocks are given by the XOR of the input blocks (that is A_0 and B_0) and the output of the MAP (that is A_4 and B_4). So, the computational complexity of the brute force attack for the target function seems to be given by $O(2^{64})$. But, we show that the computational complexity to obtain input blocks(*i.e.* plaintext) from output blocks(*i.e.* ciphertext) can be decreased to $O(2^{32})$ from $O(2^{64})$.

Let us recall equation(2.1) and (2.2) about function S-box. Any one of three values x_1, x_2 and x_3 can be determined by the other two values. From now on, this relationship must be taken into consideration.

(1) [Relationship between ciphertext and plaintext]

Let eight 1-byte blocks of ciphertext be C_0, C_1, C_2, C_3, C_4, C_5, C_6 and C_7. Then, the following equations is valid in Fig. 4, 5, 6, 7, 8, 9 and 10:

$$A_{40} = C_0 \oplus A_{00}, \tag{3.1}$$
$$A_{41} = C_1 \oplus A_{01}, \tag{3.2}$$
$$A_{42} = C_2 \oplus A_{02}, \tag{3.3}$$
$$A_{43} = C_3 \oplus A_{03}, \tag{3.4}$$
$$B_{40} = C_4 \oplus B_{00}, \tag{3.5}$$
$$B_{41} = C_5 \oplus B_{01}, \tag{3.6}$$
$$B_{42} = C_6 \oplus B_{02} \quad and \tag{3.7}$$
$$B_{43} = C_7 \oplus B_{03}. \tag{3.8}$$

Now, suppose four 1-byte blocks, *i.e.* A_{00}, B_{00}, A_{20} and B_{20}, are given. Hereafter, $lot_2(x)$ denotes a 2-bit right rotation on 8-bit data x, and $\#ai$ and $\#bi$, $(i = 1, 2, \cdots, 8)$, denote the function S-Box in Fig.9 and Fig.10.

(2) [Relation of function S-Box (#a1, #a5, #b1, #b5)]

According to each of the function S-Box denoted by #a1, #a5, #b1 and #b5, we can obtain the following equations, (3.9), (3.10),(3.11) and (3.12).

$$\#a1 : A_{21} \oplus B_{01} = lot_2(B_{00} \oplus A_{20}) - A_{00} \oplus B_{00} \tag{3.9}$$
$$\#a5 : A_{21} \oplus B_{21} \oplus A_{01} \oplus B_{01} = lot_2(A_{00} \oplus B_{00} \oplus A_{20} \oplus B_{20}) - A_{20} \tag{3.10}$$
$$\#b1 : A_{21} \oplus B_{21} \oplus C_1 \oplus A_{01} = lot_2(C_0 \oplus A_{00} \oplus A_{20} \oplus B_{20}) - B_{20} \tag{3.11}$$
$$\#b5 : B_{21} \oplus C_5 \oplus B_{01} = lot_2(B_{20} \oplus C_4 \oplus B_{00}) - C_0 \oplus A_{00} \tag{3.12}$$

By equation (3.10) and (3.11) we obtain

$$B_{01} = \{lot_2(A_{00} \oplus B_{00} \oplus A_{20} \oplus B_{20}) - A_{20}\} \oplus \{lot_2(C_0 \oplus A_{00} \oplus A_{20} \oplus B_{20}) - B_{20}\} \oplus C_1 \tag{3.13}$$

By (3.9)

$$A_{21} = \{lot_2(B_{00} \oplus A_{20}) - A_{00} \oplus B_{00}\} \oplus B_{01} \tag{3.14}$$

By (3.12)

$$B_{21} = \{lot_2(B_{20} \oplus C_4 \oplus B_{00}) - C_0 \oplus A_{00}\} \oplus C_5 \oplus B_{01} \tag{3.15}$$

By (3.11)

$$A_{01} = \{lot_2(C_0 \oplus A_{00} \oplus A_{20} \oplus B_{20}) - B_{20}\} \oplus A_{21} \oplus B_{21} \oplus C_1 \tag{3.16}$$

Thus, we obtained four 1-byte blocks, *i.e.* A_{01}, B_{01}, A_{21} and B_{21}.

(3) [Relation of function S-Box (#a2, #a3, #a6, #a7, #b2, #b3, #b6, #b7)]

According to each of the function S-Box denoted by #a2, #a3, #a6, #a7, #b2, #b3, #b6 and #b7, we can obtain the following equations, (3.17), (3.18), (3.19), (3.20), (3.21), (3.22),(3.23) and (3.24).

$$\#a2: \{A_{02} \oplus B_{02}\} \oplus A_{03} \oplus B_{03} = lot_2(A_{21} \oplus B_{01}) - A_{00} \oplus B_{00} \oplus A_{01} \oplus B_{01} - 1 \quad (3.17)$$

$$\#a3: \{A_{02} \oplus B_{02}\} \oplus A_{03} \oplus B_{03} = lot_2(A_{22} \oplus B_{02}) - A_{21} \oplus B_{01} \quad (3.18)$$

$$\#a6: \{B_{02}\} \oplus A_{22} \oplus B_{02} \oplus A_{23}$$
$$= lot_2(A_{21} \oplus B_{21} \oplus A_{01} \oplus B_{01}) - \{B_{01}\} \oplus A_{20} \oplus A_{21} \oplus B_{01} - 1 \quad (3.19)$$

$$\#a7: \{B_{02}\} \oplus A_{22} \oplus B_{02} \oplus A_{23}$$
$$= lot_2(A_{22} \oplus B_{22} \oplus A_{02} \oplus B_{02}) - A_{21} \oplus B_{21} \oplus A_{01} \oplus B_{01} \quad (3.20)$$

$$\#b2: \{C_6 \oplus B_{02}\} \oplus B_{22} \oplus C_6 \oplus B_{02} \oplus B_{23}$$
$$= lot_2(A_{21} \oplus B_{21} \oplus C_1 \oplus A_{01}) - \{C_5 \oplus B_{01}\} \oplus B_{20} \oplus B_{21} \oplus C_5 \oplus B_{01} - 1 \quad (3.21)$$

$$\#b3: \{C_6 \oplus B_{02}\} \oplus B_{22} \oplus C_6 \oplus B_{02} \oplus B_{23}$$
$$= lot_2(A_{22} \oplus B_{22} \oplus C_2 \oplus A_{02}) - A_{21} \oplus B_{21} \oplus C_1 \oplus A_{01} \quad (3.22)$$

$$\#b6: \{C_2 \oplus A_{02}\} \oplus C_3 \oplus A_{03} = lot_2(B_{21} \oplus C_5 \oplus B_{01}) - C_0 \oplus A_{00} \oplus C_1 \oplus A_{01} - 1 \quad (3.23)$$

$$\#b7: \{C_2 \oplus A_{02}\} \oplus C_3 \oplus A_{03} = lot_2(B_{22} \oplus C_6 \oplus B_{02}) - B_{21} \oplus C_5 \oplus B_{01} \quad (3.24)$$

By the combinations of equations ((3.17), (3.18)), ((3.19), (3.20)), ((3.21), (3.22)) and ((3.23), (3.24)), we can obtain the following equations.

$$\#a2, \#a3: A_{22} \oplus B_{02}$$
$$= rot_2(lot_2(A_{21} \oplus B_{01}) - A_{00} \oplus B_{00} \oplus A_{01} \oplus B_{01} - 1 + A_{21} \oplus B_{01}) \quad (3.25)$$

$$\#a6, \#a7: A_{22} \oplus B_{22} \oplus A_{02} \oplus B_{02}$$
$$= rot_2(lot_2(A_{21} \oplus B_{21} \oplus A_{01} \oplus B_{01}) - A_{20} \oplus A_{21} - 1 + A_{21} \oplus B_{21} \oplus A_{01} \oplus B_{01}) \quad (3.26)$$

$$\#b2, \#b3: A_{22} \oplus B_{22} \oplus A_{02}$$
$$= C_2 \oplus rot_2(lot_2(A_{21} \oplus B_{21} \oplus C_1 \oplus A_{01})$$
$$- B_{20} \oplus B_{21} - 1 + A_{21} \oplus B_{21} \oplus C_1 \oplus A_{01}) \quad (3.27)$$

$$\#b6, \#b7: B_{22} \oplus B_{02}$$
$$= C_6 \oplus rot_2(lot_2(B_{21} \oplus C_5 \oplus B_{01}) - C_0$$
$$\oplus A_{00} \oplus C_1 \oplus A_{01} - 1 + B_{21} \oplus C_5 \oplus B_{01}) \quad (3.28)$$

By equation (3.26) and (3.27) we obtain

$$B_{02} = \{rot_2(lot_2(A_{21} \oplus B_{21} \oplus A_{01} \oplus B_{01}) - A_{20} \oplus A_{21} - 1 + A_{21} \oplus B_{21} \oplus A_{01} \oplus B_{01})\}$$
$$\oplus \{C_2 \oplus rot_2(lot_2(A_{21} \oplus B_{21} \oplus C_1 \oplus A_{01})$$
$$- B_{20} \oplus B_{21} - 1 + A_{21} \oplus B_{21} \oplus C_1 \oplus A_{01})\} \quad (3.29)$$

By (3.25)

$$A_{22} = \{rot_2(lot_2(A_{21} \oplus B_{01}) - A_{00} \oplus B_{00} \oplus A_{01} \oplus B_{01} - 1 + A_{21} \oplus B_{01})\} \oplus B_{02} \quad (3.30)$$

By (3.28)

$$B_{22} = \{C_6 \oplus rot_2(lot_2(B_{21} \oplus C_5 \oplus B_{01}) \\ -C_0 \oplus A_{00} \oplus C_1 \oplus A_{01} - 1 + B_{21} \oplus C_5 \oplus B_{01})\} \oplus B_{02} \quad (3.31)$$

By (3.27)

$$A_{02} = \{C_2 \oplus rot_2(lot_2(A_{21} \oplus B_{21} \oplus C_1 \oplus A_{01}) \\ -B_{20} \oplus B_{21} - 1 + A_{21} \oplus B_{21} \oplus C_1 \oplus A_{01})\} \oplus A_{22} \oplus B_{22} \quad (3.32)$$

Thus, we obtained four 1-byte blocks, *i.e.* A_{02}, B_{02}, A_{22} and B_{22}.

By (3.24)

$$A_{03} = \{lot_2(B_{22} \oplus C_6 \oplus B_{02}) - B_{21} \oplus C_5 \oplus B_{01}\} \oplus C_2 \oplus A_{02} \oplus C_3 \quad (3.33)$$

By (3.22)

$$B_{23} = \{lot_2(A_{22} \oplus B_{22} \oplus C_2 \oplus A_{02}) - A_{21} \oplus B_{21} \oplus C_1 \oplus A_{01}\} \oplus B_{22} \quad (3.34)$$

By (3.20)

$$A_{23} = \{lot_2(A_{22} \oplus B_{22} \oplus A_{02} \oplus B_{02}) - A_{21} \oplus B_{21} \oplus A_{01} \oplus B_{01}\} \oplus A_{22} \quad (3.35)$$

By (3.18)

$$B_{03} = \{lot_2(A_{22} \oplus B_{02}) - A_{21} \oplus B_{01}\} \oplus A_{02} \oplus B_{02} \oplus A_{03} \quad (3.36)$$

We obtained four 1-byte blocks, *i.e.* A_{03}, B_{03}, A_{23} and B_{23}.

(4) [Relation of function S-Box (#a4,#a8,#b4,#b8)]

According to each of the function S-Box denoted by #a4, #a8, #b4 and #b8, the value of A_{02}, B_{02}, A_{22}, B_{22}, A_{03}, B_{03}, A_{23} and B_{23} must satisfy the following four equations, (3.37), (3.38),(3.39) and (3.40).

$$\#a4: A_{22} \oplus B_{02} = lot_2(A_{23} \oplus B_{03}) - A_{03} \oplus B_{03} - 1 \quad (3.37)$$
$$\#a8: A_{22} \oplus B_{22} \oplus A_{02} \oplus B_{02} = lot_2(A_{23} \oplus B_{23} \oplus A_{03} \oplus B_{03}) - A_{23} - 1 \quad (3.38)$$
$$\#b4: A_{22} \oplus B_{22} \oplus C_2 \oplus A_{02} = lot_2(C_3 \oplus A_{03} \oplus A_{23} \oplus B_{23}) - B_{23} - 1 \quad (3.39)$$
$$\#b8: B_{22} \oplus C_6 \oplus B_{02} = lot_2(B_{23} \oplus C_7 \oplus B_{03}) - C_3 \oplus A_{03} - 1 \quad (3.40)$$

This means that these equations are used for checks on the correctness of given value of A_{00}, B_{00}, A_{20}, B_{20}.

According to the above discussion, twelve 1-byte blocks, *i.e.* A_{01}, A_{02}, A_{03}, B_{01}, B_{02}, B_{03}, A_{21}, A_{22}, A_{23}, B_{21}, B_{22} and B_{23}, can be computed when one ciphertext (C_0, C_1, C_2, C_3, C_4, C_5, C_6 and C_7) is given, on the assumption that four 1-byte blocks (A_{00}, B_{00}, A_{20} and B_{20}) are given.

Therefore, taking the equations (3.1 - 3.8) into consideration, when one ciphertext (C_i), ($i = 0, \cdots, 7$), is given and that four 1-byte blocks (A_{00}, B_{00}, A_{20} and B_{20}) are presumed, we can compute twenty four 1-byte blocks, *i.e.* A_{00}, A_{01}, A_{02}, A_{03}, B_{00}, B_{01}, B_{02}, B_{03}, A_{20}, A_{21}, A_{22}, A_{23}, B_{20}, B_{21}, B_{22}, B_{23}, A_{40}, A_{41}, A_{42}, A_{43}, B_{40}, B_{41}, B_{42} and B_{43}, that satisfy the relation shown in Fig.9 and Fig.10.

3.4 Example of computation

We perform the calculation with the ciphertext given as:

$$(C_i) = (da, 8a, b9, db, 34, d1, 59, 94), \quad (i = 0, \cdots, 7).$$

under relationship of the equations shown in the previous subsection3.3.

Three patterns of the computed plaintext are obtained as follows:

$$\begin{aligned}(A_0, B_0) &= (48, 41, 53, 48, 44, 41, 54, 41)\\(A_2, B_2) &= (a5, 2d, bd, 8e, f7, df, 2e, 0b)\\(A_4, B_4) &= (92, cb, ea, 93, 70, 90, 0d, d5),\end{aligned}$$

$$\begin{aligned}(A_0, B_0) &= (4f, 1b, 01, 40, 69, 2b, 12, 08)\\(A_2, B_2) &= (d6, e2, 82, 6f, 89, 5a, 44, 32)\\(A_4, B_4) &= (95, 91, b8, 9b, 5d, fa, 4b, 9c)\end{aligned}$$

and

$$\begin{aligned}(A_0, B_0) &= (55, 20, ac, 80, 37, fa, 48, d7)\\(A_2, B_2) &= (7b, 4b, d9, 70, 82, fa, 6d, 20)\\(A_4, B_4) &= (8f, aa, 15, 5b, 03, 2b, 11, 43).\end{aligned}$$

It took 15.7 hours to compute the above case, with C program on a workstation (SUN-SPARCstation 2).

The obtained plaintext is not unique. But only a few plaintexts satisfy the algorithm of the one-way function of the MAP from (1) the fact that the MAP algorithm has the one to one correspondence[1] and (2) the result of the cycling closure test of FEAL[12], that has a very similar algorithm to this one-way function of the MAP.

4 Concluding Remarks

We studied into the method which improves FEAL algorithm against cryptanalysis based on the careful analysis of data randomizer. Since the round function f of FEAL has an inverse function f^{-1}, we studied an attack using the inverse function f^{-1} and key-independent intermediate messages. In this paper, we present an example of the application of our method to the study of ciphertext only attack for a one-way function that employs the mapping function called MAP. The algorithm of this MAP function is very similar to that of FEAL data randomizer from the view point of round function f.

Simulation results of the proposed attack have shown that the the computational complexity to obtain the plaintext of the one-way function of the MAP[1] is decreased to $O(2^{32})$ from $O(2^{64})$. Then, the inverse function of this one-way function can be practically breakable by the proposed attack.

15.7 hours are required to compute our example, with C program on a workstation(SUN-SPARCstation 2). There are three computed plaintext in this example. The obtained plaintext is not unique. But we can restrict the plaintext within narrow limits.

5 References

[1] Miyaguchi, S. : " CRITERIA FOR THE STRENGTH OF ENCIPHERMENT AND STANDARDIZATION FOR CRYPTOGRAPHIC TECHNIQUES " , Proc. of the 1986 Symposium on Cryptography and Information Security, (1986).

[2] Shimizu, A., Miyaguchi, S. :" Fast Data Encipherment Algorithm FEAL " , Advances in Cryptology - EUROCRYPT'87, p267-278 (1987).

[3] E. Biham, A. Shamir :" Differential Cryptanalysis of Feal and N-Hash " , Advances in Cryptology - EUROCRYPT'91, p1-16 (1991).

[4] H. Gilbert, G. Chasse : " A STATISTICAL ATTACK OF THE FEAL-8 CRYPTOSYSTEM", Advances in Cryptology - CRYPTO'90, p22-33 (1990).

[5] A. TARDY-CORFDIR, H. GILBERT : " A KNOWN PLAINTEXT ATTACK OF FEAL-4 AND FEAL-6 " , Advances in Cryptology - CRYPTO'91, p172-182 (1991).

[6] Kaneko, T. :" A known plaintext cryptanalytic attack on FEAL-4 " ,IEICE Tech. Report, ISEC91-25 (1991) (in Japanese).

[7] Matsui, M., Yamagishi, A. :" A Study on Known Text Attack of Involution-Type Cryptosystems" ,IEICE Tech. Report, ISEC91-26 (1991) (in Japanese).

[8] Tsujii, S., Kasahara, M. :" Cryptography and Information Security " Shoh-Kohdo (1990) (in Japanese).

[9] Matsui, K. :" Introduction of Cryptoanalysis " Morikita Publishing Co. (1990) (in Japanese).

[10] Tsunoo, Y., Okamoto, E., Doi, H. : " Analytical Known Plain-text Attack for FEAL-4 and Its Improvement " , Proc. of the 1993 Symposium on Cryptography and Information Security, SCIS93-3A (1993) (in Japanese).

[11] Okamoto, E. : " An Introduction to Theory of Cryptology " Kyoritsu Publishing Co. (1993) (in Japanese).

[12] Kazuo, N. :" Cycling closure test of FEAL " IEICE Tech. Report, ISEC90-50 (1990) (in Japanese).

Pitfalls in Designing Substitution Boxes (Extended Abstract)

Jennifer Seberry, Xian-Mo Zhang and Yuliang Zheng

Department of Computer Science
University of Wollongong, Wollongong, NSW 2522, Australia
{jennie, xianmo, yuliang}@cs.uow.edu.au

Abstract. Two significant recent advances in cryptanalysis, namely the differential attack put forward by Biham and Shamir [3] and the linear attack by Matsui [7, 8], have had devastating impact on data encryption algorithms. An eminent problem that researchers are facing is to design S-boxes or substitution boxes so that an encryption algorithm that employs the S-boxes is immune to the attacks. In this paper we present evidence indicating that there are many pitfalls on the road to achieve the goal. In particular, we show that certain types of S-boxes which are seemly very appealing do not exist. We also show that, contrary to previous perception, techniques such as chopping or repeating permutations do *not* yield cryptographically strong S-boxes. In addition, we reveal an important combinatorial structure associated with certain quadratic permutations, namely, the difference distribution table of each differentially 2-uniform quadratic permutation embodies a Hadamard matrix. As an application of this result, we show that chopping a differentially 2-uniform quadratic permutation results in an S-box that is very prone to the differential cryptanalytic attack.

1 Basic Definitions

Denote by V_n the vector space of n tuples of elements from $GF(2)$. Let $\alpha = (a_1, \ldots, a_n)$ and $\beta = (b_1, \ldots, b_n)$ be two vectors in V_n. The scalar product of α and β, denoted by $\langle \alpha, \beta \rangle$, is defined by $\langle \alpha, \beta \rangle = a_1 b_1 \oplus \cdots \oplus a_n b_n$, where multiplication and addition are over $GF(2)$. In this paper we consider Boolean functions from V_n to $GF(2)$ (or simply functions on V_n).

Let f be a function on V_n. The $(1, -1)$-sequence defined by $((-1)^{f(\alpha_0)}, (-1)^{f(\alpha_1)}, \ldots, (-1)^{f(\alpha_{2^n-1})})$ is called the *sequence* of f, and the $(0, 1)$-sequence defined by $(f(\alpha_0), f(\alpha_1), \ldots, f(\alpha_{2^n-1}))$ is called the *truth table* of f, where $\alpha_0 = (0, \ldots, 0, 0)$, $\alpha_1 = (0, \ldots, 0, 1)$, ..., $\alpha_{2^n-1} = (1, \ldots, 1, 1)$. f is said to be *balanced* if its truth table has 2^{n-1} zeros (ones).

An *affine* function f on V_n is a function that takes the form of $f = a_1 x_1 \oplus \cdots \oplus a_n x_n \oplus c$, where $a_j, c \in GF(2)$, $j = 1, 2, \ldots, n$. Furthermore f is called a *linear* function if $c = 0$. The sequence of an affine (or linear) function is called an *affine (or linear) sequence*.

The *Hamming weight* of a vector $\alpha \in V_n$, denoted by $W(\alpha)$, is the number of ones in the vector.

A $(1,-1)$-matrix H of order m is called a *Hadamard* matrix if $HH^t = mI_m$, where H^t is the transpose of H and I_m is the identity matrix of order m. A *Sylvester-Hadamard matrix* or *Walsh-Hadamard matrix* of order 2^n, denoted by H_n, is generated by the following recursive relation

$$H_0 = 1,\ H_n = \begin{bmatrix} H_{n-1} & H_{n-1} \\ H_{n-1} & -H_{n-1} \end{bmatrix},\ n = 1, 2, \ldots.$$

Now we introduce bent functions, an important combinatorial concept discovered by Rothaus in the mid 1960's, although his pioneering work was not published until some ten years later [14].

Definition 1. A function f on V_n is said to be bent if

$$2^{-\frac{n}{2}} \sum_{x \in V_n} (-1)^{f(x) \oplus \langle \beta, x \rangle} = \pm 1$$

for every $\beta \in V_n$. Here $x = (x_1, \ldots, x_n)$ and $f(x) \oplus \langle \beta, x \rangle$ is considered as a real valued function.

Bent functions can be characterized in various ways. In particular, the following statements are equivalent (see also [6]):

(i) f is bent.
(ii) $\langle \xi, \ell \rangle = \pm 2^{\frac{1}{2}n}$ for any affine sequence ℓ of length 2^n, where ξ is the sequence of f.
(iii) $f(x) \oplus f(x \oplus \alpha)$ is balanced for any non-zero vector $\alpha \in V_n$, where $x = (x_1, \ldots, x_n)$.

An $n \times s$ S-box or substitution box is a mapping from V_n to V_s, where $n \geq s$. Now we consider a nonlinearity criterion that measures the strength of an S-box against differential cryptanalysis [3, 4]. The essence of a differential attack is that it exploits particular entries in the difference distribution tables of S-boxes employed by a block cipher. The difference distribution table of an $n \times s$ S-box is a $2^n \times 2^s$ matrix. The rows of the matrix, indexed by the vectors in V_n, represent the change in the input, while the columns, indexed by the vectors in V_s, represent the change in the output of the S-box. An entry in the table indexed by (α, β) indicates the number of input vectors which, when changed by α (in the sense of bit-wise XOR), result in a change in the output by β (also in the sense of bit-wise XOR).

Note that an entry in a difference distribution table can only take an even value, the sum of the values in a row is always 2^n, and the first row is always $(2^n, 0, \ldots, 0)$. As entries with higher values in the table are particularly useful to differential cryptanalysis, a necessary condition for an S-box to be immune to differential cryptanalysis is that it does not have large values in its differential distribution table (not counting the first entry in the first row).

Definition 2. Let F be an $n \times s$ S-box, where $n \geq s$. Let δ be the largest value in differential distribution table of the S-box (not counting the first entry in the first row), namely,

$$\delta = \max_{\alpha \in V_n, \alpha \neq 0} \max_{\beta \in V_s} |\{x | F(x) \oplus F(x \oplus \alpha) = \beta\}|.$$

Then F is said to be *differentially δ-uniform*, and accordingly, δ is called the differential uniformity of f.

Obviously the differential uniformity δ of an $n \times s$ S-box is constrained by $2^{n-s} \leq \delta \leq 2^n$. Extensive research has been carried out in constructing differentially δ-uniform S-boxes with a low δ [1, 13, 2, 9, 10, 11, 12]. Some constructions, in particular those based on permutation polynomials on finite fields, are simple and elegant. However, caution must be taken with Definition 2. In particular, it should be noted that low differential uniformity (a small δ) is only a *necessary*, but not a *sufficient* condition for immunity to differential attacks. This is shown by the fact that S-boxes constructed in [1, 9], which have a flat difference distribution table, are extremely weak to differential attacks, despite that they achieve the lowest possible differential uniformity $\delta = 2^{n-s}$ [4, 5, 15]. A more complete measurement that takes into account the number of nonzero entries in the first column of a difference distribution table is the *robustness* introduced in [15].

Definition 3. Let $F = (f_1, \ldots, f_s)$ be an $n \times s$ S-box, where f_i is a function on V_n, $i = 1, \ldots, s$, and $n \geq s$. Denote by L the largest value in the difference distribution table of F, and by N the number of nonzero entries in the first column of the table. In either case the value 2^n in the first row is not counted. Then we say that F is R-robust against differential cryptanalysis, where R is defined by

$$R = (1 - \frac{N}{2^n})(1 - \frac{L}{2^n}).$$

Robustness gives more accurate information about the strength of an S-box against the differential attack than differential uniformity does. However, differential uniformity has an advantage over robustness in that the former is easier to discuss than the latter. For this reason, differential uniformity is employed as the first indicator for the strength of an S-box against the differential attack, while robustness is considered when more complete information about the strength is needed.

An $n \times s$ S-box $F = (f_1, \ldots, f_s)$ is said to be *regular* if F runs through each vector in V_s 2^{n-s} times while x runs through V_n once. S-boxes employed by a block cipher must be regular, since otherwise the cipher would be prone to statistical attacks. For a regular $n \times s$ S-box, its differential uniformity is larger than 2^{n-s} (see also Lemma 2 of [17]). The robustness of the S-box is further determined by the number of nonzero entries in the first column of the table.

We are particularly interested in $n \times s$ S-boxes that have the following property: for any nonzero vector $\alpha \in V_n$, $F(x) \oplus F(x \oplus \alpha)$ runs through half of the vectors in V_s, each 2^{n-s+1} times, but not through the other half of the vectors in V_n. With each row in the difference distribution table of such an S-box, half of its entries contain a value 2^{n-s+1} while the other half contain a value zero. For simplicity, we say such a difference distribution table to be *uniformly half-occupied*. Clearly an $n \times s$ S-box with a UHODDT or uniformly half-occupied difference distribution table achieves the differential uniformity of 2^{n-s+1}. In Theorem 3 of [17], it has been proved that for quadratic S-boxes, 2^{n-s+1} is the lower bound on differential uniformity.

Note that a differentially 2-uniform permutation is also a permutation with a UHODDT, and vice versa. These permutations have many nice properties [13, 2, 9, 10, 11, 12]. In particular, they achieve the highest possible robustness against the differential attack. The concept of $n \times s$ S-boxes with a UHODDT can be viewed as a generalization of differentially 2-uniform permutations. Hence $n \times s$ S-boxes with a UHODDT are very appealing and have received extensive research (see for instance [2]).

There are two important questions about S-boxes with a UHODDT, namely

(i) Do there exist S-boxes with a UHODDT ? If there do, how to construct them ?

(ii) What is the robustness of an S-box with a UHODDT ?

When $n = s$, the answer to the first question is "yes". It has been shown in [13, 11, 2] that certain permutation polynomials on $GF(2^n)$, n odd, have a UHODDT. So far no result has been known regarding the case of $n > s$. In Section 2, we will partially solve the problem by showing that there exist no quadratic $n \times s$ S-boxes with a UHODDT, if either n or s is even. The second question will be discussed in Section 3. We will prove that the robustness of an S-box with a UHODDT is very low.

Another important question is the synthesis of S-boxes, namely

(iii) How to construct S-boxes from existing ones ?

This question will be discussed in Section 4. We will show that many synthesis methods *which were previously taken for granted*, in fact do *not* yield strong S-boxes, even though the starting S-boxes employed are all strong ones. Section 5 is solely devoted to the investigation of combinatorial properties of the differential distribution table of an quadratic permutation. We reveal a result that is very interesting even from the point of view of pure combinatorics, namely, every uniformly half-occupied difference distribution table of a quadratic permutation embodies a Sylvester-Hadamard matrix.

2 Nonexistence of Certain Quadratic S-boxes

2.1 On Quadratic S-boxes with a UHODDT

As mentioned in the previous section, an $n \times s$ S-box with a UHODDT or uniformly half-occupied difference distribution table achieves the differential uniformity of 2^{n-s+1}, and for quadratic S-boxes, 2^{n-s+1} is the lower bound on differential uniformity. In the following we show an impossibility result, namely, there exist no quadratic S-boxes that have a UHODDT if either n or s is even.

Assume that $F = (f_1, \ldots, f_s)$ is a quadratic $n \times s$ S-box with a UHODDT, where $n > s$. We prove that neither n nor s can be even.

Recall that a vector $\alpha \in V_n$ is called a *linear structure* of a function f on V_n if $f(x) \oplus f(x \oplus \alpha)$ is a constant. The set of the linear structures of f forms a linear subspace. The dimension of the subspace is called the *linearity dimension* of f. Let $\alpha_1, \ldots, \alpha_{2^n-1}$ be the $2^n - 1$ nonzero vectors in V_n and $g_1, \ldots, g_{2^s-1}$ be the $2^s - 1$ nonzero linear combinations of $f_1, \ldots, f_s$. We construct a bipartite graph whose vertices comprise $\alpha_1, \ldots, \alpha_{2^n-1}$ on one side and $g_1, \ldots, g_{2^s-1}$ on the other side. An edge or link between α_i and g_j exists if and only if α_i is a linear structure of g_j.

Theorem 2 of [17] states that $n - \ell_i$ is even, where ℓ_i is the linearity dimension of g_i. Equivalently, n and ℓ_i must be both even or both odd. Since each g_i is balanced, it can not be bent. By Lemma 5 of [17], a quadratic function is bent if and only if it does not have linear structures. Hence we have $\ell_i \geq 1$. On the other hand, from the proof for Corollary 1 of [17], we have $\ell_i \leq n - 2$. We distinguish the following two cases:

Case 1: n is odd and ℓ_i is $1, 3, 5, \ldots,$ or $n - 2$.

Case 2: n is even and ℓ_i is $2, 4, 6, \ldots,$ or $n - 2$.

First we consider Case 1. Let p_j denote the number of ℓ_i, $1 \leq i \leq 2^n - 1$, such that $\ell_i = j$. Then we have a sequence of numbers $p_1, p_3, p_5, \ldots, p_{n-2}$. Obviously,

$$p_1 + p_3 + p_5 + \cdots + p_{n-2} = 2^s - 1. \tag{1}$$

Since F is a S-box with a UHODDT, for any nonzero vector $\alpha_k \in V_n$

$$F(x) \oplus F(x \oplus \alpha_k) = (f_1(x) \oplus f_1(x \oplus \alpha_k), \ldots, f_s(x) \oplus f_s(x \oplus \alpha_k))$$

is not regular. Thus, by Lemma 6, there exists a linear combination of $f_1(x) \oplus f_1(x \oplus \alpha_k), \ldots, f_s(x) \oplus f_s(x \oplus \alpha_k)$, say $g_j(x) \oplus g_j(x \oplus \alpha_k)$, such that $g_j(x) \oplus g_j(x \oplus \alpha_k)$ is not balanced. Since $g_j(x) \oplus g_j(x \oplus \alpha_k)$ is affine, $g_j(x) \oplus g_j(x \oplus \alpha_k)$ must be constant. This proves that any nonzero vector $\alpha_k \in V_n$ is a linear structure of a g_j, a linear combination of $f_1, \ldots, f_s$. On the other hand, by Theorem 4 of [17], for each α_k, there exists at most one g_j among $g_1, \ldots, g_{2^s-1}$ such that α_k is a linear structure of g_j. By the construction of the bipartite graph, each α_k is linked to a unique g_j. Also each g_i with $\ell_i = j$ has j linearly independent linear structures and $2^j - 1$ nonzero linear structures. Hence we have

$$(2^1 - 1)p_1 + (2^3 - 1)p_3 + (2^5 - 1)p_5 + \cdots + (2^{n-2} - 1)p_{n-2} = 2^n - 1. \tag{2}$$

From (1) and (2) we have

$$(2^1 - 2)p_1 + (2^3 - 2)p_3 + (2^5 - 2)p_5 + \cdots + (2^{n-2} - 2)p_{n-2} = 2^n - 2^s$$

or equivalently

$$(2^2 - 1)p_3 + (2^4 - 1)p_5 + \cdots + (2^{n-3} - 1)p_{n-2} = 2^{s-1}(2^{n-s} - 1) \tag{3}$$

Note that $2^k - 1$ is divisible by 3 if and only $k \geq 2$ is even. Thus the left hand side of (3) is divisible by 3. This implies that the $(2^{n-s} - 1)$ part in the right hand side of the equation is divisible by 3. Hence s must be odd. Thus there exists no quadratic $n \times s$ S-box with a UHODDT if n is odd ($n \geq 5$) and s is even.

We now consider Case 2. Let q_j denote the number of ℓ_i, $1 \leq i \leq 2^n - 1$, such that $\ell_i = j$. Similarly to Case 1, we have a sequence of numbers $q_2, q_4, q_6, \ldots, q_{n-2}$, and

$$q_2 + q_4 + q_6 \cdots + q_{n-2} = 2^s - 1,$$

$$(2^2 - 1)q_2 + (2^4 - 1)q_4 + (2^6 - 1)q_6 + \cdots + (2^{n-2} - 1)q_{n-2} = 2^n - 1.$$

By simple deduction,

$$(2^3 - 2)q_4 + (2^5 - 2)q_6 + \cdots + (2^{n-3} - 2)q_{n-2} = 2^{n-1} - 3 \cdot 2^{s-1} + 1. \quad (4)$$

It is not hard to see that the left hand side of (4) is even when $n \geq 4$, while the right hand side of (4) is always odd for $s \geq 2$. From this we can conclude that there exists no quadratic $n \times s$ S-box with a UHODDT if n is even with $n \geq 4$.

Summarizing Case 1 and Case 2, we have

Theorem 4. *For $n \geq 4$, there exists no quadratic $n \times s$ S-box with a UHODDT if either n or s is even.*

Theorem 4 can be viewed as an extension of Corollary 2 in [17], which states that there exists no differentially 2-uniform quadratic permutation on an even dimensional vector space.

By Theorem 4, $n \times s$ S-boxes with a UHODDT do not exist if either n or s is even. When n is odd and $n = s$, as mentioned before, we do have differentially 2-uniform quadratic permutation [13, 2, 11]. Thus a problem that is left open is whether there are quadratic S-boxes with a UHODDT for $n > s$, both n and s odd. It should be pointed out that an S-box which has an odd number of input bits and also an odd number of output bits may not be very useful in practice.

2.2 An Extension

The result in the previous subsection can be extended to a special kind of differentially 2^{n-s+t}-uniform quadratic S-boxes. Let F be a $n \times s$ S-box such that for any nonzero vector $\alpha \in V_n$, $F(x) \oplus F(x \oplus \alpha)$ runs through 2^{s-t} vectors in V_s, each 2^{n-s+t} times, but not through the remaining $2^s - 2^{s-t}$ vectors in V_s, where $t \geq 1$. The case when $t = 1$ has been discussed in the previous subsection. In the following we present a nonexistence result on the case when $t > 1$.

Theorem 5. *If n is odd and t is even, there exists no quadratic $n \times s$ S-boxes such that for any nonzero vector $\alpha \in V_n$, $F(x) \oplus F(x \oplus \alpha)$ runs through 2^{s-t} vectors in V_s, each 2^{n-s+t} times, but not through the remaining vectors in V_s.*

The proof will be provided in the full version.

3 Columns of a UHODDT

In the previous section we proved that there does not exist a quadratic $n \times s$ S-box with a UHODDT if either n or s is even. It is not clear whether or not higher degree S-boxes with a UHODDT exist. If there do exist such S-boxes, we would like to know whether or not they satisfy a more stringent requirement, namely high robustness. Results to be shown below give a negative answer to the question.

The following lemma is exactly the same as Theorem 1 of [17].

Lemma 6. *Let $F = (f_1, \ldots, f_s)$ be a mapping from V_n to V_s, where each f_j is a function on V_n. Then F is regular if and only if each nonzero linear combination of $f_1, \ldots, f_s$ is balanced.*

It is easy to show that the profile of the difference distribution table of an S-box is not changed by a nonsingular linear transformation on input coordinates (see for instance [2, 17]). In particular we have

Lemma 7. *Let $F = (f_1, \ldots, f_s)$ be a regular S-box with a UHODDT or uniformly half-occupied difference distribution table. Let A be a nonsingular matrix of order n and B a nonsingular matrix of order s over $GF(2)$. Then both Let $G(x) = F(xA) = (f_1(xA), \ldots, f_n(xA))$ and $H(x) = F(x)B = (f_1(x), \ldots, f_n(x))B$ are regular S-boxes with a UHODDT.*

By definition, each row in a uniformly half-occupied difference distribution table, except the first, contains an equal number of zero and nonzero entries. The following lemma shows that a similar result holds with columns in the table.

Lemma 8. *Let F be a regular $n \times s$ S-box with a UHODDT. Then each column, except the first, in the difference distribution table contains an equal number of zero and nonzero entries.*

Proof. We prove that for each nonzero $\beta \in V_s$, there exist 2^{n-1} nonzero $\alpha \in V_n$ such that $F(x) \oplus F(x \oplus \alpha) = \beta$ has solutions for x.

Fix $x_0 \in V_n$. Since the difference distribution table of F is uniformly half-occupied, $F(x_0) \oplus F(x_0 \oplus \alpha)$ runs through each nonzero $\beta \in V_s$ 2^{n-s} times while α runs through V_n. As x_0 is arbitrary, for each nonzero $\beta \in V_s$, there exist $2^n \cdot 2^{n-s}$ pairs (x, α) such that $F(x) \oplus F(x \oplus \alpha) = \beta$, where $\alpha \neq 0$. On the other hand, since the difference distribution table of F is uniformly half-occupied, $F(x) \oplus F(x \oplus \alpha) = \beta$ either has 2^{n-s+1} solutions or has no solution for x. Thus for each nonzero $\beta \in V_s$ there exist $2^n \cdot 2^{n-s}/2^{n-s+1} = 2^{n-1}$ nonzero vectors $\alpha \in V_n$ such that $F(x) \oplus F(x \oplus \alpha) = \beta$ has solutions for x.

Recall that the robustness of an S-box is determined by the largest value in the difference distribution table of the S-box, and also by the number of nonzero entries in the first column of the table. The lemma described below gives the precise number of nonzero entries in the first column of a uniformly half-occupied difference distribution table.

Lemma 9. *Let F be a regular $n \times s$ S-box with a UHODDT. Then there are $2^{n-1} - 2^{s-1}$ nonzero entries in the first column of the difference distribution table (excluding the first entry).*

As an immediate consequence of Lemma 9, we obtain the robustness of an S-box with a UHODDT:

$$R = [1 - (2^{n-1} - 2^{s-1})/2^n](1 - 2^{n-s+1}/2^n) = (1/2 + 2^{-n+s-1})(1 - 2^{-s+1}).$$

When $n = s$, we have $R = 1 - 2^{-n+1}$, which is the highest possible value for robustness. However, when s is relatively smaller than n, say $n - s > 3$, R is very close to $1/2$. For comparison, we note that the robustness of S-boxes constructed in [15] is at least $7/8$.

4 On Methods for Synthesizing S-boxes

This section is concerned with methods for constructing S-boxes from existing ones. We show that a number of techniques which were previously taken for granted do not yield good S-boxes.

4.1 Chopping Permutations

Chopping permutations which are cryptographically strong has been conceived as a promising method to construct S-boxes for DES-like encryption algorithms. For this reason, many researchers have focused their attention on permutations, especially those on a finite field [2, 9, 10, 11, 12]. Results to be present in this subsection indicate that, contrary to the common perception, this practice does not produce good S-boxes.

First we prove the following:

Theorem 10. *Let $F = (f_1, \ldots, f_s)$ be a regular $n \times s$ S-box with a UHODDT, where $n \geq s$ and each f_j is a function on V_n. The following two statements hold:*

(i) Let $1 \leq t \leq s-1$ and let G be an S-box obtained by dropping $s-t$ component functions from F, say $G = (f_1, \ldots, f_t)$. Then the difference distribution table of G is not uniformly half-occupied.

(ii) Let $n \geq t \geq s+1$ and let H be an S-box obtained by adding $t-s$ component functions to F, say $H = (f_1, \ldots, f_s, f_{s+1}, \ldots, f_t)$, where $f_{s+1}, \ldots, f_t$ are newly added. Then the difference distribution table of H is not uniformly half-occupied.

Proof. (i) Since F has a UHODDT, for any nonzero $\alpha \neq 0$, $F(x) \oplus F(x \oplus \alpha)$ runs through 2^{s-1} vectors in V_s, each 2^{n-s+1} times, but not through the other 2^{s-1} vectors in V_s, while α runs through V_n. Fix a nonzero vector, say $\gamma = (0, \beta) \in V_s$, where 0 is the zero vector in V_t and β is a nonzero vector in V_{s-t}. By Lemma 8 there exist 2^{n-1} nonzero vector α such that $F(x) \oplus F(x \oplus \alpha) = \gamma$ has solutions for x. Thus there exist 2^{n-1} nonzero vector α such that $G(x) \oplus G(x \oplus \alpha) = 0$, where

0 is the zero vector in V_t, has solutions for x. It is easy to show that G is not uniformly half-occupied. Since G is regular there exist $2^{n-1}-2^{t-1}$ nonzero vector α such that $G(x) \oplus G(x \oplus \alpha) = 0$ (see Lemma 8) if G is uniformly half-occupied.

(ii) follows (i).

From Theorem 10 chopping a regular S-box with a UHODDT does not yield a regular S-box with a UHODDT. In particular, chopping a differentially 2-uniform permutation on V_n does not produce an S-box with a UHODDT.

As quadratic permutations with a UHODDT or differentially 2-uniform quadratic permutations have been studied very extensively, an important problem is about the structure of the difference distribution table of an S-box obtained by chopping such a permutation. We will devote a single section, Section 5, to this topic.

In addition to chopping permutations, other techniques, such as linear transforms or modulo operations on inputs or outputs of differentially 2-uniform permutations, and repeating differentially 2-uniform permutations, are also conceived as possible S-box synthesis methods. In the following we show that none of these methods generates an S-box with a UHODDT.

4.2 Linear Transforms Applied on Inputs

Let F be a differentially 2-uniform permutation on V_s, B a matrix of order $n \times s$ $(n > s)$ over $GF(2)$. Set $G(y) = F(yB)$ where $y \in V_n$. Since the rank of B is s, yB runs through 2^s vectors in V_s each 2^{n-s} times while y runs through V_n. Since F is a permutation on V_s, $G(y)$ is a regular $n \times s$ S-box.

Unfortunately the difference distribution table of $G(y)$ is not uniformly half-occupied. The reason is described in the following. Since $n > s$ there exists a nonzero vector, say β, such that $\beta B = 0$, where 0 is the zero vector in V_s. Note that $G(y) \oplus G(y \oplus \beta) = F(yB) \oplus F((y \oplus \beta)B) = F(yB) \oplus F(yB \oplus \beta B) = F(yB) \oplus F(yB) = 0$, where 0 is the zero vector in V_s, for every $y \in V_n$.

4.3 Linear Transforms Applied on Outputs

Let F be a differentially 2-uniform permutation on V_s, and B a matrix of order $n \times s$ $(n > s)$ over $GF(2)$. Set $G(x) = F(x)B$. Note that the rank of B is s. Hence yB runs through 2^s vectors in V_s each 2^{n-s} times while y runs through V_n. As F is a permutation on V_n, G is a regular $n \times s$ S-box.

Since $n > s$, there exists a matrix of order $n \times (n-s)$, say D, such that the matrix $A = [BD]$ of order n is nonsingular. Set $\Psi(x) = F(x)A$. By Lemma 7, Ψ is a also a differentially 2-uniform permutation. By Theorem 10, G is not an S-box with a UHODDT.

4.4 Connecting Permutations in Parallel

Let F be a differentially 2-uniform permutation on V_s. Set

$$G(y) = (1 \oplus x_{s+1})F(x) \oplus x_{s+1}F(x \oplus \alpha)$$

where $x = (x_1, \ldots, x_s)$, $y = (x_1, \ldots, x_s, x_{s+1})$, $\alpha \in V_s$. Note that $G(x, 0) = F(x)$, $G(x, 1) = F(x \oplus \alpha)$. Since F is permutation on V_s G is a regular $(s + 1) \times s$ S-box.

Let $\beta = (\alpha, 1)$. Clearly $G(y \oplus \beta) = G(y)$ for every $y \in V_{s+1}$. Thus $G(y) \oplus G(y \oplus \beta) = 0$, where 0 is the zero vector in V_s, for every $y \in V_n$. Thus the difference distribution is very bad in this case, and $G(y)$ is not an S-box with a UHODDT.

The above discussions can be extended to the general case where F is repeated 2^k times, $k \geq 1$.

4.5 Enlarging Inputs or Reducing Outputs by Modulo Operations

Let $\alpha = (a_1, \ldots, a_n) \in V_n$. Rewrite α as $\alpha = a_1 \oplus a_2 x \oplus \cdots \oplus a_n x^{n-1}$. Thus V_n and the set of polynomials of degree at most $n-1$ over $GF(2)$ have a one-to-one correspondence. Let $\sigma(x)$ be a primitive polynomial of degree s $(s < n)$. For any $\alpha \in V_n$, we have

$$\alpha = h\sigma \oplus \overline{\alpha}$$

where the degree of h is less than or equal to $n - s - 1$, the degree of $\overline{\alpha}$ is less than s. Thus we have defined a mapping from V_n to V_s: $\alpha \rightarrow \overline{\alpha}$.

Now let ξ be a vector in V_n and $\overline{\xi}$ a vector in V_s. Let $F(\overline{\xi})$ be a differentially 2-uniform permutation on V_s. Set $G(\xi) = F(\overline{\xi})$. This gives an $n \times s$ S-box. Note that $\overline{\xi \oplus \eta} = \overline{\xi} \oplus \overline{\eta}$. This means that the mapping from V_n to V_s, $\alpha \rightarrow \overline{\alpha}$, is linear. Hence $G(\xi)$ is not an S-box with a UHODDT, although it is regular (see Subsection 5.1).

Now let $\Phi(\xi)$ be a differentially 2-uniform permutation on V_n. Set $\Psi(\xi) = \overline{\Phi(\xi)}$. Ψ is an $n \times s$ S-box. A similar argument shows that the difference distribution table of $\Psi(\xi)$ is not uniformly half-occupied.

5 Hadamard Matrices Embodied in Difference Distribution Table

In this section we reveal a very important combinatorial property of differentially 2-uniform quadratic permutations, namely, every differentially 2-uniform quadratic permutation is associated with a Sylvester-Hadamard matrix. As an application of the result, we show that chopping a differentially 2-uniform quadratic permutation results in an S-box whose difference distribution table is nearly flat. Such an S-box is very weak to the differential attack.

5.1 Difference Distribution Tables and Incidence Functions

Let $F = (f_1, \ldots, f_n)$ be a differentially 2-uniform quadratic permutation on V_n, namely, a quadratic permutation with a UHODDT or uniformly half-occupied difference distribution table. Let W_α be the set of vectors $F(x) \oplus F(x \oplus \alpha)$ runs through when x runs through V_n, namely,

$$W_\alpha = \{F(x) \oplus F(x \oplus \alpha) | x \in V_n\} \tag{5}$$

Obviously if $\alpha = 0$ then $W_\alpha = \{0\}$. Since each f_j is quadratic $f_j(x) \oplus f_j(x \oplus \alpha)$ is an affine function.

Write $f_j \oplus f_j(x \oplus \alpha) = c_{1j}x_1 \oplus \cdots \oplus c_{nj}x_n \oplus d_j$, $j = 1, \ldots, n$. Set $C_\alpha = (c_{ij})$, $\sigma_\alpha = (d_1, \ldots, d_n)$. Thus $F(x) \oplus F(x \oplus \alpha) = xC_\alpha \oplus \sigma_\alpha$ and $W_\alpha = \{F(x) \oplus F(x \oplus \alpha) | x \in V_n\} = \{xC_\alpha \oplus \sigma_\alpha | x \in V_n\}$.

Now let $\alpha \neq 0$. Since F is a permutation, $F(x) \oplus F(x \oplus \alpha) \neq 0$ for any $x \in V_n$. Hence $0 \notin W_\alpha$. Since $F(0) \oplus F(\alpha) = \sigma_\alpha$, we have $\sigma_\alpha \neq 0$. And by the definition of a UHODDT, $|W_\alpha| = 2^{n-1}$ and hence $rank(C_\alpha) = n - 1$. Thus we have

Lemma 11. *Let F be a differentially 2-uniform quadratic permutation on V_n. If $\alpha \neq 0$ then*
(i) $0 \notin W_\alpha$, (ii) $\sigma_\alpha \neq 0$, (iii) $|W_\alpha| = 2^{n-1}$, and (iv) $rank(C_\alpha) = n - 1$.

Now set $W_\alpha^0 = \{xC_\alpha | x \in V_n\}$. Then we have

Lemma 12. *Let F be a differentially 2-uniform quadratic permutation on V_n. If $\alpha \neq 0$ then $V_n = W_\alpha \cup W_\alpha^0$ and $W_\alpha \cap W_\alpha^0 = \phi$.*

Lemma 13. *Let F be a differentially 2-uniform quadratic permutation on V_n. Let $\alpha \neq 0$. Then the following statements hold:*

(i) If β, $\beta' \in W_\alpha$ then $\beta \oplus \beta' \in W_\alpha^0$,
(ii) if $\beta \in W_\alpha$, $\beta' \in W_\alpha^0$ then $\beta \oplus \beta' \in W_\alpha$,
(iii) if β, $\beta' \in W_\alpha^0$ then $\beta \oplus \beta' \in W_\alpha^0$.

Let F be a differentially 2-uniform quadratic permutation on V_n and let W_α be the same as (5). For each $\alpha \in V_n$ we define an *incidence function* φ_α as follows:

$$\varphi_\alpha(\beta) = \begin{cases} 0 \text{ if } \alpha = 0 \\ 1 \text{ if } \alpha \neq 0 \text{ and } \beta \in W_\alpha \\ 0 \text{ if } \alpha \neq 0 \text{ and } \beta \notin W_\alpha \end{cases} \tag{6}$$

As is to be proved below, each φ_α is in fact a linear function on V_n.

Lemma 14. *Let F be a differentially 2-uniform quadratic permutation on V_n. Then φ_α, defined in (6), is a linear function on V_n for every vector $\alpha \in V_n$.*

Lemma 15. *Let F be a differentially 2-uniform quadratic permutation on V_n. If $\alpha \neq \alpha'$, then $\varphi_\alpha \neq \varphi_{\alpha'}$.*

5.2 Hadamard Matrices in Difference Distribution Tables

Lemma 14 states that each row of the differential distribution table is associated with a linear function on V_n, while Lemma 15 indicates that these linear functions are all different. Hence we have

Theorem 16. *Let F be a differentially 2-uniform quadratic permutation on V_n. Then φ_α runs through all linear functions on V_n while α runs through the vectors in V_n.*

Recall that $\alpha_0, \alpha_1, \ldots, \alpha_{2^n-1}$ are all the vectors in V_n, with $\alpha_0 = (0, \ldots, 0)$, $\ldots$, $\alpha_{2^n-1} = (1, \ldots, 1)$. Let $M = (m_{ij})$ be a $(1, -1)$-matrix defined by

$$m_{ij} = (-1)^{\varphi_{\alpha_i}(\alpha_j)} \qquad (7)$$

M is called the *difference trait matrix* of F. Essentially, M is a matrix obtained from the difference distribution table of the S-box by replacing each zero entry by 1 and each nonzero entry by -1, with an exception that the first entry in the first row is replaced by 1.

Theorem 17. *Let F be a differentially 2-uniform quadratic permutation on V_n. Then M, the difference trait matrix of F, is a Sylvester-Hadamard matrix if the row-order is ignored.*

Proof. From Theorem 16, the 2^n rows of M comprise all the linear sequences of length 2^n. By Lemma 1 of [16], each linear sequence of length 2^n is a row of H_n. Thus M can be changed to H_n by re-ordering its rows.

Obviously, W_α, φ_α and M can be defined for any permutation on V_n, not restricted to quadratic ones.

Theorem 18. *Let F be a differentially 2-uniform quadratic permutation on V_n and M be the difference trait matrix of F. Then the inverse of F is also a differentially 2-uniform permutation, whose difference trait matrix is the transpose of M.*

Note that for a differentially 2-uniform quadratic permutation F based on a cubic polynomial on $GF(2^n)$, n odd, the algebraic degree of F^{-1} is larger than $(n+1)/2$. By Theorem 18, both the difference trait matrix of F and that of F^{-1} are Sylvester-Hadamard matrices (subject to re-ordering their rows).

5.3 Chopping Quadratic Permutations

Let $F = (f_1, \ldots, f_n)$ be a differentially 2-uniform permutation on V_n. Let G be an S-box obtained by chopping a component function of F, say $G = (f_2, \ldots, f_n)$. Similarly to W_α, φ and M corresponding to F (see (5), (6) and (7)), we can define

$$U_\alpha = \{G(x) \oplus G(x \oplus \alpha) | x \in V_n\},$$

where $\alpha \in V_n$, and the incidence function

$$\psi_\alpha(\beta) = \begin{cases} 0 \text{ if } \alpha = 0 \\ 1 \text{ if } \alpha \neq 0 \text{ and } \beta \in U_\alpha \\ 0 \text{ if } \alpha \neq 0 \text{ and } \beta \notin U_\alpha \end{cases}$$

where $\beta \in V_{n-1}$.

Let $\alpha_0, \alpha_1, \cdots, \alpha_{2^n-1}$ be the ordered vectors in V_n and $\beta_0, \beta_1, \cdots, \beta_{2^{n-1}-1}$ the ordered vectors in V_{n-1}. Define a $2^n \times 2^{n-1}$ (1, -1)-matrix, say $N = (n_{ij})$, where $n_{ij} = (-1)^{\psi_{\beta_i}(\beta_j)}$.

Write $M = [M_1 M_2]$ where each M_j is of order $2^n \times 2^{n-1}$, $M_1 = (m_{ij})$, and $M_2 = (m_{ij+2^{n-1}})$. It is easy to see that $\psi_\alpha(\beta) = 1$ if and only if $\varphi_\alpha(0, \beta) = 1$ or $\varphi_\alpha(1, \beta) = 1$. In other words, $n_{ij} = -1$ if and only if $m_{ij} = -1$ or $m_{ij+2^{n-1}} = -1$.

Since F is a differentially 2-uniform quadratic permutation, by Theorem 17, each row of M is a row of H_n. Now recall that $H_n = \begin{bmatrix} H_{n-1} & H_{n-1} \\ H_{n-1} & -H_{n-1} \end{bmatrix}$. Write $H_n = (h_{ij})$, $i, j = 1, \ldots, 2^n$. We can see that $-h_{ij} = h_{ij+2^{n-1}}$ if $i > 2^{n-1}$. This implies that $h_{ij} = -1$ or $m_{ij+2^{n-1}} = -1$, if $i > 2^{n-1}$. Note that M and H_n have the same set of rows. This proves that there exists 2^{n-1} nonzero $\alpha \in V_n$ such that ψ_α is constant 1. In this case $G(x) \oplus G(x \oplus \alpha)$ runs through every vector (including the zero vector) in V_{n-1}, for some 2^{n-1} nonzero vectors $\alpha \in V_n$ and hence the robustness of G is less than $\frac{1}{2}$.

To summarize the above discussions, the difference distribution table of an S-box obtained by chopping a component function of a differentially 2-uniform quadratic permutation has the following profile: it can be viewed as a folded (right to left) version of the uniformly half-occupied table of the original permutation, with half of the rows containing a value 2 in all their entries, and the remaining rows, not counting the first row, containing an equal number of 0s and 4s. Similarly, chopping two component functions from a permutation results in an S-box whose difference distribution table is almost flat: it can be viewed as a twice-folded (right to left) version of the uniformly half-occupied table of the original permutation, and three quarters of the rows contain a value 4 in all their entries, while the remaining rows, not counting the first row, have an equal number of 0s and 8s. This observation can be extended to the case when three or more component functions are chopped.

In conclusion, S-boxes obtained by chopping differentially 2-uniform quadratic permutations have an almost flat difference distribution table, which renders a DES-like encryption algorithm that employs such S-boxes very prone to the differential attack.

Acknowledgments The first author was supported in part by the Australian Research Council under the reference numbers A49130102, A49131885 and A49232172, the second author by A49130102, and the third author by A49232172. All authors were supported by a University of Wollongong Research Program grant and the first two by ATERB C010/058. The authors would like to thank the anonymous referees for Crypto'94 for their helpful comments.

References

1. Adams, C. M.: On immunity against Biham and Shamir's "differential cryptanalysis". Information Processing Letters **41** (1992) 77–80

2. Beth, T., Ding, C.: On permutations against differential cryptanalysis. In Advances in Cryptology - EUROCRYPT'93 (1994) vol. 765, Lecture Notes in Computer Science Springer-Verlag, Berlin, Heidelberg, New York pp. 65–76
3. Biham, E., Shamir, A.: Differential cryptanalysis of DES-like cryptosystems. Journal of Cryptology **Vol. 4, No. 1** (1991) 3–72
4. Biham, E., Shamir, A.: Differential Cryptanalysis of the Data Encryption Standard. Springer-Verlag New York, Heidelberg, Tokyo 1993
5. Brown, L., Kwan, M., Pieprzyk, J., Seberry, J.: Improving resistance to differential cryptanalysis and the redesign of LOKI. In Advances in Cryptology - ASIACRYPT'91 (1993) vol. 739, Lecture Notes in Computer Science Springer-Verlag, Berlin, Heidelberg, New York pp. 36–50
6. Dillon, J. F.: A survey of bent functions. The NSA Technical Journal (1972) 191–215
7. Matsui, M.: Linear cryptanalysis method for DES cipher. In Advances in Cryptology - EUROCRYPT'93 (1994) vol. 765, Lecture Notes in Computer Science Springer-Verlag, Berlin, Heidelberg, New York pp. 386–397
8. Matsui, M.: Linear cryptanalysis method for DES cipher (II). In Proceedings of 1994 Symposium on Cryptography and Information Security (Japan, 1994)
9. Nyberg, K.: Perfect nonlinear S-boxes. In Advances in Cryptology - EUROCRYPT'91 (1991) vol. 547, Lecture Notes in Computer Science Springer-Verlag, Berlin, Heidelberg, New York pp. 378–386
10. Nyberg, K.: On the construction of highly nonlinear permutations. In Advances in Cryptology - EUROCRYPT'92 (1993) vol. 658, Lecture Notes in Computer Science Springer-Verlag, Berlin, Heidelberg, New York pp. 92–98
11. Nyberg, K.: Differentially uniform mappings for cryptography. In Advances in Cryptology - EUROCRYPT'93 (1994) vol. 765, Lecture Notes in Computer Science Springer-Verlag, Berlin, Heidelberg, New York pp. 55–65
12. Nyberg, K., Knudsen, L. R.: Provable security against differential cryptanalysis. In Advances in Cryptology - CRYPTO'92 (1993) vol. 740, Lecture Notes in Computer Science Springer-Verlag, Berlin, Heidelberg, New York pp. 566–574
13. Pieprzyk, J.: Bent permutations. In Proceeding of the International Conference on Finite Fields, Coding Theory, and Advances in Communications and Computing (Las Vegas, 1991)
14. Rothaus, O. S.: On "bent" functions. Journal of Combinatorial Theory **Ser. A, 20** (1976) 300–305
15. Seberry, J., Zhang, X. M., Zheng, Y.: Systematic generation of cryptographically robust S-boxes. In Proceedings of the first ACM Conference on Computer and Communications Security (1993) The Association for Computing Machinery, New York pp. 172 – 182
16. Seberry, J., Zhang, X. M., Zheng, Y.: Nonlinearly balanced boolean functions and their propagation characteristics. In Advances in Cryptology - CRYPTO'93 (1994) vol. 773, Lecture Notes in Computer Science Springer-Verlag, Berlin, Heidelberg, New York pp. 49–60
17. Seberry, J., Zhang, X. M., Zheng, Y.: Relationships among nonlinearity criteria. Presented at *EUROCRYPT'94*, 1994

A Randomness-Rounds Tradeoff in Private Computation

Eyal Kushilevitz[1] and Adi Rosén[2]

[1] Dept. of Computer Science, Technion, Haifa, Israel*
[2] Dept. of Computer Science, Tel-Aviv University, Tel-Aviv, Israel

Abstract. We study the role of randomness in multi-party private computations. In particular, we give several results that prove the existence of a randomness-rounds tradeoff in multi-party private computation of xor. We show that with a single random bit, $\Theta(n)$ rounds are necessary and sufficient to privately compute xor of n input bits. With $d \geq 2$ random bits, $\Omega(\log n/d)$ rounds are necessary, and $O(\log n/\log d)$ are sufficient.
More generally, we show that the private computation of a boolean function f, using $d \geq 2$ random bits, requires $\Omega(\log S(f)/d)$ rounds, where $S(f)$ is the sensitivity of f. Using a single random bit, $\Omega(S(f))$ rounds are necessary.

1 Introduction

A *1-private* (or simply, *private*) protocol $\mathcal{A}$ for computing a function f is a protocol that allows n players, P_i, $1 \leq i \leq n$, each possessing an individual secret input, x_i, to compute the value of $f(\vec{x})$ in a way that no *single* player learns more about the initial inputs of other players than what is revealed by the value of $f(\vec{x})$ and its own input[1]. The players are assumed to be honest but curious. Namely, they all follow the prescribed protocol $\mathcal{A}$ but they could try to get additional information by considering the messages they receive during the execution of the protocol. Private computations in this setting were the subject of a considerable amount of work, e.g., [BGW88, CCD88, BB89, CK89, K89, B89, FY92, CK92, CGK90, CGK92]. One crucial ingredient in private protocols is the use of *randomness*. Quantifying the amount of randomness needed for computing functions privately is the subject of the present work.

Randomness as a resource was extensively studied in the last decade. Methods for saving random bits range over pseudo-random generators [BM84, Y82, N90],

* Work on this paper by the first author was supported by the E. and J. Bishop Research Fund, and by the Fund for the Promotion of Research at the Technion. Part of his research was performed while he was at Aiken Computation Laboratory, Harvard University, supported by research contracts ONR-N0001491-J-1981 and NSF-CCR-90-07677.

[1] In the literature a more general definition of t-privacy is given. The above definition is the case $t = 1$.

techniques for re-cycling random bits [IZ89, CW89], sources of weak randomness [CG88, VV85, Z91], and construction of different kinds of small probability spaces [NN90, AGHP90, S92, KM93] (which sometimes even allow to eliminate the use of randomness). A different direction of research is a quantitative study of the role of randomness in specific contexts, e.g., [RS89, KPU88, BGG90, CG90, BGS94, BSV94]. In this work, we initiate a quantitative study of randomness in private computations. We will mainly concentrate on the specific task of computing the `xor` of n input bits. However, some of our results extend to any boolean function. The task of computing `xor` was the subject of previous research due to its being a basic linear operation and its relative simplicity [FY92, CK92].

It is not difficult to show that private computation of `xor` cannot be carried out deterministically (for $n \geq 3$). On the other hand, with a single random bit this becomes possible: At the first round player P_n chooses a random bit r and sends to P_1 the bit $x_n \oplus r$. Then, in round i ($2 \leq i \leq n$) player P_{i-1} xors its bit x_{i-1} with the message it received in the previous round, and sends the result to P_i. Finally, P_n xors the message it received with the random bit r. Both the correctness and privacy of this protocol are easy to verify. The main drawback of this protocol is that it takes n rounds. Another protocol for this task computes `xor` in 2 rounds but requires a linear number of random bits: In the first round each player chooses a random bit r_i. It sends $x_i \oplus r_i$ to P_1 and r_i to P_2. In the second round P_2 xors all the (random) bits it received in the first round and sends the result to P_1 which xors all the messages it received during the protocol to get the value of the function. Again, both the correctness and privacy are not hard to verify.

In this work we prove that there is a tradeoff between the amount of randomness and the number of rounds in private computations of `xor`. For example, we show that while with a single random bit $\Theta(n)$ rounds are necessary and sufficient[2], with two random bits $O(\log n)$ rounds suffice. Namely, with a second additional random bit, the number of rounds is significantly reduced. Additional bits give a much more "modest" saving. More precisely, we prove that with $d \geq 2$ random bits $O(\log n / \log d)$ rounds suffice and $\Omega(\log n / d)$ rounds are required. Our upper bound is achieved using a new method that enables us to use linear combinations of random bits again and again (while preserving the privacy). The lower bounds are proved using combinatorial arguments, and they are strong in the sense that they also apply to protocols that are allowed to make errors, and that they actually show a lower bound on the *expected* number of rounds. In the final version of this paper, we will show that if protocols are restricted to a certain natural type (that includes, in particular, the protocol that achieves the upper bound) we can even improve the lower bound and show that $\Theta(\log n / \log d)$ rounds are necessary and sufficient.

Our lower bound techniques apply not only to the `xor` function, but in fact give lower bounds on the number of rounds for any boolean function in terms of the *sensitivity* of the function. Namely, we prove that with $d \geq 2$ random bits

[2] More precisely, $n/2$ rounds; and this can be achieved by a slight modification of the first protocol above.

$\Omega(\log S(f)/d)$ rounds are necessary to privately compute a boolean function f, whose sensitivity is $S(f)$. With a single random bit ($d = 1$) $\Omega(S(f))$ rounds are necessary.

The question whether private computations can be carried out in constant number of rounds was discussed in [BB89, BFKR91]. In light of our results, a promising approach to investigate this question may be by proving that if a constant number of rounds is sufficient then a large number of random bits is required.

The rest of the paper is organized as follows: In Section 2 we give some definitions. In Section 3 we give an upper bound on the number of rounds required to privately compute **xor**. In section 4 we give lower bounds on the number of rounds to privately compute a boolean function, in terms of its sensitivity. We conclude in Section 5 with lower bounds on the *expected* number of rounds in terms of the *average* sensitivity of the function being computed.

2 Preliminaries

We give here a description of the protocols we consider, and define the *privacy* property of protocols. More rigorous definitions of the protocols are given in Section 4.1.

A set of n players P_i ($1 \leq i \leq n$), each possessing a secret input bit x_i, collaborate in a protocol to compute a function $f(\vec{x})$. The protocol operates in rounds. In each round each player may toss some coins, and then sends messages to the other players (messages are sent over private channels so that other than the intended receiver no other player can listen to them). It then receives the messages sent to it by the other players. In addition, each player at a certain round chooses to output the value of the function. We assume that each player knows its serial number and the total number of players n.

Each player P_i receives during the execution of the protocol a sequence of messages C_i. Informally, *privacy* with respect to player P_i means that player P_i cannot learn anything (in particular, the inputs of the other players) from C_i, except what is implied by its input bit, and the value of the function computed. Formally,

Definition 1. (Privacy) A protocol $\mathcal{A}$ for computing a function f is *private* with respect to player P_i if for any two input vectors $\vec{x}$ and $\vec{y}$, such that $f(\vec{x}) = f(\vec{y})$ and $x_i = y_i$, for any sequence of messages C, and for any random coins, R_i, tossed by P_i, $Pr[C_i = C|R_i, \vec{x}] = Pr[C_i = C|R_i, \vec{y}]$, where the probability is over the random coin tosses of *all other* players.

3 Upper Bound

This section presents a protocol which allows n players to use $d \geq 2$ random bits for computing **xor** privately. This protocol takes $O(\log_d n)$ rounds. (For the case $d = 1$ a similar protocol can be presented, that uses $n/2$ rounds.)

Consider the following protocol (which we call the *basic protocol*): First organize the n players in a tree. The degree of the root of the tree is $d+1$, and that of any other internal node is d (assume for simplicity that n is such that this forms a complete tree). The computation starts from the leaves and goes towards the root by sending messages (each of them of a single bit) as follows: Each leaf player P_i sends its input bit x_i to its parent in the tree. Each internal node, after receiving messages from its d children, sums them up (modulo 2) together with its input bit x_i and sends the result to its parent. Finally, the root player sums up the $d+1$ messages it receives together with its input bit and the result is the output of the protocol.

While a simple induction shows the correctness of this protocol, and it clearly runs in $O(\log_d n)$ rounds, it is obvious that it does not maintain the required privacy. The second idea will be to "mask" each of the messages sent in the basic protocol by an appropriate random bit (constructed using the d random bits available), in a way that these masks will disappear at the end, and we will be left with the output. To do so we assign the nodes of the above tree vectors in $GF[2^d]$ as follows (the meaning of those vectors will become clear soon): Assign to the root the vector $(0,\ldots,0)$. The children of the root will be assigned $d+1$ (non-zero) vectors such that the vectors in any d-size subset of them are linearly independent and the sum of all the $d+1$ vectors is $(0,\ldots,0)$ (for example, the d unit vectors together with the $(1,\ldots,1)$ vector satisfy these requirements). Finally, in a recursive way, given an internal node which is assigned a vector v, we assign to its d children d linearly independent vectors whose sum is v (note that in particular none of these vectors is the $\vec{0}$ vector)[3].

We now show how to use the vectors we assigned to the nodes, so as to get a private protocol. We will assume that the random bits $b_1,\ldots,b_d$ are chosen by some external processor. We will later see that this assumption can be eliminated easily. Let v be the vector assigned to some player which is a *leaf* in the tree. We will give this player a single bit $r_v = v \cdot b$, where $b = (b_1,\ldots,b_d)$ is the vector consisting of the d random bits, and the product is an inner product. The players will use the basic protocol, described above, with the modification that a player in a leaf also adds to its message the bit r_v it received (the other players behave exactly as before). We claim that for every player P_i, if in the basic protocol it sends the message m when the input vector is $\vec{x}$, then in the modified protocol it sends the message $m+(v_i \cdot b)$, where v_i is the vector assigned to this player. The proof goes by induction: It is trivially true for the leaf players. For internal nodes the message is calculated by summing the incoming messages, thus, using the induction hypothesis it is $\sum_{k=1}^{d}[m^k+(v^k \cdot b)]$, where m^k is the message received from the k'th child in the basic protocol, and v^k is the vector assigned to the k'th child. Since the construction is such that v_i, the vector assigned to P_i, satisfies $v_i = \sum_{k=1}^{d} v^k$, then a simple algebraic manipulation proves the induction step.

[3] For example, such a collection of d vectors can be constructed as follows: Since $v \neq \vec{0}$ there exists an index i such that $v_i = 1$. The first $d-1$ vectors will be the $d-1$ unit vectors $e_1,\ldots,e_{i-1},e_{i+1},\ldots,e_d$. The last vector will be $v - \sum_{j \neq i} e_j$. Obviously the sum of these d vectors is v and they are linearly independent.

In particular, as the root is assigned the vector $(0,\ldots,0)$, its output equals the output of the basic protocol. Hence, the correctness follows.

We now prove the privacy property of the protocol. The leaf players do not receive any message, hence there is nothing to prove. Let P_j be an internal node in the tree. Denote by $s^1,\ldots,s^d$ the messages it receives. We claim that for every vector $w=(w_1,\ldots,w_d)\in GF[2^d]$, for any input vector, we have

$$Pr[(s^1=w_1)\wedge\ldots\wedge(s^d=w_d)] \;=\; \frac{1}{2^d},$$

where the probability is over the random choice of $b_1,\ldots,b_d$ (note that in this protocol the players do not make internal random choices). In other words, fix any specific input vector $\vec{x}$, then for every vector w, there exists exactly one choice of values for $b_1,\ldots,b_d$, such that the messages that P_j receives, when the protocol is executed with input $\vec{x}$, are the vector w. Denote by $\vec{v}^1,\ldots,\vec{v}^d$ the vectors corresponding to the d children of P_j in the tree, and let $m^1,\ldots,m^d$ be the messages they have to send in the basic protocol given a specific input vector $\vec{x}$. As claimed, for every $1\le k\le d$, the message that the k-th child sends in the modified protocol can be expressed as $s^k=m^k+(\vec{v}^k\cdot\vec{b})$. With this notation, for having $s^1=w_1,\ldots,s^d=w_d$ the following linear system has to be satisfied:

$$\begin{bmatrix}\vec{v}^1\\ \vdots\\ \vec{v}^d\end{bmatrix}\cdot\vec{b}=\begin{bmatrix}w_1-m^1\\ \vdots\\ w_d-m^d\end{bmatrix}$$

Since $\vec{v}^1,\ldots,\vec{v}^d$ are linearly independent, this system has exactly one solution, as needed.

As for the root player the same argument can be applied to any fixed d-size subset of the $d+1$ messages it receives. This gives us that given any input vector $\vec{x}$, for all d-size messages vectors $\vec{w}$,

$$Pr[(s^1=w_1)\wedge\ldots\wedge(s^d=w_d)] \;=\; \frac{1}{2^d}\,.$$

Now take two input vectors $\vec{x}$ and $\vec{y}$ such that $x_{\mathrm{root}}=y_{\mathrm{root}}$ and such that $f(\vec{x})=f(\vec{y})$. Then by the correctness of the algorithm, given a specific d-size messages-vector, the $d+1$'st message is the same for $\vec{x}$ and $\vec{y}$. Thus the privacy property holds with respect to the root too.

Finally, note that we assumed that the random choices were made by some external processor. However, we can let one of the leaf players randomly choose the bits $b_1,\ldots,b_d$ and to supply each of the leaf players with the appropriate bit r_v. As the leaf players only send messages in the protocol, the special processor that selects the random bits gets no advantage.

Note that if a player is non-honest it can easily prevent the other players from computing the correct output. However, it cannot get any additional information in the above protocol, since the only message each player gets after sending its own message is the value of the function. We have thus proved the following theorem:

Theorem 2. *The function* **xor** *can be computed privately using* $d \geq 2$ *random bits in* $O(\log n / \log d)$ *rounds.*

4 Lower Bounds

In this section we prove several lower bounds on the number of rounds required to privately compute a boolean function, given that the total number of random bits the players can toss is d. The lower bound is given in terms of the sensitivity of the function. In Section 4.1 we give some formal definitions. In Section 4.2 we present a lemma, central to our proofs, about sensitivity of functions. The proof of the lower bound appears in Section 4.3.

4.1 Preliminaries

We first give a formal definition for the protocols. A protocol operates in rounds. In each round each player P_i, based on the value of its input bit x_i, the values of the messages received in previous rounds, and the values of the coins tossed in previous rounds, tosses a certain number of additional coins, and sends messages to the other players. The values of these messages may depend on all of the above, including the coins just tossed. Then the player receives the messages sent to it by the other players. In addition, each player, at a certain round, chooses to output the value of the function as calculated by it. To define the protocol more formally we give the following definition:

Definition 3. (View)

- A time-t *partial* view of player P_i consists of its input bit x_i, the messages it has received in the first $t-1$ rounds, and the coins it tossed in the first $t-1$ rounds. We denote it by $PView_i^t$.
- A time-t view of player P_i consists of its input bit x_i, the messages it has received in the first $t-1$ rounds, and the coins it tossed in the first t rounds. We denote it by $View_i^t$.

Intuitively, the *partial view* of a player in round t determines how many coins (if at all) it will toss in round t. Then, its *view* (which includes those newly tossed coins) determines the messages it will send in round t. Formally,

Definition 4. A protocol consists of a set of functions $R_i^k(PView_i^k)$ which determine how many coins are tossed by P_i in round k, and a set of functions $M_{i \to j}^k : View_i^k \to M$, $1 \leq i, j \leq n$ (where M is a finite domain of possible message values), which determine the message sent by P_i to P_j at round k.

Definition 5. A d-random protocol is a protocol such that for any input assignment, the total number of coins tossed by all players in *any* execution is at most d.

We emphasize that the definitions allow, for example, that in different executions different players will toss the coins. This may depend on both the input of the players, and previous coin tosses.

Definition 6. A protocol to compute a function f is a protocol such that for any input vector $\vec{x}$ and every player P_i, P_i correctly outputs the value of $f(\vec{x})$ with probability 1.

Protocols that are allowed to err will be considered in Section 5.1.

We will slightly modify our view of the protocol in the following way. Fix an arbitrary binary encoding for the messages in M. We will consider a protocol where each player sends instead of a single message from M, a set of boolean messages that represent the binary encoding of the message to be sent in the original protocol. These messages are sent "in parallel" in the same round. Henceforth when we refer to messages we refer to these binary messages. Clearly, the number of rounds remains the same.

We further modify the model, with respect to its randomness. By the above definitions in a d-random protocol each player can locally toss coins, and we are assured that no more than d coin tosses occur. Assume we have an additional external agent that has a tape with d random bits. Then, we can replace the local coins tossing by a primitive with which each player communicates with this external agent, asking it for a random bit. In response, the external agent (immediately) provides the next random bit on its tape.

Given a d-random private protocol in which the players locally toss coins, there is a protocol that uses the external agent, sends the same messages, and runs in the same number of rounds, while the privacy requirements are not violated in any of the players. The external agent will never be asked for more than the d random bits it has on its tape. Note that the external agent may be able to learn things, but in our lower bound proof we will use the privacy requirement only with respect to the original players. Thus, without loss of generality, we prove our lower bounds on the number of rounds required by a d-random private protocol that uses an external agent as its source of randomness.

With this model in mind we can regard a d-random protocol as a distribution of 2^d *deterministic* protocols, each derived from the randomized protocol by a specific random tape of length d. Furthermore, $View_i^t$, for any i and t, is a function of the random tape $\vec{b}$ and the input assignment $\vec{x}$. We can thus write it as $View_i^t(\vec{x}, \vec{b})$.

Denote by $T_i(\vec{x}, \vec{b})$ the round number in which P_i outputs its result, given input assignment $\vec{x}$ and random tape $\vec{b}$ (Note that since $\vec{b}$ is an argument for T, then T is a deterministic function). The following definition defines the number of rounds of a protocol.

Definition 7. (Rounds Complexity)

- An r-round protocol to compute a function f is a protocol to compute f such that for all i, $\vec{x}$, $\vec{b}$, $T_i(\vec{x}, \vec{b}) \leq r$.

- An *expected* r-round protocol to compute a function f is a protocol to compute f such that for all i and $\vec{x}$, $E_{\vec{b}}[T_i(\vec{x}, \vec{b})] \leq r$.

We include here some definitions related to functions $f : \{0,1\}^n \rightarrow D$, where D is some finite domain.

Definition 8. (Sensitivity)

- A function f is *sensitive* to its i-th variable on assignment Y, if $f(Y) \neq f(Y^{(i)})$, where $Y^{(i)}$ is the same as Y with the i-th variable flipped.
- $\mathcal{S}_f(Y)$ is the set of variables to which the function f is sensitive on assignment Y.
- The *sensitivity* of a function f, denoted $S(f)$, is $S(f) \stackrel{\triangle}{=} \max_Y\{|\mathcal{S}_f(Y)|\}$.
- The *average sensitivity* of a function f, denoted $AS(f)$, is the average of $|\mathcal{S}_f(Y)|$. That is, $AS(f) \stackrel{\triangle}{=} \frac{1}{2^n}\sum_{Y\in\{0,1\}^n} |\mathcal{S}_f(Y)|$.
- The set of variables on which f *depends*, denoted $\mathcal{D}(f)$, is $\mathcal{D}(f) \stackrel{\triangle}{=} \{i : \exists Y \text{s.t.} i \in \mathcal{S}_f(Y)\}$. If $i \in \mathcal{D}(f)$ we say that f depends on its i-th variable.

The following claim gives a lower bound on the degree of error if we evaluate a function f by means of another function g, in terms of their average sensitivities.

Claim 9. *Consider two functions $f, g : \{0,1\}^n \rightarrow D$. Then $f(\vec{x}) = g(\vec{x})$ for at most $2^n \cdot (1 - \frac{AS(f)-AS(g)}{2n})$ input assignments $\vec{x}$.*

Proof. Consider the n-dimensional hypercube. An *f-good edge* is an edge $e = (\vec{x}, \vec{y})$ such that $f(\vec{x}) \neq f(\vec{y})$. By the definitions, the number of f-good edges is exactly $\frac{2^n AS(f)}{2}$. Therefore, there are at least $2^n \frac{AS(f)-AS(g)}{2}$ edges which are f-good but not g-good. For each such edge $e = (\vec{x}, \vec{y})$ either $f(\vec{x}) \neq g(\vec{x})$ or $f(\vec{y}) \neq g(\vec{y})$. Since the degree of each vertex in the hypercube is n there must be at least $2^n \cdot \frac{AS(f)-AS(g)}{2n}$ inputs on which f and g do not agree. □

4.2 A Lemma on Sensitivity

In this section we prove a lemma that bounds the growth of the sensitivity of a combination of functions. This lemma plays a central role in the proofs of our lower bounds, and any improvement on it will immediately improve our lower bounds.

Lemma 10. *Let $\mathcal{F} = \{f_j\}, 1 \leq j \leq m$ be a set of m functions $f_j : \{0,1\}^n \rightarrow \{0,1\}$, for some n. Assume $S(f_j) \leq C$ for all j. Define the function $F(Y) \stackrel{\triangle}{=} (f_1(Y), \ldots, f_m(Y))$. If F assumes at most 2^d different values (different vectors), then the sensitivity of F is at most $C \cdot 2^d - 1$.*[4]

[4] An obvious bound is $S(F) \leq C \cdot m$. However, for reasons that will become clear soon we are interested in bounds which are independent of m.

Proof. Assume towards a contradiction that F has a larger sensitivity. Then, there is an assignment Y such that $|\mathcal{S}_F(Y)| \geq C \cdot 2^d$. Consider Y and assume without loss of generality that $F(Y) = (0, \ldots, 0)$. We will show the existence of $2^d + 1$ different values of F, contradicting the conditions of the lemma. Pick the smallest value j_1 such that $\mathcal{S}_{f_{j_1}}(Y) \neq \emptyset$. Thus, there exists an $i \in \mathcal{S}_{f_{j_1}}(Y)$ which implies that $F(Y^{(i)}) = (0, \ldots, 0, f_{j_1}(Y^{(i)}) \neq 0, *, \ldots, *)$, where each $*$ denotes an arbitrary value. From now on we disregard all values k such that $k \in \mathcal{S}_{f_{j_1}}(Y)$ (those variables to which f_{j_1} is sensitive on Y). According to the conditions of the lemma there are at most C such variables. We now pick a new value j_2, the smallest $j_2 > j_1$ such that there exists a new variable numbered ℓ, still under consideration, and $\ell \in \mathcal{S}_{f_{j_2}}(Y)$. Since f_{j_1} is not sensitive to the ℓ'th variable on Y, we have that $F(Y^{(\ell)}) = (0, \ldots, 0, f_{j_2}(Y^{(l)}) \neq 0, *, \ldots, *)$ which is different from the previous value for F that we have built. We continue this process and at each such step we eliminate at most C variable to which F is sensitive on Y. Since we assumed that $|\mathcal{S}_F(Y)| \geq C \cdot 2^d$, we can continue this process for at least 2^d steps, creating 2^d different values for F. Together with $F(Y) = (0, \ldots, 0)$ this makes more that 2^d different values. □

4.3 Lower Bound on the Number of Rounds

In this subsection we prove the following theorem.

Theorem 11. *Let $\mathcal{A}$ be an r-round d-random ($d \geq 2$) private protocol to compute a boolean function f. Then, $r = \Omega(\log S(f)/d)$.*

Using similar arguments we can show that with a single random bit ($d = 1$) $\Omega(S(f))$ rounds are required. The proof of this case is omitted.

In the following proof we restrict our attention to a specific deterministic algorithm derived from the original protocol by a specific random tape $\vec{B}$. In such a deterministic protocol the views of the players are functions of only the input assignment $\vec{x}$.

Lemma 12. *Consider a private d-random protocol to compute a boolean function f. For a given random tape $\vec{B}$, recall that $View_i^k(\vec{y}, \vec{B})$ is the view of player P_i at round k. Then, for any P_i, $View_i^k(\vec{y}, \vec{B})$ can assume at most 2^{d+2} different values.*

Proof. Partition the input assignments $\vec{x}$ into 4 groups according to the values of x_i (0 or 1), and the value of $f(\vec{x})$ (0 or 1). We argue that the number of different values the view can assume within each such group is at most 2^d.

Fix $\vec{x}$ in one of these 4 groups and consider any other $\vec{y}$ pertaining to the same group. Since the two input assignments are in the same group, by the privacy requirement the value of $View_i^k(\vec{y}, \vec{B})$ must appear also as $View_i^k(\vec{x}, \vec{b})$ for some random tape $\vec{b}$. However, there are only 2^d different random tapes. □

Lemma 13. *Consider a private d-random protocol to compute a boolean function f, and consider a specific random tape $\vec{B}$. Then for any player P_i, the function $View_i^k(\vec{x}, \vec{B})$ (as a function of $\vec{x}$) has sensitivity of at most $T(k) \triangleq (2^{d+2})^{k-1}$.*

Proof. First note that as we consider a fixed random tape the views of the players are functions of the input assignment $\vec{x}$ only.

We prove the lemma by induction. For $k = 1$ the view of any player depends only on its single input bit. Thus, the claim is obvious. For $k > 1$ assume the claim holds for any $\ell < k$. This implies in particular that all messages received by player i and included in the view under consideration have sensitivity of at most $T(k-1)$. Denote by F the view of the player without its input bit. Assume towards a contradiction that $View_i^k$ has sensitivity strictly greater than $T(k)$. Since the view consists of the input bit of the player and the messages received, then F has sensitivity of at least $T(k)$. Using Lemma 10 with $C = T(k-1)$ we conclude that F assumes more than 2^{d+2} different values, contradicting Lemma 12. (Note that Lemma 10 allows us to give a bound which does not depend on the *number* of messages received by P_i.) □

Theorem 14. *Given a private d-random protocol ($d \geq 2$) to compute a boolean function f, consider the deterministic protocol derived from it by any given random tape $\vec{B}$. For any player P_i, there is at least one input assignment $\vec{x}$ such that $T_i(\vec{x}, \vec{B}) = \Omega(\log S(f)/d)$.*

Proof. Consider a fixed but arbitrary player P_i. Denote by t the largest round number in which P_i outputs a value, i.e. $t = \max_{\vec{x}}\{T_i(\vec{x}, \vec{B})\}$. We claim that as long as the sensitivity of the view of P_i does not reach $S(f)$, there is at least one input assignment $\vec{x}$ for which P_i cannot output the correct value $f(\vec{x})$. The reason is that the sensitivity of the view of P_i, while deciding on its output, is a bound on the sensitivity of the output of P_i. On the other hand the value $S(f)$ is obtained by some assignment Y such that the value of $f(Y)$ is different from the value of f on $S(f)$ of Y's "neighbors". Hence, the output must be wrong on either Y or on at least one of these "neighbors". Thus, t is such that $S(View_i^t(\vec{x}, \vec{B})) \geq S(f)$. By Lemma 13, $2^{(d+2)(t-1)} \geq S(f)$, i.e., $t \geq \frac{\log S(f)}{(d+2)} + 1$. □

This proves Theorem 11. Note that this proves not only that there is an input assignment $\vec{x}$ and a random tape $\vec{b}$ for which the protocol runs "for a long time", but also that for each random tape $\vec{b}$ there is such input assignment.

Corollary 15. *Let $\mathcal{A}$ be an r-round d-random private protocol ($d \geq 2$) to compute* **xor** *of n bits. Then $r = \Omega(\log n/d)$.*

As stated at the top of the section, using similar arguments we can show that with a single random bit ($d = 1$) $\Omega(n)$ rounds are required to privately compute **xor**, which is tight.

5 Lower Bounds on the Expected Number of Rounds

In this section we prove lower bounds on the *expected* number of rounds, in terms of the average sensitivity of the computed function. In particular, we prove an $\Omega(\log n/d)$ lower bound on the *expected* number of rounds required by protocols that privately compute **xor** of n bits. We further strengthen the result in the next subsection where we allow the protocol to make errors (we will formally define this notion shortly). In particular, we prove that the expected number of rounds of such protocols that compute **xor** remains $\Omega(\log n/d)$. We start with the following theorem:

Theorem 16. *Let f be a boolean function and let $\mathcal{A}$ be an* expected *r-round d-random private protocol ($d \geq 2$) to compute the function f. Then, $r = \Omega(AS(f) \log AS(f)/nd)$.*

Consider a protocol $\mathcal{A}$ and a player P_i. We say that the protocol is *late* on $\vec{x}$ and $\vec{b}$ if $T_i(\vec{x}, \vec{b}) \geq \frac{\log AS(f)}{2(d+2)} + 1$. We first show that for any deterministic protocol derived from a private protocol to compute f, not only there is at least one input on which it is late, but that this happens for a large fraction of the inputs.

Lemma 17. *Consider player P_i. For any random tape $\vec{b}$, there are at least $2^n(\frac{AS(f)-\sqrt{AS(f)}}{2n})$ input assignments $\vec{x}$ such that the protocol is late on $\vec{x}$, $\vec{b}$.*

Proof. Consider the views of P_i, $View_i^t$, given random tape $\vec{b}$. For any round t such that $t < \frac{\log AS(f)}{2(d+2)} + 1$, by Lemma 13, $S(View_i^t) < 2^{(d+2)\frac{\log AS(f)}{2(d+2)}} = \sqrt{AS(f)}$. Any function g computed from such a view can have at most the same sensitivity, and thus clearly an average sensitivity of at most $\sqrt{AS(f)}$. By Claim 9, g can have the correct value for the function f for at most $2^n(1 - \frac{AS(f)-\sqrt{AS(f)}}{2n})$ input assignments. Since we assume that $\mathcal{A}$ is correct for all input assignments, it follows that at least $2^n \frac{AS(f)-\sqrt{AS(f)}}{2n}$ input assignments are late. □

Lemma 18. *Consider player P_i. There is at least one input assignment $\vec{x}$ for which $\frac{1}{2^d} \sum_{\vec{b} \in \{0,1\}^d} T_i(\vec{x}, \vec{b}) \geq (\frac{AS(f)-\sqrt{AS(f)}}{2n})(\frac{\log AS(f)}{2(d+2)} + 1)$.*

Proof. By the previous lemma the total number of pairs $\vec{x}$, $\vec{b}$ on which the protocol is late is at least $2^d 2^n(\frac{AS(f)-\sqrt{AS(f)}}{2n})$. It follows that there is at least one input assignment $\vec{x}$ for which there are at least $2^d(\frac{AS(f)-\sqrt{AS(f)}}{2n})$ random tapes $\vec{b}$ such that the protocol is late on $\vec{x}$, $\vec{b}$. For each such tape $T_i(\vec{x}, \vec{b}) \geq \frac{\log AS(f)}{2(d+2)} + 1$. □

Theorem 16 follows from the above lemma.

Corollary 19. *Let $\mathcal{A}$ be an* expected *r-round d-random private protocol ($d \geq 2$) to compute **xor** of n bits. Then, $r = \Omega(\log n/d)$.*

Proof. Follows from Theorem 16 and the fact that $AS(\mathbf{xor}) = n$. □

5.1 Weakly Correct Protocols

We first formally define a protocol that is allowed to make a certain amount of errors. Given a protocol $\mathcal{A}$, denote by $\mathcal{A}_i(\vec{x}, \vec{b})$ the output of the protocol in player P_i, given input assignment $\vec{x}$ and random tape $\vec{b}$.

Definition 20. For $\delta < 1/2$, a $(1-\delta)$-correct protocol to compute a function f is a protocol such that for any player P_i and any input vector $\vec{x}$, $Pr_{\vec{b}}[\mathcal{A}_i(\vec{x}, \vec{b}) = f(\vec{x})] \geq (1-\delta)$.

We prove lower bounds on the number (and the expected number) of rounds of such d-random protocols.

Theorem 21. *Let f be a boolean function.*

- *Let $\mathcal{A}$ be an $(1-\delta)$-correct r-round d-random private protocol ($d \geq 2$) to compute f. If $\delta < \frac{AS(f)-\sqrt{AS(f)}}{2n}$ then $r = \Omega(\log AS(f)/d)$.*
- *Let $\mathcal{A}$ be an $(1-\delta)$-correct d-random private protocol ($d \geq 2$) to compute f. Then the* expected *number of rounds is $\Omega((1-\sqrt{2\delta}) \cdot (\frac{AS(f)-\sqrt{AS(f)}}{2n} - \sqrt{\frac{\delta}{2}}) \cdot \log AS(f)/d)$.*

Proof. We first prove a lower bound on the number of rounds, and then turn our attention to the expected number of rounds. The correctness requirement implies that for any player P_i, $Pr_{\vec{b}}[\mathcal{A}_i(\vec{x}, \vec{b}) = f(\vec{x})] \geq 1-\delta$, for all $\vec{x}$. This implies that there is at least one random tape $\vec{b}$ such that for at least $2^n(1-\delta)$ input assignments $\vec{x}$, $\mathcal{A}_i(\vec{x}, \vec{b}) = f(\vec{x})$. By the same arguments as those in the proof of Lemma 17, it follows that before round number $\frac{\log AS(f)}{2(d+2)} + 1$, the protocol can be correct on at most $2^n(1 - \frac{AS(f)-\sqrt{AS(f)}}{2n})$ inputs (with random tape $\vec{b}$). Since we require that at least $2^n(1-\delta)$ are correct, we have that at least

$$2^n(1-\delta) - 2^n(1 - \frac{AS(f) - \sqrt{AS(f)}}{2n}) = 2^n(\frac{AS(f) - \sqrt{AS(f)}}{2n} - \delta)$$

inputs are late. For a lower bound on r for an r-round protocol it is sufficient to have a single input vector $\vec{x}$ such that the execution on $(\vec{x}, \vec{b})$ is "long". For this, note that if $\delta < \frac{AS(f)-\sqrt{AS(f)}}{2n}$ then (for random tape $\vec{b}$) the number of late inputs is greater than 0. This gives us a lower bound of $r = \Omega(\log AS(f)/d)$ for any $(1-\delta)$-correct r-round d-random protocol, with δ as above.

We now prove a lower bound on the *expected* number of rounds of $(1-\delta)$-correct protocols. Again the correctness requirement implies that for any player P_i $Pr_{\vec{b}}[\mathcal{A}_i(\vec{x}, \vec{b}) = f(\vec{x})] \geq 1-\delta$, for all $\vec{x}$. By a counting argument, it follows that for at least $2^d(1-\sqrt{2\delta})$ random tapes, $\mathcal{A}_i(\vec{x}, \vec{b}) = f(\vec{x})$ for at least $2^n(1-$

$\sqrt{\frac{\delta}{2}})$ input assignments $\vec{x}$. Consider those random tapes, and the deterministic protocols derived by them. For each such protocol there are at least

$$2^n(1-\sqrt{\frac{\delta}{2}})-2^n(1-\frac{AS(f)-\sqrt{AS(f)}}{2n})=2^n(\frac{AS(f)-\sqrt{AS(f)}}{2n}-\sqrt{\frac{\delta}{2}})$$

late input assignments. Thus the total number of late pairs $\vec{x}$, $\vec{b}$ is at least

$$2^d(1-\sqrt{2\delta})\cdot 2^n(\frac{AS(f)-\sqrt{AS(f)}}{2n}-\sqrt{\frac{\delta}{2}}).$$

It follows that there is at least one input assignment $\vec{x}$ for which the number of random tapes $\vec{b}$ such that $\vec{x}$,$\vec{b}$ is late is at least $2^d(1-\sqrt{2\delta})\cdot(\frac{AS(f)-\sqrt{AS(f)}}{2n}-\sqrt{\frac{\delta}{2}})$. □

Corollary 22. *For fixed $\delta < 1/2$ let $\mathcal{A}$ be a $(1-\delta)$-correct d-random* expected *r-round private protocol to compute* **xor** *of n bits. Then $r = \Omega(\log n/d)$. (Obviously the same lower bound holds for r-round protocols.)*

Proof. Follows from Theorem 21 and the fact that $AS(\mathtt{xor}) = n$. Note that the term $(1-\sqrt{2\delta})(\frac{1}{2}-\sqrt{\frac{\delta}{2}}-\frac{1}{2\sqrt{n}})$ is greater than 0 for any $\delta < 1/2$ (and sufficiently large n). □

Acknowledgments We thank Benny Chor for useful comments.

References

[AGHP90] N. Alon, O. Goldreich, J. Hastad, and R. Peralta, "Simple constructions of almost k-wise independent random variables", Proc. of 31st FOCS, 1990, pp. 544–553.

[BB89] J. Bar-Ilan, and D. Beaver, "Non-Cryptographic Fault-Tolerant Computing in a Constant Number of Rounds", Proc. of 8th PODC, 1989, pp. 201–209.

[B89] D. Beaver, "Perfect Privacy for Two-Party Protocols", TR-11-89, Harvard University, 1989.

[BFKR91] D. Beaver, J. Feigenbaum, J. Kilian, and P. Rogaway, "Security with Low Communication Overhead", 1991.

[BGG90] M. Bellare, O. Goldreich, and S. Goldwasser, "Randomness in Interactive Proofs", Proc. of 31st FOCS, 1990, pp. 563–571.

[BGW88] M. Ben-or, S. Goldwasser, and A. Wigderson, "Completeness Theorems for Non-Cryptographic Fault-Tolerant Distributed Computation", Proc. of 20th STOC, 1988, pp. 1–10.

[BM84] M. Blum, and S. Micali "How to Generate Cryptographically Strong Sequences Of Pseudo-Random Bits", FOCS 82 and *SIAM J. on Computing,* Vol 13, 1984, pp. 850–864.

[BGS94] C. Blundo, A. Giorgio Gaggia, and D. R. Stinson, "On the Dealer's Randomness Required in Secret Sharing Schemes", *Proc. of EuroCrypt94.*

[BSV94] C. Blundo, A. De-Santis, and U. Vaccaro, "Randomness in Distribution Protocols", To appear in *Proc. of 21st ICALP*, 1994.

[CG90] R. Canetti, and O. Goldreich, "Bounds on Tradeoffs between Randomness and Communication Complexity", FOCS 90 and *Computational Complexity* Vol. 3, (1993), 141-167.

[CCD88] D. Chaum, C. Crepeau, and I. Damgard, "Multiparty Unconditionally Secure Protocols", Proc. of 20th STOC, 1988, pp. 11–19.

[CK89] B. Chor, and E. Kushilevitz, "A Zero-One Law for Boolean Privacy", STOC 89 and *SIAM J. Disc. Math.* Vol. 4, (1991), 36–47.

[CK92] B. Chor, and E. Kushilevitz, "A Communication-Privacy Tradeoff for Modular Addition", *Information Processing Letters*, Vol. 45, 1993, pp. 205–210.

[CGK90] B. Chor, M. Geréb-Graus, and E. Kushilevitz, "Private Computations Over the Integers", Proc. of 31st FOCS, 1990, pp. 335–344.

[CGK92] B. Chor, M. Geréb-Graus, and E. Kushilevitz, "On the Structure of the Privacy Hierarchy", *Journal of Cryptology*, Vol. 7, No. 1, 1994, pp. 53-60.

[CG88] B. Chor, and O. Goldreich, "Unbiased Bits from Sources of Weak Randomness and Probabilistic Communication Complexity", FOCS 85 and *SIAM J. Computing* Vol. 17, (1988), 230–261.

[CW89] A. Cohen, and A. Wigderson, "Dispersers, Deterministic Amplification, and Weak Random Sources", Proc. of 30th FOCS, 1989, pp. 14–19.

[FY92] M. Franklin, and M. Yung, "Communication complexity of secure computation", Proc. of 24th STOC, 1992, pp. 699–710.

[IZ89] R. Impagliazzo, and D. Zuckerman, "How to Recycle Random Bits", Proc. of 30th FOCS, 1989, pp. 248–253.

[KM93] D. Koller, and N. Megiddo, "Constructing Small Sample Spaces Satisfying Given Constraints", Proc. of 25th STOC, 1993, pp. 268–277.

[KPU88] D. Krizanc, D. Peleg, and E. Upfal, "A Time-Randomness Tradeoff for Oblivious Routing", Proc. of 20th STOC, 1988, pp. 93–102.

[K89] E. Kushilevitz, "Privacy and Communication Complexity", FOCS 89, and SIAM Jour. on Disc. Math., Vol. 5, No. 2, May 1992, pp. 273–284.

[NN90] J. Naor, and M. Naor, "Small-Bias Probability Spaces: Efficient Constructions and Applications", STOC 90, and *SIAM J. on Computing,* Vol 22, No. 4, 1993, pp. 838–856.

[N90] N. Nisan, "Pseudorandom Generator for Space Bounded Computation", Proc. of 22nd STOC, 1990, pp. 204–212.

[RS89] P. Raghavan, and M. Snir, "Memory vs. Randomization in On-Line Algorithms", Proc. of 16th ICALP, 1989, pp. 687–703.

[S92] L. J. Schulman, "Sample Spaces Uniform on Neighborhoods", Proc. of 24th STOC, 1992, pp. 17–25.

[VV85] U. Vazirani, and V. Vazirani, "Random Polynomial Time is Equal to Slightly-Random Polynomial Time", Proc. of 26th FOCS, 1985, pp. 417–428.

[Y82] A. C. Yao, "Theory and Applications of Trapdoor Functions" Proc. of 23rd FOCS, 1982, pp. 80–91.

[Z91] D. Zuckerman, "Simulating BPP Using a General Weak Random Source" Proc. of 32nd FOCS, 1991, pp. 79–89.

Secure Voting Using Partially Compatible Homomorphisms

Kazue Sako[1] and Joe Kilian[2]

[1] NEC Corporation, 4-1-1 Miyazaki Miyamae, Kawasaki 216, JAPAN
[2] NEC Research Institute, 4 Independence Way, Princeton, NJ 08540, USA

Abstract. We introduce a new number-theoretic based protocol for secure electronic voting. Our scheme is much more communication efficient than previous schemes of its type, and has a much lower round complexity than is currently possible using the anonymous-channel/mixer techniques. Preprocessing allows for nearly all of the communication and computation to be performed before any voting takes place. Unlike the mixer-based protocols, anyone can verify that everyone's vote has been properly counted. Also, our techniques allow for a wide variety of different schemes.

Our protocols are based on families of homomorphic encryptions which have a *partial compatibility* property, generalizing a method of Benaloh and Yung [2]. We use these functions to generate very simple interactive proofs on encrypted shares. We also develop amortization techniques yielding dramatic efficiency improvements over our simple protocols. Our protocols can be realized by current-generation PC's with access to an electronic bulletin board.

1 Introduction

Secure electronic voting is arguably the most important single application of secure multi-party computation. Yet despite extensive work on this subject, no complete solution has been found in either the theoretical or practical domains. Even the general solutions to secure multiparty protocols fail to exhibit all of the desired properties of elections (for example, the "receipt-free" property [1]). While these general solutions do have a wide breadth of security properties, and some hope of rigorous analysis, they are hopelessly impractical both in their computational and communication costs.

A number of more practical voting protocols have been proposed, with widely differing security properties. Schemes based on anonymous channels/mixers [5, 9, 12] have become very popular due to their superior efficiency and the arbitrary nature of the votes that are allowed. However, a price is paid for this efficiency. The simplest of these schemes cannot allow a voter to securely protest the omission of his vote without allowing a malicious voter to block the election. In all the schemes we know of there is a high round complexity - one round for each mixer used to implement the anonymous channel. Also, after the election each voter is typically responsible for checking that his vote was correctly tallied.

There is usually no way for an outside observer to later verify that the election was properly performed.

Another approach is the use of number theoretic techniques without anonymous channels or mixers [6, 3, 2]. The protocol has desirable security properties, but as we will discuss in detail, their communication complexity is quite high for realistic scenarios.

In this paper, we give a new number-theoretic scheme for secure electronic voting. This scheme has a number of features including moderate communication cost, low round complexity, preprocessing potential, security, universal verifiability and flexibility, which we outline below.

Communication Cost: Our protocol requires much less communication than the previous number-theoretic protocols. For one realistic setting of the parameters, we estimate that our protocol conservatively requires less than 1/20th the communication of Benaloh-Yung's scheme[2]. Furthermore, when more than one election is to be held, we can use an amortization technique that will boost our per-vote advantage to a factor of between 150 and 250. We note however that the communication complexity of each vote remains much greater than that required by the anonymous channel/mixer-based protocols. However, it is well within the range of feasibility, as we will discuss in greater detail in the Section 4.

Round Complexity: Our protocol enjoys an extremely low round complexity. Once the system has been set up, a voter can cast a vote in an election simply by posting a single message. The counting process consists of each counting center posting a single, very short message.

Preprocessing: Ideally, one would like to do the bulk of ones communication and computation in advance of an actual vote. In our scheme, one can preprocess ones vote with a single message. This preprocessing step doesn't depend on the vote that is eventually cast, and may be done at the time the voter "registers" to vote. When it comes time to actually vote, the voter can often simply post a single bit (or a bit along with a small integer in the worst case). Thus even with the signatures needed for identification, the communication cost is negligible. Similarly, after preprocessing, a PC acting as a voting center can easily count a hundred incoming votes per second.

Security: Under reasonable heuristic assumptions, no coalition of voters or centers can unfairly influence an election or significantly delay its outcome. A voter keeps his vote private as long as two of the centers he deals with are honest. The two honest center requirement can be reduced to a single honest center requirement by a simple doubling trick: each center simulates two centers of the original schemes.

While we use heuristic assumptions (such as the Fiat-Shamir method for noninteractive proofs) the only attack we know requires one to compute discrete-logarithms over the group we are working in. We can thus use elliptic curves for which the discrete-log problem is currently thought to be much harder than factoring. Previous number-theoretic approaches were based on the rth residue problem over Z_n^*, and are guaranteed breakable if one can factor n.

Universal Verifiability: Every action by a voter, whether preprocessing a vote or actually voting, is accompanied by a proof that the ballot is correctly constructed. The output of the counting center may also be easily checked for correctness. Any participant can verify that everyone's vote has been included in the tally. Once a party posts their message, it does not need to cooperate in the checking of its work. We can distribute the checking of the election over many parties, and if someone is still not satisfied they may check the results themselves. Thus, a voter has the option of minimally participating in an election by simply sending in their vote and then ceasing all involvement.

Flexibility: Our scheme is readily adaptable. For example, the number of centers can be made quite large. Voters may choose their own security/efficiency tradeoff and individually choose how many and which centers they use. Thus, we can practically support an election in which there are a 100 centers of which a typical voter chooses 10 and is protected as long as two of them are honest. For giant elections we can construct hierarchical counting schemes.

1.1 Techniques Used

New Techniques for Using Homomorphic Encryptions The main contribution of this paper is the use of a new tool: families of ***partially compatible*** homomorphic encryption functions. There are well known encryption functions with additive ($E(x+y) = E(x)E(y)$) or multiplicative $E(xy) = E(x)E(y)$ homomorphisms. These properties can be exploited to make very efficient zero-knowledge proofs, but can also work against security. For example, suppose that one has $E(x)$ and $E(y)$ and wishes to know whether $x + y$ is 1 or -1. If E is a function (as opposed to a probabilistic encryption) with an additive homomorphism then one can compute $E(x + y)$ and check if it is equal to $E(1)$ or $E(-1)$.

Benaloh and Yung [2] circumvent this difficulty by using a family of probabilistic encryption functions, $\{E_1, \ldots, E_n\}$. Each E_i probabilistically encrypts an element $x \in Z_r$, where r is a moderately large prime independent of i. While each E_i satisfies $E_i(x + y) = E_i(x)E_i(y)$, there is no obvious way of combining $E_i(x)$ and $E_j(y)$ to learn obtain an encryption of some simple function of x and y. A key requirement of their technique was that the encryptions E_i be probabilistic, a condition we weaken in the formalism below.

We consider a family of additive homomorphic, possibly deterministic encryption functions, $\{E_1, \ldots, E_n\}$. Within this family we have a single group Z_q (where q is large) such that $E_i(x+y) = E_i(x)E_i(y)$, where $x, y \in Z_q$. Thus, they are all basically compatible. However, we require that for all (i, j) the following two distributions are computationally indistinguishable:

1. $(E_i(x), E_j(y))$, where x and y are chosen uniformly from Z_q, and
2. $(E_i(x), E_j(x))$, where x is chosen uniformly from Z_q.

This implies that for any v, $(E_i(x), E_j(v-x))$ is indistinguishable from $(E_i(x), E_j(y))$. Thus, if x and y are chosen uniformly such that $x+y = v$, then seeing $(E_i(x), E_j(y))$

doesn't reveal anything about v. Similarly, if $x_1, \ldots, x_n$ are chosen uniformly to sum to v, then knowing $n-2$ of the values $\{x_1, \ldots, x_n\}$ tells one nothing about v.

We call such families of encryption functions with this property *partially compatible* homomorphic encryption functions. We don't know how to reduce the existence of such families to any well-known algebraic assumption. However, there are a number of candidates for such a family of encryption functions. For example, let primes $p_i = k_i q + 1$, where q is a prime, let g_i be a randomly (or pseudorandomly) chosen generator for $Z^*_{p_i}$ and let $\alpha_i = g_i^{k_i}$. If we define $E_i(x) = \alpha_i^x$ then we know of no way of obtaining any information about $x_1 + x_2$ given $E_1(x_1)$ and $E_2(x_2)$ without computing discrete logarithms. The only weakness we know of is that if $p_1 = p_2$ and ℓ such that $\alpha_2 = \alpha_1^\ell$ is known, then one can compute $E_1(x_1+x_2)^\ell = E_1(x_1)^\ell E_2(x_2)$, which allows one to determine if x_1+x_2 is equal to a given number. We know of no such attack when $p_1 \neq p_2$.

We can also incorporate encryption functions based on elliptic curves or other groups. Furthermore, we can mix which types of groups we use arbitrarily. For ease of exposition, we will use multiplicative notation for the cyclic group generated by α_i, regardless if the group is normally treated as multiplicative or additive.

Using these families of partially compatible homomorphic encryption functions, we construct very efficient interactive proofs for assertions such as:

- $x_1 + \cdots + x_n = a + b$, given the encryptions for these values, and
- $x + y \in \{1, -1\}$, given the encryptions for x and y.

Because of the efficiency of these proofs, we can afford to run them many times, and use the heuristic trick of Fiat-Shamir [7] to make them noninteractive.

Reduced Costs Through Amortization Our improved proof techniques brings the complexity of the number-theoretic techniques down to the point where they can be used by a PC that can post messages to a bulletin board. However, when we use very strong confidence parameters (2^{-40} or even 2^{-60} error rates are recommended when using the Fiat-Shamir trick) and allow the voter to protect his vote by using many (e.g., 10) of the available centers, our costs are at the outer margin of usability. Hence, we have developed techniques for making these burdens easier to bear.

By allowing nearly all of the work to be done in advance of any election, we can amortize the computational and communication burden over a much larger period of time and still have very fast elections. To lower the computational burden of a proof, we use table lookup techniques to reduce the number of group operations required. Finally, we show how a voter can process many votes using much less communication and computation than would be required to process them individually.

We note that the use of amortization in cryptography is not new. Kurosawa and Tsujii [11] construct a zero-knowledge proof for two assertions that is more efficient than simply concatenating the zero-knowledge proofs for each assertion.

Boyar, Brassard and Peralta [4] and Kilian [10] consider the problem of achieving ultra-high confidence zero-knowledge proofs for NP using less communication than is required by simple sequential repetition. Franklin and Yung [8] show how to implement k instances of a secure multiparty computation much less expensively than k times the cost of a single secure computation.

1.2 Comparison to Related Work

Our work has been motivated by the work of Benaloh [6] and the later work of Benaloh and Yung [2]. The latter protocol enjoys most of desirable security properties we obtain, and is based on partially compatible homomorphisms of the form $E_i(x) = y_i^x \cdot g_i^r \bmod n_i$. The technical advances made by our work include:

1. Greater generality: The encryption used by [2] was tuned to the factoring problem. Each center i had the prime factorization of n_i as part of its secret information. This secret information complicated the protocol in that the voters needed to verify the correctness of the centers' public information and the correctness of their subtallies through interactive protocols. Also, we can apply our method to "discrete-log type" problems - we don't know how to attack our scheme without computing discrete logarithms.
2. Amortization techniques: Unlike most previous work in voting, we consider how to run multiple elections more efficiently. Since there are usually many voters, and checking each vote involves many subchecks, we can effectively use amortization techniques to speed up single elections as well.
3. Improved zero-knowledge proofs: We give direct and efficient protocols to show, for example, that $x + y$ is either 1 or -1, without conveying which is the case. These proofs are more efficient than the *cryptographic capsule* (c.f. [3]) methods used in [6, 2].

Also, we incorporate techniques, such as the Fiat-Shamir heuristic for removing interaction, that were not available at the time of [2]. Some of these techniques can also be applied to the original [2] protocol (with varying degrees of difficulty and utility). By using more modern techniques we can better realize the basic approach laid out by Benaloh and Yung.

2 The Basic Scheme

We now describe the basic voting scheme. For simplicity, we assume that there are only two centers counting the votes, and that a single yes/no vote is being held. This scenario is solely to help understand the protocol given in Section 3. We note that this basic scheme does not protect privacy of one's vote against the centers. This issue will be overcome in Section 3, where we involve more than 2 centers.

2.1 Preliminaries and Initial Preparation

We denote the two centers by C_1 and C_2. Each vote v will be broken into shares x_1 and x_2, where x_i is a member of Z_q, and q is a prime. Before being posted, each share x_i is encrypted using encryption function E_i, where $\{E_1, E_2\}$ form a family of partially compatible homomorphic encryption functions.

As part of the setup process, which need only be done once for all time, the parties agree on $\{E_1, E_2\}$. Note that implementations based on discrete-log functions, there is no trapdoor information that need be kept hidden. Thus, for example a few bits from some global source can be fed into a pseudorandom bit generator and these random bits could be used to choose the moduli and generators needed to specify the desired functions. Heuristically, anyone can provide the seed to the pseudorandom generator, and it is unlikely that they will make it output a weak set of functions.

Along with setting up the family of encryption functions, we assume that basic primitives such as public-key cryptography and secure bit-string commitment have already been established. Let $H(x)$ denote a possibly probabilistic hash function that commits the sender to x without giving away any useful information about x.

2.2 The Election Procedure

The basic election runs in three stages: vote preparation, vote casting and vote counting.

Vote Preparation Each voter i chooses his vote v_i, 1 for yes-vote and -1 for no-vote. He uniformly generates $x_i^{(1)}$ and $x_i^{(2)}$ such that

$$x_i^{(1)} + x_i^{(2)} = v_i \bmod q.$$

He then posts $E_1(x_i^{(1)}) = \alpha_1^{x_i^{(1)}}$ and $E_2(x_i^{(2)}) = \alpha_2^{x_i^{(2)}}$ and proves $x_i^{(1)} + x_i^{(2)} \in \{1, -1\}$ without disclosing $x_i^{(1)}, x_i^{(2)}$ nor v_i. We discuss this proof in Section 2.3.

Voting Casting Each voter i encrypts $x_i^{(1)}$ and $x_i^{(2)}$ using the public keys of C_1 and C_2 respectively. Each center j computes $E_j(x_i^{(j)})$ and checks that it agrees with the previously posted value.

Counting Votes Each center j sums up $x_i^{(j)}$ modulo q for all voters i and posts sub-tally, t_j. Each voter verifies that

$$E_j(t_j) = \prod_j E_j(x_i^{(j)}),$$

and computes $T = t_1 + t_2$, which is equal to the number of "yes" votes minus the number of "no" votes.

2.3 Proving $\mathbf{x_1 + x_2 \in \{1, -1\}}$

In Figure 1 we exhibit a simple protocol for proving that $x_1+x_2 \in \{1,-1\}$ mod q given $E_1(x_1)$ and $E_2(x_2)$.

prove±1(x_1, x_2) /* Given $E_1(x_1), E_2(x_2)$, prove that $x_1 + x_2 \in \{1, -1\}$ mod q */

1. The prover uniformly chooses $r \in Z_q$ and $s \in \{1, -1\}$. He computes $R = H(r)$, the secure commitment for r, computes

$$Y_1 = E_1(s(x_1 + r)) = (E_1(x_1)E_1(r))^s \text{ and}$$
$$Y_2 = E_2(s(x_2 - r)) = \left(E_2(x_2)E_2(r)^{-1}\right)^s,$$

and posts (Y_1, Y_2, R).

2a. With probability $\frac{1}{2}$, the verifier asks the prover to reveal r and s. The verifier checks that r is consistent with R and that the above identities for Y_1 and Y_2 hold.

2b. With probability $\frac{1}{2}$, the verifier asks the prover to reveal $s(x_1 + r)$ and $t = s(x_1 + r) + s(x_2 - r) \in \{1, -1\}$. He then checks that $Y_1 = (E_1(x_1)E_1(r))^s$ and that $Y_2 = E_2(t - s(x_1 + r))$.

Fig. 1. Protocol for proving validity of shares.

Each execution of the above protocol will catch a cheating prover with probability $\frac{1}{2}$. We note that the distribution of (Y_1, Y_2) is easy to simulate given $(E_1(x_1), E_2(x_2))$. Indeed, if R is a perfect zero-knowledge bit commitment then the protocol is perfect zero-knowledge. We note that a conceptually more simple protocol would have the prover reveal $s(x_2 - r)$ in Step 2b. We chose the above protocol for its reduced communication complexity. Both s and t could also be eliminated by having the verifier check both possibilities, but this would save only 2 bits.

While we give this protocol in terms of a verifier, a more round efficient solution is to use the Fiat-Shamir method of eliminating interaction. First, the protocol is run many times (on the order of 40 or 60) in order to make the probability of withstanding all of the challenges vanishingly small. Then the verifier is replaced by a suitably "random looking" hash function which generates the challenges from the prover's posting in Step 1 of the protocol. If the prover is trying to prove an incorrect statement, then heuristically his only strategy is to run different postings through the hash function until he finds one whose challenges he can meet. However, the cost of this attack is prohibitive if the error probability is truly small (2^{-40} or 2^{-60}).

3 Many Centers and Many Votes

In the basic scheme we outline above, there were only two centers and a single yes-no vote. However, in more practical scenarios a voter will want to divide his vote among as many centers as possible - the more centers the more private the vote. Also, a voter is likely to participate in many elections and a given election is likely to have many yes/no votes. For example, Benaloh points out that approval voting (where a voter may cast a vote for any number of the given candidates) is really just a case of several independent yes/no votes[3]. In this section, we show how to split many votes over many centers with substantial amortized savings over preparing each vote separately.

For simplicity, we assume that there are only n centers and that each voter will split his votes over all n centers. For each center i we have an encryption function E_i from this family. Following the basic scheme, the voter breaks his vote $v \in \{1, -1\}$ into shares $X^{(1)}, \ldots, X^{(n)}$ such that $v = X^{(1)} + \ldots + X^{(n)}$, and then proves that these shares are correctly constructed. The most straightforward solution is to adapt protocol prove±1 to handle more than two shares. Instead, we break the proof into two stages. First, the prover randomly generates a, b such that $v = a + b$ and proves that $X^{(1)} + \ldots + X^{(n)} = A + B$. Then he uses prove±1 to prove that $v = a + b$. This gives us chance to handle multiple votes efficiently as shown in Subsection 3.2.

3.1 Proving $X^{(1)} + X^{(2)} + \cdots + X^{(n)} = A + B$

Figure 2 gives our protocol, prove-sum, for reducing a sum of n encrypted shares to a sum of 2 shares. We assume that the encryptions $E_i(X^{(i)})$, $E_a(A)$ and $E_b(B)$ are known, and that

$$(E_1, \ldots, E_n, E_a, E_b)$$

is a family of partially compatible homomorphic encryptions with domain Z_q.

If the summation assertion is not true, then in each iteration of the protocol the prover will fail a check with probability at least $\frac{1}{2}$. As before, we lower this error rate to a very small value by repeated repetition, and then use the Fiat-Shamir heuristic to make the proof noninteractive.

3.2 Proving Equations Simultaneously

The bulk of computation and communication required for the full n-party scheme is taken up by the proof of the reduction to the 2-share sum. By combining many of these proofs into a single proof, the voter can efficiently prepare many "yes/no" ballots at once with significant savings in the amortized computation and communication required.

Suppose that the voter wants to prove that the following equations hold.

$$X_1^{(1)} + X_1^{(2)} + \cdots + X_1^{(n)} = A_1 + B_1$$
$$X_2^{(1)} + X_2^{(2)} + \cdots + X_2^{(n)} = A_2 + B_2$$

prove-sum$(X^{(1)}, \ldots, X^{(n)}, A, B)$
/* Given $E_1(X^{(1)}), \ldots, E_n(X^{(n)}), E_a(A), E_b(B)$, prove that $X^{(1)} + X^{(2)} + \cdots X^{(n)} = A + B$ */

1. For $0 \le k \le n$ the prover uniformly chooses $r_i \in Z_q$. He computes the commitment $R_i = H(r_i)$ and computes $Y_1, \ldots, Y_n, Y_a, Y_b$ by

$$Y_i = E_i(X^{(i)} + r_i) \text{ for } 1 \le i \le n,$$
$$Y_a = E_a(A + r_0) \text{ and}$$
$$Y_b = E_b(B + \left(\sum_{i \ge 1} r_i\right) - r_0).$$

The prover posts $R_1, \ldots, R_n$ and $Y_1, \ldots, Y_n, Y_a, Y_b$.

2a. With probability $\frac{1}{2}$, the verifier challenges the prover to reveal $r_0, \ldots, r_n$. The verifier checks that

$$R_i = H(r_i) \text{ for } 1 \le i \le n,$$
$$Y_i = E_i(X^{(i)}) E_i(r_i) \text{ for } 1 \le i \le n,$$
$$Y_a = E_a(A) E_a(r_0), \text{ and}$$
$$Y_b = E_b(B) E_b\left(-r_0 + \sum_{i=1}^{n} r_i\right).$$

2b. With probability $\frac{1}{2}$ the verifier challenges the prover to reveal $(A + r_0), (B - r_0 + \sum_{i=1}^{n} r_i)$ and $(X^{(i)} + r_i)$ for $1 \le i \le n$. The verifier checks that

$$Y_i = E_i(X^{(i)} + r_i) \text{ for } 1 \le i \le n,$$
$$Y_a = E_a(A + r_0) = \alpha_a^{(A+r_0)} \text{ and}$$
$$Y_b = E_b(B + \left(\sum_{i \ge 1} r_i\right) - r_0) = \alpha_b^{(B + (\sum_{i=1}^{n} r_k) - r_0)}.$$

Finally, the verifier checks that

$$\sum_{i=1}^{n} (X^{(i)} + r_k) = (A + r_0) + \left(B - r_0 + \sum_{i=1}^{n} r_i\right).$$

Fig. 2. Protocol for proving summation assertions.

$$\vdots$$
$$X_m^{(1)} + X_m^{(2)} + \cdots + X_m^{(n)} = A_m + B_m$$

and the values of $E_i(X_j^{(i)})$, $E_a(A_j)$ and $E_b(B_j)$ are known for $1 \le i \le n$ and $1 \le j \le m$. Let coefficients $c_1, \ldots, c_m \in Z_q$ be chosen at random, and consider the following linear equation:

$$\bar{X^{(1)}} + \bar{X^{(2)}} + \cdots \bar{X^{(n)}} = \bar{A} + \bar{B}, \text{ where}$$

$$\bar{X^{(i)}} = \sum_{j=1}^{m} c_j X_j^{(i)}$$

$$\bar{A} = \sum_{j=1}^{m} c_j A_j \text{ and}$$

$$\bar{B} = \sum_{j=1}^{m} c_j B_j.$$

By a simple probability argument, the following facts hold:

1. If all of the original linear equations were true, then the new linear equation will also be true, and
2. If at least one of the original linear equations is false, then the new linear equation will be false with probability $1 - 1/q$.

Thus, a proof of the new equation will suffice as a proof of *all* of the original equations.

It remains to show how to generate the encryptions for the new variables and how to choose the coefficients. The encryptions are given by

$$E_i(\bar{X^{(i)}}) = \prod_{j=1}^{m} E_i(X_j^{(i)})^{c_j}$$

$$E_a(\bar{A}) = \prod_{j=1}^{m} E_a(A_j)^{c_j} \text{ and}$$

$$E_b(\bar{B}) = \prod_{j=1}^{m} E_b(A_j)^{c_j}.$$

We can view the c_i coefficients as challenges. As before, we can use the Fiat-Shamir trick and generate the c_i by a hash function of the original encryptions. Note that in this case, we do not have to perform the operation multiple times, since for a random setting of the coefficients an error in the original set of equations will result in an error in the final equation with all but vanishing probability. Indeed, for computational efficiency it suffices to choose c_i from $\{1, \ldots, 2^{60}\}$, which will greatly speed up the exponentiations.

We give a summary of the election scheme in Figure 3.

m votes n centers Election Scheme

Precomputation (In the following description, i ranges over $1, \ldots, n$, j ranges over $1, \ldots, m$, and k ranges over the indices of all valid voters V_k.)

1. The vote counters, $C_1, C_2, \cdots, C_n$ agree on a randomly selected family of partially compatible encryption functions, E_a, E_b, $\{E_i\}$ and post them.
2. Each voter V_k randomly chooses his "masking vote" $v_{k,j} \in \{1, -1\}$ and then randomly chooses two representations for $v_{k,j}$,

$$v_{k,j} = X_{k,j}^{(1)} + X_{k,j}^{(2)} + \cdots + X_{k,j}^{(n)} = A_{k,j} + B_{k,j}.$$

   He posts $E_i(X_{k,j}^{(i)})$, $E_a(A_{k,j})$ and $E_b(B_{k,j})$.
3. V_k executes the reduction given in Section 3.2 and obtains a single equation

$$\bar{X_k^{(1)}} + \bar{X_k^{(2)}} + \cdots + \bar{X_k^{(n)}} = \bar{A_k} + \bar{B_k},$$

   equality of which implies that the given two representations of masking vote $v_{k,j}$ are equal for all j. V_k then proves the validity of the equation using prove-sum.
4. V_k executes protocol prove±1 to show that $A_{k,j} + B_{k,j} \in \{1, -1\}$ for all j.
5. V_k encrypts $X_{k,j}^{(i)}$ using C_i's public encryption algorithm, for all j and i, and posts the encryptions.

Vote Casting

1. To use his jth vote, voter k computes $s_{k,j} \in \{-1, 1\}$ such that his actual vote is equal to $s_{k,j} v_{k,j}$, and posts j and $s_{k,j}$. Note that often j is known and thus need not be sent. We assume that the correct j for each voter is known, and henceforth drop j from our notation.

Vote Counting

1. Each center C_i decrypts $X_k^{(i)}$ for all k and j, and verifies whether it is consistent with $E_i(X_k^{(i)})$. He calculates his subtally $t^{(i)} = \sum_k s_k \cdot X_k^{(i)}$ and posts $t^{(i)}$.
2. Each person checking the vote verifies that $\prod_k (E_i(X_k^{(i)}))^{s_k}$ is equal to $E_i(t^{(i)})$. If so, they accept $\sum_i t^{(i)}$ as the tally of the vote.

If precomputation is not used, then v_k can be used as the actual vote and s_k is omitted.

Fig. 3. Overview of the general protocol.

4 Efficiency

4.1 Communication Cost

In this subsection, we roughly estimate communication cost of our scheme. While there are many possible variations of our scheme, a good understanding of their complexity can be had by analyzing the cost of splitting a vote into encrypted shares and proving that the shares are well formed.

A number of security parameters come into play in this analysis. First, let us assume that our encryption functions are based on modular exponentiation over $Z_{p_i}^*$, and let k be an upper bound on the length of p_i (if different moduli are uses, then they will not exactly the same size). Let h be the output of the hash function H used for commitments and let l be the security parameter that effectively denotes how many times the proofs are run.

We consider the most general case of splitting m votes to n centers. Note that for m large, we achieve a higher amortized efficiency due to the method used in Section 3.2. Not counting the cost of the proof, representing these pieces along with the additional 2 shares used in our reduction requires $(n+2)km$ bits. The cost of proving the correctness of the combined equation (from Section 3.2) is $[2(n+2)k+(n+1)h]l$ bits. At this point, the voter has proved that each set of n shares representing a vote is equal to the two auxiliary shares. The proof that the two auxiliary shares sum to 1 or -1 costs $[3k+h]lm$ bits. The cost of revealing these shares to the proper counting authorities is approximately nkm bits. Altogether, this gives a total of $2(n+1)mk+[(2n+2)k+(n+1)h+(3k+h)m]l$ bits. Some of the resulting numbers are shown in Table1. If one uses the "center doubling" trick so as to need only one good center instead of two, then the costs are all doubled.

4.2 Computation Cost for Voters

In this subsection, we roughly estimate computation cost for the voters. In the scheme, the costly computations are mainly modular multiplication and modular exponentiation. Note that many modular exponentiations with the same base are being done. We can exploit this fact by computing lookup tables that will reduce the number of multiplications required by the exponentiations. For example, we can precompute α_j^i for all i's that are powers of 2. This will reduce the average number of multiplications needed to compute $\alpha_x \bmod p$ from $\frac{3}{2}k$ to $\frac{1}{2}k$, requiring a table size of $(n+2)k^2$ bits. Using a more sophisticated table, we can obtain a further factor of 3 for the typical number ranges.

We again consider the case of splitting m votes into n shares each. Splitting m votes to mn pieces requires $\frac{1}{2}(n+2)km$ multiplications. $\frac{1}{2}(n+2)kl$ multiplications are needed for proving the reduction to the the reduced 2-share representation of a vote. klm multiplications are needed to complete the proof that the votes are well formed. Verifying the subtallies of each center requires $(\frac{1}{2}k+[\#\text{ of voters}])nm$ modular multiplications.

Altogether, our protocol requires approximately

$$\frac{1}{2}[(2n+2+2l)m+(n+2)l]k+[\#\text{ of voters}]nm$$

modular multiplications. A PC running at 33MHz can executes 768 multiplications in a second. Based on this, some of the resulting numbers are shown in Table 1.

Note that these figures are only approximate. However, the cost of the other modular addition or such operations as computing hash functions is comparatively negligible.

4.3 Computation Cost for Verifying Other Voters

In this subsection, we roughly estimate the computation cost needed for verification. Here again, k is the length of p, and l is a security parameter which determines maximum probability of cheating. c is the length of coefficients used in the protocol which can be set small. Also, modular exponentiation can exploit the previously mentioned table lookup techniques.

We again consider the case of splitting m votes into n shares each. $\frac{1}{2}(n+2)|c|(m-1)$ multiplications are needed for generating the encryption of the shares, including their representation. $\frac{1}{2}(n+2)(k+1)l$ multiplications are needed to verify that the combined equation is correct. $(k+1)lm$ multiplications are needed to complete the proofs that the shares are well formed. Altogether, this yields $\frac{1}{2}[((n+2)|c|+2l(k-1))m+(n+2)(kl-l-|c|)]$ modular multiplications for each voter.

We can reduce this number by using techniques for verifying many modular exponentiations, resulting in a factor of 4 improvement over actually computing the exponentiations.

4.4 Efficiency Comparisons with Previous Work

The work of [2] gave the first protocol where votes are divided into pieces and the verifiable subtally yields total outcome of voting. However, their scheme suffers from large communication complexity and seems not yet practical for implementation on existing networks. One of the reasons they need large communication complexity is that each centers i generate secret prime factors of their public key N_i. Therefore the scheme involves an interactive protocol to detect possible cheating at the setting of the public keys, together with an interactive protocol to show detected cheat was not due to a malicious voter. Also, since extra information of subtally may reveal these secret primes, an interactive protocol was necessary to prove the correctness of subtally. For above reasons, their protocol needed $(4l^2+5l+2)kn$ bits for communication, where k is the size of the public keys of the n centers, and l is a security parameter.

The computation complexity is rather small for each iteration of their scheme, since the computation is based on $y^e x^r \bmod n$ where e and r are much smaller than n. However, since this interactive proof takes place many times, the total cost does not remain so small. We estimate their total computation assuming that they use the same trick constructing a table of $y^i \bmod n_j$ that requires nrk bits. Then, there will be $3(l^2+3l+1)\lg rn + 2(l^2+l+1)n + [\# \text{ of voters}]n$ n bit modular multiplication in total, where k is the size of public keys of the n centers, r determines number of voters and l is a security parameter. We give a rough numerical comparison in Table 1.

	1 vote, $l = 40$	1 vote, $l = 20$	100 votes, $l = 40$
Proposed Scheme	56K bytes 2.5 min.	28K bytes 1.5 min	1M bytes 58 min.
Benaloh & Yung (1000 voters)	4M bytes 11 min	1M bytes 3 min	400M bytes 19 hrs

All with $n = 10$ centers.

Table 1. Comparisons with the Benaloh - Yung Scheme

Acknowledgments

We would like to thank Mr. Masumoto of Central Labs for pointing out a vulnerability in our earlier scheme and for supplying us with data on the computational capabilities of various machines. We would further like to thank Josh Benaloh, Moti Yung and anonymous referees for their valuable comments.

References

1. Benaloh, J. C., Tuinstra, D.: Receipt-free secret-ballot elections. In STOC 94 (1994) pp. 544–553
2. Benaloh, J., Yung, M.: Distributing the power of a government to enhance the privacy of voters. In ACM Symposium on Principles of Distributed Computing (1986) pp. 52–62
3. Benaloh, J.: Verifiable secret-ballot Elections. PhD thesis Yale University 1987. YALEU/DCS/TR-561
4. Boyar, J., Brassard, G., Peralta, R.: Subquadratic zero-knowledge. In FOCS 91 (1991) pp. 69–78
5. Chaum, D.: Untraceable electronic mail, return addresses, and digital pseudonyms. In Communications of the ACM. ACM 1981 pp. 84–88
6. Cohen, J., Fischer, M. J.: A robust and verifiable cryptographically secure election scheme. In FOCS 85 (1985) pp. 372–382.
7. Fiat, A., Shamir, A.: How to prove yourself: Practical solutions to identification and signature problems. In Advances in Cryptology –Crypto '86. Springer-Verlag 1986 pp. 186–199
8. Franklin, M., Yung, M.: Communication complexity of secure computation. In STOC 92 (1992) pp. 699–710
9. Fujioka, A., Okamoto, T., Ohta, K.: A practical secret voting scheme for large scale elections. In Advances in Cryptology –Auscrypt '92 (1992) pp. 244–251
10. Kilian, J.: A note on efficient zero-knowledge proofs and arguments. In STOC 92 (1992) pp. 722–732
11. Kurosawa, K., Tsujii, S.: Multi-language zero knowledge interactive proof systems. In Advances in Cryptology –Crypto '90 (1991) pp. 339–352
12. Park, C., Itoh, K., Kurosawa, K.: All/Nothing election scheme and anonymous channel. In Advances in Cryptology –EUROCRYPT '93 (1993)pp. 248–259

Maintaining Security in the Presence of Transient Faults

Ran Canetti[1]* and Amir Herzberg[2]

[1] Department of Applied Mathematics and Computer Science, Weizmann Institute, Israel, canetti@wisdom.weizmann.ac.il.
[2] IBM T.J. Watson Research Center, amir@watson.ibm.com.

Abstract. Consider a multiparty system where parties may occasionally be "infected" by malicious, coordinated agents, called viruses. After some time the virus is expelled and the party wishes to regain its security. Since the leaving virus knows the entire contents of the infected party's memory, a source of "fresh" randomness seems essential for regaining security (e.g., for selecting new keys). However, such an "on-line" source of randomness may not be always readily available.
We describe a scheme which, using randomness only at the beginning of the computation, supplies each party with a new pseudorandom number at each round of communication. Each generated number is unpredictable by an adversary controlling the viruses, even if the party was infected in previous rounds. Our scheme is valid as long as in each round there is at least *one* noninfected party, and some of the communication links are secure.
We describe an important application of our scheme to secure sign-on protocols.

1 Introduction

Traditionally, cryptography was focused on protecting interacting parties (i.e., computers) against *external* malicious entities. Such cryptographic tasks include private communication over insecure channels, authentication of parties, unforgeable signatures, and general multiparty secure computation. An inherent property of all these scenarios is that once a party is "corrupted" it remains this way.

However, as computers become more complex, *internal* attacks on computers (i.e., attacks that corrupt components within a computer) have become an even more important security threat [LE93, Sto88]. Such attacks may be performed by internal (human) fraud, operating system weaknesses, or Trojan horse software (e.g. viruses). Security administrators often find internal attacks more alarming than external attacks, such as line tappings. An important property of internal

* Part of this research was done while visiting IBM T.J. Watson Research Center. Supported by grant no. 92-00226 from the United States — Israel Binational Science Foundation, Jerusalem, Israel.

attacks is that they are often temporary, or transient [ER89]. Thus the paradigm of "bad once means bad forever" does not hold here.

Still, almost no known solution to internal attacks allows recovery from faults. We briefly outline two main approaches underlying the known solutions to internal attacks. One approach to reducing the vulnerability of computers to internal attacks is to minimize the use of data that has to be kept secret. Most notable in this approach is the use of public key cryptosystems, where only the private key has to be kept secret (see, for instance, Novell NetWare 4.0). Still, the public key approach requires a secure, infallible certification entity. A different approach follows the popular paradigm of 'not putting all eggs in one basket'. That is, critical components are multiplied, and the overall security of the system is ensured as long as the attacker is unable to break all, or a large number, of the components.

However, in all these solutions, since the faults are assumed to be non-transient, there is no mechanism for taking advantage of a possible *recovery* of a component. This approach is contrasted with the traditional approach of fault-tolerance, which relies heavily on the fact that faults are transient, and on the reuse of recovered components. We believe that the idea of recovering and reusing components that have once been corrupted can be extremely useful also for cryptographic purposes. This idea, and in particular the recovery process, is the focus of this work.

Note that we may not know whether a particular component has been corrupted, and when the attackers have left; thus, the recovery process will be invoked periodically, regardless of whether we identified an attack. We call such an approach proactive, and say that such systems provide proactive security.

The goal of the recovery process is to ensure that once a component is recovered, it will again contribute to the overall security of the system. This goal is somewhat tricky. Even after the attacker loses control of a component, it still knows the (possibly modified) state of the component (e.g., the private cryptographic keys). Thus, a first step in the recovery process *must* be to somehow hand the recovering component some new secrets unknown to the attacker. These secrets can then be used to, say, choose new keys. The obvious way to generate such secrets is to use some source of "fresh", physical randomness. However, such a source may not be readily available or beneficial to use. In this paper we show that new secrets can be generated *without* fresh randomness. Instead, we use other, non-corrupted components in the system.

We consider a system (network) of components (parties) where every two parties are connected via a communication channel. (We elaborate below on the security requirements from the channels.) Parties may be temporarily corrupted (or infected) by malicious agents, called viruses. We assume that the viruses are controlled by an adversary. The adversary may choose to infect different parties at different times (i.e., communication rounds), as long as the *number* of infected parties is limited. We stress that there may be no party that has never been infected! This model was suggested by Ostrovsky and Yung [OY91], who also coined the terms 'infected' and 'viruses'.

We assume that, even if the faults are Byzantine, once the virus has left the party resumes executing its original *code* of the protocol (while the state may be corrupted). This assumption is explained as follows. If the virus can control the code after it leaves, then there is little meaning to recovery and it would be impossible to regain security. Furthermore, in practice there are reasonable ways to ensure that the code is not modified, such as physical read-only storage or comparison against backup copies. These techniques are used regularly in many systems.

In this work, we describe a scheme in which the parties use randomness *only at the beginning of the computation.* At each round, the scheme supplies each noninfected party with a "fresh" pseudorandom number, unpredictable by the adversary, even if this party was infected in previous rounds, and if the adversary knows all the other pseudorandom numbers supplied to any party at any round. In particular, these pseudorandom numbers can be used by a recovering party just as fresh random numbers (e.g., for regaining security). We call such a scheme a proactive pseudorandomness (PP) protocol.

Our implementation is simple, using pseudorandom functions [GGM86, GGM84]. We require the following very weak conditions. First, we assume that the adversary is computationally bounded. Next, we require that in each round of computation there is at least *one* secure party. A party is secure at a given round if it is noninfected at this round, and it has a secure (i.e., authenticated and private) link to a party that was secure in the previous round. We note that this link has to be secure only at this round.

1.1 Reconstructability and its Application to Secure Sign-On

Reconstructability. Pseudorandom generators, being deterministic functions applied to a random seed, have the following advantage over truly random sources. A pseudorandom sequence is reconstructible, in the sense that it is possible to generate exactly the same sequence again by using the same seed. This property is very useful for several purposes, such as repeatable simulations and debugging, and is used in the construction of pseudo-random functions [GGM86]. Our application to secure sign-on protocols also makes use of this property.

In our setting, we say that a PP protocol is reconstructible if the value generated within each party at each round depends only on the seeds chosen by the parties at the beginning of the computation. In particular, these values should not depend on the random choices of the adversary.

Reconstructability is not easily achieved for proactive pseudorandomness protocols. In particular, the basic protocol described in this paper is reconstructible only if the faults are passive (namely, "eavesdropping" only). Crash faults (and also Byzantine faults, at the price of slightly compromising the security) could be tolerated by simple modifications described in section 5.

Application of Reconstructability to Secure Sign-On. Unix and other operating systems provide security for the passwords by storing only a one-way function of the passwords on disk [MT79]. This technique allows authentication of the users, secure against eavesdropping the password file. Session security is not provided if the communication links are not secure.

In secure LAN systems, it is not realistic to assume that the communication is secure. Security mechanisms, therefore, avoid sending the password "on the clear". Instead, they use the user's password to derive a *session key*, with which they secure the communication. In both Kerberos [MNSS87] and NetSP / KryptoKnight[BGH+93a], this is done by using the password as a key for exchanging a random session key; this method also allows NetSP / KryptoKnight to authenticate the user automatically to additional systems ('single sign on').

However, this mechanism implies that some server must be able to compute the session key itself, using some secret (e.g. the password). This in turn implies that the server has to maintain the password file *secret*. This secrecy requirement is a major 'Achilles heel' of any security system. Indeed, NetWare 4.0 provides a more complicated and expensive solution, where the server keeps, for each user, an RSA private key encrypted using the user's password. The encrypted private key is sent to the workstation, which decrypts it using the password, and then uses it to derive a session key. This solution does not require the password file to be secret (but it is assumed that the password file is not modified).

We show how a reconstructible proactive pseudorandomness protocol can be used to overcome this weakness, without compromising efficiency. Our solution uses several proactive sign-on servers. The servers run a different copy of our PP protocol for each user. The initial seed of each server P_i is a pseudorandom value derived from the user's password, e.g. $f_{PW}(i)$. Each server sets its key for each time period to be the current output of the PP protocol. The user, knowing all the servers' inputs of this deterministic computation, can simulate the computation and compute each server's key at any time period *without need for any communication.* Thus, a user can always interact with the server of his choice. The security of our PP protocol makes sure that a mobile adversary does not know the key currently used by a secure server, as long as in each round there exists at least one secure server.

This solution does not require public key mechanisms, and is secure even if the attacker can modify the login files kept by the servers.

1.2 Related Works

Previous works have addressed the issue of transient faults in different ways. Reischuk [Rei85] designed a Byzantine agreement protocol which is able to tolerate faults covering a fraction of the network, if they remain stationary for a given interval of time. His results have been recently been extended by Garay [Gar94], who has also noted that randomization could provide a solution better in (expected) time by using the coin tossing techniques of [FM88]. Ostrovsky and Yung [OY91] showed how to perform secret sharing and how to compute any function in the presence of transient viruses. Their work required the parties to have a trusted random source, as well as complete security of the links; furthermore, their protocols tolerate only a small fraction of infected parties at each round. In [FY93], Franklin and Yung present a graph theoretic approach to privacy in a system with transient eavesdroppers.

We note that our PP protocol can be used to provide randomness to both [OY91] and [FM88].

Our application to single sign-on provides protection for both server and user against impersonation, even when attacker may break into servers. Traditional unix security [MT79] keeps only a hashing of the password, which allows an attacker (which reads the hashed-password file) to spoof the user. This has been extended by [BM93] to protect weak passwords from dictionary attack.

Organization. In Section 2 we define PP protocols, and recall the definition of pseudorandom function families. In Sections 3 and 4 we describe our PP protocol and prove its correctness. In Section 5 we describe some modifications to our application to secure sign-on protocols. In Appendix A we offer an alternative definition of PP protocols, and show that it is implied by our first definition.

2 Definitions

In this section we describe the communication model, the assumptions on the adversary, and define proactive pseudorandomness protocols.

Communication network. We consider a network where every two parties are connected via a secure (i.e., private and authenticated) communication channel. (In Section 4.1 we describe a relaxation of this security requirement on the links.) The parties are synchronized. Namely, the communication proceeds in rounds and all parties know the current round number. For simplicity, we also assume that at the end of each round, each party can send a message to each other party; these messages are received at the beginning of the next round. It is simple to extend our results to more realistic communication and synchronization models.

The Adversary. We assume that parties are occasionally being infected by some external, malicious entities, called viruses. Upon infecting a party, the entire contents of the party's memory becomes known to the virus. Furthermore, the virus can alter the party's memory and program. After some time the virus is discovered and removed from the party. Once a virus is removed, the party returns to execute its original program; however, its memory may have been altered. Assuming that the rate in which new parties are infected is roughly equal to the rate in which the viruses are being removed, we may model this scenario as follows. We assume a limit, t, on the number of viruses. The adversary may decide which parties are infected at each round, as long as at each round the number of infected parties is at most t.

We assume that the viruses cooperate. That is, there exists an entity that accumulates all the data gathered by the viruses and coordinates their activity. We call this entity a mobile adversary. We say that a mobile adversary is t-limited if in each round of the computation at most t parties are infected. (We stress that there may exist no party that has never been infected!)

A definition of proactive pseudorandomness. Consider a network of parties performing some computation in the presence of a mobile adversary. We limit the parties' access to randomness to the beginning of the computation. Once the interaction starts no additional randomness is available. Still, the parties may need secret, "fresh" random input at different rounds of the computation (e.g., in order to recover from a virus).

For this purpose, the parties will run a special deterministic protocol; this protocol will generate a new value within each party at each round. Given that the parties' initial inputs of this protocol are randomly chosen, the value generated within each party at each round will be indistinguishable from random from the point of view of a mobile adversary, *even if the values generated within all the other parties, at all rounds, are known.* We call such a protocol a proactive pseudorandomness (PP) protocol.

We stress that at each round the adversary may know, in addition to the data gathered by the viruses, the outputs of all the parties (including the noninfected ones) at all the previous rounds. Still, it cannot distinguish between the current output of a noninfected party and a random value.

More specifically, consider the following attack, called an on-line attack, with respect to an n-party PP protocol. Let the input of each party be taken at random from $\{0,1\}^k$, where k is a security parameter (assume $n < k$). Furthermore, each party's output at each round is also a value in $\{0,1\}^k$.

On-line attack: *The protocol is run in the presence of a mobile adversary for m rounds (m is polynomial in n and k), where in addition to the data gathered by the viruses, the adversary knows the outputs of all the parties at all the rounds. At a certain round, l (chosen "on-line" by the adversary), the adversary chooses a party, P, out of the noninfected parties at this round. The adversary is then given a test value, v, instead of P's output at this round. The execution of the protocol is then resumed for rounds $l+1,\ldots,m$.* (Our definition will require that the adversary be unable to say whether v is P's output at round l, or a random value.)

For an n-party protocol π, and a mobile adversary A, let $A(\pi,\mathrm{PR})$ (respectively, $(A(\pi,\mathrm{R}))$ denote the output of A after an on-line attack on π, and when the test value v given to A is indeed the output of the specified party (respectively, when v is a random value). Without loss of generality, we assume that $A(\pi,\mathrm{PR}) \in \{0,1\}$.

Definition 1. Let π be a deterministic n-party protocol with security parameter k. We say that π is a t-resilient proactive pseudorandomness protocol (PP) if for every t-limited polynomial time mobile adversary A, for all $c > 1$ and all large enough k we have

$$|\mathrm{Prob}(A(\pi,\mathrm{PR})=1) - \mathrm{Prob}(A(\pi,\mathrm{R})=1)| \leq \frac{m}{k^c}$$

where m is the total number of rounds of protocol π, and the probability is taken over the parties' inputs of π and the choices of A. (We stress that m is polynomial in k.)

We say that π is efficient if it uses resources polynomial in n and k.

An alternative definition. Using Definition 1 above, it can be shown that the following property holds for any randomized application protocol α run by the parties. Consider a variant, α', of α that runs a PP protocol π along with α, and uses the output of π as random input for α at each round. Then, The parties' outputs of α and α' are indistinguishable; furthermore, running α' the adversary gains no knowledge it did not gain running α. In fact, this property may serve as an alternative definition for PP protocols. In Appendix A we present a more precise definition of this property.

We can now state our main result:

Theorem 2. *If one-way functions exist, then for all $n \in \mathrm{N}$ there exists an efficient, $(n-1)$-resilient PP protocol for n parties.*

Pseudorandom function families. Our constructions make use of pseudorandom functions families. We briefly sketch the standard definition.

Let $\mathcal{F}_k$ denote the set of functions from $\{0,1\}^k$ to $\{0,1\}^k$. Say that an algorithm D with oracle access distinguishes between two random variables f and g over $\mathcal{F}_k$ with gap $s(k)$, if the probability that D outputs 1 with oracle to f differs by $s(k)$ from the probability that D outputs 1 with oracle to g. Say that a random variable f over $\mathcal{F}_k$ is $s(k)$-pseudorandom if no polynomial time (in k) algorithm with oracle access distinguishes between f and $g \in_R \mathcal{F}_k$ with gap $s(k)$. (Throughout the paper, we let $e \in_R D$ denote the process of choosing an element e uniformly at random from domain D.)

We say that a function family $F_k = \{f_\kappa\}_{\kappa \in \{0,1\}^k}$ (where each $f_\kappa \in \mathcal{F}_k$) is $s(k)$-pseudorandom if the random variable f_κ where $\kappa \in_R \{0,1\}^k$ is $s(k)$-pseudorandom. A collection $\{F_k\}_{k \in \mathbf{N}}$ is pseudorandom if for all $c > 0$ and for all large enough k, the family F_k is $\frac{1}{k^c}$-pseudorandom. We consider pseudorandom collections which are efficiently constructible. Namely, there exists a polytime algorithm that on input $\kappa, x \in \{0,1\}^k$ outputs $f_\kappa(x)$.

Pseudorandom function families and their cryptographic applications were introduced by Goldreich, Goldwasser and Micali [GGM86, GGM84]. Applications to practical key distribution and authentication protocols were shown by Bellare and Rogaway [BR93]. In [GGM86] it is shown how to construct pseudorandom functions from any pseudo-random generator, which in turn could be constructed from any one-way function [HILL93]. However, practitioners often trust and use much simpler constructions based on DES or other widely available cryptographic functions.

3 The Protocol

In this section we describe the basic protocol. Several modifications useful for the application to secure sign-on are described in Section 5.

Consider a network of n parties, $P_1, \ldots, P_n$, having inputs $x_1, \ldots, x_n$ respectively. Each input value x_i is uniformly distributed in $\{0,1\}^k$, where k is a security parameter. We assume that parties have agreed on a predefined pseudorandom function family $F = \{f_\kappa\}_{\kappa \in \{0,1\}^k}$, where each $f_\kappa : \{0,1\}^k \to \{0,1\}^k$.

In each round l each party P_i computes an internal value (called a key), $\kappa_{i,l}$, in a way described below. P_i's output at round l, denoted $r_{i,l}$, is set to be $r_{i,l} = f_{\kappa_{i,l}}(0)$, where 0 is an arbitrary fixed value.

The key $\kappa_{i,l}$ is computed as follows. Initially, P_i sets its key to be its input value, namely $\kappa_{i,0} = x_i$. At the end of each round $l \geq 0$, party P_i sends $f_{\kappa_{i,l}}(j)$ to each party P_j. Next, P_i *erases* its key for round l and sets its key for round $l+1$ to the bitwise exclusive or of the values received from all the parties at this round:

$$\kappa_{i,l+1} = \oplus_{j=1}^{n} f_{\kappa_{j,l}}(i) \quad (1)$$

We stress that it is crucial that the parties *erase* the old keys. In fact, if parties *cannot erase* their memory, proactive pseudo-randomness is impossible. In particular, once each party has been infected in the past, the adversary has complete information on the system at, say, the first round. Now the adversary can predict all the subsequent outputs of this deterministic protocol.

4 Analysis

We first offer some intuition for the security of our protocol. This intuition is based on an inductive argument. Assume that, at round l, the key of a noninfected party is pseudorandom from the point of view of the adversary. Therefore, the value that this party sends to each other party is also pseudorandom. Furthermore, the values received by different parties seem unrelated to the adversary; thus, the value that each party receives from a noninfected party is pseudorandom from the point of view of the adversary, even if the values sent to other parties are known. Thus, the value computed by each noninfected party at round $l+1$ (being the bitwise exclusive or of the values received from all the parties) is also pseudorandom.

Naturally, this argument serves only as intuition. The main inaccuracy in it is in the implied assumption that we do not lose any pseudo-randomness in the repeated applications of pseudorandom functions. A more rigorous proof of correctness (using known techniques) is presented below.

Theorem 3. *Our protocol, given a pseudorandom function family, is an efficient, $(n-1)$-resilient PP protocol.*

Proof. Let π denote our protocol (run for m rounds). Assume there exists a polytime mobile adversary A such that

$$|\text{Prob}(A(\pi, \text{PR}) = 1) - \text{Prob}(A(\pi, \text{R}) = 1)| > \frac{m}{k^c}$$

for some constant $c > 0$ and some value of k. For simplicity we assume that A infects exactly $n-1$ parties at each round, and that A always runs the full m rounds before outputting its guess. The proof can be easily generalized to all A. We show that F_k is not pseudorandom. Specifically, we construct a distinguisher D that distinguishes with gap $\frac{1}{2k^c}$ between the case where its oracle is taken at random from F_k and the case where its oracle is a random function in $\mathcal{F}_k$.

In order to describe the operation of D, we define hybrid probabilities as follows. First, define $m+1$ hybrid protocols, $H_0, \ldots, H_m$, related to protocol π. Protocol H_i instructs each party P_s to proceed as follows.

- In rounds $l \leq i$ party P_s outputs a random value and sends a random value to each other party P_t (instead of $f_{\kappa_{s,l}}(0)$ and $f_{\kappa_{s,l}}(t)$, respectively). In other words, P_s uses a random function from $\mathcal{F}_k$ instead of $f_{\kappa_{s,l}}$ for his computations.
- In rounds $l > i$ party P_s executes the original protocol, π.

Ofcourse, whenever a party is infected it follows the instructions of the adversary.

Distinguisher D, given oracle access to function g, operates as follows. First, D chooses at random a round number $l_0 \in_R [0, \ldots, m-1]$. Next, D runs adversary A on the following simulated on-line attack on a network of n parties. The infected parties follow the instructions of A. The (single) noninfected party at each round l, denoted $P_{*(l)}$, proceeds as follows.

1. In rounds $l < l_0$, party $P_{*(l)}$ outputs a random value and sends a random value to each other party (as in the first steps of the hybrid interactions).
2. In round l_0, party $P_{*(l_0)}$ uses the oracle function g to compute its output and messages. Namely, it outputs $g(0)$ and sends $g(j)$ to each other party P_j.
3. In rounds $l > l_0$, party $P_{*(l)}$ follows protocol π.

(Note that D knows which parties are infected by A at each round.)

Once a round is completed, D reveals all the parties' outputs of this round to A (as expected by A in an on-line attack). When A asks for a test value v, D proceeds as follows. First, D chooses a bit $b \in_R \{0,1\}$. If $b=0$, then D sets v to the actual corresponding output of the party chosen by A. Otherwise, D sets v to a random value. Finally, if $b=0$ then at the end of the simulated interaction D outputs whatever A outputs. If $b=1$ then D outputs the *opposite value* to whatever A outputs.

The operation of D can be intuitively explained as follows. It follows from a standard hybrids argument that there must exist an i such that either $|\text{Prob}(A(H_i, \text{PR}) = 1) - \text{Prob}(A(H_{i+1}, \text{PR}) = 1)|$ is large or $|\text{Prob}(A(H_i, \text{R}) = 1) - \text{Prob}(A(H_{i+1}, \text{R}) = 1)|$ is large. Thus, if D chooses the "correct" values for l_0 and b it can use the output of A to distinguish between the two possible distributions of its oracle. We show that a similar distinction can be achieved if l_0 and b are chosen at random.

We analyze the output of D as follows. Let $\text{PR}_i \triangleq \text{Prob}(A(H_i, \text{PR}) = 1)$. (Namely, PR_i is the probability that A outputs 1 after interacting with parties running protocol H_i and when the test value given to A is indeed the corresponding output value of the party that A chose.) Similarly, let $\text{R}_i \triangleq \text{Prob}(A(H_i, \text{R}) = 1)$. Let ρ (resp., ϕ) be a random variable distributed uniformly over $\mathcal{F}_k$ (resp., over F_k). Assume that D is given oracle access to ρ. Then, at round l_0 party $P_{*(l_0)}$ outputs a random value and sends random values to all the other parties.

Thus, the simulated interaction of A is in fact an on-line attack of A on protocol H_{l_0}. Therefore, if $b = 0$ (resp., if $b = 1$), then D outputs 1 with probability PR_{l_0} (resp., $1 - \mathrm{R}_{l_0}$). Similarly, D is given oracle access to ϕ then the simulated interaction of A is an on-line attack of A on protocol H_{l_0+1}. In this case, if $b = 0$ (resp., if $b = 1$), then D outputs 1 with probability PR_{l_0+1} (resp., $1 - \mathrm{R}_{l_0+1}$). Thus,

$$\mathrm{Prob}(D^{\rho} = 1) - \mathrm{Prob}(D^{\phi} = 1) = \frac{1}{2m}\sum_{i=0}^{m-1}(\mathrm{PR}_i + 1 - \mathrm{R}_i) - \frac{1}{2m}\sum_{i=0}^{m-1}(\mathrm{PR}_{i+1} + 1 - \mathrm{R}_{i+1})$$

$$= \frac{1}{2m}\sum_{i=0}^{m-1}[(\mathrm{PR}_i - \mathrm{R}_i) - (\mathrm{PR}_{i+1} - \mathrm{R}_{i+1})]$$

$$= \frac{1}{2m}[(\mathrm{PR}_0 - \mathrm{R}_0) - (\mathrm{PR}_m - \mathrm{R}_m)].$$

Clearly, H_0 is the original protocol π. Thus, by the contradiction hypothesis, $|\mathrm{PR}_0 - \mathrm{R}_0| > \frac{m}{k^c}$. On the other end, in protocol H_m the parties output random values in all the m rounds, thus $\mathrm{PR}_m - \mathrm{R}_m = 0$. We conclude that $|\mathrm{Prob}(D^{\rho} = 1) - \mathrm{Prob}(D^{\phi} = 1)| > \frac{1}{2m} \cdot \frac{m}{k^c} = \frac{1}{2k^c}$. □

4.1 Insecure Links

When describing the model, we assumed that all the communication links are secure (i.e., private and authenticated). Here, we discuss the effect of insecure links on our protocol. We note that the protocol remains a PP protocol even if in each round l only a single noninfected party P_i has a single link which is secure in round l to a party P_j that wasn't infected in round $l-1$ (and P_j had in round $l-1$ a secure link to a noninfected party, etc.).

This security requirement on the links is minimal in the following sense. If no randomness is allowed after the interaction begins then a mobile adversary that sees the entire communication can continue simulating each party that has once been infected, even after the virus has left this party. Thus, after few rounds, the adversary will be able to simulate all parties and predict the output of each party at each subsequent round.

5 On the Application to Secure Sign-On

In subsection 1.1 we discussed reconstructible protocols and described an application of our PP protocol to proactive secure sign-on, using its reconstructability. However, as mentioned there, the protocol described in Section 3 is reconstructible only if all the parties (servers) follow their protocols at all times (that is, the adversary is only eavesdropping).

In this section we describe modifications of our protocol, aimed at two goals: one goal is to make the protocol more efficient for the user; the other goal is to maintain the reconstructability property for the case where the servers don't follow their protocols.

We start by describing a variant of the protocol which is more efficient for the user. Using the protocol described in Section 3, the user had to simulate the computation performed by the servers step by step. In case that many rounds have passed since the last time the user updated its keys, this may pose a considerable overhead. Using this variant, denoted ρ, the user can compute its updated key simulating only *one* round of computation of the servers. On the other hand, this variant has a weaker resilience property: it assures that the servers' keys be unpredictable by the adversary only if there exists a server that has never been infected.

The variant is similar in structure to the original protocol with the following modification. Each P_i has a master key which is never erased. This master key is set to be the initial key, $\kappa_{i,0}$ (derived, say, from the password). The master key is used as the index for the function at all the rounds. Namely, the key $\kappa_{i,l}$ at round l is computed as follows:

$$\kappa_{i,l} = \oplus_{j=1}^{n} f_{\kappa_{j,0}}(i)$$

It is also possible to combine the original protocol with the variant described above, in order to reach a compromise between efficiency and security. We define a special type of round: a major round. (For instance, let every 10th round be a major round.) The parties now update their master keys, using the original protocol, only at major rounds. In non-major rounds, the servers use their current master key as the index for the function.

This combined protocol has the following properties. On one hand, the user has one tenth of the rounds to simulate than in the original scheme. On the other hand, we only need that in any period of 10 rounds there exists a server that has not been infected. We believe that such versions may be a more reasonable design for actual implementations.

Next, we describe additions to the protocol, aimed at maintaining the reconstructability property for the case where the servers don't follow their protocols. We note that it is possible to withstand crash failures of servers, if the servers cooperatively keep track of which servers crashed at each round (this coordination can be done using standard consensus protocols).

The following addition to the protocol handles the case of Byzantine faults, if variant ρ described above is used. The only way an adversary controlling party P_j could interfere with the reconstructability of variant ρ is by sending a wrong value instead of $f_{\kappa_{j,0}}(i)$, to some server P_i. However, P_i can, when unable to authenticate a user, compare each of the $f_{\kappa_{j,0}}(i)$ values with the user, e.g. using the 2PP protocol [BGH+93b], without exposing any of the values. If there are more than half of the values which match, the server and the user may use the exclusive or of these values only. This technique requires that at any round, the majority of the servers are non-faulty (otherwise the server may end up using values which are all known to the adversary). We note that this idea does not work if the basic protocol (that of Section 3) is used instead of variant ρ.

Conclusions and Open Problems

We believe that incorporating proactive security (i.e., the repeated, periodic attempt at recovery from potential break-ins, using other components in the system) in the design of systems can greatly inhance their security. In particular, our proactive pseudorandomness protocol may prove very useful in the design of proactive security protocols, supplying them with pseudorandomness (with some reconstructibility properties).

We propose the following open problem as a particularly challenging example for the applicability of proactive security to many security tasks. One of the major drawbacks of identity based cryptosystems is their reliance on a *single, trusted* key-generation facility, which knows the keys of all parties. It would be much better if one could design a distributed key-generation facility composed of several, potentially less trusted servers, but having proactive security.

Another challenging and important problem is to find a PP protocol with full reconstructibility against byzantine faults without giving up on security (preliminary results have been obtained by [CH94]).

Acknowledgments

Many thanks to Oded Goldreich and Hugo Krawczyk, who helped us with many constructive comments and suggestions. Thanks also to Chee-Seng Chow, Juan Garay and Moti Yung, for their support and comments.

References

[BGH+93a] Ray Bird, Inder Gopal, Amir Herzberg, Phil Janson, Shay Kutten, Refik Molva, and Moti Yung. A family of light-weight protocols for authentication and key distribution. Submitted to IEEE T. Networking, 1993.

[BGH+93b] Ray Bird, Inder Gopal, Amir Herzberg, Phil Janson, Shay Kutten, Refik Molva, and Moti Yung. Systematic design of a family of attack-resistant authentication protocols. *IEEE Journal on Selected Areas in Communications*, 11(5):679–693, June 1993. Special issue on Secure Communications. See also a different version in *Crypto 91*.

[BM93] Steven M. Bellovin and Michael Merritt. Augmented encrypted key exchange: a password based protocol secure against dictionary attacks and password file compromise. In *1st ACM Conference on Computer and Communications Security*, pages 244–250, November 1993.

[BR93] Mihir Bellare and Phil Rogaway. Entity authentication and key distribution. In *Advances in Cryptology: Proc. of Crypto 93*, pages 232–249, August 1993.

[CH94] Chee-Seng Chow and Amir Herzberg. A reconstructible proactive pseudorandomness protocol. Work in progress, June 1994.

[ER89] M. Elchin and J. Rochlis. With microscope and tweezers: An analysis of the internet virus of november 1988. In *IEEE Symp. on Security and Privacy*, pages 326–343, 1989.

[FM88] P. Feldman and S. Micali. Optimal algorithms for byzantine agreement. In *Proceedings of the* 20th *Annual ACM Symposium on Theory of Computing*, pages 148–161, May 1988.

[FY93] Mat Franklin and Moti Yung. Eavesdropping games: A graph-theoretic approach to privacy in distributed systems. In 34th *Annual Symposium on Foundations of Computer Science*, pages 670–679, November 1993.

[Gar94] Juan A. Garay. Reaching (and maintaining) agreement in the presence of mobile faults. To be presented in WDAG, 1994.

[GGM84] Oded Goldreich, Shafi Goldwasser, and Silvio Micali. On the cryptographic applications of random functions. In *Advances in Cryptology: Proc. of Crypto 84*, pages 276–288, 1984.

[GGM86] Oded Goldreich, Shafi Goldwasser, and Silvio Micali. How to construct random functions. *J. ACM*, 33(4):792–807, 1986. Extended abstract in FOCS84.

[HILL93] John Hästad, Russell Impagliazzo, Leonid Levin, and Mike Luby. Construction of pseudo-random generator from any one-way functions. Manuscript, see preliminary versions by Impagliazzo et al. in *21st STOC* and Hästad in *22nd STOC*, 1993.

[LE93] Thomas A. Longstaff and Schultz E. Eugene. Beyond preliminary analysis of the wank and oilz worms: A case of study of malicious code. *Computers and Security*, 12(1):61–77, 1993.

[MNSS87] S. P. Miller, C. Neuman, J. I. Schiller, and J. H. Saltzer. Kerberos authentication and authorization system. In *Project Athena Technical Plan*. Massachusetts Institute of Technology, July 1987.

[MT79] R. H. Morris and K. Thompson. Unix password security. *Comm. ACM*, 22(11):594–597, November 1979.

[OY91] Rafail Ostrovsky and Moti Yung. How to withstand mobile virus attacks. In *Proceedings of the* 10th *Annual ACM Symposium on Principles of Distributed Computing, Montreal, Quebec, Canada*, pages 51–59, 1991.

[Rei85] R. Reischuk. A new solution to the byzantine generals problem. *Information and Control*, pages 23–42, 1985.

[Sto88] Cliff Stoll. How secure are computers in the u.s.a.? *Computers and Security*, 7(6):543–547, 1988.

Appendix A: An alternative definition of PP

We offer an alternative definition of a PP. This definition follows from Definition 1. Borrowing from the theory of secure encryption functions, we call Definition 1 "PP in the sense of indistinguishability" (or, in short, PP), whereas Definition 5 below is called "semantic PP" (or, in short, SPP). We first recall the standard definition of polynomial indistinguishability of distributions.

Definition 4. Let $\mathcal{A} = \{A_k\}_{k \in \mathbf{N}}$ and $\mathcal{B} = \{B_k\}_{k \in \mathbf{N}}$ be two ensembles of probability distributions. We say that $\mathcal{A}$ and $\mathcal{B}$ are polynomially indistinguishable if there exists a constant $c > 0$ such that for every polytime distinguisher D and for all large enough k,

$$|\text{Prob}(D(A_k) = 1) - \text{Prob}(D(B_k) = 1)| < \frac{1}{k^c}.$$

We colloquially let $A_k \approx B_k$ denote "$\{A_k\}$ and $\{B_k\}$ are polynomially indistinguishable".

Let α be some distributed randomized protocol, which is resilient against t-limited mobile adversaries, for some value of t. We wish to adapt protocol α to a situation where no randomness is available once the interaction starts. Namely, we want to construct a protocol α' in which the parties use randomness only before the interaction starts, and the parties' outputs of protocol α' are "the same as" their outputs of protocol α.

A general framework for a solution to this problem proceeds as follows. The parties run protocol α along with another *deterministic* protocol, π. Each party's local input of protocol π is chosen at random at the beginning of the interaction. In each round, each party sets the random input of protocol α for this round to be the current output of protocol π. We call π a semantically secure proactive pseudorandomness(SPP) protocol. We refer to protocol α as the application protocol.

We state the requirements from a SPP protocol π. Informally, we want the following requirement to be satisfied for every application α. Whatever an adversary can achieve by interacting with α combined with protocol π as described above, could also be achieved by interacting with the original protocol α when combined with a truly random oracle. More formally,

- for an n-party randomized application protocol α, a mobile adversary A, an input vector $\mathbf{x} = x_1, \ldots, x_n$, and $1 \leq i \leq n$, let $\alpha(\mathbf{x}, A)_i$ denote party P_i's output of protocol α with a random oracle when party P_j has input x_j and in the presence of adversary A. Let $\alpha(\mathbf{x}, A)_0$ denote A's output of this execution. Let $\boldsymbol{\alpha}(\mathbf{x}, A) \stackrel{\triangle}{=} \alpha(\mathbf{x}, A)_0, \ldots, \alpha(\mathbf{x}, A)_n$.
- For a randomized protocol α and a deterministic protocol π for which each party has an output at each round, let $\alpha\pi$ denote the protocol in which α and π are run simultaneously and each party at each round sets the random input of α to be the current output of π.

Definition 5. We say that an n party deterministic protocol π is a t-resilient SPP protocol if for every (randomized) application protocol α and every t-limited mobile adversary A there exists a t-limited mobile adversary A' such that for every input vector $\mathbf{x}$ (for protocol α) we have

$$\boldsymbol{\alpha\pi}(\mathbf{x}, A') \approx \boldsymbol{\alpha}(\mathbf{x}, A).$$

where the probabilities are taken over the inputs of π and the random choices of A and of the oracle of α.

Theorem 6. *If a protocol is a t-resilient PP protocol then it is a t-resilient SPP protocol.*

Author Index

Springer-Verlag and the Environment

We at Springer-Verlag firmly believe that an international science publisher has a special obligation to the environment, and our corporate policies consistently reflect this conviction.

We also expect our business partners – paper mills, printers, packaging manufacturers, etc. – to commit themselves to using environmentally friendly materials and production processes.

The paper in this book is made from low- or no-chlorine pulp and is acid free, in conformance with international standards for paper permanency.

Lecture Notes in Computer Science

For information about Vols. 1–762
please contact your bookseller or Springer-Verlag

Vol. 763: F. Pichler, R. Moreno Díaz (Eds.), Computer Aided Systems Theory – EUROCAST '93. Proceedings, 1993. IX, 451 pages. 1994.

Vol. 764: G. Wagner, Vivid Logic. XII, 148 pages. 1994. (Subseries LNAI).

Vol. 765: T. Helleseth (Ed.), Advances in Cryptology – EUROCRYPT '93. Proceedings, 1993. X, 467 pages. 1994.

Vol. 766: P. R. Van Loocke, The Dynamics of Concepts. XI, 340 pages. 1994. (Subseries LNAI).

Vol. 767: M. Gogolla, An Extended Entity-Relationship Model. X, 136 pages. 1994.

Vol. 768: U. Banerjee, D. Gelernter, A. Nicolau, D. Padua (Eds.), Languages and Compilers for Parallel Computing. Proceedings, 1993. XI, 655 pages. 1994.

Vol. 769: J. L. Nazareth, The Newton-Cauchy Framework. XII, 101 pages. 1994.

Vol. 770: P. Haddawy (Representing Plans Under Uncertainty. X, 129 pages. 1994. (Subseries LNAI).

Vol. 771: G. Tomas, C. W. Ueberhuber, Visualization of Scientific Parallel Programs. XI, 310 pages. 1994.

Vol. 772: B. C. Warboys (Ed.),Software Process Technology. Proceedings, 1994. IX, 275 pages. 1994.

Vol. 773: D. R. Stinson (Ed.), Advances in Cryptology – CRYPTO '93. Proceedings, 1993. X, 492 pages. 1994.

Vol. 774: M. Banâtre, P. A. Lee (Eds.), Hardware and Software Architectures for Fault Tolerance. XIII, 311 pages. 1994.

Vol. 775: P. Enjalbert, E. W. Mayr, K. W. Wagner (Eds.), STACS 94. Proceedings, 1994. XIV, 782 pages. 1994.

Vol. 776: H. J. Schneider, H. Ehrig (Eds.), Graph Transformations in Computer Science. Proceedings, 1993. VIII, 395 pages. 1994.

Vol. 777: K. von Luck, H. Marburger (Eds.), Management and Processing of Complex Data Structures. Proceedings, 1994. VII, 220 pages. 1994.

Vol. 778: M. Bonuccelli, P. Crescenzi, R. Petreschi (Eds.), Algorithms and Complexity. Proceedings, 1994. VIII, 222 pages. 1994.

Vol. 779: M. Jarke, J. Bubenko, K. Jeffery (Eds.), Advances in Database Technology — EDBT '94. Proceedings, 1994. XII, 406 pages. 1994.

Vol. 780: J. J. Joyce, C.-J. H. Seger (Eds.), Higher Order Logic Theorem Proving and Its Applications. Proceedings, 1993. X, 518 pages. 1994.

Vol. 781: G. Cohen, S. Litsyn, A. Lobstein, G. Zémor (Eds.), Algebraic Coding. Proceedings, 1993. XII, 326 pages. 1994.

Vol. 782: J. Gutknecht (Ed.), Programming Languages and System Architectures. Proceedings, 1994. X, 344 pages. 1994.

Vol. 783: C. G. Günther (Ed.), Mobile Communications. Proceedings, 1994. XVI, 564 pages. 1994.

Vol. 784: F. Bergadano, L. De Raedt (Eds.), Machine Learning: ECML-94. Proceedings, 1994. XI, 439 pages. 1994. (Subseries LNAI).

Vol. 785: H. Ehrig, F. Orejas (Eds.), Recent Trends in Data Type Specification. Proceedings, 1992. VIII, 350 pages. 1994.

Vol. 786: P. A. Fritzson (Ed.), Compiler Construction. Proceedings, 1994. XI, 451 pages. 1994.

Vol. 787: S. Tison (Ed.), Trees in Algebra and Programming – CAAP '94. Proceedings, 1994. X, 351 pages. 1994.

Vol. 788: D. Sannella (Ed.), Programming Languages and Systems – ESOP '94. Proceedings, 1994. VIII, 516 pages. 1994.

Vol. 789: M. Hagiya, J. C. Mitchell (Eds.), Theoretical Aspects of Computer Software. Proceedings, 1994. XI, 887 pages. 1994.

Vol. 790: J. van Leeuwen (Ed.), Graph-Theoretic Concepts in Computer Science. Proceedings, 1993. IX, 431 pages. 1994.

Vol. 791: R. Guerraoui, O. Nierstrasz, M. Riveill (Eds.), Object-Based Distributed Programming. Proceedings, 1993. VII, 262 pages. 1994.

Vol. 792: N. D. Jones, M. Hagiya, M. Sato (Eds.), Logic, Language and Computation. XII, 269 pages. 1994.

Vol. 793: T. A. Gulliver, N. P. Secord (Eds.), Information Theory and Applications. Proceedings, 1993. XI, 394 pages. 1994.

Vol. 794: G. Haring, G. Kotsis (Eds.), Computer Performance Evaluation. Proceedings, 1994. X, 464 pages. 1994.

Vol. 795: W. A. Hunt, Jr., FM8501: A Verified Microprocessor. XIII, 333 pages. 1994.

Vol. 796: W. Gentzsch, U. Harms (Eds.), High-Performance Computing and Networking. Proceedings, 1994, Vol. I. XXI, 453 pages. 1994.

Vol. 797: W. Gentzsch, U. Harms (Eds.), High-Performance Computing and Networking. Proceedings, 1994, Vol. II. XXII, 519 pages. 1994.

Vol. 798: R. Dyckhoff (Ed.), Extensions of Logic Programming. Proceedings, 1993. VIII, 362 pages. 1994.

Vol. 799: M. P. Singh, Multiagent Systems. XXIII, 168 pages. 1994. (Subseries LNAI).

Vol. 800: J.-O. Eklundh (Ed.), Computer Vision – ECCV '94. Proceedings 1994, Vol. I. XVIII, 603 pages. 1994.

Vol. 801: J.-O. Eklundh (Ed.), Computer Vision – ECCV '94. Proceedings 1994, Vol. II. XV, 485 pages. 1994.

Vol. 802: S. Brookes, M. Main, A. Melton, M. Mislove, D. Schmidt (Eds.), Mathematical Foundations of Programming Semantics. Proceedings, 1993. IX, 647 pages. 1994.

Vol. 803: J. W. de Bakker, W.-P. de Roever, G. Rozenberg (Eds.), A Decade of Concurrency. Proceedings, 1993. VII, 683 pages. 1994.

Vol. 804: D. Hernández, Qualitative Representation of Spatial Knowledge. IX, 202 pages. 1994. (Subseries LNAI).

Vol. 805: M. Cosnard, A. Ferreira, J. Peters (Eds.), Parallel and Distributed Computing. Proceedings, 1994. X, 280 pages. 1994.

Vol. 806: H. Barendregt, T. Nipkow (Eds.), Types for Proofs and Programs. VIII, 383 pages. 1994.

Vol. 807: M. Crochemore, D. Gusfield (Eds.), Combinatorial Pattern Matching. Proceedings, 1994. VIII, 326 pages. 1994.

Vol. 808: M. Masuch, L. Pólos (Eds.), Knowledge Representation and Reasoning Under Uncertainty. VII, 237 pages. 1994. (Subseries LNAI).

Vol. 809: R. Anderson (Ed.), Fast Software Encryption. Proceedings, 1993. IX, 223 pages. 1994.

Vol. 810: G. Lakemeyer, B. Nebel (Eds.), Foundations of Knowledge Representation and Reasoning. VIII, 355 pages. 1994. (Subseries LNAI).

Vol. 811: G. Wijers, S. Brinkkemper, T. Wasserman (Eds.), Advanced Information Systems Engineering. Proceedings, 1994. XI, 420 pages. 1994.

Vol. 812: J. Karhumäki, H. Maurer, G. Rozenberg (Eds.), Results and Trends in Theoretical Computer Science. Proceedings, 1994. X, 445 pages. 1994.

Vol. 813: A. Nerode, Yu. N. Matiyasevich (Eds.), Logical Foundations of Computer Science. Proceedings, 1994. IX, 392 pages. 1994.

Vol. 814: A. Bundy (Ed.), Automated Deduction—CADE-12. Proceedings, 1994. XVI, 848 pages. 1994. (Subseries LNAI).

Vol. 815: R. Valette (Ed.), Application and Theory of Petri Nets 1994. Proceedings. IX, 587 pages. 1994.

Vol. 816: J. Heering, K. Meinke, B. Möller, T. Nipkow (Eds.), Higher-Order Algebra, Logic, and Term Rewriting. Proceedings, 1993. VII, 344 pages. 1994.

Vol. 817: C. Halatsis, D. Maritsas, G. Philokyprou, S. Theodoridis (Eds.), PARLE '94. Parallel Architectures and Languages Europe. Proceedings, 1994. XV, 837 pages. 1994.

Vol. 818: D. L. Dill (Ed.), Computer Aided Verification. Proceedings, 1994. IX, 480 pages. 1994.

Vol. 819: W. Litwin, T. Risch (Eds.), Applications of Databases. Proceedings, 1994. XII, 471 pages. 1994.

Vol. 820: S. Abiteboul, E. Shamir (Eds.), Automata, Languages and Programming. Proceedings, 1994. XIII, 644 pages. 1994.

Vol. 821: M. Tokoro, R. Pareschi (Eds.), Object-Oriented Programming. Proceedings, 1994. XI, 535 pages. 1994.

Vol. 822: F. Pfenning (Ed.), Logic Programming and Automated Reasoning. Proceedings, 1994. X, 345 pages. 1994. (Subseries LNAI).

Vol. 823: R. A. Elmasri, V. Kouramajian, B. Thalheim (Eds.), Entity-Relationship Approach — ER '93. Proceedings, 1993. X, 531 pages. 1994.

Vol. 824: E. M. Schmidt, S. Skyum (Eds.), Algorithm Theory – SWAT '94. Proceedings. IX, 383 pages. 1994.

Vol. 825: J. L. Mundy, A. Zisserman, D. Forsyth (Eds.), Applications of Invariance in Computer Vision. Proceedings, 1993. IX, 510 pages. 1994.

Vol. 826: D. S. Bowers (Ed.), Directions in Databases. Proceedings, 1994. X, 234 pages. 1994.

Vol. 827: D. M. Gabbay, H. J. Ohlbach (Eds.), Temporal Logic. Proceedings, 1994. XI, 546 pages. 1994. (Subseries LNAI).

Vol. 828: L. C. Paulson, Isabelle. XVII, 321 pages. 1994.

Vol. 829: A. Chmora, S. B. Wicker (Eds.), Error Control, Cryptology, and Speech Compression. Proceedings, 1993. VIII, 121 pages. 1994.

Vol. 830: C. Castelfranchi, E. Werner (Eds.), Artificial Social Systems. Proceedings, 1992. XVIII, 337 pages. 1994. (Subseries LNAI).

Vol. 831: V. Bouchitté, M. Morvan (Eds.), Orders, Algorithms, and Applications. Proceedings, 1994. IX, 204 pages. 1994.

Vol. 832: E. Börger, Y. Gurevich, K. Meinke (Eds.), Computer Science Logic. Proceedings, 1993. VIII, 336 pages. 1994.

Vol. 833: D. Driankov, P. W. Eklund, A. Ralescu (Eds.), Fuzzy Logic and Fuzzy Control. Proceedings, 1991. XII, 157 pages. 1994. (Subseries LNAI).

Vol. 834: D.-Z. Du, X.-S. Zhang (Eds.), Algorithms and Computation. Proceedings, 1994. XIII, 687 pages. 1994.

Vol. 835: W. M. Tepfenhart, J. P. Dick, J. F. Sowa (Eds.), Conceptual Structures: Current Practices. Proceedings, 1994. VIII, 331 pages. 1994. (Subseries LNAI).

Vol. 836: B. Jonsson, J. Parrow (Eds.), CONCUR '94: Concurrency Theory. Proceedings, 1994. IX, 529 pages. 1994.

Vol. 837: S. Wess, K.-D. Althoff, M. M. Richter (Eds.), Topics in Case-Based Reasoning. Proceedings, 1993. IX, 471 pages. 1994. (Subseries LNAI).

Vol. 838: C. MacNish, D. Pearce, L. Moniz Pereira (Eds.), Logics in AI. Proceedings, 1994. IX, 413 pages. 1994. (Subseries LNAI).

Vol. 839: Y. G. Desmedt (Ed.), Advances in Cryptology - CRYPTO '94. Proceedings, 1994. XII, 439 pages. 1994.

Vol. 840: G. Reinelt, The Traveling Salesman. VIII, 223 pages. 1994.

Vol. 841: I. Prívara, B. Rovan, P. Ružička (Eds.), Mathematical Foundations of Computer Science 1994. Proceedings, 1994. X, 628 pages. 1994.

Vol. 842: T. Kloks, Treewidth. IX, 209 pages. 1994.

Vol. 843: A.Szepietowski, Turing Machines with Sublogarithmic Space. VIII, 115 pages. 1994.